T0233831

Lecture Notes in Computer Science   **9643**

Commenced Publication in 1973
Founding and Former Series Editors:
Gerhard Goos, Juris Hartmanis, and Jan van Leeuwen

More information about this series at http://www.springer.com/series/7409

Shamkant B. Navathe · Weili Wu
Shashi Shekhar · Xiaoyong Du
X. Sean Wang · Hui Xiong (Eds.)

Database Systems
for Advanced Applications

21st International Conference, DASFAA 2016
Dallas, TX, USA, April 16–19, 2016
Proceedings, Part II

 Springer

Editors
Shamkant B. Navathe
Georgia Institute of Technology
Atlanta, GA
USA

Weili Wu
University of Texas at Dallas
Richardson, TX
USA

Shashi Shekhar
University of Minnesota
Minneapolis, MN
USA

Xiaoyong Du
Renmin University
Beijing
China

X. Sean Wang
Fudan University
Shanghai
China

Hui Xiong
Rutgers, The State University of New Jersey
New Brunswick, NJ
USA

ISSN 0302-9743   ISSN 1611-3349   (electronic)
Lecture Notes in Computer Science
ISBN 978-3-319-32048-9   ISBN 978-3-319-32049-6   (eBook)
DOI 10.1007/978-3-319-32049-6

Library of Congress Control Number: 2016934671

LNCS Sublibrary: SL3 – Information Systems and Applications, incl. Internet/Web, and HCI

This Springer imprint is published by Springer Nature
The registered company is Springer International Publishing AG Switzerland

Preface

Welcome to the proceedings of the 21st International Conference on Database Systems for Advanced Applications (DASFAA)! The DASFAA conference is held in varying locations throughout the world, and the 2016 DASFAA conference was held in Dallas, Texas, USA. DASFAA is an annual international database conference, which showcases state-of-the-art R&D activities in database systems and their applications. It provides a forum for technical presentations and discussions among database researchers, developers, and users from academia, business, and industry.

The DASFAA conference is truly an international forum. During its 21-year history, the conference has been held in more than 12 countries around the world. This year's conference continued this global trend: Our Organizing and Program Committee (PC) members represent 10 countries, and authors submitted papers from 24 different countries/regions.

This year's conference was competitive. A total of 183 papers were submitted for review. Each paper was reviewed by at least three PC members (except for a few reviewed by two PC members) and the selection was made on the basis of discussion among the reviewers and the program co-chairs. This year, 61 papers were accepted for presentation, representing an acceptance rate of about 33 %. In keeping with the goal of advancing the state of the art in databases, paper topics span numerous active and emerging topic areas including big data, crowdsourcing, Web applications, cloud data management, data archive and digital library, data mining, data model and query language, data quality and credibility, data semantics and data integration, data streams and time-series data, data warehouse and OLAP, databases for emerging hardware, database usability and HCI, graph data management, index and storage systems, information extraction and summarization, multimedia databases, parallel, distributed and P2P systems, probabilistic and uncertain data, query processing and optimization, real-time data management, recommendation systems, search and information retrieval, security and privacy, Semantic Web and knowledge management, sensor data management, social network analytics, statistical and scientific databases, temporal and spatial databases, transaction management, Web information systems, and XML and semi-structured data.

Reviewing and selecting papers from such a large set of research groups required the coordinated effort of many individuals. We want to thank all 83 members of the Program Committee and 90 external reviewers, who provided insightful feedback to the authors and helped with this selection process. In addition to the technical presentations, our program also included two invited speakers: Aidong Zhang, Jian Pei and 10-year best paper "Probabilistic Similarity Join on Uncertain Data" which appeared in DASFAA 2006 (written by Hans-Peter Kriegel, Peter Kunath, Martin Pfeifle, Matthias Renz). In addition, a set of four workshops completed the program.

Organizing the DASFAA 2016 program required the time and expertise of numerous contributors. We are grateful for the tremendous help of Hong Gao, Jinho Kim, and Yasushi Sakurai, who organized the workshops, Shaojie Tang, who was this year's publication chair, Ming Wang and Jun Liang, who organized the indusrial/practitioners track, Latifur Khan, Lidong Wu, and Dingzhu Du, who served as local organization co-chairs, Sang Won Lee, Jin Soung Yoo, and Jiaofei Zhong, who served as publicity co-chairs, Jing Yuan, who served as the registration chair, and Lei Cui and Jing Yuan, who served as webmasters. In addition, the guidance of the Steering Committee liaison, Xiaofang Zhou, was invaluable throughout each step of the conference organization and we thank him for his tireless efforts as well.

Finally, we thank the DASFAA community for their support of this international conference. We hope you enjoy the DASFAA conference and that you are inspired by the ideas found in these papers.

January 2016

Shamkant B. Navathe
Weili Wu
Shashi Shekhar
Xiaoyong Du
X. Sean Wang
Hui Xiong

Organization

Organizing Committee

General Co-chairs

Shamkant B. Navathe	Georgia Tech, USA
Weili Wu	University of Texas at Dallas, USA
Shashi Shekhar	University of Minnesota, USA

Program Committee Co-chairs

Xiaoyong Du	Renmin University, China
X. Sean Wang	Fudan University, China
Hui Xiong	Rutgers, The State University of New Jersey, USA

Workshop Co-chairs

Hong Gao	Harbin Institute of Technology, China
Jinho Kim	Kangwon National University, South Korea
Yasushi Sakurai	Kumamoto University, Japan

Industrial/Practitioners Track Co-chairs

Ming Wang	Google, USA
Jun Liang	SAP, USA

Publication Co-chair

Shaojie Tang	University of Texas at Dallas, USA

Local Organization Co-chairs

Latifur Khan	University of Texas at Dallas, USA
Lidong Wu	University of Texas at Tyler, USA
Dingzhu Du	University of Texas at Dallas, USA

Publicity Co-chairs

Sang Won Lee   Sungkyunkwan University, Korea
Jin Soung Yoo   Indiana University - Purdue University Fort Wayne, USA
Jiaofei Zhong   California State University, USA

Registration Chair

Jing Yuan   University of Texas at Dallas, USA

Webmasters

Lei Cui   University of Texas at Dallas, USA
Jing Yuan   University of Texas at Dallas, USA

Steering Committee Liaison

Xiaofang Zhou   University of Queensland, Australia

Program Committee

Toshiyuki Amagasa   University of Tsukuba, Japan
Spiridon Bakiras   John Jay College, CUNY, USA
Zhifeng Bao   RMIT University, Australia
Boualem Benatallah   University of New South Wales, Australia
Zhipeng Cai   Georgia State University, USA
K. Selcuk Candan   Arizona State University, USA
Jianneng Cao   Institute for Infocomm Research, A* Singapore
Varun Chandola   State University of New York at Buffalo, USA
Lei Chen   Hong Kong University of Science and Technology,
                           SAR China
Yi Chen   New Jersey Institute of Technology, USA
James Cheng   Chinese University of Hong Kong, SAR China
Reynold Cheng   The University of Hong Kong, SAR China
Bin Cui   Peking University, China
Ugur Demiryurek   University of Southern California, USA
Prasad Deshpande   IBM Research, India
Bowen Du   Beihang University, China
Hong Gao   Harbin Institute of Technology, China
Yunjun Gao   College of Computer Science, Zhejiang University, China
Yong Ge   The University of North Carolina at Charlotte, USA
Vikram Goyal   IIIT-Delhi, India
Le Gruenwald   The University of Oklahoma, USA
Ralf Hartmut Güting   Fernuniversität Hagen, Germany
Takahiro Hara   Graduate School of Information Science and Technology,
                           Osaka University, Japan

Haibo Hu	Hong Kong Baptist University, SAR China
Yan Huang	University of North Texas, USA
Yoshiharu Ishikawa	Nagoya University, Japan
Bo Jin	Dalian University of Technology, China
Donghyun Kim	North Carolina Central University, USA
Wang-Chien Lee	The Pennsylvania State University, USA
Cuiping Li	Renmin University of China
Guoliang Li	Tsinghua University, China
Siming Li	eBay, USA
Yingshu Li	Georgia State University, USA
Zhanhuai Li	Northwestern Polytechnical University, China
Zhongmou Li	Rutgers University, USA
Lipyeow Lim	University of Hawaii at Manoa, USA
Chuanren Liu	Rutgers Business School, USA
Jiaheng Lu	Renmin University of China
Zaixin Lu	Marywood University, USA
Jun Luo	Nanyang Technological University, Singapore
Qiong Luo	Hong Kong University of Science and Technology, SAR China
Jun Miyazaki	Tokyo Institute of Technology, Japan
Yasuhiko Morimoto	Hiroshima University, Japan
Shinsuke Nakajima	Kyoto Sangyo University, Japan
Wilfred Ng	Hong Kong University of Science and Technology, SAR China
Sarana Nutanong	City University of Hong Kong, SAR China
Makoto Onizuka	Osaka University, Japan
Jian Pei	Simon Fraser University, Canada
Zhiyong Peng	State Key Lab. of Software Engineeering, China
Weining Qian	East China Normal University, China
Zbigniew Ras	University of North Carolina, USA
Kun Ren	Yale University, USA
Chiara Renso	KDDLAB, ISTI-CNR, Pisa, Italy
Markus Schneider	University of Florida, USA
Shuo Shang	China University of Petroleum, China
Xuequn Shang	Northwestern Polytechnical University, China
Yan Shi	University of Wisconsin-Platteville, USA
Kyuseok Shim	Seoul National University, South Korea
Yu Suzuki	Nara Institute of Science and Technology, Japan
Keishi Tajima	Kyoto University, Japan
Nan Tang	Qatar Computing Research Institute, Qatar
Yong Tang	South China Normal University, China
Dimitri Theodoratos	New Jersey Institute of Technology, USA
Goce Trajcevski	Northwestern University, USA
Vincent S. Tseng	National Chiao Tung University, Taiwan
Ranga Raju Vatsavai	North Carolina State University, USA
Bin Wang	North Eastern University of China

Hongzhi Wang	Harbin Institute of Technology, China
Jianmin Wang	Tsinghua University, China
Jie Wang	Indiana University, USA
Li Wang	Taiyuan University of Technology, China
Peng Wang	Fudan University, China
Wei Wang	University of New South Wales, Australia
Jia Wu	University of Technology, Sydney, Australia
Keli Xiao	Stony Brook University, USA
Yanghua Xiao	Fudan University, China
Xike Xie	Aalborg University, Denmark
Jianliang Xu	Hong Kong Baptist University, Kowloon Tong, SAR China
Jeffery Xu Yu	Chinese University of Hong Kong, SAR China
Xiaochun Yang	University of California, Irvine, USA
Jian Yin	Sun Yat-sen University, China
Ge Yu	Northeastern University, China
Dayu Yuan	Google, USA
Zhongnan Zhang	Xiamen University, China
Fay Zhong	CSUEB, USA
Aoying Zhou	East China Normal University, China
Yuanchun Zhou	Computer Network Information Center, Chinese Academy of Sciences, China
Yuqing Zhu	California State University Los Angeles, USA
Roger Zimmermann	National University of Singapore
Lei Zou	Peking University, China

External Reviewers

Chunyu Ai	Shumo Chu	Zhipeng Huang
Ibrahim Almubark	Ananya Dass	Hui-Ju Hung
Daichi Amagata	Atreyee Dey	Atsushi Keyaki
Moshe Chai Barukh	Aggeliki Dimitriou	Huayu Li
Favyen Bastani	Zhaoan Dong	Jinfeng Li
Seyed-Mehdi-Reza Beheshti	Yixiang Fang	Mingda Li
	Xiaoyi Fu	Yafei Li
Hongzhi Chen	Chuancong Gao	Xinsheng Li
Jinchuan Chen	Li Gao	Yusan Lin
Lei Chen	Yanjun Gao	Huan Liu
Lu Chen	Yash Garg	Sicong Liu
Xilun Chen	Kazuo Goda	Yanchi Liu
Yueguo Chen	Xiaotian Hao	Zhi Liu
Ji Cheng	Juhua Hu	Wei Lu
Wenliang Chen	Shengyu Huang	Siqiang Luo
Lingyang Chu	Xiangdong Huang	Xiangbo Mao

Xiaoye Miao
Vinicius Monteiro de Lira
Bin Mu
Phuc Nguyen
Jan Kristof Nidzwetzki
Konstantinos
  Nikolopoulos
Kenta Oku
Silvestro Poccia
Bo Qin
Jianbin Qin
Raheem Sarwar
Yuya Sasaki
Caihua Shan
Yingxia Shao
Hiroaki Shiokawa
Masumi Shirakawa
Hong-Han Shuai

Jinhe Shi
Souvik Sinha
Shaoxu Song
Haiqi Sun
Jing Sun
Yifang Sun
Weitian Tong
Erald Troja
Fabio Valdés
Chiemi Watanabe
Huanhuan Wu
Liang Wu
Xiaoying Wu
Fan Xia
Chuan Xiao
Cheng Xu
Jianqiu Xu
Yong Xu

Yuto Yamaguchi
Fan Yang
Jingyuan Yang
Yu Yang
Yingjie Wang
Chenyun Yu
Jianwei Zhang
Qizhen Zhang
Zhipeng Zhang
Hongke Zhao
Jinwen Zhao
Shuai Zhao
Yunjian Zhao
Yudian Zheng
Huan Zhou
Hengshu Zhu
Huaijie Zhu
Tao Zhu

Contents – Part II

Contents – Part I

Social Networks

When Peculiarity Makes a Difference: Object Characterisation in Heterogeneous Information Networks

Wei Chen[1,2(✉)], Feida Zhu[1], Lei Zhao[2], and Xiaofang Zhou[2,3]

[1] School of Information Systems,
Singapore Management University, Singapore, Singapore
`fdzhu@smu.edu.sg`
[2] School of Computer Science and Technology, Soochow University, Jiangsu, China
`wchzhg@gmail.com, zhaol@suda.edu.cn, zxf@itee.uq.edu.au`
[3] School of ITEE, The University of Queensland, Brisbane, Australia

Abstract. A central task in heterogeneous information networks (HIN) is how to characterise an entity, which underlies a wide range of applications such as similarity search, entity profiling and linkage. Most existing work focus on using the main features common to all. While this approach makes sense in settings where commonality is of primary interest, there are many scenarios as important where uncommon and discriminative features are more useful. To address the problem, a novel model COHIN (Characterize Objects in Heterogeneous Information Networks) is proposed, where each object is characterized as a set of feature paths that contain both main and discriminative features. In addition, we develop an effective pruning strategy to achieve greater query performance. Extensive experiments on real datasets demonstrate that our proposed model can achieve high performance.

1 Introduction

The recent boom of social network services of all kinds has brought excitement to the research community with a wide range of interesting yet challenging topics. Among them one area of particular research interest and practical value is the characterisation, comparison and linkage of entities, especially user identity, across different platforms. Recent advances along this line include [1–4] which focus on connecting users across different social platforms with structural and semantic information, [5–7] addressing entity resolution, and [4,8,9] investigating similarity between objects of the same or different types. Despite their multiplicity and diversity, one common task central to all these work is to decide how to select and prioritise the features in consideration. The answers to these questions, unfortunately, are hardly straightforward as they depend on the nature of the application. One large class of applications look for the commonality among entities where entities are characterised with the most similar features among them. Naturally, solutions to these tasks focus on frequent patterns, or the "main"

S.B. Navathe et al. (Eds.): DASFAA 2016, Part II, LNCS 9643, pp. 3–17, 2016.
DOI: 10.1007/978-3-319-32049-6_1

features of the entities, and it is along this line that most existing approaches have been identified with. However, it is crucial to notice that, there are also sufficient, and equally important, application settings where "uncommon", which we refer to as *discriminative*, features actually play a more important role. In these situations, we seek to identify what are unique about an entity, in order to distinguish it from others in tasks of summarisation, comparison and linkage. We further illustrate with the following examples.

Example 1: Entity Profiling. Profiling researchers from a heterogeneous information network (HIN) such as DBLP is useful for academia. While it is important to summarise the main features of a researcher, it is also valuable to identify their unique aspects. This is because for many leading researchers, their profiles would appear highly similar if we only focus on their main features. For example, Philip S. Yu and Jiawei Han, both prominent researchers in data mining, have published a great number of papers in the same venues, including KDD, ICDM, SDM, etc., for many of which they are even co-authors. An informative and insightful profiling algorithm should in this case be able to identify not only what are similar between them but, perhaps more importantly, what are the distinct research aspects of each as well, as further illustrated in the following conceptual example.

Consider the example in Fig. 1, suppose Jim, Lee and Tom are researchers who have published papers in several venues with different terms. As seen in Table 1(c), while they share the same main features, they do exhibit their uniqueness if we take the discriminative features into account — Jim connects SIGIR which has accepted his paper with the term IR, ICDE has accepted Lee's paper with the term DB, and Tom has published a paper in AAAI. Yet, as shown in Table 1(a), since the authors have the same main features, frequent-pattern-based solutions [10] would not work. Similarily, meta-path-based methods [9] hardly help from Table 1(b). It is clear that to comprehensively profile entities in HINs, existing methods based on frequent patterns and meta-paths, such as [8,9] would fall short by neglecting the discriminative features.

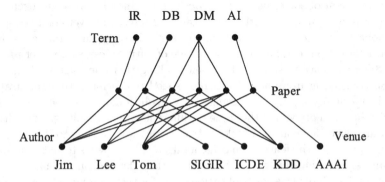

Fig. 1. An example of characterizing objects

Table 1. Characterize authors with venue

(a) Frequent pattern		(b) Meta path		(c) Main and discriminative features	
Author	Venue	Author	Venue	Author	Venue
Jim	KDD	Jim	KDD	Jim	KDD, SIGIR
Lee	KDD	Lee	KDD	Lee	KDD, ICDE
Tom	KDD	Tom	KDD	Tom	KDD, AAAI

Example 2: User Identity Linkage. The problem of *User Identity Linkage* (UIL), which aims to identify the accounts of the same user across different social platforms, has recently been attracting an increasing amount of attention and effort due to both the significant research challenges and the immense practical value of the problem [4]. One popular concept widely applied as intuition in this problem is that of "homophile", which essentially states that similar people share similar traits, e.g., friends. In this setting, it translates into the logic that, if two accounts u_A and u_B on two social platforms A and B respectively, we can use the number of their common friends to measure how likely u_A and u_B belong to the same user. Correct as it may be, this observation does not provide the most effective and efficient clue to the identity linkage problem due to the fact that people from the same context (workplace, school, neighbourhood) would naturally share many common friends, which means it is hard to distinguish two such people by this observation. On the other hand, what could really help quickly identify the linkage is to look for those unique and discriminative friends, few there maybe, that u_A and u_B both have. Similarly we can take advantage of users' discriminative interests or behaviour patterns, the availability of which are well supported by the long-tail observation for social platforms.

We propose in this paper a novel model COHIN to extract both the main and discriminative features that can be applied to a wide range of applications including user linkage, similarity search and entity profiling. Lying at the heart of these applications is how to measure similarity between two objects and find two objects with the minimum distance. We also propose a pruning strategy to improve query performance instead of enumerating all candidates. We summarise our major contributions as follows.

- We identify the importance of discriminative features in characterizing objects in heterogeneous information networks, and propose the COHIN model.
- We develop novel algorithms to extract features for objects across different platforms. Furthermore, a novel pruning strategy is proposed with the goal of achieving better query performance.
- We conduct extensive experiments on real datasets, and the experimental results demonstrate that the proposed model COHIN can achieve high efficiency and accuracy.

The rest of paper is organized as follows. We briefly review existing work related to our problem in Sect. 2 and formulate the problem in Sect. 3. In Sect. 4, we introduce the baseline approach, which is followed by the optimization in Sect. 5. We develop a pruning strategy in Sect. 6. Experiment results are reported in Sect. 7, followed by the conclusion in Sect. 8.

2   Related Work

2.1   Heterogeneous Information Networks Analysis

As the basic mining functions for heterogeneous information networks, clustering and classification have received great attentions during the last decade [11–13]. Sun et al. [11] propose the RankClus to address the problem of generating clusters with ranking information. The following work [12] develops a novel ranking-based clustering method called NetClus, different from [11] that focuses on bi-typed heterogeneous network, a start network schema is proposed for the clustering of multi-typed heterogeneous network.

Meta path-based similarity search and mining also play an important role in analyzing the heterogeneous information networks, and the existing studies [8,9,14] have made significant contributions. The meta path-based similarity measure called PathSim [8] is able to find peer objects with the same type, however it is not applicable for measuring relatedness between different types of objects. Followed by this work, a novel measure HeteSim [9] is proposed, with the goal of measuring similarity between objects with the same type or different types based on the given meta paths.

2.2   User Linkage and Entity Resolution

Connecting corresponding identities across communities is a challenging problem in heterogeneous information networks, which is firstly introduced in [15]. In the following work, [1,3,4] pay attentions to user linkage with more abundant information. The purpose of [3] is to address the cross-media user identification problem, where a behavioral model is proposed by considering the user names, language and writing styles. The framework HYDRA [4] investigates the problem of large-scale social identity linkage across different social networks by integrating all social information associated with a user.

Entity resolution is another concern in social networks [6,16,17]. [16] focuses on finding identifiers referring to the same real-world entities, where several adaptive techniques are developed for clustering and matching. As it is rather hard to automatically select appropriate similarity functions, [6] defines "how similar is similar", by which inappropriate similarity functions are pruned.

2.3   Mining of Discriminative Features

Recently people pay attentions to mine the discriminative features for graph pattern recognition [18–20]. A mining framework, called LEAP, is proposed in [18] to find the most discriminative subgraph. [19] studies the problem of supervised feature selection among frequent subgraphs, where an approach called CORK is designed to optimize a submodular quality criterion for subgraph mining. [20] develops a diversified discriminative feature selection method for graph classification, where discriminative score is used to select frequent subgraph features, and a new diversified discriminative score is introduced to select features that have a higher diversity.

3   Problem Statement

In this section, we present several definitions and the notations used throughout the paper, and formulate the problem.

Definition 1 Information Network [8]. *Given a network schema $S = (\mathcal{A}, \mathcal{R})$, where $\mathcal{A} = \{A\}$ is a set of object types and $\mathcal{R} = \{R\}$ is a set of relations, a heterogeneous information network is defined as a directed graph $G = (V, E)$ with an object type mapping function $\phi : V \rightarrow \mathcal{A}$ and a link type mapping function $\psi : E \rightarrow \mathcal{R}$. For each object $v \in V$, it belongs to one particular object type $\phi(v) \in \mathcal{A}$, and each link $e \in E$ belongs to a particular relation $\psi(e) \in R$. When the types of objects $|\mathcal{A}| > 1$ or the types of relations $|\mathcal{R}| > 1$, the network is called* **heterogeneous information network;** *otherwise, it reduces to a* **homogeneous information network.**

$$\mathcal{F}_{Jim}^{APC} = Jim \xrightarrow{(1,0.03)} Paper \xrightarrow{(0.308,0.03)} Venue \begin{cases} \xrightarrow{(0.25,0.67)} SIGIR \\ \xrightarrow{(0.75,0.042)} KDD \end{cases} \tag{1}$$

$$\mathcal{F}_{Lee}^{APC} = Lee \xrightarrow{(1,0.061)} Paper \xrightarrow{(0.3,0.06)} Venue \begin{cases} \xrightarrow{(0.33,0.67)} ICDE \\ \xrightarrow{(0.67,0.083)} KDD \end{cases} \tag{2}$$

Definition 2 Meta Path [8]. *A meta path $\mathcal{P}$ is a path defined on a schema $S = (\mathcal{A}, \mathcal{R})$, and is denoted in the form of $A_1 \xrightarrow{R_1} A_1 \xrightarrow{A_2} \cdots \xrightarrow{R_{l-1}} A_l$, which defines a composite relation $R = R_1 \circ R_2 \circ \cdots \circ R_{l-1}$ between type A_1 and A_l, where $\circ$ represents the composition operator on relations.*

The well-known bibliographic information network DBLP is a typical heterogeneous information network. As presented in the Fig. 2(a), the network schema of DBLP dataset contains objects from four types of entities: authors (A), papers (P), venues (C) and terms (T). The types of links connecting two objects are defined by the relations between them. For instance, the links between different papers denote citing or cited-by relations. Given two types A_1 and A_2, we use $A_1 \xrightarrow{R} A_2$ to denote the relation R from A_1 to A_2.

For each meta path $\mathcal{P}$, the length of $\mathcal{P}$ is the number of relations in it. Furthermore, the type names are used to represent the meta path if there exist no multiple relations between the same pair of types: $\mathcal{P} = (A_1 A_2 \cdots A_l)$. Seen from Fig. 2(b), the length-2 meta path $A \xrightarrow{writes} P \xrightarrow{published} C$ means authors publish

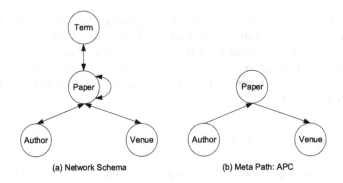

Fig. 2. Bibliographic information network schema and a meta path

papers in conferences, short as APC. Given a concrete path $p = (a_1 a_2 \cdots a_l)$ between a_1 and a_l in network G, the path is a instance of $\mathcal{P}(p \in \mathcal{P})$, if for each a_i we have $\phi(a_i) = A_i$ and each link $e_i = \langle a_i, a_{i+1} \rangle$ belongs to the relation R_i.

Definition 3 Relation Matrix. *Given a meta path* $\mathcal{P} = (A_1 A_2 \cdots A_l)$, *a matrix* $\mathcal{M_P} = U_{A_1 A_2} \cdots U_{A_{l-1} A_l}$ *is defined as the relation matrix of* $\mathcal{P}$, *where* $U_{A_i A_{i+1}}$ *is the adjacency matrix between type* A_i *and type* A_{i+1}, *and each element* $\mathcal{M_P}(i,j)$ *denotes the number of concrete paths between node* $x_i \in A_1$ *and node* $y_j \in A_l$, *and* $|\mathcal{M_P}(i,\cdot)|$ *is used to represent the number of paths from* $x_i \in A_1$ *to the last type* A_l *following the meta path* $\mathcal{P}$.

As seen in Eq. (3), $\mathcal{M}_{APC}$ represents the relation matrix of the meta path $\mathcal{P} = APC$ in Fig. 1. The element $\mathcal{M}(1,3) = 3$ denotes the first author Jim has published 3 papers in the third conference KDD.

$$\mathcal{M}_{APC} = U_{AP} \bullet U_{PC} = \begin{pmatrix} 1 & 0 & 3 & 0 \\ 0 & 1 & 2 & 0 \\ 0 & 0 & 3 & 1 \end{pmatrix} \tag{3}$$

Definition 4 Feature Path. *Given a meta path* $\mathcal{P} = (A_1 A_2 \cdots A_l)$ *and an object* v_i, *a path* $\mathcal{F}^{\mathcal{P}}_{v_i} = v_i \xrightarrow{(\omega,\delta)} A_2 \xrightarrow{(\omega,\delta)} \cdots \xrightarrow{(\omega,\delta)} A_l \xrightarrow{(\omega,\delta)} \theta$ *is defined as a feature path following the meta path* $\mathcal{P}$, *where* v_i *is an object of* A_1, ω *and* δ *denote the main and discrimination score of a link respectively, and* θ *is a set of objects contained by the entity type* A_l.

For each feature path, the computation of ω and δ is threefold:
(1) Computing ω and δ base on the Eq. (4) for the first link $v_i \rightarrow A_2$,

$$\begin{cases} \omega(v_i, A_2) = \dfrac{|\mathcal{M}_{A_1 A_2}(i,\cdot)|}{\sum\limits_{B_k \in S} |\mathcal{M}_{A_1 B_k}(i,\cdot)|} \\ \\ \delta(v_i, A_2) = \dfrac{||\mathcal{M}_{A_1 A_2}(i,\cdot)| - \dfrac{1}{m}\sum\limits_{j=1}^{m} |\mathcal{M}_{A_1 A_2}(j,\cdot)||}{\sum\limits_{j=1}^{m} |\mathcal{M}_{A_1 A_2}(j,\cdot)|} \end{cases} \tag{4}$$

where $\omega(v_i, A_2)$ denotes the main score of the link $v_i \rightarrow A_2$, $\delta(v_i, A_2)$ represents the discriminative score of this link compared with other links $v_i \rightarrow A_k$, and $\mathcal{S}$ is a set of types that link A_1 directly.

(2) For any two adjacent types A_{t-1} and A_t ($3 \leq t \leq l$), we use the following equations to compute ω and δ,

$$
\begin{cases}
\omega(A_{t-1}, A_t) = \dfrac{|\mathcal{M}_{A_1 \cdots A_t}(i,\cdot)|}{\sum\limits_{B_k \in \mathcal{S}} |\mathcal{M}_{A_1 \cdots A_{t-1} B_k}(i,\cdot)|} \\[20pt]
\delta(A_{t-1}, A_t) = \dfrac{||\mathcal{M}_{A_1 \cdots A_t}(i,\cdot)| - \dfrac{1}{m}\sum\limits_{j=1}^{m} |\mathcal{M}_{A_1 \cdots A_t}(j,\cdot)||}{\sum\limits_{j=1}^{m} |\mathcal{M}_{A_1 \cdots A_t}(j,\cdot)|}
\end{cases}
\tag{5}
$$

where $\mathcal{S}$ is a set of types that link type A_{t-1} directly, and $\dfrac{1}{m}\sum\limits_{j=1}^{m} |\mathcal{M}_{A_1 \cdots A_t}(j,\cdot)|$ is the average number of paths of all objects from A_1 to A_t following the meta path $\mathcal{P} = A_1 \cdots A_t$.

(3) For the last type A_l ($A_l \xrightarrow{(\omega,\delta)} \theta$), Eq. (6) is used to compute $\omega(A_l, v_j)$ and $\delta(A_l, v_j)$ between A_l and each object $v_j \in A_l$.

$$
\begin{cases}
\omega(A_l, v_j) = \dfrac{\mathcal{M}(i,j)}{\sum\limits_{k=1}^{m} \mathcal{M}(i,k)} \\[20pt]
\delta(A_l, v_j) = \dfrac{|\mathcal{M}(i,j) - \dfrac{1}{m}\sum\limits_{k=1}^{m} \mathcal{M}(i,k)|}{\sum\limits_{k=1}^{m} \mathcal{M}(i,k)}
\end{cases}
\tag{6}
$$

Consider the example in Fig. 1, given the meta path $\mathcal{P}_1 = APC$ for author Jim, we get the feature path $\mathcal{F}_{Jim}^{\mathcal{P}_1}$ presented in Eq. (1). Obviously, the venue KDD is the main research domain of Jim due to the main score $\omega(Venue, KDD) = 0.75$. However, SIGIR is also a significant property, which distinguishes Jim from other authors with discriminative score $\delta(Venue, IR) = 0.67$. Compared with Jim, Lee has the same main research domain KDD. However, we can distinguish them easily, due to the discriminative feature $\delta(Venue, ICDE) = 0.67$ of Lee. From the example we know the main and discriminative features are necessary while characterizing objects in heterogeneous information networks.

Problem Formalization. Given a heterogeneous information network $G = (V, E)$ with a schema $\mathcal{S}$, each object v in V is characterized as a set of feature paths $\mathcal{D}_v^{\mathcal{F}} = (\mathcal{F}_v^{\mathcal{P}_1}, \mathcal{F}_v^{\mathcal{P}_2}, \cdots, \mathcal{F}_v^{\mathcal{P}_m})$, where each feature path $\mathcal{F}_v^{\mathcal{P}_i}$ follows a certain meta path $\mathcal{P}_i$.

4   Baseline Method of COHIN

4.1   Calculation of Feature Paths

In this section, we propose a baseline method to calculate feature paths for objects in V, where each feature path is computed based on a given meta path.

Note that, objects are often compared with the same properties in real applications. For example, in DBLP, we measure the similarity between two authors in their research domain with the meta path APT, and the meta path APA is used to compare their co-authors. Consequently, we characterize objects with feature paths that follow the same meta paths. Given a set of meta paths $\mathcal{P} = \{\mathcal{P}_1, \mathcal{P}_2, \cdots, \mathcal{P}_m\}$, we can obtain $\mathcal{D}_v^{\mathcal{F}} = (\mathcal{F}_v^{\mathcal{P}_1}, \mathcal{F}_v^{\mathcal{P}_2}, \cdots, \mathcal{F}_v^{\mathcal{P}_m})$ for each object v based on Eqs. (4), (5) and (6).

4.2   Measure Similarity

The proposed model is applicable for many applications, such as entity match, user linkage and similarity search. And the key element of these applications is how to measure the similarity between any two objects with feature paths. Given two objects v and v', let $\mathcal{D}_v^{\mathcal{F}} = \{\mathcal{F}_v^{\mathcal{P}_1}, \cdots, \mathcal{F}_v^{\mathcal{P}_m}\}$ and $\mathcal{D}_{v'}^{\mathcal{F}} = \{\mathcal{F}_{v'}^{\mathcal{P}_1}, \cdots, \mathcal{F}_{v'}^{\mathcal{P}_m}\}$, we use $Dis(v, v')$ to denote the distance between v and v', which is given as follows:

$$Dis(v, v') = \sum_{k=1}^{m} Dis(\mathcal{F}_v^{\mathcal{P}_k}, \mathcal{F}_{v'}^{\mathcal{P}_k}) \tag{7}$$

where $Dis(\mathcal{F}_v^{\mathcal{P}_k}, \mathcal{F}_{v'}^{\mathcal{P}_k})$ denotes the distance between $\mathcal{F}_v^{\mathcal{P}_k}$ and $\mathcal{F}_{v'}^{\mathcal{P}_k}$, which is discussed in the sequel.

Given a feature path $\mathcal{F}_v^{\mathcal{P}_k} = v \xrightarrow{(\omega, \delta)} A_2 \xrightarrow{(\omega, \delta)} \cdots \xrightarrow{(\omega, \delta)} A_l \xrightarrow{(\omega, \delta)} \theta$ following meta path $\mathcal{P}_k$, and another feature path $\mathcal{F}_{v'}^{\mathcal{P}_k} = v' \xrightarrow{(\omega, \delta)} A_2 \xrightarrow{(\omega, \delta)} \cdots \xrightarrow{(\omega, \delta)} A_m \xrightarrow{(\omega, \delta)} \theta'$ following the same meta path $\mathcal{P}_k$, the distance between $\mathcal{F}_v^{\mathcal{P}_k}$ and $\mathcal{F}_{v'}^{\mathcal{P}_k}$ is defined as follows:

$$Dis(\mathcal{F}_v^{\mathcal{P}_k}, \mathcal{F}_{v'}^{\mathcal{P}_k}) = |\omega(v, A_2) - \omega(v', A_2)| + |\delta(v, A_2) - \delta(v', A_2)| + Dis(\theta, \theta')$$
$$+ \sum_{k=2}^{l-1} |\omega_v(A_k, A_{k+1}) - \omega_{v'}(A_k, A_{k+1})|$$
$$+ \sum_{k=2}^{l-1} |\delta_v(A_k, A_{k+1}) - \delta_{v'}(A_k, A_{k+1})| \tag{8}$$

where the $Dis(\theta, \theta')$ denotes the distance between θ and θ'. Let $\theta = (v_1, v_2, \cdots, v_m)$ and $\theta' = (v'_1, v'_2, \cdots, v'_n)$, $Dis(\theta, \theta')$ is given as follows:

$$Dis(\theta, \theta') = \sum_{v_k \in \theta \cap \theta'} (|\omega_v(A_l, v_k) - \omega_{v'}(A_l, v_k)| + |\delta_v(A_l, v_k) - \delta_{v'}(A_l, v_k)|) \tag{9}$$
$$+ \sum_{v_k \in \theta - \theta \cap \theta'} (\omega_v(A_l, v_k) + \delta_v(A_l, v_k)) + \sum_{v_k \in \theta' - \theta \cap \theta'} (\omega_{v'}(A_l, v_k) + \delta_{v'}(A_l, v_k))$$

Example 3: Given meta paths APC, APT and APA to calculate feature paths for authors in Fig. 1, the distances between them are presented in Table 2. Based on Eq. (7), we have $Dis(Jim, Tom) = Dis(\mathcal{F}_{Jim}^{APC}, \mathcal{F}_{Tom}^{APC}) + Dis(\mathcal{F}_{Jim}^{APT}, \mathcal{F}_{Tom}^{APT})$

Table 2. Distance between objects

(a) APC				(b) APT				(c) APA		
	Jim	Lee	Tom		Jim	Lee	Tom		Jim	Lee	Tom
Jim	0	2.104	1.834	Jim	0	2.092	1.834	Jim	0	1.253	1.4
Lee	2.104	0	2.104	Lee	2.092	0	2.092	Lee	1.253	0	1.253
Tom	1.834	2.104	0	Tom	1.834	2.092	0	Tom	1.4	1.253	0

$+Dis(\mathcal{F}_{Jim}^{APA}, \mathcal{F}_{Tom}^{APA})$=5.068. Similarily, we can obtain $Dis(Jim, Lee)$=5.449 and $Dis(Tom, Lee)$=5.449. Obviously, Jim and Tom are the most similarity objects with respect to research interest and co-authorship, since $Dis(Jim, Tom) < Dis(Jim, Lee)$ and $Dis(Jim, Tom) < Dis(Lee, Tom)$.

5 Optimization

The baseline method discussed in Sect. 4 can achieve high performance while characterizing objects with feature paths that follow the same meta paths. However, this method may become unuseful if we characterize objects across different social platforms, due to the diversity of network schemas across different platforms, and it is difficult to characterize objects with same meta paths. In order to tackle the problem, we design more general approaches here.

Algorithm 1. Compute a Feature Path

Input: a network G with a schema $\mathcal{S}$, an object v, λ
Output: a feature path $\mathcal{F}_v^{\mathcal{P}}$
1: $A_v \leftarrow v$;
2: **repeat**
3:     add link $A_o \xrightarrow{(\omega,\delta)} A_i$ into the feature path $\mathcal{F}_v^{\mathcal{P}}$ with the function *choose* (A_o);
4:     $A_o \leftarrow A_i$;
5: **until** the length of $\mathcal{F}_v^{\mathcal{P}}$ is larger than λ;
6: **for** $v_j \in A_l$ **do**
7:     compute $\omega(A_l, v_j)$ and $\delta(A_l, v_j)$ based on Eq.(6); //A_l is the last node of $\mathcal{F}_v^{\mathcal{P}}$
8: **end for**
9: **return** the feature path $\mathcal{F}_v^{\mathcal{P}}$

10: **Procedure** *choose* (A_o)
11: **for** each type A_i linked with A_o **do**
12:     select A_i with the maximum $\omega(A_i, A_o) + \delta(A_i, A_o)$;
13: **end for**
14: **return** the link $A_o \xrightarrow{(\omega,\delta)} A_i$;

As presented in Algorithm 1, we develop a depth-first method to characterize objects with main and discriminative features, where A_o denotes the node that

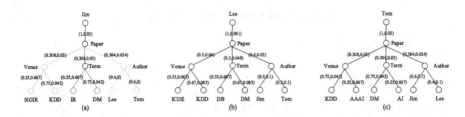

Fig. 3. Characterizing objects with feature paths (Color figure online)

need to be visited next, the function *choose* (A_o) is designed for selecting a proper link $A_o \xrightarrow{(\omega,\delta)} A_i$ for each process, and λ denotes a given threshold of the length of feature path.

Example 4: Continue the example in Fig. 1, given $\lambda = 3$, we get results presented in Fig. 3, where each tree is composed of a set of feature paths. In Fig. 3(a), the blue part is a feature path that follows the mate path APA and the red part follows the meta path APT. Compared with the baseline method, the results here contain more significant information.

Next, we develop novel approaches to measure similarity between different objects. As discussed in [21, 22], measuring similarity between two trees contains two components. (1) Transform a tree to a string. We construct an inverted labeled Prüfer sequence (IPS) for a tree, where the post-order is used to number tree nodes. By extending IPS, we construct FIPS (extended IPS), which is composed of a sequence of tuples in the form of $(type, \omega, \delta)$. Seen from Table 3, we construct IPS and FIPS for Jim, Lee and Tom. As we just consider the tree nodes with ω and δ, the root node is omitted for FIPS. (2) Compute the edit distance between two strings.

Given two trees with corresponding strings FIPS=$(\alpha_1, \cdots, \alpha_m)$, FIPS$' = (\beta_1, \cdots, \beta_n)$, a matrix $\mathcal{G}$ is proposed such that its element $\mathcal{G}[i,j]$ denotes the distance between substrings FIPS$_i = (\alpha_1, \cdots, \alpha_i)(1 \leq i \leq m)$ and FIPS$'_j = (\beta_1, \cdots, \beta_j)(1 \leq j \leq n)$. The element $\mathcal{G}[i,j]$ is given as follows:

Table 3. IPS and FIPS

Author	IPS and FIPS
Jim	IPS: JimPaperPaperAuthorPaperTermPaperVenueAuthorTomAuthorLeeTermDMTermIRVenueKDDVenueSIGIR
	FIPS: (Paper,1,0.03)(Paper,1,0.03)(Author,0.384,0.024)···(Venue,0.308,0.03)(SIGIR,0.25,0.667)
Lee	IPS: LeePaperPaperAuthorPaperTermPaperVenueAuthorTomAuthorJimTermDMTermDBVenueKDDConfICDE
	FIPS: (Paper,1,0.061)(Paper,1,0.061)(Author,0.4,0.03)···(Venue,0.3,0.06)(ICDE,0.33,0.667)
Tom	IPS: TomPaperPaperAuthorPaperTermPaperVenueAuthorLeeAuthorJimTermAITermDMVenueAAAIConfKDD
	FIPS: (Paper,1,0.03)(Paper,1,0.03)(Author,0.384,0.024)···(Venue,0.308,0.03)(KDD,0.75,0.042)

$$\mathcal{G}[i,j] = min \begin{cases} \mathcal{G}[i-1,j] + \alpha_i.\omega + \alpha_i.\delta + 1 \\ \mathcal{G}[i,j-1] + \beta_j.\omega + \beta_j.\delta + 1 \\ \mathcal{G}[i-1,j-1] + |\alpha_i.\omega - \beta_j.\omega| + |\alpha_i.\delta - \beta_j.\delta|, if\ \alpha_i.type = \beta_j.type \\ \mathcal{G}[i-1,j-1] + \alpha_i.\omega + \beta_j.\omega + \alpha_i.\delta + \beta_j.\delta + 1, if\ \alpha_i.type \neq \beta_j.type \end{cases}$$

(10)

Note that the element $\mathcal{G}[m,n]$ holds the distance between the given trees. In Fig. 3, we can obtain $Dis(Jim, Lee)=\mathcal{G}[19,19] = 5.46$, $Dis(Jim, Tom) = \mathcal{G}[19,19] = 5.05$, and $Dis(Lee, Tom) = \mathcal{G}[19,19] = 5.37$.

6   Feature Score Based Prune Strategy

Given an object v and a set of objects V, a straightforward way to find the top-k objects with the minimum distance $Dis(v, v')$ is to calculate the distance for each object v' in V. However, this is very time consuming, especially for a large $|V|$. With the goal of achieving high query performance, a feature score based strategy is proposed to prune search space.

Firstly, we calculate the feature score for each object v' in V based on Eq. (11).

$$FS_{v'} = \sum_{p \in FIPS_{v'}} p.\omega + p.\delta$$

(11)

Theorem 1. *Given two objects v and v' with $FIPS_v = (\alpha_1, \cdots, \alpha_m)$ and $FIPS_{v'} = (\beta_1, \cdots, \beta_n)$, it holds that $Dis(v, v') = \mathcal{G}[m,n] \geq |FS_v - FS_{v'}|$.*

Proof. Let $m \leq n$, $Dis(v, v')$ gets the minimum value when IPS_v is a part of $IPS_{v'}$ after removing the first node, which means that $\forall \alpha_k \in FIPS_v$, there exists $\beta_t \in FIPS_{v'}$ such that $\alpha_k.tpye = \beta_t.type$, then $Dis(v, v') = \mathcal{G}[m,n] \geq=$ $\sum_{\alpha_k.tpye=\beta_t.type}^{m} (|\alpha_k.\omega - \beta_t.\omega| + |\alpha_k.\delta - \beta_t.\delta|) + \sum(\beta_j.\omega + \beta_j.\delta) = |FS_v - FS_{v'}|$, where $\sum(\beta_j.\omega + \beta_j.\delta)$ denotes the sum of ω and δ for all β_j in remained $FIPS_{v'}$.   □

Secondly, we obtain a sorted set V_s by sorting objects in V according to feature scores. Note that, we keep track of top-k objects with the minimum distance $Dis(v, v')$, and the maximum value in current top-k results is denoted as $Dis(v, v_k)$.

Theorem 2. *Given an object v, for all unvisited objects v' in V_s, they are pruned if $|FS_v - FS_{v'}| \geq Dis(v, v_k)$.*

Proof. According to Theorem 1, for all unvisited objects, it holds that $Dis(v, v') \geq |FS_v - FS_{v'}|$. If $|FS_v - FS_{v'}| \geq Dis(v, v_k)$, we have $Dis(v, v') \geq Dis(v, v_k)$. Consequently, these objects should be pruned.   □

During the process of finding object v' with the minimum distance $Dis(v, v')$, we calculate $Dis(v, v')$ based on Eq. (10) and update $Dis(v, v_k)$ if necessary. The process is terminated if $|FS_v - FS_{v'}| \geq Dis(v, v_k)$, according to Theorem 2.

7   Experiments

We conduct extensive experiments on real datasets to evaluate the performance of the model through case studies, entity match and real-time query. All algorithms are implemented on a Core i5-4570 3.2 GHz machine with 16 GB memory.

7.1   Datasets

We use the following three datasets: (1) **DBLP**: This dataset is collected from DBLP website, which contains 1217512 authors with 2907314 papers. In order to study the most powerful authors, we sort all authors with corresponding number of papers and select the top 1015 authors with 742858 papers. (2) **Renren**: Renren is an important social network for users to share comments, pictures, videos, etc. The platform contains 240 million users in China. (3) **Sina Weibo**: Sina is another popular social network of China with more than 200 million registered users, and the number of active users is 89 million per month.

In order to study the performance of the proposed methods while collecting objects from different platforms, we select 2679 users from Renren and Sina as the ground truth. Note that we use the following algorithms to compare performance and efficiency. (1) HeteSim: which is proposed in [9]. (2) The baseline method MFP, details of which are presented in Sect. 4. (3) The second method is OFP, details of which are discussed in Sect. 5. (4) The final approach PFP, which uses the prune strategy to reduce the number of candidates.

7.2   Summary of Objects

We compare the performance of HeteSim, MFP and OFP, while characterizing objects in DBLP. As venues and co-authors are of critical importance to authors in DBLP, we use the meta paths APA and APC to characterize authors for HeteSim and the results are presented in Table 4, where the top-5 most similar authors and venues to author "Jiawei Han" are returned. The same meta paths are also used to calculate feature paths for MFP and the characterizations are presented in Fig. 4. Different from HeteSim that only focuses on the main features, MFP takes main and discriminative features into account and it contains more significant information, where the main and discriminative scores are used to denote the importance of a property. Note that OFP performs even better than the baseline methods. As seen from Fig. 5, it contains the most abundant information compared with the baseline methods. More importantly, there is no need to give meta paths for OFP, as which characterizes objects automatically with depth-first strategy. Obviously, OFP is the optimal approach to characterize objects in heterogeneous information.

7.3   Performance on Entity Match

Entity match is an important application of the proposed model COHIN, we randomly select 1K, 1.5K, 2K, 2.5K users from the ground truth to investigate the

Table 4. Top-5 similar authors and top-5 related venues to "Jiawei Han" in DBLP

(a) Meta path: APA

Rank	Author
1	Xifeng Yan
2	Philip S. Yu
3	Jing Gao
4	Yizhou Sun
5	Xin Jin

(b) Meta path: APC

Rank	Venue
1	KDD
2	ICDM
3	SDM
4	TKDE
5	SIGMOD

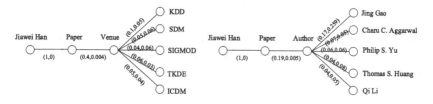

Fig. 4. Characterize objects with MFP

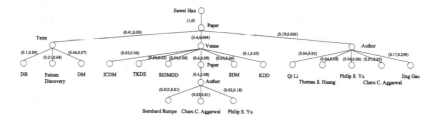

Fig. 5. Characterize objects with OFP

performance of MFP and OFP on entity match, by comparing the corresponding recall and precision. We conduct the experiments 30 times and report the average result in Fig. 6(a) and (b). Without surprise, OFP outperforms MFP with higher recall and precision, since it contains more significant information. As shown in Figs. 4 and 5, OFP contains more main and discriminative features while characterizing the same object. Consequently, OFP is more likely to have higher recall and precision. Note that, the recall and precision of these two approaches do not change obviously with the increase of the number of objects to be characterized, which means that they have good scalability.

7.4 Query Efficiency

In order to investigate the efficiency of the proposed methods OFP and PFP, we randomly select 1K, 1.5K, 2K, 2.5K users from Sina to find the most similar objects in Renren. Each experiment is repeated 20 times and the average time cost of OFP and PFP is presented in Fig. 6(c). Without surprise, PFP performs

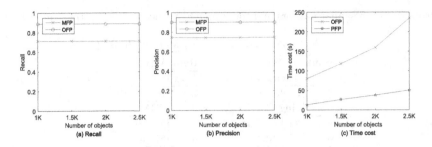

Fig. 6. Performance on entity match and efficiency (Color figure online)

better than OFP since many candidates are pruned during the process by the proposed strategy, i.e., we can achieve higher query efficiency with the proposed prune strategy.

8   Conclusion

We study a problem of characterizing objects in heterogeneous information networks. Different from traditional studies that focus on the main features of an object, we have proposed a novel model COHIN to characterize objects with main and discriminative features. The proposed model has many applications, such as similarity search, entity resolution, user linkage, etc. In oder to achieve higher query performance, we develop a prune strategy to reduce search space. Experiment results demonstrate that COHIN can achieve high performance.

Acknowledgments. This work was partially supported by the Singapore National Research Foundation under its International Research Centre @ Singapore Funding Initiative and administered by the IDM Programme Office, Media Development Authority (MDA) and the Pinnacle Lab at Singapore Management University, Natural Science Foundation of China (Grant No. 61572335), and Natural Science Foundation of Jiangsu Province, China (Grant No. BK20151223).

References

1. Vosecky, J., Hong, D., Shen, V.Y.: User identification across multiple social networks. In: First International Conference on Networked Digital Technologies NDT2009, pp. 360–365 (2009)
2. Iofciu, T., Fankhauser, P., Abel, F., Bischoff, K.: Identifying users across social tagging systems. In: ICWSM (2011)
3. Zafarani, R., Liu, H.: Connecting users across social media sites: a behavioral-modeling approach. In: SIGKDD, pp. 41–49 (2013)
4. Liu, S., Wang, S., Zhu, F., Zhang, J., Krishnan, R.: Hydra: large-scale social identity linkage via heterogeneous behavior modeling. In: SIGMOD, pp. 51–62 (2014)
5. Zheng, R., Li, J., Chen, H., Huang, Z.: A framework for authorship identification of online messages: writing-style features and classification techniques. JASIST **57**(3), 378–393 (2006)

6. Wang, J., Li, G., Yu, J.X., Feng, J.: Entity matching: how similar is similar. Proc. VLDB Endow. **4**(10), 622–633 (2011)
7. Peled, O., Fire, M., Rokach, L., Elovici, Y.: Entity matching in online social networks. In: Social Computing, pp. 339–344 (2013)
8. Sun, Y., Han, J., Yan, X., Yu, P.S., Wu, T.: Pathsim: meta path-based top-k similarity search in heterogeneous information networks. In: VLDB (2011)
9. Shi, C., Kong, X., Yu, P.S., Xie, S., Wu, B.: Relevance search in heterogeneous networks. In: Proceedings of the 15th International Conference on Extending Database Technology, pp. 180–191 (2012)
10. Aggarwal, C.C., Han, J.: Frequent Pattern Mining. Springer, Heidelberg (2014)
11. Sun, Y., Han, J., Zhao, P., Yin, Z., Cheng, H., Wu, T.: Rankclus: integrating clustering with ranking for heterogeneous information network analysis. In: Proceedings of the 12th International Conference on Extending Database Technology: Advances in Database Technology, pp. 565–576 (2009)
12. Sun, Y., Yu, Y., Han, J.: Ranking-based clustering of heterogeneous information networks with star network schema. In: SIGKDD, pp. 797–806 (2009)
13. Sun, Y., Norick, B., Han, J., Yan, X., Yu, P.S., Yu, X.: Integrating meta-path selection with user-guided object clustering in heterogeneous information networks. In: SIGKDD, pp. 1348–1356 (2012)
14. Gao, M., Lim, E.-P., Lo, D., Zhu, F., Prasetyo, P.K., Zhou, A.: CNL: collective network linkage across heterogeneous social platforms. In: ICDM, pp. 757–762 (2015)
15. Zafarani, R., Liu, H.: Connecting corresponding identities across communities. In: ICWSM (2009)
16. Cohen, W.W., Richman, J.: Learning to match and cluster large high-dimensional data sets for data integration, pp. 475–480 (2002)
17. Tejada, S., Knoblock, C.A., Minton, S.: Learning domain-independent string transformation weights for high accuracy object identification. In: KDD, pp. 350–359 (2002)
18. Yan, X., Cheng, H., Han, J., Yu, P.S.: Mining significant graph patterns by leap search. In: SIGMOD, pp. 433–444 (2008)
19. Thoma, M., Cheng, H., Gretton, A., Han, J., Kriegel, H.-P., Smola, A.J., Song, L., Philip, S.Y., Yan, X., Borgwardt, K.M.: Near-optimal supervised feature selection among frequent subgraphs. In: SDM, pp. 1076–1087 (2009)
20. Zhu, Y., Yu, X.J., Cheng, H., Qin, L.: Graph classification: a diversified discriminative feature selection approach. In: Proceedings of the 21st ACM International Conference on Information and Knowledge Management, pp. 205–214 (2012)
21. Yang, R., Kalnis, P., Tung, A.K.: Similarity evaluation on tree-structured data. In: SIGMOD, pp. 754–765 (2005)
22. Li, G., Liu, X., Feng, J.-H., Zhou, L.: Efficient similarity search for tree-structured data. In: Ludäscher, B., Mamoulis, N. (eds.) SSDBM 2008. LNCS, vol. 5069, pp. 131–149. Springer, Heidelberg (2008)

STH-Bass: A Spatial-Temporal Heterogeneous Bass Model to Predict Single-Tweet Popularity

Yan Yan[1], Zhaowei Tan[1], Xiaofeng Gao[1(✉)], Shaojie Tang[2], and Guihai Chen[1]

[1] Shanghai Key Laboratory of Data Science, Department of Computer Science and Engineering, Shanghai Jiao Tong University, Shanghai, China
gao-xf@cs.sjtu.edu.cn
[2] Department of Information Systems,
The University of Texas at Dallas, Richardson, USA

Abstract. Prediction in social networks attracts more and more attentions since social networks have become an important part of people's lives. Although a few topic or event prediction models have been proposed in the past few years, researches that focus on the single tweet prediction just emerge recently. In this paper, we propose STH-Bass, a Spatial and Temporal Heterogeneous Bass model derived from economic field, to predict the popularity of a single tweet. Leveraging only the first day's information after a tweet is posted, STH-Bass can not only predict the trend of a tweet with favorite count and retweet count, but also classify whether the tweet will be popular in the future. We perform extensive experiments to evaluate the efficiency and accuracy of STH-Bass based on real-world Twitter data. The evaluation results show that STH-Bass obtains much less APE than the baselines when predicting the trend of a single tweet, and an average of 24 % higher *precision* when classifying the tweets popularity.

Keywords: The Bass model · Predicting popularity · Social network

1 Introduction

Twitter, which is centered by users and communications, is one of the best-known social networks all over the world. What differs Twitter from other popular social networks is the asymmetric friend relationship, where a user is able to see tweets posted by his followings, while his tweets can be seen by his followers. Because of this characteristic, Twitter is more suitable for users who want to know celebrities' lives compared to symmetric social network such as Facebook,

This work has been supported in part by the China 973 project (2012CB316200), National Natural Science Foundation of China (Grant number 61202024, 61472252, 61133006, 61272443, 61473109), the Opening Project of Key Lab of Information Network Security of Ministry of Public Security (The Third Research Institute of Ministry of Public Security) Grant number C15602, and the Opening Project of Baidu (Grant number 181515P005267).

© Springer International Publishing Switzerland 2016
S.B. Navathe et al. (Eds.): DASFAA 2016, Part II, LNCS 9643, pp. 18–32, 2016.
DOI: 10.1007/978-3-319-32049-6_2

WeChat, etc. In Twitter, users tweet not only to express their emotions, but also to share their lives with followers. In other words, users tweet for what are worth tweeting, usually things that can lead to people's interest, attention or discussion. Once users have posted these tweets, they will have an expectation that these tweets will become popular, gaining a large number of retweets and favorites.

In recent years, prediction in social networks attracts more and more attention from both academia and industry since social networks have become parts of people's lives. Foresight is always users' favorite, and prediction satisfies the demand. Plenty of contents in social networks are worth predicting, such as user's personality [1], top10 news [2], popular stories [3], and controversial events [4]. Even a film's box office which seems to have nothing to do with social networks can be predicted through contents posted by users [5]. Most of these predictions use methods of machine learning such as Support Vector Machine, Naive Bayesian, Neural Network, etc., plus statistical model to do classification or regression.

Although researchers have done plenty of works on prediction, there are few works about predicting the popularity of a single tweet. Single tweets are components of topics and events. For users, from the perspective of psychology, posting popular tweets will bring satisfaction, and drive them to post more tweets. For Twitter and other official accounts, if they are capable of predicting which single tweet will become popular in advance, then they get an opportunity to lead the trend, and even come up with new hot topics or hashtags generating from that popular tweet. For third-party companies other than Twitter, predicting whether a single tweet will become popular and learning what makes a single tweet popular are useful as well, in that they are likely to create a popular tweet artificially, helping propagate their products and so on. Last but not least, abnormal popular tweets can set alarm for disaster, criminal, or catastrophe.

However, adopting existing topic or event prediction models cannot obtain satisfactory results. Topics or events usually consist of multiple tweets, but the popularity of one topic or event cannot well represent a single tweet's popularity. Actually, predicting the popularity of a single tweet requires different models. Compared with topics or events, which have more information generated from all tweets that they are made up of for prediction, predicting a single tweet's popularity can only use its own textual information and user's information with time line. Besides, the lifespan of a single tweet is mush less than a topic or event. Most of tweets are out of sight after posted for one week, which leaves us less time to compare the prediction with real trend, let alone correct the prediction. In addition, traditional machine learning methods with large training set does not work for single tweets prediction, since each tweet is a unique new target to predict. Information of those historical popular tweets helps little, in that we cannot learn a new model with unique parameters for each new tweet.

When it comes to the prediction of online contents popularity, most of related works focus on predicting the popularity of topics, events, or news. To the best of our knowledge, few researches discuss the prediction of a single tweet [6,7].

Zhang et al. [6] only predicts whether a tweet will be retweeted, which is a 2-class classification problem, leaving aside the prediction of the tweet's trend, which is a regression problem. The accuracy of regression is harder to ensure, in that the trend of a tweet changes greatly everyday after it is posted, and an efficient regression model have to predict as much as possible random changes of the trend. On the other hand, the model in [7] is impractical, in that it requires too many features that are difficult to obtain for predicting the trend of a tweet, such as the post time of all retweets relative to the original tweet, and the number of followers of each retweet user. Currently, there exists no reported study of predicting contents in social network using models modified from economics fields.

In this paper, we design STH-Bass, a *Spatial and Temporal Heterogeneous Bass Model* to predict the future of a single tweet. The Bass model [8] is one of the most widely applied models in management science, and the "Bass Model" paper is one of the Top 10 Most Influential Papers published in the 50-year history of *Management Science* [9]. The model is originally used in economic field to model the sales of a newly put-on-market product. To make up for its deficiency, we work more on spatial and temporal heterogeneity. We would like to predict the trend of a single tweet. In specific, according to the information observed after a single tweet posted for one day, we can use STH-Bass model to predict its favorite count and retweet count later of its whole life cycle. We further predict whether a single tweet will be popular using the results of trend prediction. Our model does not need large training set like traditional machine learning methods. In addition, there is no need of the topology of the social network such as following and follower relationships. Our experiments using real-world Twitter data validate the efficiency and accuracy of the trend prediction, with less absolute percent error and better classification detection.

We summarize our contributions as follow:

- We predict the trend of a single tweet after it is posted for a day, and whether the tweet will be hot in the end. Our model only needs attributes about the tweet to predict and its poster. Large training set and the topology of the network are not needed.
- We are the first to use the Bass model which is a famous statistical model from management science, in social network content prediction. In addition, we combine spatial and temporal heterogeneity into the Bass model, proposing a practical model which can be used in predicting a single tweet's popularity.
- Our heterogeneous Bass model is not only suitable for predicting a single tweet's popularity in Twitter, but also suitable for other social networks which have an asymmetric following-follower relationship, such as Weibo and Digg.
- We use real-world Twitter data to examine the efficiency of STH-Bass model, and the simulation results well exhibit the efficiency and accuracy of the trend prediction, with less absolute percent error and better classification detection.

The rest of the paper is organized as follows. In Sect. 2, we introduce the related work in the field of social network prediction. In Sect. 3, we give some preliminaries about our problem statement and data analysis, which is the basis for parameters of our proposed model in Sect. 4. In Sect. 5, we illustrate our experiments on Twitter data set with discussions. Finally, we conclude our work in Sect. 6.

2   Related Work

As soon as social network was brought to people's eyesight, researchers began to explore it from different perspectives. They first analyzed social network, discovered and found popular things [10,11] and influenced users [12]. Then, they classified things [13,14], and finally they recommended things to users [15], and predicted the future [5].

Prediction in social network goes to different directions as well. Social network is consist of users and contents posted by users. Therefore, one direction of prediction goes to people [1,16]. Another direction goes to contents, which can be classified specifically into predicting events [4,17], topics [18], news [2,19,20] or activities [21]. Most of contents prediction were at collective level. They paid attention to predicting things that a group of people took part in, not things that created by an individual.

Collective prediction such as the prediction of topic and event has grown relatively mature recently. These predictions are mostly classification problems, predicting whether a topic or event will be popular in future. Deng et al. [18] used a probability method of Bayesian combined with generative learning method. Furthermore, they divided time into continuous time intervals and predicted in which interval a topic will be popular. Zhang et al. [17] firstly detected events from burst word clustering, then used linear spread model to predict event popularity. All these topic or event models need extra tools to generate topics or events at first step, then use self-designed model with machine learning methods to reach their goals.

In each social network, there are special contents to predict, such as images on Flickr [22], stories on Digg [3], or hashtags on Twitter [23–25]. Kong et al. [25] is an representative example of predicting bursts and popularity of hashtags. It is smart to classify the life cycle of a hashtag into different statuses. Hashtag is a great innovation to help annotate topics in tweets. It helps predict popularity of online contents. Chang [26] brought in *Diffusion of Innovation* (DOI) theory, which was first applied in the researches in social network. It regarded hashtag as a kind of innovation, and showed that the Bass model [8] is feasible to hashtag prediction. Cui et al. [27] and Yang et al. [28] mentioned the DOI theory to discover popular tweets, but did not apply any model of DOI to predict the popularity of online contents.

For works related to ours most, predicting the popularity of a single story on Digg [3] is the first relevant work we know that paid attention to contents of individual level. It took into account individual behaviors to generate social dynamic model to obtain good performance on Digg. When it comes to Twitter, retweet behavior prediction [6] used a hierarchical Dirichlet process to predict whether a tweet would gain retweets. It was a classification problem which cannot provide a quantitative outcome of how many retweets a tweet will finally gain. Hong et al. [29] predicted popular messages in Twitter using methods of binary classification and multi-class classification. The newly published [7] is one of the most relevant work that predicts the popularity of a single tweet. However, besides the shortages mentioned in Sect. 1, their model had to obtain the specific information of each user who retweeted, which is hard to track when generating the data set.

3 Preliminary

3.1 Problem Statement

After a tweet w is posted by user u, the followers of u can favorite the tweet, which increases its favorite count $f(t)$, or retweet, which increases its retweet count $r(t)$. Once a follower retweets u, the tweet can be seen by the follower's followers. Therefore, $f(t)$ and $r(t)$ keep increasing until everyone who have seen the tweet stopped favoriting it or retweeting it. The final count of favorite count is denoted as fc, and the final count of retweet count is denoted as rc. In order to state our task clearly, we give some definitions first.

Definition 1 *(Stable Time). The time T that satisfies $f(T) \geq \nu fc$ and $r(T) \geq \nu rc$ is the **stable time** of a tweet. We set $\nu = 0.95$ in our experiments.*

We set parameter ν here because not all tweets can reach their final counts given a fixed period of time. Some of the most popular tweets will continue to be retweeted and favorited for over a month. Therefore, we use ν to ensure that the favorite counts and retweet counts of tweets in our data set can finally reach a stable state. This will be explained later in Sect. 3.2.

Definition 2 *(Popularity Count). Both favorite count and retweet count represent a tweet's popularity. We define the **popularity count** $Y(t)$ as: $Y(t) = \mu f(t) + (1 - \mu)r(t)$, where $0 < t \leq T$, and μ is a coefficient to maintain the balance of favorite count and retweet count.*

Therefore, predicting the trend of tweet w turns into predicting $Y(t)$, given $0 < t \leq T$. At the same time, $Y(T)$ is the approximate final popularity count of a tweet. If $Y(T) > \gamma$, where γ is a threshold, then we regard the tweet as **popular**.

3.2 Data Analysis

Our data set is collected through Twitter API[1]. In order to efficiently crawl as many tweets as possible, and select the features we need, we first randomly crawl a set of tweets (time from 16th July, 2015 to 23rd July, 2015; with a quantity of $102,756$ tweets), and do some data analysis (including Tables 1, 2, and Figs. 1, 2, 3, 4, 5 and 6). We find several characteristics of single tweet:

---

[1] https://dev.twitter.com/.

Table 1. Tweets' popular count distribution

Popular count	0	1–9	10–99	100–999	1000+
Potion (%)	74.60 %	24.06 %	1.20 %	0.12 %	0.02 %

- From Table 1 we can see that 74.6 % of original tweets receive no favorite count or retweet count. This is because the users do not have enough followers, or the contents of these tweets are not the type which can be favorited or retweeted, e.g., something sad or meaningless. The rest 25 % non-zero tweets mostly gain 1–10 popular counts. Only 1.34 % tweets gain over 10 popular counts.
- The popular count of a single tweet increases rapidly during the first several hours after it is posted. As we can see from Fig. 1, 75 % tweets' favorite counts and retweet counts remain unchanged since they were posted. Referring to the characteristic above we know that nearly all of these tweets have no favorite count or retweet count. It is also possible that 0.4 % of these tweets gain several popular counts, and soon reach their stationary. There is a huge increase at 100 h (4–5 days after posted). At this time, 90 % tweets reach their final counts, and nearly 98 % tweets become stable after 240 h (10 days after posted). The rest 1 % tweets are with high probability to become popular tweets, which still receive favorites and retweets, although at a very slow rate.
- We can also see from Fig. 1 that favorite count reaches its stationary much faster than retweet count, which makes sense. It takes little time to click your mouse to favorite a tweet you see. But retweet action includes something like chatting. For example, user A retweets user B's tweet, and B retweets back with something A is interested in. Then A will retweet again, and B replies back. Finally, the retweet count will keep increasing until their conversation ends.

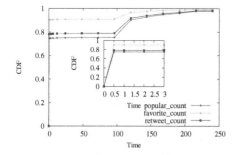

Fig. 1. The CDF of tweets' stable time (Color figure online)

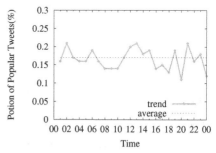

Fig. 2. When are popular tweets created (Color figure online)

Based on the above characteristics, we use Twitter stream to stochastically crawl tweets which has just been posted on Twitter, and record their favorite counts and retweet counts at $t = 0$, $0.5\,\text{h}$, $1\,\text{h}$, $2\,\text{h}$, $3\,\text{h}$, $6\,\text{h}$. For tweets with $Y(t = 6\,\text{h}) > 10$, we track down their favorite counts and retweet counts at the following $t = 12\,\text{h}$, $1\,\text{d}$, $2\,\text{d}$, ...$10\,\text{d}$. For the rest tweets, we only record their popularity count at $t = 10\,\text{d}$ as the final popularity count. Finally, our second data set is from 24th July, 2015 to 24th August, 2015, with a quantity of 569,149 tweets. We just crawl information about original tweets (instead of retweets), since the original tweets must be popular if some of its retweets are popular. Besides, we crawl the features about tweets' posters, including their number of followings, followers, tweets they have posted and the favorites they have obtained, etc.

According to our second data set, we dig deeper about the correlation between features of tweets and popularity of tweets. Apart from semantic and emotion features which need extra tools to generate, we study the effect of creating time and length of a tweet, which is simple but worthwhile.

Figure 2 shows the creating time of tweets and the number of popular tweets during 24 hours in a day. The truth is, at what time a tweet is posted does have an impact on its popularity. The dash line gives the average proportion of popular tweets per hour. It is obvious that 1 a.m. – 2 a.m., 11 a.m. – 1 p.m. and 8 p.m. – 9 p.m. are best time periods during which there is a large proportion of popular tweets. However, 7 a.m. – 10 a.m. in the morning is not a good time for posting a tweet, so as 4 p.m. – 6 p.m. in the afternoon, because people are busy going to work and going back home. This is a general analysis, because the creating time of a tweet is not strongly related to its popularity: a popular tweet is born to be popular according to its inner property. Nevertheless, without the right posting time, a popular tweet may submerge under large quantities of tweet streams. In a word, choosing an appropriate time to post helps accelerate the speed of popularity.

Figure 3 shows the relationship between the length of a tweet and its final popular count. Intuitively the length of a tweet is a useful feature for prediction, for that popular tweets always come up with clearly declarative or logical

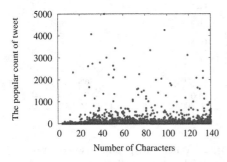

Fig. 3. Number of characters and popular counts of tweets

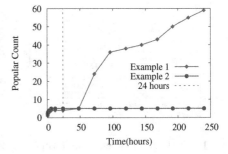

Fig. 4. An example of two tweets which had the same trend at the beginning, but ended differently (Color figure online)

Table 2. Popular tweet examples with different length

Content	Length	Favorite count	Retweet count
I can not deal with people commenting on my shit. I'm a human being, stop leaving comments on my pictures like I don't have eyes to read it	138	814	3442
I can remember song lyrics from 2006 but not whatever maths formula we were learning yesterday	94	2112	3019
Her smile puts stars in my eyes	31	857	3212
GOODMORNING	11	73	334

statements, which are all in need of quite a few words. However, it turns out that the contribution of the length to a popular tweet is not significant. Table 2 gives some examples of popular tweets with different length. However, whether the length can be regarded as one of predicting features remains to be examined in experiments later.

Tweets in Fig. 4 have similar trend in the observation period during the first 24 h after posted. However, one gets popular and another stays unpopular.

We also classify the trends of tweets mainly into four (Fig. 5): (1) gets high popular count immediately after posted, and stays popular (Example 2); (2) slowly increases and finally gets high popular count (Example 1); increases at a high rate at beginning but are not popular in the end (Example 3); slowly increases and stays slow in the end (Example 4). The threshold of being popular is 100, meaning that a tweet with final popular count greater than 100 will be classified as popular. Since Example 3 and Example 4 get small number of popular count, their trends have been zoomed in in Fig. 6.

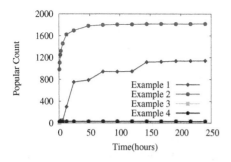

Fig. 5. Different trends of tweets (Color figure online)

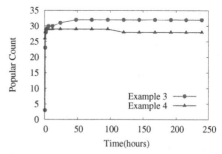

Fig. 6. A clearer illustration of Examples 3 and 4 in Fig. 5 (Color figure online)

4   The Heterogeneous Bass Model

The Bass model [8] is one of the most widely applied models in management science. The model was proposed to predict the sales of a new product when it is lunched on the market. The model is different from traditional machine learning method in that it does not need large numbers of training set. Given the first several days or months of sales of a new product, we can easily predict the performance of the product later via only two parameters.

The standard formulation of a diffusion process is:

$$h(t) = (p + qH(t))(1 - H(t)) \qquad (1)$$

Where p is the coefficient of innovators (or the impact of factors outside the population), q is the coefficient of imitators (or the impact of contacts within the population). $h(t)$ is the hazard rate of adoption, which is also known as the likelihood of purchase at time t. $1 - H(t)$ is the probability that one has not yet adopted at time t. Obviously, $H(t) = \int_0^t h(x)dx$. Here t is represented in *hours*.

Assuming the size of potential buyers is fixed as m, the number of purchase at t is:

$$S(t) = mh(t) = pm + (q - p)Y(t) - \frac{q}{m}[Y(t)]^2 \qquad (2)$$

Where $Y(t) = \int_0^t S(x)dx$ is the accumulative number of sales. Then Eq. (2) is the formulation of standard Bass model.

Although the Bass model is an excellent model in economic fields, it has many drawbacks as well. It only has two parameters, which gives little chance to add feature of social network into the model. In addition, the standard Bass model assumes spatial and temporal homogeneity, leading to no distinction of individuals. Therefore, we can combine features of Twitter with the original model, and relax it to individual-level heterogeneity. Due to the limitation of Twitter features, we focus on incorporating spatial heterogeneity, which allows everyone to have different possibilities to favorite or retweet a tweet. According to [30], there are 3 kinds of terms to reflect spatial heterogeneity: the intrinsic probability of adoption, the susceptibility to intra population linkages and the infectiousness of adopters.

Endowing the features we can get from Twitter, we derive the model as below:

$$S(t) = pm + (q - p)Y(t) - \frac{q}{m}[Y(t)]^2 + \boldsymbol{\alpha x} + \boldsymbol{\beta y} \qquad (3)$$

Here $\boldsymbol{x}$ is a vector of variables representing user features, such as the number of followings and followers of a user, the number of tweets a user posted since the creation of his account, the number of favorites the users collected. $\boldsymbol{y}$ is a vector of single tweet features including the creating time of a tweet, the number of URLs appearing in a tweet, and the number of characters in the text. Although the number of characters is not the main component contributing to popular count based on our data analysis in Sect. 3.2, we still take it into consideration here. It should be noted that not all of these features contribute to the final result of our experiments. Those nonsignificant features will be ruled out by **PCA** (Principal Component Analysis).

$$
\boldsymbol{x} = \begin{bmatrix} \log num\ Of\ Followings \\ \log num\ Of\ Followers \\ \log num\ Of\ Tweets \\ \log num\ Of\ Favorites \end{bmatrix} \tag{4}
$$

$$
\boldsymbol{y} = \begin{bmatrix} \log num\ Of\ Creating\ Time \\ \log num\ Of\ URLs \\ \log num\ Of\ Characters \end{bmatrix} \tag{5}
$$

In order to find $Y(t)$ we must solve the non-linear differential equation:

$$
\frac{dY}{dt} = pm + (q - p)Y(t) - \frac{q}{m}[Y(t)]^2 + \boldsymbol{\alpha x} + \boldsymbol{\beta y} \tag{6}
$$

For simplicity, let $V = pm + \boldsymbol{\alpha x} + \boldsymbol{\beta y}$. Then we have:

$$
\frac{mdY}{q[Y(t)]^2 + m(p-q)Y(t) - mV} = -dt \tag{7}
$$

Factoring the denominator on the left of Eq. (7), we have:

$$
\frac{mdY}{(Y - y_1)(Y - y_2)} = -dt \tag{8}
$$

with $y_1 = \frac{m(q-p)+\sqrt{\Delta}}{2q}$, $y_2 = \frac{m(q-p)-\sqrt{\Delta}}{2q}$, and $\Delta = m^2(p-q)^2 + 4mqV$.
Change Eq. (8) into:

$$
(\frac{1}{Y - y_1} - \frac{1}{Y - y_2})\frac{mdY}{y_1 - y_2} = -dt \tag{9}
$$

Then we can do integration on both sides of the equation:

$$
\int_0^T (\frac{1}{Y - y_1} - \frac{1}{Y - y_2})\frac{mdY}{y_1 - y_2} = \int_0^T -dt \tag{10}
$$

The solution is:

$$
Y(t) = \frac{y_2 e^{-\frac{\sqrt{\Delta}}{mq}t+C} + y_1}{1 + e^{-\frac{\sqrt{\Delta}}{mq}t+C}} \tag{11}
$$

Because $Y(0) = 0$, constant C generated by the integration can be solved:

$$
C = \ln(-\frac{y_1}{y_2}) \tag{12}
$$

And:

$$Y(t) = \frac{y_2 e^{-\frac{\sqrt{\Delta}}{mq}t + \ln(-\frac{y_1}{y_2})} + y_1}{1 + e^{-\frac{\sqrt{\Delta}}{mq}t + \ln(-\frac{y_1}{y_2})}} \qquad (13)$$

Hence we get the spatial and temporal heterogeneous Bass model. Using **Least Square Method**, which is one of the most recommending mathematical methods to fit Eq. (13). Then we can predict the popular count of a tweet at any time t, where $1d < t \leq T$.

5   Experiments

Our experiments are divided into two parts. The first part is the trend predicting, and the second part is the popularity predicting. Due to the fact that the favorite count and retweet count of a tweet varies a lot, some tweets gain large amount of favorite count but much fewer retweet count, while some tweets are on the contrary. Therefore, we set $\mu = 0.5$ to treat the number of favorite count and retweet count equally. In addition, we set $\nu = 10d$ and $\gamma = 100$ according to our data analysis. Furthermore, we treat follower count and friend count as invariant during the period of our experiments. As a matter of fact, these features do not change much for a mature user who has already set up his relationship network.

5.1   Predicting Trends

Different from traditional machine learning methods, which usually have a training set to train parameters and a test set to evaluate trained models, our heterogeneous Bass model divides each tweet into training part and predicting part. The training part includes the first 7 sample points at $t = 0.5\,\mathrm{h}$ 1 h, 2 h, 3 h, 6 h, 12 h, 24 h after a tweet posted. We use **Least Square Method** on these sample points of a single tweet, and the parameters in Eq. (13) are solved. Each tweet has a set of unique parameters to predict its trend in the nearly future. To test the efficiency of our model, we sample 9 time points at $t = 48\,\mathrm{h}(2\,\mathrm{d})$, $72\,\mathrm{h}(3\,\mathrm{d})$,... $240\,\mathrm{h}(10\,\mathrm{d})$. According to our data analysis in Sect. 3, most of popular counts of tweets tend to be stable after posted for 240 h. Therefore, our sample points can well test the whole life cycle of a tweet.

We give the comparison of the predicting final popular count and the real final popular count, as illustrated in Fig. 7. The dash line shows the correlation between the predicted value and the real value, with slope 1.3. Besides several outliers which are far away from the dash line, most of our predicting values are close to real values. When the final popular count is less than 200, our model performs best. As the real popular count becomes larger, our prediction emerges its conservative. There is a high probability that our predicting value is smaller than the real ones. However, it does not affect the precision of predicting popularity.

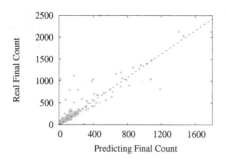

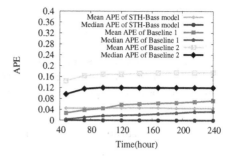

Fig. 7. The predicting final count VS the real final count

Fig. 8. The APE of our model and baselines (Color figure online)

Our STH-Bass model only need each tweet's text information (favorite count, retweet count, created time, length of the tweet, etc.) and it's poster's information (friends count, followers count, etc.), which can be easily obtained from Twitter API. For baselines, we initially intend to use the SEISMIC Model [7] as one of our baselines, since it is the most relevant study compared to ours. However, the SEISMIC Model needs extra information (such as time and follower counts) about user i who contributed to the ith retweet of a certain tweet. It is difficult to crawl these extra features from Twitter. Without these extra information, the SEISMIC model will lose its accuracy. Hence, we give up on the SEISMIC model, and propose another two baselines, which need a few features about each tweet and its author like our STH-Bass model:

- **Naive Model (Baseline1).** Since there are not many literature about predicting the trend of a single tweet, we come up with a naive model based on the characteristic (analyzed in Sect. 3.2) of our data set. We only use the popular count of $t = 24$ h to predict the trend of next 9 days. The Naive Model performs rather competitive because 75 % of tweets in our data set reach their final count after posted for one day.
- **Log Linear Regression Model (Baseline2).** Szabo and Huberman [31] analyzed there are strong correlations between early and later times of the logarithmical transformed popularity:

$$\ln Y(t_r) = \ln r(t_i, t_r) + \ln N(t_i) + \xi(t_i, t_r) \tag{14}$$

Therefore, Log Linear Regression Model can also predict well.

We calculate the **Absolute Percentage Error (APE)** for our evaluation. For each tweet w at time t, the formulation of APE is as follows:

$$APE(t) = \frac{|Y(t) - popularCount(t)|}{popularCount(t)} \tag{15}$$

Figure 8 shows the Mean APE and Median APE of our model and two baselines. We test 9 time points from 48 h (2 *days* after posted) to 240 h (10 *days* after posted). Both Mean and Median APE of our STH-Bass model change little during the whole period. The Mean APE is around 4 %, and the Median APE is around 1 %. Baseline 2 is far less accurate than STH-Bass in terms of Mean APE and Media APE. Baseline 1 is competitive at first several time points with STH-Bass, but tends to be less accurate later, indicating that there are still plenty of increases of popular counts from 48 h to 240 h.

Table 3. The performances of different methods

Model	Precision	Recall	F1-score
Our model	0.997	0.796	0.886
Baseline1	0.949	0.792	0.863
Baseline2	0.757	0.950	0.843

5.2 Predicting Popularity

According to Table 1, tweets which finally gain 0–9 popular count has a proportion of over 98 %. It is easy for our model and two baselines to get accuracy rates of over 95 %, because tweets receiving 0–9 popular count do not change much in their trends, and most of them already have reached stable after posted for one day. Therefore, we focus on the rest 2 % tweets, whose trends change relatively a lot, and are hard to reach their final counts after posted for one day. Therefore, we set $\gamma = 100$, which indicates that a tweet with $Y(T) > 100$ is popular.

Once we have predicted the final count of a tweet, we can immediately decide whether a tweet is popular. Table 3 shows the performance of predicting whether a tweet becomes popular by different methods. STH-Bass model gets the highest *Precision* and *F1-score*, while Baseline 2 gets the highest *Recall*. Comprehensively, STH-Bass model is 3 % better than Baseline 1 and 5 % better than Baseline 2 in terms of *F1-score*. In addition, STH-Bass model is 5 % better than Baseline 1 and 24 % better than Baseline 2 in terms of *Precision*, meaning that unpopular tweets are seldom classified as popular ones. As a matter of fact, classification is easier than regression. Although the advantage of STH-Bass is not as obvious as predicting trend in Sect. 5.1, STH-Bass ensures that those popular tweets which the model has classified, will be popular in the end with a high probability.

6 Conclusion

In this paper, we propose STH-Bass, a spatial and temporal heterogeneous Bass model to predict the popularity of a single tweet. STH-Bass uses the data features of a single tweet from the first day it has been posted, and can successfully predict the whether this tweet can be popular in the future. More specifically, STH-Bass can well depict the trend of a single tweet during its life cycle. Our model can even distinguish the tweets which have similar beginning popular count in first 24 h, but gain extremely different popular count in the end. We also use real-world Twitter data set to examine the performance of STH-Bass and compare the results with several baseline strategies. The simulation results validate the efficiency and accuracy of STH-Bass model with much less **APE** than baselines when predicting trend of a single tweet, and higher *Precision* and *F*1-*score* than one of our baselines when classifying the popularity.

References

1. Golbeck, J., Robles, C., Turner, K.: Predicting personality with social media. In: ACM CHI Conference on Human Computer Interaction, pp. 253–262 (2011)
2. Kong, L., Jiang, S., Yan, R., Xu, S., Zhang, Y.: Ranking news events by influence decay and information fusion for media and users. In: ACM International Conference on Information and Knowledge Management (CIKM), pp. 1849–1853 (2012)
3. Lerman, K., Hogg, T.: Using a model of social dynamics to predict popularity of news. In: International Conference on World Wide Web (WWW), pp. 621–630 (2010)
4. Popescu, A.-M., Pennacchiotti, M.: Detecting controversial events from twitter. In: ACM International Conference on Information and Knowledge Management (CIKM), pp. 1873–1876 (2010)
5. Asur, S., Huberman, B.: Predicting the future with social media. In: IEEE International Conference on Web Intelligence and Intelligent Agent Technology (WI-IAT), vol. 1, pp. 492–499 (2010)
6. Zhang, Q., Gong, Y., Guo, Y., Huang, X.: Retweet behavior prediction using hierarchical dirichlet process. In: AAAI Conference on Artificial Intelligence (AAAI), pp. 403–409 (2015)
7. Zhao, Q., Erdogdu, M.A., He, H.Y., Rajaraman, A., Leskovec, J.: Seismic: a self-exciting point process model for predicting tweet popularity. In: ACM SIGKDD International Conference on Knowledge Discovery andData Mining (SIGKDD), pp. 1513–1522 (2015)
8. Bass, F.M.: A new product growth for model consumer durables. Manag. Sci. **15**(5), 215–227 (1969)
9. Bass, F.M.: Comments on "a new product growth for model consumer durables the Bass model". Manag. Sci. **50**(12), 1833–1840 (2004)
10. Lerman, K., Ghosh, R.: Information contagion: an empirical study of the spread of news on digg and twitter social networks. In: The International AAAI Conference on Web and Social Media (ICWSM), vol. 10, pp. 90–97 (2010)
11. Kwak, H., Lee, C., Park, H., Moon, S.: What is Twitter, a social network or a news media? In: ACM International Conference on World Wide Web (WWW), pp. 591–600 (2010)
12. Weng, J., Lim, E.-P., Jiang, J., He, Q.: Twitterrank: finding topicsensitive inuential twitterers. In: ACM International Conference on Web Searchand Data Mining (WSDM), pp. 261–270 (2010)
13. Yang, S.-H., Kolcz, A., Schlaikjer, A., Gupta, P.: Large-scale high-precision topic modeling on Twitter. In: ACM SIGKDD International Conference on Knowledge Discovery and Data Mining (SIGKDD), pp. 1907–1916 (2014)

14. Zhou, D., Chen, L., He, Y.: An unsupervised framework of exploring events on Twitter: filtering, extraction and categorization. In: AAAI Conferenceon Artificial Intelligence (AAAI), pp. 2468–2474 (2015)

15. de Macedo, A.Q., Marinho, L.B.: Event recommendation in eventbasedsocial networks. In: Hypertext, Social Personalization Workshop, pp. 3130–3131 (2014)

16. Akram, H.A.A.L., Mahmood, A.: Predicting personality traits, gender and psychopath behavior of Twitter users. Int. J. Technol. Diffus. (IJTD) **5**(2), 1–14 (2014)

17. Zhang, X., Chen, X., Chen, Y., Wang, S., Li, Z., Xia, J.: Event detection and popularity prediction in microblogging. Neurocomputing **149**, 1469–1480 (2015)

18. Deng, Z.-H., Gong, X., Jiang, F., Tsang, I.W.: Effectively predicting whether and when a topic will become prevalent in a social network. In: AAAI Conference on Artificial Intelligence (AAAI), pp. 210–216 (2015)

19. Castillo, C., El-Haddad, M., Pfeffer, J., Stempeck, M.: Characterizing the life cycle of online news stories using social media reactions. In: ACM Conference on Computer Supported Cooperative Work and Social Computing (CSCW), pp. 211–223 (2014)

20. Bandari, R., Asur, S., Huberman, B.A.: The pulse of news insocial media: forecasting popularity. In: International AAAI Conference on Web and Social Media (ICWSM), pp. 26–33 (2012)

21. Huang, S., Chen, M., Luo, B., Lee, D.: Predicting aggregate social activities using continuous-time stochastic process. In: ACM international Conference on Information and Knowledge Management (CIKM), pp. 982–991 (2012)

22. McParlane, P.J., Moshfeghi, Y., Jose, J.M.: Nobody comes here anymore, it's too crowded; predicting image popularity on flickr. In: ACM International Conference on Multimedia Retrieval (ICMR), pp. 385–391 (2014)

23. Kamath, K.Y., Caverlee, J.: Spatio-temporal meme prediction: learning what hashtags will be popular where. In: ACM International Conference on Information Knowledge Management (CIKM), pp. 1341–1350 (2013)

24. Zaman, T.R., Herbrich, R., Van Gael, J., Stern, D.: Predicting information spreading in Twitter. In: Workshop on Computational Social Science and the Wisdom of Crowds, Annual Conference on Neural Information Processing Systems (NIPS), pp. 599–601. Citeseer (2010)

25. Kong, S., Mei, Q., Feng, L., Ye, F., Zhao, Z.: Predicting bursts and popularity of hashtags in real-time. In: International ACM SIGIR Conference on Research and Development in Information Retrieval, pp. 927–930 (2014)

26. Chang, H.-C.: A new perspective on Twitter hashtag use: diffusion of innovation theory. J. Am. Soc. Inf. Sci. Technol. **47**(1), 1–4 (2010)

27. Cui, A., Zhang, M., Liu, Y., Ma, S., Zhang, K.: Discover breaking events with popular hashtags in Twitter. In: ACM International Conference on Information and Knowledge Management (CIKM), pp. 1794–1798 (2012)

28. Yang, L., Sun, T., Zhang, M., Mei, Q.: We know what@ you# tag: does the dual role affect hashtag adoption? In: ACM International Conference on World Wide Web (WWW), pp. 261–270 (2012)

29. Hong, L., Dan, O., Davison, B.D.: Predicting popular messages in Twitter. In: ACM International Conference Companion on World Wide Web (WWW), pp. 57–58 (2011)

30. Strang, D., Tuma, N.B.: Spatial and temporal heterogeneity in diffusion. Am. J. Soc. **99**(3), 614–639 (1993)

31. Szaba, G., Huberman, B.A.: Predicting the popularity of online content. Commun. ACM **53**(8), 80–88 (2010)

Closeness and Structure of Friends Help to Estimate User Locations

Zhi Liu$^{(\boxtimes)}$ and Yan Huang

Computer Science and Engineering, University of North Texas, Denton, TX, USA
zhiliu@my.unt.edu, huangyan@unt.edu

Abstract. A tremendous amount of information is being shared every day on social media sites such as Facebook, Twitter or Google+. However, only a small portion of users provide their location information, which can be helpful in targeted advertising and many other services.Current methods in location estimation using social relationships consider social friendship as a simple binary relationship. However, social closeness between users and structure of friends have strong implications on geographic distances. In this paper, we introduce new measures to evaluate the social closeness between users and structure of friends. We propose models that use them for location estimation. Compared with the models which take the friend relation as a binary feature, social closeness can help identify which friend of a user is more important and friend structure can help to determine significance level of locations, thus improving the accuracy of the location estimation models. A confidence iteration method is further introduced to improve estimation accuracy and overcome the problem of scarce location information. We evaluate our methods on two different datasets, Twitter and Gowalla. The results show that our model can improve the estimation accuracy by 5 %–20 % compared with state-of-the-art friend-based models.

1 Introduction

A tremendous amount of information is being shared every day on social media platforms such as Facebook, Twitter, or Google+. For example, more than 241 million active Twitter users have published 300 million tweets worldwide, and this number continues to increase at a rate of 5,700 per second [17]. Oftentimes these messages include geo-information that is valuable to others, such as activities (*e.g.*, art fairs, jazz festivals, and social gatherings), natural disaster occurrences (*e.g.*, tornadoes, earthquakes), or other incidents (*e.g.*, traffic jams). The goal of this paper is to develop effective algorithms to estimate user home location. The results will help on the event detection from social media which in turn can assist the assimilation of social media information of interest for application domains such as smart transportation, disaster relief and recovery, and national security.

In this paper, we develop several models to estimate home locations of users on social media. Our goal is to locate users whose locations are unknown by

© Springer International Publishing Switzerland 2016
S.B. Navathe et al. (Eds.): DASFAA 2016, Part II, LNCS 9643, pp. 33–48, 2016.
DOI: 10.1007/978-3-319-32049-6_3

their social network and those located users. We face the following challenges: (1) People can share their location information more easily nowadays. Paradoxically, the problem of lacking location information still exists. Only 30 % of users provide their location information to at least one social media account and 46 % of teen app users have turned off the location tracking feature on their cell phone[1]; (2) User behavior varies greatly between different social networking services. In the datasets used in this paper, 27 % of friend pairs on Gowalla locate within 100 km of each other. In Twitter, however, this ratio is only 12 %. This because Twitter is not mainly a location-based social networking service and users tend to follow various media sources that are far away; and (3) Social media information is noisy and mixed with meaningless information. For example, 40 % tweets are not associated with a particular subject [16]. Additionally, some users are global travelers and have many friends from many cities from the world.

Previous methods use the social network, user location, and the content information for location estimation. However, the content information is not always available on different social media tools. In this paper, we focus on the analysis of social network and how the user locations affect the social connections. Our study shows that features such as friend structure of a user are important in improving the accuracy of the location estimation. This paper makes the following contributions: (1) We study the geographic features of social networks. We propose measures of social closeness between friends and the tightness of the friend structure of a user. We study the relationship between the closeness/tightness and the geographic distance. Existing algorithms consider friend relation as a binary relationship. However, finer level features such as friend co-location can help determine the social closeness of two friends. Friend co-location is an index measuring how overlapped two users' friend distributions are. Statistics shows that the friend co-location has a significant influence on the probability of two friends located close to each other. A user typically has a tight social structure among his friends in his home location. Local social coefficient measures the tightness of a user's friends in an area and can be a good indicator to measure if the user is located at that area. (2) We propose three user location estimation models which take social closeness and tightness of friend structure into consideration. (3) To deal with the challenge of location sparsity, we propose a confidence-based iteration model in location estimation which significantly improved the estimation accuracy. (4) We evaluate our models using two real world datasets. Extensive experimental results show that our methods improve the estimation accuracy by 5 %–20 % compared with the state-of-the-art algorithm.

2   Related Work

Recent research on location estimation in social networks follows two directions based on the data used: content-based and social-network-based. Content-based prediction models extract location information, like venue signals, from content provided by users. Cheng et al. [4] proposed and evaluated a city-level location

---

[1] http://www.pewinternet.org/2013/09/12/location-based-services/.

estimation model of Twitter users purely by taking the location related words in tweet content as features and applying a classification method. Chandra et al. [3] improved the content based method by using user interactions and exploiting the relationship between different tweet message types. They also provided the estimation of the top-K probable cities for a user. Another similar method is proposed in [8]. When a user declares a place, it will be checked in gazetteer to see if it corresponds to a city name. Location information will then be applied to infer the user location by the Twitter network. Content-based methods have also been studied to solve the web page geotagging problem [1]. The authors extract the toponyms from the web page to predict its location.

Our work is closely related to the work by Backstrom et al. [2]. By modeling the relation between distance and probability of being a friend, the authors proposed a general formula to calculate the probability of a user located at a specific place. Places with the maximal probability will be estimated as the location of the user. Both [15] and [5] aimed to build a user mobility model by using the location of their friends. Cho et al. [5] used several factors in their probability model, including check-in records, social network, friends' location, and time. Sadilek et al. [15] applied a machine learning method with similar information. Li et al. [13] proposed the $\mathcal{UDI}$ (unified discriminative influence) framework to combine content and friends' location analysis in a unique model to profile users' home location. In [14], the authors extract features from users' tweets content, social relation, and other behaviors like geotag to infer the home location on different level. Based on the geotagged information, they also built a classifier to predict whether a user is traveling. In [11], the author applied the geometric median between users, Oja's Simplex Median, and the transitivity of social network in the method. The goal of the algorithm is to select the nearest neighbor as the estimation result. In [7], the authors use the number of @mentions to indicate the social ties and take it as the weight ω_{ij} between users u_i and u_j. For all the connected users, they define the total variation as $\sum_{ij} \omega_{ij} d_{ij}$ where d_{ij} is the geographic distance between them. When estimating the users' locations, they seek the solution that makes the sum as small as possible. However, all of those models take the friend relation as a binary feature, i.e., being friends or not. This premise does not allow those models to take advantage of all the information from the social network. In our model, we take the social relation as a continuous feature by introducing the concept of social closeness. Network structure and locations are taken into consideration compared with our previous work [12] to help to achieve better results. By studying and using the relation between social closeness and geographic distance, our model: (1) significantly improves the estimation accuracy, especially when people have a small number of friends, (2) overcomes the problem that only a small number of users provide their location information.

3   Network Structure and Geographic Features

In this section, we analyze how location affects user behaviors. We study the geographic features of the social network from three aspects: the geographic

Table 1. Notations

u_i, l_i	User u_i and his/her location
U	All users in the social network
E	Set of friend relation in the social network
Γ_i	Friend set of user u_i
A_k	Set of users located in city k

Table 2. Features of the data sets. $|U'|$: number of located users, $|\Gamma|$: average number of friends, d: average distance between users, d_f: average distance between friend pairs.

Dataset	$|U'|$	$|\Gamma|$	d	d_f
Twitter	148,860	29.4	2,207	1,124 km
Gowalla	99,563	4.8	1,361	536 km

distribution of friends of an individual, the friend co-location, and the social structure of an individual's friends in the same city. Table 1 lists the notions used in this paper. In this paper, we use d_{ij} to refer to the geographic distance between users u_i and u_j, and all the distance used in this paper means the geographic distance.

3.1   Friend Distribution of an Individual

Data Preparation: We collected our data from two different social media platforms, Gowalla and Twitter. Gowalla is a location-based social network, and users are able to check in at "spots" in their local vicinity. The Gowalla dataset [5] was collected from February 2009 to October 2010, which contains 196,591 users' friendship network and 6,442,890 check-in records. We use 99,563 of those users who have at least one check-in records in our experiment. Since there is no user profile, we use the same method as that in [5] and take the center of the 25 km × 25 km area with the most number of check-ins as the home location. We then collected user profiles from Twitter, an on-line social networking and microblogging service which allows users to follow each other; as well as post and read "tweets". The user IDs and social network come from [19]. There are 660,000 distinct user IDs in total together with their social relations. We collected the profiles of these users using Twitter API[2]. We obtained locations of 148,860 users by converting the address in their profiles into geographic coordinates by the Google Maps Geocoding API[3]. The data was collected from April 14 to April 28, 2013. We define the friend relation in the same way as [10], i.e., users u_i and u_j have friend relation if they follow each other. Table 2 gives more information about these two data sets.

---

[2] https://dev.twitter.com/rest/public.
[3] https://developers.google.com/maps/documentation/geocoding/intro.

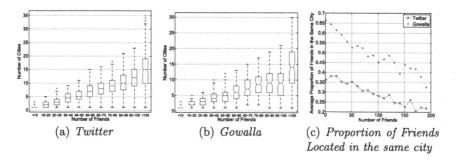

Fig. 1. The distribution of number of cities a user has friends located at and the average proportion of friends located in the same city.

Figure 1(a) and 1(b) show the city distribution of friends on Twitter and Gowalla social networks. For most users with less than 70 located friends, their friends will be located in no more than 10 cities. When a user has more than 100 located friends, his/her friends may be located in 10 to 20 cities. The rapidity and ease of modem transportation and communication present a great opportunity of meeting new friends in different places, leading to a wide distribution of friends, which challenges social network based location estimation. Figure 1(c) shows the average proportion of friends in the same city of a user with respect to the number of friends of a user. The result shows that with an increasing of the number of friends, the probability of friends located in the same city decreases quickly.

Friend locality is the average distance to the friends of the user u_i:

$$F_{u_i} = \frac{1}{|\Gamma_i'|} \sum_{u_j \in \Gamma_i'} d_{ij} \tag{1}$$

Here we use Γ_i' to denote the located friends of user u_i. In Fig. 2, it not surprising that friend locality increases with respect to the number of friends of a user. On one hand, not having enough friends will make location estimation difficult and on the other hand, having too many friends is not helping as well.

3.2 Friend Distribution of Connected Pairs

Friend-based methods were widely used in research literature [2,8,15]. By analyzing the distribution of friends' location or mobility, probability models can be built to predict locations of users. In these works, friend relation was taken as a binary feature: being friend or not. However, in the real world, depending on the social relation and other user behaviors, friends of a user can be very different. In this section, we propose the friend co-location index $(FCoI)$ to measure the "closeness" of friends on the social network.

We represent the social network as an undirected graph $G = (U, E)$, where U represents the user set, and edges in E exist between two users if they have a friend relation. There are two kinds of nodes in U. U' is the set of located

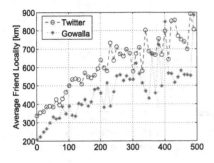
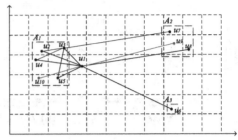

Fig. 2. The relationship between average distance and number of friends.

Fig. 3. Estimating location of user u_1. Here $u_2, u_3, \ldots, u_8$ are located friends, and u_1, u_9, and u_{10}'s locations are unknown

user and U^- represents the others, so $U = U' \cup U^-$. Before performing the location estimation, we first cluster a user u_i's friends Γ_i by their location. Friends in the same city will be put in the same set and we represent those sets as $\mathscr{A} = \{A_1, A_2...\}$. City is selected because of its natural definition of activity concentration by human geography. An example is shown in Fig. 3, users u_1 has 7 located friends. These friends distribute in three different cities and form three sets of friends A_1, A_2 and A_3.

To measure the closeness of two users on the social network, we propose the Friend Co-location Index ($FCoI$) that takes both the social connection and the location into account. The key idea is to measure the correlation of the friends' geographic distribution of two users. For a pair of friends u_i and u_j, we firstly generate two vectors for each of them to describe the friend distributions as:

$$v_k^i = |A_k^i|/|\Gamma_i'| \tag{2}$$

The $|\Gamma_i'|$ is the total number of located friends of user u_i and $|A_k^i|$ is the number of friends of user u_i located in city A_k. For example, in Fig. 3, u_3 has 4 located friends where three of them are in city A_1, one is in A_2, and none in city A_3. So the vector v^3 is [0.75 0.25 0]. Similarly, $v^1 = [0.57\ 0.29\ 0.14]$. After getting the distribution vectors, we can define the friend co-location index between user u_i and u_j as:

$$FCoI(u_i, u_j) = \frac{\sum_{k=1}^{m} min\{v_k^i, v_k^j\}}{\sum_{k=1}^{m} max\{v_k^i, v_k^j\}} \tag{3}$$

Here m is the total number of the cities. When the friends of two users have the same distribution, the distribution vectors of them will be also the same and the result of $FCoI$ is 1. On the other hand, if the distributions are completely different, e.g., all the friends of u_i are located in city A_1 and friends of u_j are located in A_2, the result will be 0. For example, in Fig. 3, the friend co-location index between u_1 and u_3 will be: $(0.57 + 0.25 + 0)/(0.75 + 0.29 + 0.14) = 0.69$. The reason we choose the index as the sum of min over the sum over max is to give more priority to a few cities where both users have a large number of friends over many cities where both users have a small number of friends. From the observation of the dataset, we can see that many users tend to have a small number of friends in many cities but only friends located in the same city tend to have a large number of friends in a few cities.

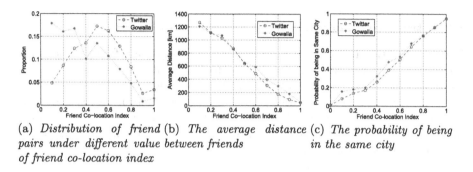

(a) *Distribution of friend pairs under different value of friend co-location index* (b) *The average distance between friends* (c) *The probability of being in the same city*

Fig. 4. Statistics on friend co-location index

Then we study whether the friend co-location index can reflect the geographic distance between two users effectively. We carry out these investigations on the Twitter and Gowalla users located in the North America to avoid the effect of oceans. Firstly, we calculate the proportion of friend pairs under different values of $FCoI$ in Fig. 4(a). The distributions of the two datasets are significantly different. More friend pairs in Gowalla have friend co-location index under 0.4, but the peak of the distribution of Twitter dataset is between 0.4 and 0.6. Another observation is that only a few friend pairs have the $FCoI$ more than 0.8 on both Twitter and Gowalla, which tells us that people have friends far away nowadays, and the social connections of each individual are quite different.

In Fig. 4(b), we show the relationship between the average geographic distance and different value of friend co-location index between friend pairs. Obviously when friends have higher $FCoI$, they tend to be geographic close. The results of Twitter and Gowalla data sets are very similar to each other. Figure 4(c) shows the relationship between $FCoI$ and the probability of the friend pairs located in the same city. When friend pairs have friend co-location index higher than 0.9, the probability of they located in the same city can be higher than 85 %. The probability increases with the increasing of the value of $FCoI$. From the investigations above, we can see that $FCoI$ can be a good index to reflect the geographic relationship between friends.

Friend Co-location Model: Our first model will be based on the investigations above. We firstly calculate the probability $P(FCoI(u_i, u_j))$. It denotes the probability of users u_i and u_j located at the same city when the value of friend co-location index between them is $FCoI(u_i, u_j)$. We can get the value of $P(FCoI(u_i, u_j))$ based on the statistics shown in Fig. 4(c). So for a user u_i in U^-, we can calculate the probability of u_i located at a city A_k as:

$$P(u_i, A_k^i) = 1 - \prod_{u_j \in A_k^i, (u_i, u_j) \in E} (1 - P(FCoI(u_i, u_j))) \qquad (4)$$

The city A_k with the maximum value of the probability will be chosen as the estimated location of user u_i.

3.3 Structure of an Individual's Friends in a City

Friend co-location index can help to identify more important friends in location estimation. However, when user pairs which contain at least one user who has more than 50 friends, the value of friend co-location index will be in a small range (0 to 0.2) and the influence of friend co-location index on location distribution is not distinct. In this case, the effectiveness of friend co-location index in finding important friends is weakened. We introduce the concept of the local social coefficient to deal with this problem.

Local Social Coefficient. The new measurement, local social coefficient, is similar to modularity used in community detection [6], but incorporates location information. The local social coefficient of a user u_i in city A_k is defined as:

$$LSoC(A_k^i) = \sum_{u_p, u_q \in A_k \wedge (u_i, u_p) \in E \wedge (u_i, u_q) \in E} [a_{pq} - \frac{|\Gamma_p||\Gamma_q|}{2|E|}] \qquad (5)$$

Here $|E|$ is the total number of edges in the social network G and $a_{pq} = 1$ if $(u_p, u_q) \in E$ and otherwise $a_{pq} = 0$. Intuitively, local social coefficient measures how tight the friends of u_i in city A_k are compared with expected number of friend connections from a random friendship formation as measured by $\frac{|\Gamma_p||\Gamma_q|}{2|E|}$.

One can choose the city with the highest value of the local social coefficient as the estimated location. The method may work well when a user has many friends and his/her friends can form structures in one or more cities. However, when a user has a small number of friends and his/her friends do not form structures in any cities, this method does not work well. Fortunately, this method performs better when the friend co-location based method fails (when a user has too many friends). In this paper, we propose a method to leverage the advantages of these two approaches and achieve overall much better performance than the state-of-the-art.

Local Social Coefficient Model: There are nearly 28 percent of users in Twitter and 8 percent of users in Gowalla who have only less than 10 percent of friends located within 100 km with them, and many of them have more friends in another city. The friend co-location based method can be very helpful when the estimated users don't have enough friends, but when the number of friends becomes larger, the effect of friend co-location index will be weakened. The local social coefficient becomes more helpful when users have more than 30 friends. In our investigation on Twitter and Gowalla data sets, the probabilities of two friends u_1 and u_2 located within 100 km are 0.2 (Twitter) and 0.27 (Gowalla). However, if there exists another user u_3 who can form a size-3 clique with u_1 and u_2 in the social network, the probability of u_1 u_2 located together can increase to 0.32 (Twitter) and 0.37 (Gowalla). Moreover, if u_1 and u_2 are located within 100 km, the probability of u_3 located close to them is 0.69 (Twitter) and 0.58 (Gowalla). In this method, we will calculate the local social coefficient of the groups of friends in each city as formula 5 and use the city which has the largest value of $LSoC$ as the estimation result.

4   Social Closeness and Social Structure-Based Model (SoSS)

We have introduced a friend co-location based model, in which we want to find more important friends in location estimation, and the local coefficient based model, which has better results when a user have many friends. Scale-free network theory [18] states that in a social network, a large portion of users has a small number of friends and the locations of these users are difficult to estimate by the location of their friends. So we propose a confidence-based iteration method to overcome the problem of sparsity of location information. In the Social Closeness and Social Structure-Based Model ($SoSS$), we combine these two models (friend co-location based and local social coefficient based models) together to overcome the disadvantages of them. After getting the two results from the two models above, we combine these two models by following the logistic regression based method in [9] as in formula 6. The parameters α, β_1, β_2, and β_3 can be trained by methods based on maximum likelihood. Since the number of friends have a great impact on the location estimation models, $|\Gamma_i|$ is chosen as a feature.

$$g(u_i, A_k) = \frac{exp(\alpha + \beta_1 LSoC(A_k) + \beta_2 P(u_i, A_k) + \beta_3 |\Gamma_i|)}{1 + exp(\alpha + \beta_1 LSoC(A_k) + \beta_2 P(u_i, A_k) + \beta_3 |\Gamma_i|)} \qquad (6)$$

4.1   Iteration with Confidence-Based Improvement

In location estimation, one of the main problem is the sparseness of location information. By the investigation of [2], only 6 % users provide their home address in their Facebook profiles. Our experiment shows that there are two main challenges in the location estimation: (1) some users only have a few number of friends. For these users, the accuracy can be very low since we cannot get enough location information from their friends, and (2) most of the users do not provide their location information in profiles. So we try to apply an iteration method to use the estimated locations. In the iteration steps, the estimated location will be taken as a friend's real location. However, there is a problem that the incorrect estimation results may lead to the decreasing of accuracy. So we propose a confidence-based iteration method.

Confidence-based method means that when we use the estimated location, we will judge the result with a confidence value and only use the results with high reliability of being correct in the iteration. Our investigation shows that the most helpful information is the aggregation of location distribution of friends. So we use an entropy-like method to measure the friends aggregate:

$$C_{u_i} = -\sum \frac{|A_k^i|}{|\Gamma_i|} \log \frac{|A_k^i|}{|\Gamma_i|} \tag{7}$$

Here, $|A_k^i|$ is the number of friends of user u_i in city A_k. So each time after we finish estimating of a user, we will calculate this entropy of him/her. In our model, we only use 66 % estimated locations on Twitter dataset and 74 % on Gowalla dataset with lower C_{u_i} in the iteration process. We get this ratio by analyzing the estimating accuracy on the training data and with this ratio, the iteration method can achieve the best accuracy. Over filtering will decrease the estimation accuracy since it may delete many useful correct estimating locations.

4.2   Complexity

Assume there are totally n users in the social network and each user has m friends on average. In the friend co-location based model, for each user u_i, we need to calculate the value of $FCoI(u_i, u_j)$ and $1 - P(FCoI(u_i, u_j))$ for every friend u_j. So the cost of this step is $O(m^2)$ and the complexity of the model is $O(nm^2)$. For the local social coefficient based model, we need to calculate the local social coefficient of each user at each city where he/she has friends located at. The worst case is that all the friends located at the same city and the complexity is $O(m^2)$. In the second step, we only need to choose the city with the highest value of the local social coefficient as the estimated result, so the complexity of this model is $O(nm^2)$. To combine the two models together, we need to repeat the calculation above at first and the complexity is $O(nm^2)$. The cost of the combination step depends on the number of cities and the cost will be less than or equal to $O(m)$. So the complexity of the $SoSS$ model is $O(nm^2)$.

5   Experiments

In this section, we evaluate our user location estimation models in comparison with existing state-of-the-art methods on two data sets, i.e., the Gowalla and the Twitter data sets introduced before. We first show the estimation accuracy of the Friend Co-location Model and Local Social Coefficient Model with respect to the number of friends which illustrates why we combine these two models into our Social Closeness and Social Structure-Based Model. Then we test the effect of different parameters including the percentage of users who provide their locations, the number of friends, and the number of iterations, and the error distance.

Evaluation Metrics. We first define the error distance of the estimation of user u_i as $Err(u_i)$, which represents the distance between the estimated location and the actual location of the user u_i. We consider the estimated locations with error distance less than 100 km as correct estimations. So the estimation accuracy can be represented as $\frac{|u_i|u_i \in U \wedge Err(u_i) \leq 100km|}{|U|}$.

5.1   Methods

The models, we tested in the experiment, are shown as follows:

- **Friend-based method** (FB): We take the friend-based method (FB) provided in [2] as the baseline method. In [2], the authors estimated a user location by his friend locations based on the relationship between distance and the probability of being friends. They proposed their estimation model as: $\prod_{(u_i, u_j) \in E} P(|l_i - l_j|) \prod_{(u_i, u_j) \notin E} (1 - P(|l_i - l_j|))$. Here $P(|l_i - l_j|)$ represents the probability of user u_i and u_j located with the distance of $|l_i - l_j|$ and E is the set of friend relation. Then they optimize the formula as: $\prod_{(u_i, u_j) \in E} \frac{P(|l_i - l_j|)}{1 - P(|l_i - l_j|)}$.
- **Social relation based model** (SR): This method is proposed in [11]. The authors apply three important methods to select the nearest friend: (1) the geometric median; (2) the minimum area formed by users and two of his friends (Oja's Simplex Median); and (3) there exists friend relationship between three users which is referred as the Triangle Heuristic.
- **Friend Co-location Based Model**: Our method.
- **Local Social Coefficient Based Model**: Our method.
- **Social Closeness and Social Structure Based Model** ($SoSS$): Our method.

We also test the FB, SR, and $SoSS$ methods combined with our confidence-based iteration model, which are noted as FB_I, SR_I, and $SoSS_I$. The default value for the percentage of location withheld is 25 % for all experiments when not specified. At each time, the parameters used in our algorithms, like the probabilities $P(FCoI(u_i, u_j))$ in the Friend Co-location Model, will be retrained on the other 75 % users. All accuracies reported is based on 3-time average on random sampling.

5.2   The Results of Experiment

Friend Co-location Model vs. Local Coefficient Model. Firstly, we will show the performance of friend co-location model and local coefficient model separately and explain why we combine them together. Figure 5(a) and 5(b) show the estimation accuracy of these two models on Twitter and Gowalla networks. From the figure, we can see that when users have less than 20 friends, the friend co-location based model can be much more useful than the local social coefficient model. It can help us to find out more important friends by the analysis of the relevance of friends distribution. The local social coefficient model performs worse because when a user has less than 20 friends, it is likely that there is no friend relation between his/her friends. With the increasing of the number of friends, the local social coefficient based model will perform better. The accuracy can be more than 60 percent on Twitter and 80 on Gowalla, which indicates that the analysis of the local tightness can be helpful. Since these two models perform quite differently, we combine them together to make sure that our model can work in different cases.

Effect of Percentages of Unknown Locations. We then test those models under different settings of the two datasets. We randomly withhold 0 %, 25 %, 50 %, and 75 % users' location information of the two datasets and compare the performances of different models. Table 3 shows the results of each method.

From these results, we can see that our models can improve the estimation accuracy by 5 to 20 percent compared with the baseline methods. The *SoSS* model combined with the confidence-based iteration achieves the best performance among them. This phenomenon demonstrates that the friend co-location

Table 3. Effect of percentage of unknown locations

% locations withheld, by platform	FB	FB_I	SR	SR_I	$SoSS$	$SoSS_I$
Twitter 75 %	35.7	43.5	35.9	43.7	40.9	**48.6**
Twitter 50 %	41.8	46.5	43.1	48.0	47.3	**52.5**
Twitter 25 %	48.0	48.5	48.3	50.2	54.6	**55.2**
Twitter 0 %	48.3	–	51.8	–	**55.6**	–
Gowalla 75 %	47.1	62.7	44.9	60.8	50.1	**70.8**
Gowalla 50 %	64.7	69.3	61.5	69.0	68.7	**77.5**
Gowalla 25 %	73.0	73.5	71.2	73.1	79.6	**80.3**
Gowalla 0 %	74.2	–	72.9	–	**80.6**	–

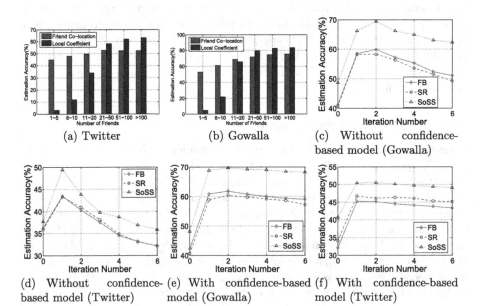

(a) Twitter   (b) Gowalla   (c) Without confidence-based model (Gowalla)

(d) Without confidence-based model (Twitter) (e) With confidence-based model (Gowalla) (f) With confidence-based model (Twitter)

Fig. 5. The difference between the friend co-location method and the local social coefficient method and the influence of confidence-based iteration process.

and the local social coefficient can help us to detect the more important friend or group of friends effectively in the estimation.

Another observation is that the confidence-based iteration process contributes greatly in the estimation accuracy, especially in the cases when only a small portion of users have their location information known in the dataset. From the estimation results, the iteration process can improve the accuracy by 1 to 3 percent when we withhold 25 % users' location information. This ratio increases to 5 to 22 percent when 75 % of users' location are unknown. When more users' location information are unknown (from 25 % to 75 %), the estimation accuracy of those models without iteration process (FB, SR, and $SoSS$) will decrease by more than 15 percent on both Twitter and Gowalla datasets. However, at the same time, the decreasing of accuracy of the models with iteration process is just about 7 percent on Twitter dataset and 10 percentage on Gowalla dataset. This phenomenon tells that the confidence-based iteration model can help us to overcome the problem of sparsity of user location.

The models work better on Gowalla dataset. We can explain this phenomenon by analyzing the difference of user behavior on Gowalla and Twitter. Users who use Gowalla tend to add friends who live close to their locations (27 % within 100 km), and Twitter users will add many users which may live far away from them and only 12 % friend pairs are located within 100 km.

Effect of Confidence-Based Iteration. In this experiment, we investigate how the iteration process improves the estimation accuracy and explain why we introduce the confidence-based method. Here we withhold 75 % user location information and estimate their locations. Figure 5(c) and 5(d) show the estimation accuracy of the iteration method without the confidence based selection. When the iteration number is 0, it is the accuracy of the basic models. The iteration process can improve the estimation accuracy by nearly 20 percent on the Gowalla dataset and 10 percent on the Twitter dataset. However, as the iteration process continues, the estimation accuracy begins to decrease. Prominently, the accuracy will decrease by nearly 10 percent in the experiment on the Twitter dataset after the sixth iteration. So we introduce the confidence-based iteration method as shown in Fig. 5(e) and 5(f). The confidence-based method can prevent the estimation accuracy from decreasing and keep it to stable and higher values of accuracy. In most cases, this process can achieve the best accuracy in three times iteration.

Effect of Number of Friends. Then we investigate the influence of the number of friends. Figure 6 gives the summary of the estimation accuracy of groups of users with different number of friends. The estimation accuracies of all the models without the iteration step are very low when the users have few number of friends. That is because: (1) if a user only has a few friends, none of his/her friends may have their location information known. So we cannot perform the estimation by their friends' location. (2) It is difficult to decide which friend may locate closer to the user if his/her friends do not cluster. The performances of $SoSS$ model are better than the FB and SR models on both the datasets. This

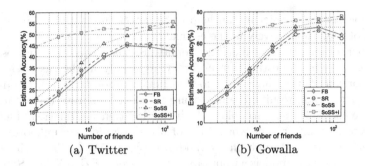

(a) Twitter   (b) Gowalla

Fig. 6. Influence of number of friends.

is because our *SoSS* model distinguishes friends based on friend co-location and local coefficient indexes. Choosing socially close friends for location estimation improves the results.

On the other hand, the confidence-based iteration method can help to improve the estimation accuracy by more than 20 percent for the users who have few number of friends. With the iteration process, more users who don't have location information in the social network will be given an estimated location, and those estimated location can be used in estimating others' location to help to overcome the first problem above. So after we combine those models with the confidence based iteration model, many of the users who have a few friends can also be estimated correctly. When the users have more than 30 friends, the confidence based iteration model cannot help much on the accuracy.

Influence of Error Distance. We test the estimation accuracy on different values of error distance and show the results in Fig. 7. The accuracy of each method will be close to 1 when the error distance is more than 3,000 km. Our method performs better when we set a smaller error distance. When the error distance is larger than 200 km, the accuracy of different models will get close.

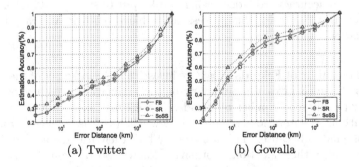

(a) Twitter   (b) Gowalla

Fig. 7. The estimation accuracy under different error distances.

6   Conclusion

In this paper, we investigate the relationship between social closeness/friend structure and geographic distance. Based on these investigations, we develop two user location estimation models, the friend co-location based model and a local social coefficient model. We further combine these two models to avoid the disadvantages of each. To improve the estimation accuracy, we also propose a confidence-based iteration method. Finally, we test these models on two different datasets and demonstrate how the models work under different situations. The results of the experiment show that our method can improve the estimation accuracy by 5 %–20 % compared with the baseline algorithms.

Acknowledgments. This work is supported in part by USDOD. We would like to thank the scientists from USDOD, Dr. James Kang and Dr. Joshua Trampier, for their insights and detailed feedback on this work.

References

1. Amitay, E., Har'El, N., Sivan, R., Soffer, A.: Web-a-where: geotagging web content. In: SIGIR, pp. 273–280. ACM (2004)
2. Backstrom, L., Sun, E., Marlow, C.: Find me if you can: improving geographical prediction with social and spatial proximity. In: WWW, pp. 61–70. ACM (2010)
3. Chandra, S., Khan, L., Muhaya, F.B.: Estimating twitter user location using social interactions-a content based approach. In: SocialCom, pp. 838–843. IEEE (2011)
4. Cheng, Z., Caverlee, J., Lee, K.: You are where you tweet: a content-based approach to geo-locating twitter users. In: CIKM, pp. 759–768. ACM (2010)
5. Cho, E., Myers, S.A., Leskovec, J.: Friendship and mobility: user movement in location-based social networks. In: SIGKDD, pp. 1082–1090. ACM (2011)
6. Clauset, A., Newman, M.E.J., Moore, C.: Finding community structure in very large networks. Phys. Rev. E **70**(6), 066111 (2004)
7. Compton, R., Jurgens, D., Allen, D.: Geotagging one hundred million twitter accounts with total variation minimization (2014). arXiv preprint arxiv:1404.7152
8. Davis Jr., C.A., Papa, G.L., de Oliveira, D.R.R., de L Arcanjo, F.: Inferring the location of twitter messages based on user relationships. Trans. GIS **15**(6), 735–751 (2011)
9. Ho, T.K., Hull, J.J., Srihari, S.N.: Decision combination in multiple classifier systems. IEEE Trans. Pattern Anal. Mach. Intell. **16**(1), 66–75 (1994)
10. Huberman, B.A., Romero, D.M., Fang, W.: Social networks that matter: Twitter under the microscope (2008). CoRR, abs/0812.1045
11. Jurgens, D.: That's what friends are for: inferring location in online social media platforms based on social relationships. ICWSM **13**, 273–282 (2013)
12. Kong, L., Liu, Z., Huang, Y.: Spot: locating social media users based on social network context. In: Proceedings of the VLDB Endowment, vol. 7, (13), pp. 1681–1684 (2014)
13. Li, R., Wang, S., Deng, H., Wang, R., Chang, K.C.-C.: Towards social user profiling: unified and discriminative influence model for inferring home locations. In: SIGKDD, pp. 1023–1031. ACM (2012)

14. Mahmud, J., Nichols, J., Drews, C.: Home location identification of twitter users (2014). CoRR, abs/1403.2345
15. Sadilek, A., Kautz, H., Bigham, J.P.: Finding your friends and following them to where you are. In: WSDM, pp. 723–732. ACM (2012)
16. Sankaranarayanan, J., Samet, H., Teitler, B.E., Lieberman, M.D., Sperling, J.: Twitterstand: news in tweets. In: ACM SIGSPATIAL, pp. 42–51. ACM (2009)
17. SocialMediaToday (2013). http://socialmediatoday.com/irfan-ahmad/1854311/twitter-statistics-ipo-infographic
18. Wang, X.F., Chen, G.: Complex networks: small-world, scale-free and beyond. IEEE Circuits Syst. Mag. **3**(1), 6–20 (2003)
19. Wang, X., Liu, H., Zhang, P., Li, B.: Identifying information spreaders in twitter follower networks. Technical report TR-12-001, School of Computing, Informatics, and Decision Systems Engineering, Arizona State University (2012)

Efficient Influence Maximization in Weighted Independent Cascade Model

Yaxuan Wang$^{(\boxtimes)}$, Hongzhi Wang, Jianzhong Li, and Hong Gao

Harbin Institute of Technology, Harbin, China
{wangyaxuan,wangzh,lijzh,honggao}@hit.edu.cn

Abstract. Influence maximization (IM) problem which aims to find the most influential seed set in a social network plays an important role in viral marketing. However, previous solutions pay all attention to the structure of network, which causes trouble in real-word applications.

D. Kempe et al. [8] presented that a non-negative weight can be attached to each node to extend the applicability of traditional models. Although this idea is much applicable in practice, there is little research based on this opinion. Thus, we develop substantial study about this issue. We extend the Independent Cascade (IC) model and present Weighted IC (WIC) model. The IM problem in WIC model is NP-hard. To solve this problem, we present a basic greedy algorithm and Weight Reset (WR) algorithm. Moreover, we propose Bounded WR (BWR) algorithm, a Fully Polynomial-Time Approximation Scheme (FPTAS).

Experimentally, WIC model outperforms IC model in nearly 90 % in weighted IM problem. Moreover, BWR achieves excellent approximation and efficiency which is faster than greedy algorithm more than four orders of magnitude. Especially, BWR can handle huge networks with millions of nodes in several tens of seconds while keeping high accuracy. This result demonstrates the effectiveness and efficiency of BWR.

1 Introduction

Viral marketing requires to select the initial crowd to make most people, who are interested in a specific topic, receive the product information and generate the largest value [12]. Such requirement involves Influence Maximization (IM), one of the most popular research topics in social network. The general IM problem is to find k initial seeds in a network to achieve the greatest propagation.

Even though existing solutions could solve this problem in many scenarios, they pay all attention to the connectivity of nodes and ignore other attributes of nodes. This defect may cause distress in many practical applications. Consider a usual scenario, an automobile manufacturing company wants to promote the sale for its luxury cars by providing test drive chances to a small crowd. Our target is maximizing the value of the propagation process (such as selling most cars) instead of influencing the largest population. Thus, WIC model is more preferable than IC model in such case. But little research on WIC model has been developed. It is an arduous task to revise existing solutions to cater on the

© Springer International Publishing Switzerland 2016
S.B. Navathe et al. (Eds.): DASFAA 2016, Part II, LNCS 9643, pp. 49–64, 2016.
DOI: 10.1007/978-3-319-32049-6_4

extra node attributes since they neglect the properties of node itself. Motivated by this idea, we attempt to present exclusive solutions to solve the IM problem on WIC model by taking the node attributes into consideration.

The IM problem on WIC model is not trivial. It is an NP-hard problem. In addition to connectivity estimating, designing a criterion which considers both the networking structure and independent attributes of nodes is the main target of this paper. By designing an elaborate mechanism, our solution can select the most valuable nodes according to both independent attribute and connectivity.

1.1 Our Contribution

In this paper, we first start substantial research about the IM problem on WIC model. We prove that the IM problem in this model is NP-hard. Then, we present a basic greedy algorithm to solve the IM problem in our WIC model.

Considering that basic greedy algorithm may have intrinsic trouble to be scalable in large graphs, we present WR algorithm to tackle the efficiency issue. Our WR algorithm can return a $(1 - 1/e)$-approximation and its expected running time is $O(kn_{in} \cdot n)$. Moreover, we propose BWR algorithm, an FPTAS, to make further effort to improve the efficiency. The experimental results show that our BWR algorithm is effective and scalable in both IC and WIC model. More importantly, our algorithm achieves both efficiency and effectiveness. It is scalable to handle large networks with millions of nodes in several tens of seconds while its performance is close to the best outcomes in polynomial time.

In summary, our main contributions in this paper are as follows.

1. We first apply the idea of node weight and present WIC model to provide a more applicable solution for IM problem, which could maximize the value of influence instead of the amount of the influenced nodes.
2. We propose a basic greedy algorithm which can achieve a $(1-1/e-\epsilon)$ approximation in polynomial time. For efficiency issues, we design WR algorithm with a similar approximation ratio whereas the time complexity is narrowed from $O(knRm)$ to $O(kn_{in} \cdot n)$. To accelerate the algorithm further, we add a branching strategy and present BWR algorithm.
3. We conduct extensive experiments on different real-world social networks to prove that our WIC model outperforms IC model in terms of IM problem in practice. Experimental results also show that BWR algorithm is better than other existing algorithms. Its running time outperforms greedy about four orders of magnitude with little sacrifice, which illustrates the high efficiency of our BWR algorithm in gigantic networks.

1.2 Related Work

IM problem has been extensively studied. In [4], Domingos et al. defined the basic problem and presented a fundamental algorithm for digging a network from the data. Kempe et al.[8] believed that the issue of choosing influential sets was a discrete optimization problem. He proved that this problem is NP-hard and designed three kinds of cascade models: IC model, WC model and LT model. He proposed a greedy algorithm framework which can guarantee a 63 % accuracy bound in three models. More models which are integrated with other factors such as time [11,17] or location [7] are explored.

Moreover, Leskovec et al. [10] optimized basic greedy algorithm by avoiding evaluating the expected spreads. This approach was enhanced in [5] with 50 % additional improvements in efficiency. Recently, TIM algorithm whose node selection phase is similar to RIS [5], was presented [16]. Chen et al. [2] proposed PMIA and Wang et al. [18] identified influential nodes from different small communities individually. However, the lack of considering extra attributes of nodes makes them ineffective under some practical circumstances.

Paper Organization. Section 2 introduces WIC model, its hardness and a basic greedy algorithm. Section 3 proposes our WR algorithm as well as its extended version, BWR algorithm. In Sect. 4, we show our experimental results. We draw conclusions and discuss future directions for our topic in Sect. 5.

2   WIC Model and Its Greedy Algorithm

In this section, we formally define our WIC model and present a basic greedy algorithm with the best performance accuracy in WIC model.

2.1   Problem Definition

Definition 1. *(WIC model) Given a directed graph* $G = (V, E)$, *let each edge* $e \in E$ *have a propagation probability* $p_{u,v} \in [0,1]$. *For each node* $v \in V$, *there is a non-negative weight* w_v *which is independent of the network structure.*

For node u, the predecessors of u are the nodes which can arrive u in finite steps and the nodes which u can arrive are the successors of u. Any social network can be modeled as WIC model. For each node v, w_v shows its uniform weight. In WIC model, the steps of a time-stamped influence process are as follows.

1. At timestamp 0, all nodes in $G = (V, E)$ are inactive.
2. At timestamp 1, we activate a set of nodes called Seed Set S_1 while other nodes are still inactive.
3. At timestamp i $(i > 1)$, we assume the nodes in S_{i-1} are activated in step $i - 1$. For each node u in S_{i-1} and edge $e_{u,v} \in E$ with v as an inactive node, v is activated with probability p_{uv}. If v is activated in this step, v is added to S_i. For any $j < i$, $S_i \cap S_j = \emptyset$.
4. The process halts when in some step t, $S_t = \emptyset$.

WIC model provides an attribute-based node selection mechanism to maximize the profit of the influence. The profit is defined as the values of all nodes which are activated by seed set S, denoted by V_S. Given a seed set S_1, $\sigma(S_1)$ denotes the expectation of influence value generated by S_1. That is,

---

Algorithm 1. BasicGreedy(G, k)

1: Initialize a set $S = \phi$
2: **for** $i = 1$ *to* k **do**
3:      **for** each node $v \in V \backslash S$ **do**
4:         $\text{sum}_v = 0$
5:         **for** $j = 1$ *to* R **do**
6:            $\text{sum}_v + = |RanCas(S \cup v)|$
7:         $\text{sum}_v = \text{sum}_v / R$
8:      $S = \{S \cup arg\max_{v \in V \backslash S}\{s_v\}\}$
9: **return** S

$$\sigma(S) = \sum_{u \in S}(\sum_{v \in V} w_v \cdot p_r(u, v) + w_u) \tag{1}$$

$p_r(u, v)$ is the comprehensive probability of reachability from u to v including all reachable paths. Obviously, the target of weighted IM(WIM) is to select the seed set S_1 to maximize $\sigma(S_1)$. Therefore, the WIM problem is defined as follows:

Problem 1. Given a non-negative integer k and graph $G = (V, E)$, the WIM problem is to find a node set $S^* = arg\max_{S \subseteq V}\{\sigma(S) \mid |S| = k\}$ where $S^* \subseteq V$.

Theorem 1. *The weighted influence maximization problem (WIM) is NP-hard.*

The WIM problem is more general than IM problem [8]. According to [3], the WIM can be also reduced from a classic NP-hard problem, Set Cover problem [6]. For the interest of space, we omit the detail of the proof. If we set the value of every node equally, the WIC model can be simplified into IC model. Therefore, WIM is a generalization of IM and solutions of WIC can be adopted in IC model.

2.2 The Basic Greedy Algorithm

In this section, we propose the greedy algorithm and its accuracy guarantees.

The strategy of our greedy algorithm is to choose the node which can make maximal marginal gain for σ. Algorithm 1 shows the general basic algorithm based on hill-climbing strategy. In each round, the algorithm computes the additional influence spread of each node v if node $v \notin S$ is activated. The function $RanCas(S \cup \{v\})$ is a random process and repeated R times (Line 6) to simulate the process of real propagation. Then the node with max marginal gain is added to the selected set S (Line 8). Thus the time complexity of Algorithm 1 is $O(knRm)$, where n and m are the total number of the nodes and edges. The following lemma explores the property of the value function $\sigma(.)$.

Lemma 1. *The value function $\sigma(S)$ is submodular and monotone.*

Proof. For all $v \in V$ and all subsets of V where $S \subseteq T \subseteq V$, we define the successors of a node v ($v \notin T$) as $R(v)$. The probability of reachability from set S and T to v_1 is p_{S,v_1} and p_{T,v_1}. Then, according to (1), we can obtain $\sigma(S \cup v) - \sigma(S) = \sum_{v_1 \in R(v)} w_{v_i} \cdot p_{vv_1}(1 - p_{S,v_1})$.

Similarly, for T, the relation is $\sigma(T \cup v) - \sigma(T) = \sum_{v_1 \in R(v)} w_{v_i} \cdot p_{vv_1}(1 - p_{T,v_1})$.

The only difference between these two equations is p_{S,v_1} and p_{T,v_1}. Since $S \subseteq T$, $p_{S,v_1} \leq p_{T,v_1}$, $\sigma(S \cup v) - \sigma(S) \geq \sigma(T \cup v) - \sigma(T)$ holds. Thus, non-negative real valued function σ is *submodular*. Moreover, since $\sigma(S)$ is the expectation of influence value generated by S, $\sigma(\emptyset) = 0$ and the marginal increase of σ always > 0. If $S \subseteq T$, $\sigma(S) \leq \sigma(T)$. Therefore, the value function $\sigma(S)$ is monotone. □

Since σ is a submodular and monotone function, maximizing $\sigma(S)$ can be approximated by maximizing the marginal gain [6]. In Algorithm 1, R is large enough to eliminate deviations from random processes.

Theorem 2. *Algorithm 1 yields $(1 - 1/e - \epsilon)$-approximate solutions where e is the base number of the natural logarithm and ϵ is any real number which $\epsilon \geq 0$.*

Proof. According to Lemma 1, the objective function $\sigma(.)$ is submodular and monotone. Let S be the outcome of Algorithm 1 and S* be an optimal set that maximizes the value of $\sigma(.)$. According to [13], $\sigma(S) \geq (1 - 1/e) \cdot \sigma(S^*)$. Therefore, Algorithm 1 achieves a $(1 - 1/e - \epsilon)$-approximation. □

3   Weight Reset Algorithm

The time complexity of Greedy prevents it from scaling to large graph, we design a novel algorithm to reduce time expenditure. In this section, we present WR algorithm to estimate the influence spread by resetting the weights of nodes. For ease of understanding, Table 1 summarizes the notations used.

At a high level, WR contains following two phases.

1. **Pre-treatment.** This phase computes $p_r(u, v)$ for each pair of reachable nodes and organizes proper data structures to facilitate node selection.
2. **Node Selection.** This phase selects k nodes with the largest marginal value of $\sigma(.)$ iteratively. Once a node is selected, its weight is reset and the value of relevant nodes are updated.

3.1   Pre-treatment

Given a WIC model, the reachability from node u to v is $p_r(u, v) = 1 - \prod_{i=1}^{r_{uv}}(1 - p_i(u, v))$. Intuitively, $p_r(u, v)$ is the probability that u activates v through all possible paths from u to v. Since each node has *weight* as an additional attribute, we need to develop a particular mechanism to estimate V_u. Moreover, we should estimate the influence of u on its predecessors and successors if u is selected.

For a node u, the estimations of both V_u and W_u require to access all successors and predecessors of u, respectively. Firstly, to estimate V_u, we organize all successors of u into a tree. By Breadth-First-Search(BFS) with u as the root,

Table 1. Frequently used notations.

Notation	Description
p_{uv}	The probability of edge $e_{u,v}$
$p_r(u,v)$	The probability that u and v are reachable
$p_i(u,v)$	The ith path from u to v
r_{uv}	The sum of the paths from u to v
w_u	The weight of node u
$IVT(u)$	The influence value tree of u
$WDT(u)$	The weight discount tree of u
V_u	The value of node u
W_u	The total value created by u if u's neighbors activate u
O_u	The set of successors of u
I_u	The set of predecessors of u
θ	Bound parameter of BWR algorithm
α	Steps which influence can propagate
β	Performance bound of BWR

if there is an edge $e_{u,v}$, v is added into this tree as one child of u. If a node is visited multiple times during the traversal due to multiple paths between a pair of nodes, we only update $p_r(u,v)$ rather than add another edge to prevent rings in this tree. According to discussion above, we define such tree as follows.

Definition 2. *Influence Value Tree(IVT) For a node $u \in V$, the $IVT(u)$, is a weighted tree (O_u, E_u, w), where O_u is the set of all successors of u, E_u is the set of all edges from u to v if $v \in O_u$, and w is the weight set of all v. The expected value of a node V_u, is $E[V_u] = \sum_{i=1}^{|O_u|} p_r(u,v_i) \cdot w_{v_i}$.*

Correspondingly, We organize u's predecessors $WDT(u)$. To build WDT for each node, we add the node v where $p_r(v,u) > 0$ with u as the root and ignoring rings. However, different from IVT, if there is an edge $e_{v,u}$, v is added into the WDT as a child of u. Thus, in $WDT(u)$, a child node points to its father node to represent the direction of $e_{v,u}$.

Definition 3. *Weight Discount Tree(WDT) For a node $u \in V$, $WDT(u)$ is a weighted tree (I_u, E_u, w), where I_u is the set of all predecessors of u, E_u is the set of all edges from v to u if $v \in I_u$, and w is the weight set of all v. The expected W_u is $E[W_u] = \sum_{i=1}^{|I_u|} p_r(v_i,u) \cdot w_u$.*

The pseudo code of computing $p_r(v,u)$ is shown in Algorithm 2, which is a recursive algorithm. u is the initial node and v is the neighbor of u. PathList is a list of nodes between u and v. If v is not in current $IVT(u)$, v is added into $IVT(u)$ (Line 3). For each new path from u to v, we update $p_r(u,v)$ (Line 6). We do not have to keep IVT and WDT for each node since they can be built rapidly

Algorithm 2. $genPr(u, v, pathList)$

1: **if** $v \in pathList$ **then**
2:    **return**
3: add v into pathList
4: **if** $v \notin IVT(u)$ **then**
5:    add v into $IVT(u)$
6: $p_r(u, v) = 1 - (1 - p_r(u, v))(1 - p_r(u, v') \cdot p_{v'v})$
7: /* where v' is the previous node of v on pathList*/
8: **for** each out-neighbor w of v **do**
9:    $genPr(u, w, pathList)$

Algorithm 3. NodeSelection(G, k)

1: Initialize $S = \phi$
2: **for** $i = 1$ to k **do**
3:    select $u = arg \max_{v \in V \setminus S} V_v$; add u into S; $w_u = 0$;
4:    /* update the value of others nodes*/
5:    **for** each node $v \in WDT(u)$ **do**
6:       recompute the V_v
7:    **for** $v' \in IVT(u)$ **do**
8:       $w_{v'} = (1 - p_r(u, v')) \cdot w_{v'}$
9:    update value V for each node
10: **return** S

according to $p_r(u, v)$. So we just build IVT and WDT on demand to save space. In our implementations, we keep $p_r(u, v)$ as key-value pair where $p_r(u, v) > 0$. Given that most nodes are not reachable, more space is saved.

3.2   Node Selection

Node selection process chooses the node to make maximal marginal increase of $\sigma(\cdot)$. Algorithm 3 presents WR's node selection algorithm which contains k iterations (Line 2–10). In each iteration, the algorithm selects a node u with the largest IVT value. After k round iterations, S is returned as the final result. The IVT value updating is the core of this algorithm.

Updating WDT and IVT. Once u is added into S, we should estimate the expected influence generated by u on WDT(u) and IVT(u).

To update WDT(u), once a node u is selected, w_u is reset into 0 (Line 3). Then we calculate all V_v if $v \in WDT(u)$ again (Line 5–6). This solution can perfectly reduce all the value increment on V_v caused by u where v is a predecessor of u.

Approximation Ratio Bound. The WIM problem can be reduced into Set Cover problem (Theorem 1). Moreover, WR algorithm is based on hill-climbing strategy. Thus, we can conclude the approximation ratio bound of WR algorithm.

Theorem 3. *WR achieves* $(1 - 1/e)$*-approximate ratio for WIM problem.*

For the interests of space, we omit the proof, which is similar to Theorem 2.

Time and Space Complexity. Assuming that $n_{in} = \max_{u \in V}\{|WDT(u)|\}$, building IVT and WDT can be viewed as BFS from each node. So, average running time of building IVT and WDT for a node(Algorithm 2) is far less than $O(kn_{in} \cdot n)$. We keep $p_r(u,v)$ for all pairs of reachable nodes since the computation of IVT and WDT need $p_r(u,v)$ all the time. It costs $O(n)$ to keep V_u of each node u. Thus, the space complexity in pre-treatment is $O(n^2)$.

In each round of node selection, it costs $O(n)$ to select a node with maximal value in V_u and $O(2n_{in})$ to update IVT and WDT, and $O(n_{in}^2)$ to update V_v whose $IVT(v)$ has changed. Thus, the running time of Node Selection is $O(k(n + 2n_{in} + n_{in}^2))$. This phase requires no extra space since it just updates the outdated node value. Therefore, the total time complexity of WR algorithm is $O(kn_{in} \cdot n)$.

n_{in} is related to structure of networks. If a network is dense, the IVTs and WDTs could be very large which means WR will be inefficient. To handle high-density networks within tiny loss in accuracy, we present bounded WR algorithm.

3.3   Bounded Weighted Reset Algorithm

In this subsection, we improve the practical performance of WR algorithm. According to Sect. 3.2, Pre-treatment process is costly. Since n is fixed, we reduce n_{in} to increase the efficiency. Moreover, after several steps in each iteration in Algorithm 2, $p_r(u,v)$ may get too small to influence the node selection order. Based on this observation, we use a threshold θ to bound the volume of IVT and WDT to achieve a high performance even in dense networks.

Definition 4. *(Bounded IVT and WDT) For a node $u \in V$, the bounded IVT of u is $BIVT(u, \theta) = \{v | v \in V, p_r(u,v) > \theta\}$ and $BWDT(u, \theta) = \{v | v \in V, p_r(v,u) > \theta\}$.*

The pre-treatment of BWR is shown in Algorithm 4. After computing $p_r(u,v)$, we anticipate the $p_r(u,w)$ for the next node (Line 8–10). If $p_r(u,w) < \theta$, we assume that u and w are not reachable and stop the iteration. This is because $p_r(v,w)$ gets too small to influence the node selection order. Furthermore, the estimation of these tiny differences is extremely costly since the number of reachable neighbors grows exponentially. In each iteration, we pre-compute the $p_r(u,w)$ of the next iteration (Line 9) to decide whether we start the next iteration, which can further save more unnecessary calculations. With θ, we can bound IVT and WDT in a small size since the nodes with low $p_r(u,v)$ in original IVT and WDT are cut out. By keeping n_{in} small, we save much running time.

BWR is shown as Algorithm 5. In initialization step, we set each node as the initial node u (see Algorithm 4) to obtain all $p_r(u,v)$ where $p_r(u,v) > \theta$ (Line 3–4). Then, node selection process (Line 6–15) starts. In each round, Algorithm 5 selects a node u with maximal V_u. For each node v in $BWDT(u)$, the expectation value of activating u is eliminated (Line 9–10). For each node v' in $BIVT(u)$, their weights are reset (Line 11–12). Furthermore, it resets the value of u to 0 (Line 13). The values of nodes in $BIVT(u)$ are recomputed (Line 14–15).

Algorithm 4. $genPr(u, v, pathList, \theta)$

1: **if** $v \in pathList$ **then**
2:     **return**
3: add v into pathList
4: **if** $v \notin IVT(u)$ **then**
5:     add v into $IVT(u)$
6: $p_r(u, v) = 1 - (1 - p_r(u, v))(1 - p_r(u, v^{'}) \cdot p_{v^{'}v})$
7: **for** each neighbor w of v **do**
8:     **if** $p_r(u, v) \cdot p_{v,w} > \theta$ **then**
9:         $genPr(u, w, pathList, \theta)$
10:     **else**
11:         **return**

Algorithm 5. $BWR(G, k, \theta)$

1: /*Initialization*/
2: set $S = \phi$; set each $p_r(u, v) = 0$;
3: **for** each node $v \in V$ **do**
4:     $genPr(v, new\ pathList, v, \theta)$
5: /*main loop*/
6: **for** $i = 1$ *to* k **do**
7:     select $u = arg\max_{v \in V \setminus S} V_v$; add u into S;
8:     /* update the value of others nodes*/
9:     **for** each node $v \in BWDT(u)$ **do**
10:         $V_v = V_v - (w_u \cdot p_r(v, u))$; /* remove the expectation value of u*/
11:     **for** $v^{'} \in BIVT(u)$ **do**
12:         $w_{v^{'}} = (1 - p_r(u, v^{'})) \cdot w_{v^{'}}$
13:     $w_u = 0$
14:     **for** each node $v^{'}$ where $BIVT(v^{'})$ has changed **do**
15:         $V_{v^{'}} = \sum_{i=1}^{O_{v^{'}}} p_r(v^{'}, v_i) \cdot w_{v_i}$; /*update value V for each node*/
16: **return** S

Considering the WIC model, V_v is a random quantity. p_e is the probability of edge e, d_i is the out-degree of node v_i and w_i is the weight of node v_i. In order to simplify this expression, we set v_{i_j} as the j hops neighbor of v_i; p_{i_j} as the possibility of edge $e_{v_{i_{j-1}}, v_{i_j}}$; d_{i_j} as the out-degree of v_{i_j} and w_{i_j} as the weight of v_{i_j}. Assuming the probability p_e, d_i and w_i are independent to each other, we can estimate the expectation of V_v as $E[V_v] = \sum_{i_1=1}^{d_v} E[p_{v_{i_1}}]E[w_{i_1}] + \cdots + \sum_{i_\alpha=1}^{d_{i_{\alpha-1}}} E[p_{i_1}] \cdot E[p_{i_2}] \cdots E[p_{i_\alpha}] \cdot E[w_{i_\alpha}]$.

Supposing nodes are independent, by treating p_i, d_i and w_i as random variables, all nodes share same $E[p_i]$, $E[d_i]$ and $E[w_i]$, which are denoted by p, d and w, respectively. Thus, the expectation of V_v is $E[V_v] = (p \cdot d + p^2 \cdot d^2 + \cdots + p^\alpha \cdot d^\alpha) \cdot w$. Since we bound the value V_v by θ, there exists $\alpha^{'}$ such that $E[\alpha^{'}] = E[log_{p_i}\theta] = log_p\theta$. Similarly, the expectation of bounded value $V_v^{'}$ is $E[V_v^{'}] = (p \cdot d + p^2 \cdot d^2 + \cdots + p^{log_p\theta} \cdot d^{log_p\theta}) \cdot w$.

Lemma 2. *The excepted solution of WR, y^*, and the excepted solution of BWR, z^*, satisfy $\dfrac{y^*}{z^*} \leq \dfrac{1 - (pd)^\alpha}{1 - (pd)^{\alpha'}}$.*

Proof. According to the analysis above, z^* is $E[\sum_{v \in S^*} V_v]$ and y^* is $E[\sum_{v \in S} V_v]$ with S denoting the result set of WR and S^* denoting the result set of BWR. We have $E[\sum_{v \in S} V_v] \geq E[\sum_{v \in S^*} V_v'] \geq E[\sum_{v \in S} V_v']$. Comparing z^* and y^*, we can get:

$$\frac{y^*}{z^*} = \frac{E[\sum_{v \in S} V_v]}{E[\sum_{v \in S^*} V_v]} \leq \frac{E[\sum_{v \in S} V_v]}{E[\sum_{v \in S} V_v']} = \frac{\sum_{v \in S} E[V_v]}{\sum_{v \in S} E[V_v']} \leq \frac{V_v}{V_v'} = \frac{1 - (pd)^\alpha}{1 - (pd)^{\alpha'}} \qquad (2)$$

$\square$

Theorem 4. *BWR algorithm is a fully polynomial-time approximation scheme for the influence maximization problem in WIC model.*

Proof. We set the parameter θ of BWR as follows,

$$\theta \leq (1 - \frac{1 - (pd)^\alpha}{(1 - 1/e)(1 + \epsilon)})^{\frac{1}{1 + \frac{1}{\log_d p}}}, \qquad (3)$$

where e is the base number of the natural logarithm. Thus, θ is only relevant to ϵ since other symbols in (3) are constants. Then by transforming (3), we can get

$$\frac{x^*}{z^*} \leq \frac{1}{(1 - \frac{1}{e})} \cdot \frac{1 - (pd)^\alpha}{\frac{1 - (pd)^\alpha}{(1 - 1/e)(1 + \epsilon)}} = 1 + \epsilon.$$

Now, we complete the analysis of the approximation ratio. In WR algorithm, the time complexity is $O(kn_{in} \cdot n)$. In BWR algorithm, the size of BIVT and BWDT is limited by θ. We can estimate n_{in}' as $O(d^{\alpha'})$. Thus, the time complexity is $O(kd^{\alpha'} \cdot n)$. According to (3), we get: $\alpha' = (1 + \frac{1}{\log_d p}) \cdot \log_p (1 - \frac{1 - (pd)^\alpha}{1 + \epsilon})$. To simplify this expression, we denote $A = 1 - \frac{1}{e}, B = 1 - (pd)^\alpha, C = \frac{1}{1 + \frac{1}{\log_d p}}$. Thus, the expression of $d^{\alpha'}$ is:

$$d^{\alpha'} = (d^{\log_p 1 - \frac{B}{A(1 + \epsilon)}})^{1 + \frac{1}{\log_d p}} = (1 - \frac{B}{A(1 + \epsilon)})^{\frac{1}{C \cdot \log_d p}} \qquad (4)$$

This bound is polynomial in the input, which is in turn polynomial in d, p, α and in $1/\epsilon$. Since the running time of BWR is polynomial in n, k and $d^{\alpha'}$, BWR is a fully polynomial-time approximation scheme. $\square$

Since $d^{\alpha'}$ is much smaller than $O(|V| + |E|)$, the space complexity is decreased as well. Considering that many pairs of nodes with low $p_r(u, v)$ are ignored, the space complexity is $O(n + n \cdot d^{\alpha'})$. Therefore, we provide a free space for users to make a trade-off between cost and accuracy.

4   Experiments

4.1   Experimental Settings

The experiments are performed on a PC with an Intel Core i5-3470 CPU and 8 GB memory, running 64 bit Ubuntu 12.04. The TIM+ algorithm is implemented in C++ while others are implemented in JAVA 8.

Datasets. We use three real-world networks [9] from various areas shown as Table 2. The Gnutella is a sequence of snapshots of the Gnutella peer-to-peer file sharing network from August 2002. The second is the Amazon product co-purchasing network. The last is a road network of California. By evaluating the experiment results, we attempt to show the broad application areas of BWR.

Table 2. Dataset characteristics.

Dataset	Gnutella	Amazon	RoadNet-CA
#Nodes	6 K	262 K	2.0 M
#Edges	21 K	1.2 M	2.8 M
Average degree	6.6	9.4	2.8
Largest component size	6299	262 K	2.0 M
Diameter	9	32	849

Propagation Models. We use TRIVALENCY model in [2]. That is, on each edge $e_{u,v}$, we select a probability from the set $\{0.001, 0.01, 0.1\}$ randomly to represent weak, medium and strong connection. We generate a random integer weight w where $w \in [1, 10]$ for each node to represent the weight of each node.

Algorithms. We compare our BWR algorithm with basic greedy algorithm (Sect. 2.2) and other related algorithms. For these comparisons, we have two main targets. One is to compare the applicability of IC and WIC model. The other is to test the performance of BWR in WIC model. The setup and implementation details of these algorithms are as follows.

- **BWR(θ):** We implement BWR algorithm on both IC and WIC model.
- **TIM+:** It is a near-optimal time complexity algorithm [16] which can achieve almost best performance in IC model. Due to the limitation of the memory of our computer, we set $\epsilon = 0.1$ in first graph whereas as small as possible to ensure its accuracy in larger graphs.
- **Greedy Algorithm** [8] **for IC model:** The influence simulations are repeated $20,000$ times to obtain an optimal seed set S.
- **Greedy Algorithm for WIC model:** We implement Algorithm 1 in Sect. 2.2 and the simulation times R is also $20,000$.

- **PageRank for IC model:** We implement PageRank [1] as a baseline. Along edge $e_{u,v}$, the transition probability is $p(u,v)/\sum_{i=1}^{O_u} p(u,v_i)$. The damping factor d is 0.85. By power iteration[1], it stops when the iterations are more than 10,000 times or the outcome between two iterations is less than 0.001.
- **PageRank for WIC model:** Nodes with high weights own more votes. Thus, the node which can activate valuable nodes will obtain more votes.
- **Random:** As a baseline, we randomly select k nodes.

4.2   Comparison Between IC and WIC

We compare the performance of IC and WIC model by comparing the expectation of influence spread. The seed set size k is 50. We only implement Greedy in first graph since other graphs are too large for Greedy algorithm.

The experimental results are shown in Table 3. All algorithms can produce better solutions in WIC except TIM+ and Random. More precisely, Greedy in WIC performs 89.87 % better than in IC. BWR in WIC also performs 23.8 %, 53.75 % and 60.68 % better than IC. Considering the significant difference between two models, WIC model is more effective than IC model in WIM problem.

Table 3. Influence spread of two models.

Graph and model		Greedy	BWR	TIM+	PageRank	Random
Gnutella	IC	420.11	508.47	502.55	371.73	291.84
	WIC	797.67	629.49	502.55	375.08	291.84
Amazon	IC	N/A	400.95	402.06	272.15	338.61
	WIC	N/A	616.45	402.06	282.40	338.61
RoadNet	IC	N/A	369.88	348.90	310.45	289.60
	WIC	N/A	594.31	348.90	315.50	289.60

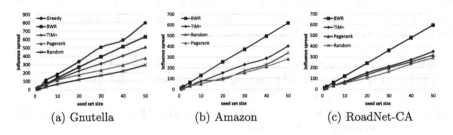

(a) Gnutella   (b) Amazon   (c) RoadNet-CA

Fig. 1. Influence spread results on WIC model

---

[1] In mathematics, the power iteration is an eigenvalue algorithm: given a matrix A, the algorithm will produce a number λ(the eigenvalue) and a nonzero vector v(the eigenvector), such that $Av = \lambda v$. This algorithm is also known as the Von Mises iteration [15].

4.3   Comparison of Algorithms

We run these algorithms on both IC and WIC model. The range of seed set size k are $1, 2, 5, 10, 20, 30, 40, 50$ and $\theta = 1/10^4$.

Figure 1 shows the experimental results in WIC model while Fig. 2 shows the results in IC model. In Fig. 2, we add the number of influenced nodes of BWR in WIC model to compare the outcomes of BWR between two models. Figure 3 shows the running time of three graphs for $k = 50$.

Gnutella. In Fig. 1(a), Greedy achieves the best result. BWR outperforms other algorithms except Greedy. BWR is 37.8 % and 40.4 % better than TIM+ and PageRank. In Fig. 2(a), Greedy, BWR and TIM+ produce same results in IC model. Although nodes activated by BWR on WIC model are less than TIM+, BWR produces more influence spread value. Such result indicates that previous algorithms are not suitable for WIC model whereas BWR solve the WIC problem effectively in both models. The efficiency results are similar in both models. Greedy is the slowest, taking more than 8,000 s while BWR is faster than Greedy four orders of magnitude. PageRank and Random are faster than BWR whereas their accuracy are not comparable with other three algorithms.

Amazon. In WIC model, BWR has a great winning margin over other algorithms: it outperforms TIM+ 53.2 % and any other algorithms. In IC model, BWR also performs well. BWR is only 5.7 % less than TIM+ but better than PageRank and Random. For running time, even though BWR loses 5.7 % accuracy, it is 6 times faster than TIM+. The results of PageRank and Random are

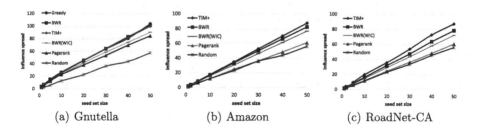

(a) Gnutella   (b) Amazon   (c) RoadNet-CA

Fig. 2. Influence spread results on general IC model

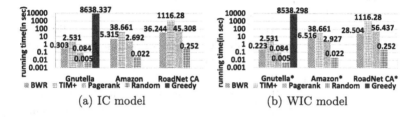

(a) IC model   (b) WIC model

Fig. 3. Running time of different algorithms

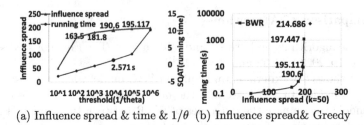

(a) Influence spread & time & $1/\theta$ (b) Influence spread& Greedy

Fig. 4. Relation between θ and maximal influence and running time

similar. This is because there is a giant component[2] and influential nodes gather together [14]. So, it is probably to activate numerous nodes by random selection.

Road-CA. According to Figs. 1(c) and 2(c), BWR produces undoubtedly the best results (at least 70.7 % stronger than others) contrasts with any other algorithms in WIC. In IC model, TIM+ is slightly better than BWR whereas its running time is 30 times more than BWR. BWR is much better than PageRank and Random in terms of the accuracy.

4.4 The Impact of θ

To investigate the impact of threshold θ, we explore the relationship between running time and influence spread in Gnutella graph with all $p_{uv} = 0.1$ and all $V_u = 1$. In Fig. 4(a), when θ gets less than $1/10^4$, the influence spread grows slowly. This tendency of accuracy coincides with Lemma 2. According to (2), when $p \cdot d < 1$, the denominator of (2) grows slower and slower. Thus, the additional performance bound decreases tardily with the growth of α', which is linear with $(\log_p 1/\theta)$. Moreover, with the decrease of θ, the running time increases almost linearly with $\lg 1/\theta$. With the decrease of θ, the sizes of BIVT and BWDT grow quadratically. However, since different graphs have different densities, the running time is not always linear with $(\lg 1/\theta)$.

In Fig. 4(b), there is an inflection on the result curve of BWR. As the influence spread increases, the running time grows immediately where $\theta = 1/10^5$. On the inflection, we can obtain a proper trade-off of accuracy and efficiency. When $\theta = 1/10^4$ or $1/10^5$, the performance is very well whereas the running time is extremely smaller than Greedy algorithm. With the increase of θ, the influence spread hardly changes whereas the running time grows quadratically. Hence for this graph, $\theta = 1/10^4$ is ideal. We observe similar situations in other datasets.

Furthermore, we give a bottom line of the accuracy with particular θ. According to (2), we can estimate $y^*/x^* \geq 0.7125$. According to experiment settings, $\theta = 1/10^4$ and all $p_r(u,v) = 0.1$ and $\alpha' = 3$. The approximation ratio of BWR is $87.88\% \cdot (1-1/e)$. It is better than the ratio bound calculated by (2) significantly,

---

[2] In network theory, a giant component is a connected component, of a large scale connected graph, which contains most of the nodes in the network.

since we assume the numerator of (2) equals 1. That means all nodes are reachable which is impossible in fact.

5   Conclusion

In this paper, we present WIC model which is more practical in various application scenarios. Then we present a basic greedy algorithm which is 89.87 % more accurate than previous greedy algorithm in IC model. To improve the efficiency, we design BWR algorithm to make a free trade-off between accuracy and efficiency. Extensive experimental results show that BWR can handle a million-node graph on a usual PC within tens of seconds. Such brilliant performance make BWR to be an excellent solution for practical applications.

A potential research direction based on this paper is integrating IM and social relationships. Data mining of social relations from real online social network is a valuable aspect. Combining influence maximization and social influence relationship together could achieve better prevalent viral marketing effectiveness and mine latent and invisible information.

Acknowledgement. This paper was supported by NGFR 973 grant 2012CB316200, NSFC grant U1509216,61472099,61133002 and National Sci-Tech Support Plan 2015BAH10F01.

References

1. Brin, S., Page, L.: Reprint of: the anatomy of a large-scale hypertextual web search engine. Comput. Netw. **56**(18), 3825–3833 (2012)
2. Chen, W., Wang, C., Wang, Y.: Scalable influence maximization for prevalent viral marketing in large-scale social networks. In: KDD, pp. 1029–1038. ACM (2010)
3. Chen, W., Wang, Y., Yang, S.: Efficient influence maximization in social networks. In: ACM SIGKDD, pp. 199–208. ACM (2009)
4. Domingos, P., Richardson, M.: Mining the network value of customers. In: KDD, pp. 57–66. ACM (2001)
5. Goyal, A., Lu, W., Lakshmanan, L.V.: Celf++: optimizing the greedy algorithm for influence maximization in social networks. In: WWW, pp. 47–48. ACM (2011)
6. Hochba, D.S.: Approximation algorithms for NP-hard problems. ACM SIGACT News **28**(2), 40–52 (1997)
7. Cai, J.L.Z., Yan, M., Li, Y.: Using crowdsourced data in location-based social networks to explore influence maximization. In: The 35th Annual IEEE International Conference on Computer Communications (INFOCOM 2016) (2016)
8. Kempe, D., Kleinberg, J., Tardos, É.: Maximizing the spread of influence through a social network. In: KDD, pp. 137–146. ACM (2003)
9. Leskovec, J.: Stanford large network dataset collection. http://snap.stanford.edu/data
10. Leskovec, J., Krause, A., Guestrin, C., Faloutsos, C., VanBriesen, J., Glance, N.: Cost-effective outbreak detection in networks. In: ACM SIGKDD, pp. 420–429. ACM (2007)

11. Han, M., Yan, M., Cai, Z., Li, Y.: An exploration of broader influence maximization in timeliness networks with opportunistic selection. J. Netw. Comput. Appl. (2016, in press)
12. Nail, J.: The consumer advertising backlash, Forrester Research and Intelliseek Market Research Report, 137, May 2004
13. Nemhauser, G.L., Wolsey, L.A., Fisher, M.L.: An analysis of approximations for maximizing submodular set functions–I. Math. Program. **14**(1), 265–294 (1978)
14. Newman, M.E.: The structure of scientific collaboration networks. Proc. Nat. Acad. Sci. **98**(2), 404–409 (2001)
15. Siegmund-Schultze, R.: Richard von mises (1883–1953): a pioneer of applied mathematics in four countries. Newslett. Eur. Math. Soc. **73**, 31–34 (2009)
16. Tang, Y., Xiao, X., Shi, Y.: Influence maximization: near-optimal time complexity meets practical efficiency. In: SIGMOD, pp. 75–86. ACM (2014)
17. Shi, T., Wan, J., Cheng, S., Cai, Z., Li, Y., Li, J.: Time-bounded positive influence in social networks. In: International Conference on Identification, Information and Knowledge in the Internet of Things (2015)
18. Wang, Y., Cong, G., Song, G., Xie, K.: Community-based greedy algorithm for mining top-k influential nodes in mobile social networks. In: ACM SIGKDD, pp. 1039–1048. ACM (2010)

Complex Queries

ListMerge: Accelerating Top-k Aggregation Queries Over Large Number of Lists

Shile Zhang[1,3], Chao Sun[2,3], and Zhenying He[2,3]($\boxtimes$)

[1] School of Software Engineering, Fudan University, Shanghai, China
shilezhang14@fudan.edu.cn
[2] School of Computer Science, Fudan University, Shanghai, China
{chaosun14,zhenying}@fudan.edu.cn
[3] Shanghai Key Laboratory of Data Science, Fudan University, Shanghai, China

Abstract. Sorted list is widely used to feature indexing in a variety of applications, such as multimedia database and information retrieval. Answering top-k aggregation queries on a set of lists plays an increasingly important role in these domains. Unfortunately the existing solutions, such as threshold-style (TA-style) algorithms, do not guarantee superior performance on a large number of lists. In this paper, we introduce a merge-based strategy, called ListMerge, to accelerating TA-style algorithms. ListMerge exploits a critical observation to TA-style algorithms: if aggregation functions are monotone and distributive, it is much more efficient that *merging several lists together, then applying a TA-style algorithm*. This observation also inspires the development of our cost model, which can evaluate the best number of merged lists. Experimental results show that ListMerge could outperform the baseline algorithms up to 4–20 times in synthetic datasets generated by various distributions.

1 Introduction

Top-k aggregation queries, as a means of retrieving a ranked set of the k most interesting objects based on the uniform aggregation function, have attracted considerable attention in many analytic applications. Threshold-style (TA-style) algorithms are widely used because they can guarantee superior performance on a small number of lists. These algorithms benefit from reducing the number of objects accessed.

Take text analysis for example, a typical scenario is to find top k word count in a set of documents satisfying certain query criteria. Each document can be viewed as a list of $\langle word, count \rangle$ pairs sorted in descending order. TA-style algorithms need to fetch words in every sorted list. However, when the number of queried documents is large, TA-style algorithms may fetch considerable number of objects in each round. As a result, execution cost would still be expensive.

The work was partially supported by the National Natural Science Foundation of China (No. 61370080, No. 61170007) and Science and Technology Commission of Shanghai Municipality (No. 14511106802).

S.B. Navathe et al. (Eds.): DASFAA 2016, Part II, LNCS 9643, pp. 67–81, 2016.
DOI: 10.1007/978-3-319-32049-6_5

In this paper, we study the problem of answering top-k aggregation queries on a large number of lists, which is motivated by the fact that high dimensional data is usually stored as a number of lists. We propose a merge strategy, called ListMerge, by exploiting dimension-reduction for high dimensional data so that much less objects are accessed in each round. Furthermore, as lists are merged, objects with high aggregation score are more likely to rank higher in merged lists, and thus the algorithm can stop much sooner. We also propose an execution cost model to determine the number of lists to be merged so that execution cost can be minimized. The contributions of this paper are described as follows:

- This paper studies the problem of answering top-k aggregation queries over a large number of lists. To the best of our knowledge, this problem is not discussed by existing TA-style algorithms.
- We propose a merge-based strategy – ListMerge, which is applicable to a class of top-k aggregation query algorithms, i.e. TA-style. The execution cost of ListMerge can be greatly reduced.
- An uniform cost model is developed to estimate best merge strategy for TA-style algorithms.
- ListMerge can be integrated into existing TA-style algorithms easily.

The rest of this paper is organized as follows: Sect. 2 introduces related works and clarifies the applicable boundary of our strategy. In Sect. 3, we define some common concepts and terminology that will be used in this paper. Sections 4 and 5 present the merge strategy and cost model respectively. In Sect. 6, we give a performance evaluation of our merge strategy. Finally we conclude this paper in Sect. 7.

2   Related Works

Top-k query processing has always been receiving constant attention from different fields, such as multimedia retrieval [4,9,14,22], P2P networks [1,3,5] and relational database management systems [17,19,20].

Most of the algorithms require pre-processing steps to construct a materialized data structure. These data structures can be roughly categorized into three classes: *layer-based*, *view-based* and *list-based*. Our method targets at list-based data structure algorithms in high dimension situation. To clarify our applicable scope, we give a rough description to each category.

Layer-based structure views data objects in a hierarchical way. Each data object belongs to a specific layer. The layer in which data objects belongs to is determined by the probability of certain rank that object may be in. The final effect this structure guarantees is that top-k objects exists in the first k layers. Convex hull [8,10,15] and skyline [18,24,25,27] are often used to realize layer-based structure.

View-based structure materializes historical query results. For a given new top-k query under a score function f, [11,16,23,26] tries to answer the query by utilizing previous query results whose score function is similar to f, with the purpose of accessing less data objects.

List-based structure views data as lists. First important paper proposed using this structure is [12], called Fagin's Algorithm (FA). Later, Several groups [13,14,22] discover TA independently, which is much efficient than FA under all circumstances. Fagin et al. [13] also proved that TA is instance optimal. To guarantee bounded memory consumption, TA does not remember position it has seen under random access, thus one object may be accessed many times during execution. With the help of position information, Best Position Algorithm (BPA) [2] is proposed. In the distributed environment, data is vertically divided over nodes. Three-Phase Uniform Threshold (TPUT) algorithm [7] improves distributed TA by ensuring finding top k objects in three phase of communication. Later, KLEE [21] is proposed to give approximate top-k results instead of exact ones. With only small penalties in result quality, this algorithm can enjoy significant performance benefits.

However, none of these type of algorithms considers performance issues in high dimension situation. Although TPUT does an experiment on high dimension web data, it doesn't give detailed study of effects different data distribution has on performance, nor does it optimize this case.

3 Preliminaries

In this section, we present the data model and basic related concepts that will be used in this paper. TA is also introduced to show how this algorithm utilize the data model. The symbols used in this paper are described in Table 1.

ListMerge is applicable for TA-style algorithms under the same precondition except that aggregation function should be both monotone and distributive.

Definition 1 (Monotone Function). *An aggregation function $f()$ is monotone if $f(x_1, \ldots, x_l) \leq f(x'_1, \ldots, x'_l)$ whenever $x_i \leq x'_i$, for every i.*

Definition 2 (Distributive Function). *Aggregation function $f()$ is distributive if there is a function $g()$ such that $f(x_1, \ldots, x_l) = g(\{f(\{x_i | i \in S_j\}) | j = 1, 2, \ldots, J\})$ where $S_m \cap S_n = \emptyset$ for $\forall m \neq n \wedge m, n \in \{1, \ldots, J\}$ and $\cup_{m=1}^{J} S_m = \{1, \ldots, l\}$.*

3.1 Data Model

We adopt the notion in [13] to depict sorted list data model used by TA-style algorithms. Data set consists of N *data objects*. Each data object O has l *fields* (or *attributes*) $x_1, \ldots, x_l$ where $x_i \in \mathbb{R}$ $\forall i \in [1, l]$, and we refer the value of x_i as the *grade* of the object. List data model views above data set as l sorted lists $L_1, \ldots, L_l$, each of which has N *data items*. For any *list i*, the *data items* are in the form of $\langle O, x_i \rangle$ and sorted in descending order by x_i value. Each list corresponds to a *field (attribute)*, so we will use *field, attribute* or *dimension* to refer to list interchangeably.

There are two forms of data access. The first is *sorted (sequential) access*, which obtains the grade of an object in one of the sorted lists by proceeding through the list sequentially from the top. The other form is *random access*. Given an object O, we can retrieve the list entry $\langle O, x_i \rangle$ from list i in one *random access*.

We denote the cost of sorted and random access as α and β respectively and *execution cost* is defined as $\alpha * n_s + \beta * n_r$ (called middleware cost in [13]) where n_s and n_r is the number of sorted and random access performed respectively.

Table 1. Definition of symbols

Symbol	Description
α	Sorted access cost
β	Random access cost
γ	Sort cost
l	Number of lists
m	Number of lists to merge
n_s	Number of sorted access
n_r	Number of random access
f	Aggregation function that is monotone and distributive
L_i	List i
O_i	Object i
N	Number of objects
k	Number of retrieved objects
Y	Ordered set to contain top-k $\langle object, score \rangle$ pair

3.2 Threshold Algorithm

TA works as follows:

1. Do sorted access in parallel to each of the l sorted lists L_i. As an object O is seen under sorted access in some list, do random access to the other lists to find the grade x_i of object O in every list L_i. Then compute overall score using monotone aggregation function $f(O) = f(x_1, \ldots, x_l)$ of object O. If this score is one of the k highest overall score seen so far, add object O and its score to ordered set Y.
2. For each list L_i, let $\underline{x}_i$ be the attribute of the last object seen under sorted access. Define the threshold value $\delta = f(\underline{x}_1, \ldots, \underline{x}_l)$. If Y has k data items whose overall scores are higher than or equal to δ, then halt. Otherwise go to step 1.
3. Return Y.

4 Merge Strategy for TA-style Algorithms

Suppose there are l lists and we are going to merge per m lists, the aggregation function $f()$ is distributive and monotone. A TA-style algorithm adopting our merge strategy can first perform merge per m lists. For the merged l/m lists, we can then apply the original algorithm to them. Our merge strategy works as follows:

1. Do sorted access in parallel to *list* j where $j \in \{\, i \mid i \bmod m = 1,\ i \in L,\ L = 1, 2, \ldots, l\}$, we denote this set of lists as M and call them *merge root* lists. For each *list* j in M, as an object O is seen under sorted access, do random access to list $(j+1), \ldots, (j+m-1)$ to find the corresponding grade of O. Then compute *merged grade* $f(O) = f(x_j, \ldots, x_{j+m-1})$ and update grade of *list* j with *merged grade*.
2. Sort each *list* j in M according to the *merged grade*.
3. Running a TA-style algorithm on sorted lists in M.

Let's illustrate merge strategy using the example in Fig. 1. We assume $l = 6$, $m = 3$ and aggregation function is sum, which satisfies distributive and monotone property. Original lists are present in Fig. 1(a). Now sorted access are run in parallel to L_1 and L_4. In the first round, object O_3 is seen under sorted access in L_1, so we retrieve grade 52 and 95 of O_3 from L_2 and L_3 respectively. Merged grade of O_3 is computed as $f(O_3) = 98 + 52 + 95 = 245$. Then merged grade for O_2 in L_4 is computed from L_5 and L_6. In the second round, merged grade of O_5 in L_1 and L_4 are calculated. After all objects in L_1 and L_4 are updated by their merged grade, we sort those two lists by scores and the resulting lists is shown in Fig. 1(b).

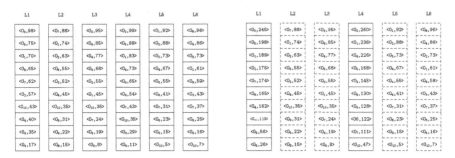

(a) Sorted lists before merge   (b) Sorted lists after merge

Fig. 1. Example illustrate how merge works.

5 Cost Model

In this section, a cost model is proposed to estimate the best merge number so that the execution cost can be minimized. Section 5.1 defines a uniform cost model that is composed of merge execution cost and algorithm execution cost for a TA-style algorithm. While merge cost is the same for all algorithms, parameters of algorithm execution cost require estimation for different algorithms case by case. In Subsect. 5.2, we present merge cost. Subsection 5.3 uses TA as a representative of TA-style algorithms to show how to refine parameters of examine algorithm execution.

5.1   Cost Model for TA-style Algorithms Using Merge Strategy

As with [13] and [2], we take the grade of objects as precomputed, so that we are taking only access cost into account and ignoring internal computation cost. Total execution cost for a TA-style algorithm is estimated as $TotalCost = MergeExecutionCost + AlgoExecutionCost$. Since merge strategy is applicable for TA-style algorithms, *merge execution cost* model can be expressed using a uniform formula (See Eq. 2). *Algorithm execution cost* model can be expressed as $\alpha * n_{s\_est} + \beta * n_{r\_est}$, where n_{s_est} and n_{r_est} denote the estimated number of sorted and random access which have algorithm-specific indicators for different TA-style algorithms. We formulate our cost model as follows:

$$TotalCost = \frac{l}{m} \left[\alpha n + (m - 1) \beta n + \gamma n \log n \right] + \alpha n_{s\_est} + \beta n_{r\_est}. \quad (1)$$

We use TA as a representative of TA-style algorithms to give a rough understanding of how execution cost changes under different number of merged lists. For centralized TA, execution cost can be measured using response time because each access is executed sequentially. We generate 20 lists using independent uniform distribution. Each list has 1000 objects, and the range of grade of these objects is [0, 3000]. Our aggregation function is $sum()$. To accurately estimate the cost, we eliminate the effect of cache miss and the JIT[1] optimization of Java.

As shown in Fig. 2, we can see that when dimension is high and no merge is performed (merged list number is 20), TA takes about 1541 ms to run. However, execution cost can drop 86.6 % percent to 195 ms if we merge to 3 lists. Merge execution time decreases and TA execution time increases as merged list number (l/m) increases, i.e. merge number m decreases. The trade-off lies in that cost incurred by merging can be covered by the cost declined because of dimension reduction for a TA-style algorithm.

5.2   Merge Cost

Since merge cost is composed of three parts: sorted access cost, random access cost and sort cost. Adding three parts together, we formulate *merge cost* as follows:

$$MergeCost = \frac{l}{m} \left[\alpha n + (m - 1) \beta n + \gamma n \log n \right]. \quad (2)$$

---

[1] Just-in-time (JIT) compilation, also known as dynamic translation, is compilation done during execution of a program.

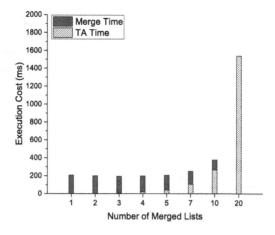

Fig. 2. Total cost under different number of merged lists.

where n denotes number of objects and γ represents the cost of one sort step.

Let's consider *merge root list i*. Using our merge strategy in Sect. 4, we need to do n sorted access to *merge root i*, and for each object seen under sorted access, $(m-1)$ random access is performed to list $i+1, \ldots, i+m-1$. So *sorted access cost* and *random access cost* are αn and $(m-1)\beta n$ respectively. Number of sort steps to perform is estimated as $n \log n$ when there are n objects in a list, which is typical for sort algorithms. There are in total l/m merge root to consider, so *merge execution cost* is $\frac{l}{m}[\alpha n + (m-1)\beta n + \gamma n \log n]$.

Equation 2 can be easily transformed to $(\gamma \log n + \alpha - \beta)nl/m + \beta ln$. Since γ, n, α, β and l are all constant, we can view it as a liner function with respect to variable l/m, i.e. number of merged lists. If $\beta \geq (\gamma \log n + a)$, merge cost decreases as the number of merged lists gets larger, and vice versa. If we consider the execution cost of sorted access, random access and sort separately, they are linear to l/m with slope αn, $-\beta n$ and $\gamma n \log n$ respectively.

To prove the correctness of our merge cost model, we plot total execution cost, sorted access cost, random access cost and sort cost against merged list number l/m in Fig. 3. As our R^2 statistics suggest, experimental data is highly in accord with the cost model analysis in previous paragraph. Values of α, β and γ can be estimated using least square method.

5.3 Algorithm Execution Cost

When it comes to TA, n_{s_est} and n_{r_est} in *algorithm execution cost* can be estimated using *access depth d*, which denotes the depth reached in lists under sorted access. So TA execution cost is estimated as follows:

$$TAExecutionCost = \alpha d\frac{l}{m} + \beta d\frac{l}{m}(\frac{l}{m} - 1). \tag{3}$$

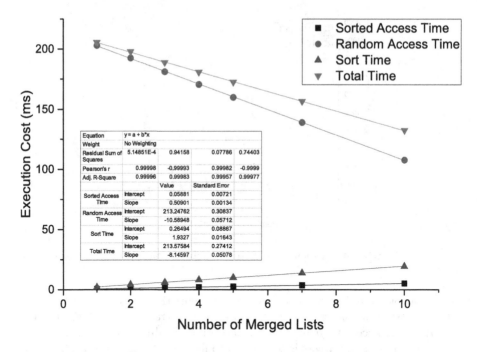

Fig. 3. Composition of merge execution cost

Suppose we have l/m sorted merged lists and have correctly estimated *access depth* d for those merged lists. Recall how TA works in Subsect. 3.2, for each sorted access, we need to perform $l/m - 1$ random access to other merged lists to retrieve the same object seen under sorted access. Each round requires l/m sorted access to merged lists, so we need to perform $l/m * (l/m - 1)$ random access in the same round. Since an estimated d rounds are necessary to finish TA processing, our estimated sorted and random access cost are $\alpha dl/m$ and $\beta dl/m(l/m - 1)$ respectively.

Estimating Access Depth. The next question is: how do we estimate access depth? Let S_i denotes the random variable representing the grade of the object in list i. Recall that the grade of the object in a list is uniformly generated and scores in each list are independent and obeys the same distribution. So we can view S_i as a uniform distribution variable with mean value μ and standard deviation δ. Total score of object can be represented as $S = \sum_{i=1}^{l} S_i$. We also define merged grade random variable $M_i = \sum_{j=0}^{m-1} = S_{i+j}$ where $i \in M$, the *merge root lists* set as is described in Sect. 4. We now present the *central limit theorem* and discuss in detail how to estimate access depth d.

Theorem 1 (The Central Limit Theorem). *If random variable X is defined as the sum of n independent and identically distributed (i.i.d.) random variables, $X_1, X_2, \ldots, X_n$; with mean μ, and standard deviation δ. Then, for large* *enough* n, X *is approximately normally distributed with parameters:* $\mu_X = n\mu$ *and* $\delta_X = \sqrt{n}\delta$.

According to Theorem 1, $S_1, S_2, \ldots, S_l$ are i.i.d random variables, so S obeys normal distribution with mean value $n\mu$ and standard deviation $\sqrt{n}\delta$, namely

$$\frac{S - n\mu}{\sqrt{n}\delta} \sim \mathcal{N}(0, 1). \tag{4}$$

For the same reason, we have

$$\frac{M_i - m\mu}{\sqrt{m}\delta} \sim \mathcal{N}(0, 1). \tag{5}$$

Suppose a database contains n objects, these n objects can be considered as n samples taken from S. Now we need two steps to approximate d. (1) Estimate the total score s_k of $k'th$ largest object. (2) Estimate the depth d at which threshold value of merged lists can drop below s_k. For the first step, we can use cumulative distribution function (cdf) to estimate s_k as:

$$\phi\left(\frac{x_k - n\mu}{\sqrt{n}\delta}\right) = 1 - \frac{k}{n}. \tag{6}$$

n, μ, δ and k are known constant, so we can look up the standard normal distribution table for the value of $\frac{x_k - n\mu}{\sqrt{n}\delta}$ corresponding to $1 - \frac{k}{n}$. And then compute s_k. In the second step, each M_i follows identical independent normal distribution with mean value $m\mu$ and standard deviation $\sqrt{m}\delta$. So on average, threshold value τ_{m_i} of each merged list should be $x_k/(l/m) = mx_k/l$. The percentage that objects with a *merged grade* higher than τ_{m_i} can be formulated as:

$$Pr\left(M_i \leq \tau_{m_i}\right) = Pr\left(\frac{M_i - m\mu}{\sqrt{m}\delta} \leq \frac{\tau_{m_i} - m\mu}{\sqrt{m}\delta}\right) = \phi\left(\frac{\tau_{m_i} - m\mu}{\sqrt{m}\delta}\right). \tag{7}$$

So estimated depth $d = n\phi\left(\frac{\tau_{m_i} - m\mu}{\sqrt{m}\delta}\right)$.

If S_i is not i.i.d, different methods should be applied to estimate s_k and d. Adopting the idea in [6], we can use sampling to estimate s_k and histogram to approximate threshold value of certain depth d.

6   Performance Evaluation

We next report an extensive evaluation of our merge strategy over various conditions. The rest of this section is organized as follows. We first describe our experimental setup. Then, we evaluate our result from three aspects, the accuracy of our cost model, the accelerating effect on TA-style algorithms under diverse distribution and performance gain for high dimensional data. Finally, we summarize the performance result.

6.1 Experimental Setup

Environment. All experiments in this paper are implemented by Java. In centralized environment, i.e. TA and ListMerge+TA, experiments are conducted on a computer running Windows 10 with an Intel i7 3.4 GHz and 8 GB RAM. Experiments in distributed environment are deployed on a 10-machine cluster, each machine has a twelve 2.1 GHz Intel Xeon processor running 64-bit Ubuntu server 14.04 with 16 GB RAM. We turn off the JIT features of JVM and eliminate the cache miss impact to make our experiments universally applicable.

Queries. We consider queries retrieving top-10 highest score objects ($k = 10$), and we test the aggregation function of $sum()$ since it is widely used.

Data Sets. To evaluate the performance of merge strategy over common cases, we generate data sets covering independent and correlated situation under various distributions. Table 2 summarizes the synthetic data sets used in our experiments. Lists of $unif1000$ data set are uniformly generated with grade from 1 to 1000, and object values between lists are independent. For $corr0.01$, first list is generated the same way as $unif1000$, values of other lists are produced by randomly adding a integer from range $[-5, 5]$ to objects in the first list. As for $gaus(500, 167)$ and $zipf5$, object values in each list obey Gaussian distribution with $\mu = 500, \delta = 167$ and Zipfian distribution with $\lambda = 5$ respectively.

Table 2. Data distributions used in experiment

Distribution	Description
$unif1000$	Random(1,1000)
$corr0.01$	$unif1000$ + Random(-5,5)
$gaus(500, 167)$	Gaussian distribution with $\mu = 500$ and $\delta = 167$
$zipf5$	Zipfian distribution with $\lambda = 5$

Metrics. To compare the performance of algorithms before and after applying our merge strategy, we measure the following metrics:

- Execution cost. As is introduced in Subsect. 3.1, execution cost is calculated as $\alpha * n_s + \beta * n_r$ where n_s and n_r are the number of sorted and random access. α is the cost of a sorted access and β is the cost of a random access. α and β are measured by time consumed per access. Execution cost can be viewed as an indicator of total CPU time consumption. For centralized algorithm, execution cost can be used interchangeably with response time because each access is performed sequentially. For distributed systems, object accesses may run in parallel so response time can be much shorter than execution cost. A more common metric for distributed algorithms is the next one.

- Number of object transmission. This metric measures the total number of objects transferred. In the distributed environment where message size is small, transmission number between nodes is the dominating factor to measure communication cost.
- response time. This is the total time an algorithm takes to find the top-k data items. For centralized algorithms, this metric is the same as execution cost. However, in distributed settings, response time can be much shorter than execution cost due to parallel accesses.

6.2   Performance Study

Evaluating Cost Model. Adopting the initial experiment setup in Subsect. 5.1, we carry out a more comprehensive and in-depth study to evaluate the accuracy of our proposed cost model. Still, data in each list is generated independently, object value in each list follow the same uniform distribution in range $[1, 3n]$, where n denotes the number of objects in data set.

We first estimate α and β with $l = 60$ and $n = 1000$. Based on the estimated α and β, we then estimate execution cost for different n and l value and compare estimated cost with actual running results in Table 3. Columns of "Actual Best" indicate the results of actual best execution cost can be achieved using ListMerge. Results exploiting the estimated cost model are described in the columns of "Estimated Best". In most case, i.e. when $l = 40, 60, 80$, the estimated merge number is just the actual best case. When $l = 20$, although the estimated best merge number is 3 while the actual one is 2, the running cost (118.1) is close to the best (117.6).

Table 3. Actual best cost v.s. estimated best cost

List number	Actual best		Estimated best		
	Cost (ms)	Merged #	Estimated (ms)	Actual (ms)	Merged #
20	117.6	2	114.4	118.1	3
40	241.8	3	239.8	241.8	3
60	369.2	3	365.1	369.2	3
80	495.9	3	490.5	495.9	3

Results on Various Distributions. For distributed algorithms, execution cost is not of much importance so we only consider the accelerating effect on response time and total number of object transmission. We run distributed TA and TPUT on data sets generated with $l = 200$ and $n = 5000$ using distribution in Table 2. The result is presented in Tables 4 and 5. For each algorithm, we compare the cost (response time or number of object transmission) without applying our merge strategy (the *origin* column) with the best cost that can be achieved using different merge number (the *best run* column), we also record the number of merged list (the *merged number* column) at which best cost is reached. For each best run, we keep track of the transmission number in that run. We also record the best transmission number and origin transmission number for comparison.

For distributed TA, response time using merge is much shorter than that without merge, giving at least 10 times performance improvements (525 times improvements for *unif*1000 data set). The corresponding object transmission number of best run usually reduces two orders of magnitude, though not optimal but close to minimal transfer amount, with respect to origin except for *corr*0.01 data distribution. The reason why transmission amount of *corr*0.01 is two times that of origin is that data are so correlated that we only need about 10 rounds to stop in original lists, merge incurs too many data transfer that can't be balanced out in TA.

As for TPUT, response time sees on average 5 times performance gain. Although transmission amount is higher compared with origin, we think it is worthy to sacrifice such little bandwidth in exchange for such huge running time reduction. Just like distributed TA, TPUT transfers twice as much objects under minimal response time over *corr*0.01 distribution because highly correlated data can filter out many unnecessary objects in the first phase even without merge.

Table 4. Response time (ms) for different distribution

Data set	TA		Distributed TA		TPUT	
	Origin	Best run (merged #)	Origin	Best run (merged #)	Origin	Best run (merged #)
*unif*1000	598548	6230(4)	1126505	2144(4)	10706	2524(5)
*corr*0.01	4364	4364(200)	23160	1772(6)	6559	2276(5)
gaus(500, 167)	565616	6215(4)	1266102	2691(2)	11827	2340(5)
*zipf*5	408628	6318(30)	766509	2432(4)	9747	2314(5)

Results for Large Number of Lists. We now present the accelerating effect of the merge strategy in various high dimension situation with respect to response time in Table 6. The number of words per list is fixed to 5000 and data distribution is uniform. For each algorithm, we list the response time without merge and best possible response time under different merge number (along with at which merged number response time reaches its minimal).

Response time can be much shorter compared with no merge condition. As number of lists grows, acceleration ratio increases from 21 times to 97 times for TA and from 132 times to 525 times for distributed TA. TPUT sees on average 4 times improvement in response time, with a small increase from 3.99 to 4.24 times.

Results When Varying *n* and *l*. From Table 7 we can see that the execution cost using ListMerge is much better than that of using orginal TA for different

Table 5. Object transmission number

Data Set	Distributed TA			TPUT		
	Origin	Trans # of BR	Best trans #	Origin	trans # of BR	Best trans #
$unif1000$	71799200	980360	964144	990156	1000070	990156
$corr0.01$	517400	990240	517400	506387	985300	506387
$gaus(500, 167)$	83341200	990190	975000	1002489	1000115	995041
$zipf5$	49033600	980504	968176	878367	1000049	878367

Table 6. Response time (ms) for different dimension

List number	TA		Distributed TA		TPUT	
	Origin	Best run (merged #)	Origin	Best run (merged #)	Origin	Best run (merged #)
50	32791	1532 (3)	166875	1257 (2)	4410	1103 (2)
100	142584	3124 (2)	488954	1761 (4)	6437	1969 (1)
150	333168	4693 (3)	745417	2127 (3)	9031	2183 (3)
200	606544	6279 (3)	1126505	2144 (4)	10706	2524 (5)

Table 7. Response time (ms) when varying n and l

List number	$n = 1000$		$n = 2000$		$n = 4000$	
	Origin	Best	Origin	Best	Origin	Best
20	964.8	117.6	1828.1	232.6	3634.9	463.1
40	4277.8	241.8	8279.5	480.9	15810.8	968.3
60	10002.9	369.2	19652.9	728.7	38330.6	1459.0
80	18291.5	495.9	36530.3	1003.8	71332.8	1971.9

n and l. The running costs for both TA and ListMerge grow almost linearly as n or l increasing. Furthermore, the slope of ListMerge is much smaller than that of original TA.

7   Conclusion

To answer top-k aggregation queries, TA-style algorithms running on sorted lists have been well-studied recently. However, when the number of lists is large (the dimension is high), there is still performance issues regarding response time. In this paper, we propose a merge strategy that is applicable to list-based data model to perform dimensionality reduction for high dimensional data so that execution cost can be greatly reduced. We also propose a cost model to help determine the best number to merge lists to get the minimal response time. We have done extensive experiment under different list settings. The results of the experiments show considerable improvements in response time.

References

1. Akbarinia, R., Pacitti, E., Valduriez, P.: Reducing network traffic in unstructured p2p systems using top-k queries. Distrib. Parallel Databases 19(2–3), 67–86 (2006)
2. Akbarinia, R., Pacitti, E., Valduriez, P.: Best position algorithms for top-k queries. In: Proceedings of the 33rd International Conference on Very Large Data Bases, pp. 495–506. VLDB Endowment (2007)
3. Akbarinia, R., Pacitti, E., Valduriez, P.: Processing top-k queries in distributed hash tables. In: Kermarrec, A.-M., Bougé, L., Priol, T. (eds.) Euro-Par 2007. LNCS, vol. 4641, pp. 489–502. Springer, Heidelberg (2007)
4. Balke, W.T., Kießling, W.: Optimizing multi-feature queries for image databases. In: VLDB, pp. 10–14, September 2000
5. Balke, W.T., Nejdl, W., Siberski, W., Thaden, U.: Progressive distributed top-k retrieval in peer-to-peer networks. In: 21st International Conference on Data Engineering, 2005, ICDE 2005, Proceedings, pp. 174–185. IEEE (2005)
6. Bruno, N., Wang, H.: The threshold algorithm: from middleware systems to the relational engine. IEEE Trans. Knowl. Data Eng. 19(4), 523–537 (2007)
7. Cao, P., Wang, Z.: Efficient top-k query calculation in distributed networks. In: Proceedings of the Twenty-Third Annual ACM Symposium on Principles of Distributed Computing, pp. 206–215. ACM (2004)
8. Chang, Y.C., Bergman, L., Castelli, V., Li, C.S., Lo, M.L., Smith, J.R.: The onion technique: indexing for linear optimization queries. In: ACM SIGMOD Record, vol. 29, pp. 391–402. ACM (2000)
9. Chaudhuri, S., Gravano, L., Marian, A.: Optimizing top-k selection queries over multimedia repositories. IEEE Trans. Knowl. Data Eng. 16(8), 992–1009 (2004)
10. Cheema, M.A., Shen, Z., Lin, X., Zhang, W.: A unified framework for efficiently processing ranking related queries. In: EDBT, pp. 427–438 (2014)
11. Das, G., Gunopulos, D., Koudas, N., Tsirogiannis, D.: Answering top-k queries using views. In: Proceedings of the 32nd International Conference on Very Large Data Bases, pp. 451–462. VLDB Endowment (2006)
12. Fagin, R.: Combining fuzzy information from multiple systems. In: Proceedings of the Fifteenth ACM SIGACT-SIGMOD-SIGART Symposium on Principles of Database Systems, pp. 216–226. ACM (1996)
13. Fagin, R., Lotem, A., Naor, M.: Optimal aggregation algorithms for middleware. J. Comput. Syst. Sci. 66(4), 614–656 (2003)
14. Güntzer, U., Balke, W.T., Kießling, W.: Towards efficient multi-feature queries in heterogeneous environments. In: International Conference on Information Technology: Coding and Computing, 2001, Proceedings, pp. 622–628. IEEE (2001)
15. Heo, J.S., Cho, J., Whang, K.Y.: The hybrid-layer index: A synergic approach to answering top-k queries in arbitrary subspaces. In: 2010 IEEE 26th International Conference on Data Engineering (ICDE), pp. 445–448. IEEE (2010)
16. Hristidis, V., Koudas, N., Papakonstantinou, Y.: Prefer: A system for the efficient execution of multi-parametric ranked queries. In: ACM SIGMOD Record, vol. 30, pp. 259–270. ACM (2001)

17. Ilyas, I.F., Aref, W.G., Elmagarmid, A.K.: Joining ranked inputs in practice. In: Proceedings of the 28th International Conference on Very Large Data Bases, pp. 950–961. VLDB Endowment (2002)
18. Lee, J., Cho, H., Hwang, S.W.: Efficient dual-resolution layer indexing for top-k queries. In: 2012 IEEE 28th International Conference on Data Engineering (ICDE), pp. 1084–1095. IEEE (2012)
19. Li, C., Chang, K.C.C., Ilyas, I.F., Song, S.: RankSQL: query algebra and optimization for relational top-k queries. In: Proceedings of the 2005 ACM SIGMOD International Conference on Management of Data, pp. 131–142. ACM (2005)
20. Li, C., Chen-Chuan Chang, K., Ilyas, I.F.: Supporting ad-hoc ranking aggregates. In: Proceedings of the 2006 ACM SIGMOD International Conference on Management of Data, pp. 61–72. ACM(2006)
21. Michel, S., Triantafillou, P., Weikum, G.: Klee: A framework for distributed top-k query algorithms. In: Proceedings of the 31st International Conference on Very Large Data Bases, pp. 637–648. VLDB Endowment (2005)
22. Nepal, S., Ramakrishna, M.: Query processing issues in image (multimedia) databases. In: 15th International Conference on Data Engineering, 1999, Proceedings, pp. 22–29. IEEE (1999)
23. Ryeng, N.H., Vlachou, A., Doulkeridis, C., Nørvåg, K.: Efficient distributed top-k query processing with caching. In: Yu, J.X., Kim, M.H., Unland, R. (eds.) DASFAA 2011, Part II. LNCS, vol. 6588, pp. 280–295. Springer, Heidelberg (2011)
24. Vlachou, A., Doulkeridis, C., Nørvåg, K.: Distributed top-k query processing by exploiting skyline summaries. Distrib. Parallel Databases 30(3–4), 239–271 (2012)
25. Vlachou, A., Doulkeridis, C., Nørvåg, K., Vazirgiannis, M.: On efficient top-k query processing in highly distributed environments. In: Proceedings of the 2008 ACM SIGMOD International Conference on Management of Data, pp. 753–764. ACM (2008)
26. Xie, M., Lakshmanan, L.V., Wood, P.T.: Efficient top-k query answering using cached views. In: Proceedings of the 16th International Conference on Extending Database Technology, pp. 489–500. ACM (2013)
27. Zou, L., Chen, L.: Pareto-based dominant graph: an efficient indexing structure to answer top-k queries. IEEE Trans. Knowl. Data Eng. 23(5), 727–741 (2011)

Approximate Iceberg Cube
on Heterogeneous Dimensions

Dan Yin[1][(✉)], Hong Gao[1], Zhaonian Zou[1], Jianzhong Li[1], and Zhipeng Cai[2]

[1] Harbin Institute of Technology, Harbin, China
{yindan,honggao,znzou,lijzh}@hit.edu.cn
[2] Georgia State University, Atlanta, USA
zcai@gsu.edu

Abstract. Heterogeneous information networks contain heterogeneous types of nodes and edges, e.g., social networks and knowledge graphs. A meta-path is a path connecting nodes through a sequence of heterogeneous edges, representing different kinds of semantic relations among nodes. Meta-paths are good mechanisms to improve the quality of graph analysis on heterogeneous information networks. This paper presents an iceberg cube framework for heterogeneous information networks based on meta-paths. To the best of our knowledge, there is no such proposal in the past. (1) We use meta-paths to measure the similarities of nodes, and prove the problem is NP-hard. (2) An optimal solution is proposed for the strict case. We develop the variant of slice tree to aggregate networks hierarchically. (3) To improve the scalability, a general approximate algorithm is provided for fast aggregation, where random walk on meta-paths is employed to measure the similarities. (4) Two pruning strategies are designed for reducing search space when the aggregate function is monotonic. (5) Experiments on both real-world and synthetic networks demonstrate the effectiveness and efficiency of the algorithms.

1 Introduction

With the rapid development of social media networks and knowledge graphs, graphs have become increasingly popular and content-rich. Heterogeneous information networks are such graphs which contain multiple types of nodes and edges, and each type of nodes has a set of attributes.

In heterogeneous information networks, a meta-path is a path connecting nodes through a sequence of heterogeneous edges, representing different kinds of semantic relations among nodes. We give a real world example to illustrate the heterogeneous information networks and meta-paths.

Figure 1 is an IMDb network[1]. It contains four types of nodes: *Movie (M)*, *Actor (A)*, *Director (D)* and *Movie Studio (S)*. Edges exist between actors and actors by the Relation *Coorperate*, between actors and movies by *Play*, between movies and directors by *Direct*, between movies and studios by *Publish*.

---

[1] http://www.imdb.com/.

S.B. Navathe et al. (Eds.): DASFAA 2016, Part II, LNCS 9643, pp. 82–97, 2016.
DOI: 10.1007/978-3-319-32049-6_6

In IMDb network, $A - M - D$ is a meta-path denoting the relation between an actor and a director who directs the movie he plays, and $A - M - S$ is a meta-path denoting the relation between an actor and a movie studio which produces the movie the actor plays. Meta-paths are good mechanisms to improve the quality of graph analysis on heterogeneous information networks. The meta-path framework provides a powerful mechanism to select close nodes.

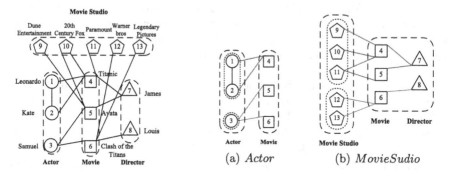

Fig. 1. IMDb network   **Fig. 2.** Example cuboids of IMDb network

Figure 2 shows two cuboids of IMDb network, where nodes are aggregated according to a meta-path. Figure 2(a) gives a cuboid on actors, which are aggregated based on meta-path $A - M - A$, e.g., Leonardo and Kate are aggregated because they are connected by path Leonardo-Titanic-Kate. In the same way, Fig. 2(b) is a cuboid on movie studios, which are aggregated based on the edge sequence of studios, movies and directors, which can be represented by meta-path $S - M - D - M - S$.

The meta-paths of a knowledge graph are numerous, for the types of nodes and edges are complicated. For example, a single knowledge graph could have more than 10K types of nodes. The aggregate graphs based on different meta-paths represent various semantics. For example, the cuboid based on meta-path $A - M - D - M - A$ represents the actors who play the movies directed by the same directors, and the cuboid based on meta-path $A - M - S - M - A$ represents the actors who play the movies produced by the same studios. Due to the large size of meta-paths, the number of cuboids is explosive. We need an approach to discover the interesting cuboids.

In IMDb network, if user wants to find the cuboids which aggregates more nodes together. The threshold of iceberg cube is set to $\theta = 3$. Figure 2(a) is not an iceberg cuboid, since the maximum number of actors aggregated together is 2 (nodes 1, 2), smaller than θ. Figure 2(b) is an iceberg cuboid, since 3 movie studios aggregated (nodes 9, 10, 11), which is equal to θ.

The research work on iceberg cube in large graphs is a new topic. The traditional multi-dimensional iceberg cube analysis can't be directly applied to graph data due to the lack of dimensionality of graphs. The only existing study on iceberg cube analysis in graphs is [7], which focuses on the iceberg nodes for which the aggregation of an attribute in their vicinities as above a given threshold. However, it cannot handle the problem of finding iceberg aggregate graphs in heterogeneous information networks.

To overcome the challenges, we propose an iceberg cube model for heterogeneous information networks. We use meta-paths to measure the similarities of nodes. The variant of slice tree [10] is designed to aggregate networks hierarchically. An optimal solution is proposed for the strict case. To relax the restriction on node similarities, we develop an approximate algorithm, where random walk on meta-path is employed. We propose two novel approaches to effectively prune the search space when the aggregate function is monotonic. The main contributions of this paper are summarized as follows:

1. The iceberg cube problem based on meta-paths in heterogeneous information networks is first investigated, where meta-paths are employed to measure the similarities among nodes. The problem is proved to be NP-hard.
2. An optimal model is introduced for the strict case. The variant of slice tree is employed for hierarchical aggregation.
3. A general approximate algorithm is proposed for fast aggregation, which devises random walk on meta-paths to estimate the similarities.
4. Two effectively pruning strategies are designed for limiting the search space of candidate iceberg cuboids, when the aggregate function is monotonic.
5. Experiments over real world and synthetic networks demonstrate the proposed algorithms effective and efficient.

2   Preliminaries

2.1   Heterogeneous Information Networks

Definition 1 *(**Heterogeneous   information   network**). A heterogeneous information network is defined as a graph $G = (V, E, T, R, A, \phi_V, \phi_E, \phi_A)$, where V is the node set, E is the edge set, T is the set of node types, R is the set of edge types and A is the attribute set of nodes. $\phi_V : V \to T$ is the node type mapping function, $\phi_E : E \to R$ is the edge type mapping function and $\phi_A : T \to A$ is the mapping function from nodes types to attributes.*

Definition 2 *(**Graph partition**). Given a network G and $T_i \in T$, the graph partition $G_{T_i} = \{G_1, G_2, \cdots, G_p\}$ of G on T_i is a family of disjoint subgraphs, where $G_j = (V_j, E_j)$ and $p = |G_{T_i}|$, iff it satisfies*

1. $\forall G_j \in G_{T_i}$, $V_{T_i} = \cup_{j=1}^{p} V_j$, where $V_{T_i} = \{v | v \in V, \phi_V(v) = T_i\}$;
2. $\forall G_j \in G_{T_i}$, $V_j \subseteq V_{T_i}$, $E_j = \{(u,v) | u, v \in V_j, (u,v) \in E\}$;
3. $\forall G_j, G_k \in G_{T_i}$, $j \neq k$, $V_j \cap V_k = \emptyset$.

2.2   Meta-path

Definition 3 *(Meta-path). A meta-path P is denoted in the form of $T_1 - T_2 - \ldots - T_{l+1}$, which defines a composite relation between types T_1 and T_{l+1}.*

We say a path $p = (a_1 a_2 \ldots a_{l+1})$ between a_1 and a_{l+1} follows meta-path P, if $\forall i,\ \phi_V(a_i) = T_i$. We call these paths as path instances of P. Further, we say a meta-path is symmetric if the relation R defined by it is symmetric. For example, $A - M - D - M - A$ is a length-4 symmetric meta-path denoting the actors who play in the movies directed by the same directors. In this paper, we employs the symmetric meta-paths to aggregate close nodes.

Definition 4 *(Similarity).   Given a symmetric meta-path P, the similarity between the same type of nodes v and u is,*

$$ sim(v, u) = \frac{2 \times |\{p_{v \rightsquigarrow u} : p_{v \rightsquigarrow u} \in P\}|}{|\{p_{v \rightsquigarrow v} : p_{v \rightsquigarrow v} \in P\}| + |\{p_{u \rightsquigarrow u} : p_{u \rightsquigarrow u} \in P\}|} $$

Where $p_{v \rightsquigarrow u}$ is a path instance between v and u following P, $p_{v \rightsquigarrow v}$ is that between v and v following P, and $p_{u \rightsquigarrow u}$ is that between u and u following P.

Given a meta-path P, $sim(v, u)$ is defined in terms of two parts: (1) their connectivity defined by the number of path instances between them following P; and (2) the balance of their visibility, where the visibility is defined as the number of path instances between themselves. From the view of structures, the more similar two nodes are, the more likely they are connected by a meta-path.
   The similarity properties are:

1. Symmetric. $s(v, u) = s(u, v)$, when G is an undirected graph;
2. Self-maximum. $s(v, u) \in [0, 1]$, and $s(v, v) = 1$. (Proof omitted)

2.3   Cuboid

Definition 5 *(Cuboid). Given a network G, a symmetric meta-path $T_1 - T_2 - \ldots - T_2 - T_1$, The cuboid G_c of G on P satisfies:*

1. *$G_c = \{G_1, G_2, \ldots, G_k\}$ is a graph partition of T_1;*
2. *For $\forall u, v \in G_i$, $sim(u, v) \geq \tau$;*
3. *The aggregate value of G_c is $\Psi(\varphi(G_1), \varphi(G_2), \ldots, \varphi(G_k))$, where φ is the aggregate function of subgraphs and Ψ is the aggregate function of cuboid.*

We give some explanations for **cuboid** as follows:

(1) The aggregate function φ on subgraphs can be average degree, centrality, diameter, containment and so on, besides the traditional functions.
(2) The aggregate function Ψ is a function on $\{\varphi(G_1), \varphi(G_2), \ldots, \varphi(G_k)\}$. $\Psi(\varphi)$ constructs the composite aggregate function on cuboid.
(3) $\tau \in (0, 1]$, the larger τ is, the more similar the nodes are. When $\tau = 1$, nodes in the same groups have the exact path instances.

2.4   Iceberg Cube

For a cuboid G_c, if its aggregate function is above a certain threshold θ, G_c is called an iceberg cuboid, and all the iceberg cuboids are called iceberg cube.

Problem 1 **(IceCubeH)**. Given a network G and the iceberg threshold θ, the goal is to find the iceberg cube *w.r.t θ*.

Theorem 1. *The hardness of* **IceCubeH** *is NP-hard.*

Proof. We consider a special case of the problem. We set the aggregate functions to $\psi = MAX$ and $\varphi = COUNT(V)$. Specifically, the problem is the maximal clique problem and decide if the size of maximal clique is larger than θ, which is NP-hard. Thus the hardness of the problem is NP-hard.   □

The greatest challenges in solving **IceCubeH** are: (1) The problem is NP-hard, which is difficult to solve in a short time; (2) The meta-paths are numerous, since different types of nodes have different referenced meta-paths; (3) The large search space causes explosive computation. For challenge 1, we propose an approximate algorithm to tackle the problem in Sect. 4. For challenge 2 and 3, we propose two pruning strategies to address the challenges in Sect. 5.

3   Optimal Model for Strict Case

We introduce the optimal algorithm for the strict case when $\tau = 1$. We make the nodes in the same groups have the same structures correspondence with specific meta-paths. Motivated by the *Slice Tree*, which is a hierarchical decomposition of the network, we propose a network aggregation strategy to compute the cuboid based on meta-paths. A *slice* partitions a set of nodes into two parts: one composed of nodes with the same path instances following meta-path P and the other containing the remaining nodes.

Definition 6 *(Slice)*. *Given a network G, nodes $X \subseteq V$, $\forall u, v \in X$, $\phi_u = \phi_v$, meta-path P, a slice $s(v, X, P)$ partitions X into $B = \{u \in X | \forall u, v \in X, sim(u, v) = 1\}$ and $X \backslash B$.*

The Slice function is given in Algorithm 1, which partitions the node set X into two parts and guarantees the node set B be maximum. We outline a greedy procedure which select the largest size of B, such that the partition of X can stop earlier. Optimal-Slice first computes the $p_{v \leadsto v}$ for each v (Line 3). Then finds the other nodes in X which has the same structures with v (Line 4–7). Finally select the slice which leads to the maximum $|B|$ (Line 8–9).

Next, we propose a cuboid computation algorithm based on the slice function. A greedy optimal algorithm, which aggregates the network by selecting consecutive the best slices is designed. At beginning, the node set of T_1 is considered as the node set X. Call the Optimal-Slice function to partition X into two groups B and $X \setminus B$. B is the first subgraph in G_c, and $X \setminus B$ is the second subgraph

Algorithm 1. Optimal-Slice

Input: A network G, meta-path P, a set of nodes X
Output: Best slice t

1: Initialize $num = 0$, $max = 0$
2: **for** each $v \in X$ **do**
3:    Compute $p_{v \rightsquigarrow v}$
4:    **for** each $u \in X$ **do**
5:       Compute $p_{v \rightsquigarrow u}$
6:       **if** $sim(v, u) = 1$ **then**
7:          num++
8:    **if** $num > max$ **then**
9:       $t \leftarrow s(v, X, B)$, $max = num$
10: **return** t

which is to be partitioned in next iteration. Then the iterative partitions don't stop until $X \setminus B = \emptyset$.

The pseudo-code of the algorithm for computing cuboid is described in Algorithm 2. The algorithm consider the being partitioned group as X. Initialize the node set of T_1 as the first partitioned node set X (Line 2–3). Then for each partitioned subgraph G_i, call Optimal-Slice to slice G_i into two parts: B and $X \setminus B$. Delete the node set $X \setminus B$ from G_i and construct a new subgraph G_{i+1} whose node set is $X \setminus B$. G_{i+1} is the next partitioned subgraph (Line 4–10). Finally construct cuboid (Line 11).

Algorithm 2. Greedy-Optimal-Cuboid

Input: A network G, meta-path P, $\tau = 1$
Output: Cuboid $G_c = \{G_1, G_2, \dots\}$

1: Initialize hash table ξ for mapping from nodes to groups
2: Initialize $V_1 = \{v | \phi_V(v) = T_1\}$, $E_1 = \{(u, v) | u, v \in V_1, (u, v) \in E\}$
3: Initialize $G_c = \{G_1\}$, $i = 1$
4: **repeat**
5:    $X \leftarrow V_i$
6:    Let B and $X \setminus B$ be the partitions produced by Optimal-Slice(G, P, X)
7:    **if** $X \setminus P \neq \emptyset$ **then**
8:       Construct a new subgraph G_{i+1} into ξ
9:       $V_i \leftarrow B$, $V_{i+1} = X \setminus B$, i++
10: **until** $X \setminus P = \emptyset$
11: Construct $G_c = (G_1, G_2, \dots)$
12: **return** G_c

Theorem 2. *The time complexity of Greedy-Optimal-Cuboid algorithm is:* $O(|V_{G_P}|^2 |E_{G_P}|)$, *where G_P is the induced subgraph of G by P, V_{G_P} is the node set and E_{G_P} is the edge set.*

Proof. A slice s over X produces subgraphs B and $X \setminus B$. s applies $|X|$ BFSs on G_P, which costs $O(|X||E_{G_P}|)$ time. In the worst case, s generates P and $X \setminus P$ such that $|P| = 1$ and $|X \setminus P| = |X| - 1$, resulting in a complexity: $O(\Sigma_{i=0}^{|V_{G_P}|}(||V_{G_P} - i|)(O(|E_{G_P}| - 1))) = O(|V_{G_P}|^2 |E_{G_P}|)$. $\square$

Algorithm 3. IceCube

Input: A network G, $\tau = 1$, threshold of iceberg cube θ, a set of meta-paths S
Output: Iceberg cube

1: Initialize queue Q for iceberg cube
2: **for** each $P \in S$ **do**
3:    $G_c^P \leftarrow$ Greedy-Optimal-Cuboid(G, P)
4:    Compute aggregate value $\psi(G_c^P)$
5:    **if** $\psi(G_c^P) \geq \theta$ **then**
6:        Insert G_c^P into Q
7: **return** Q

3.1 Iceberg Cube

The iceberg cube algorithm is displayed in Algorithm 3. For each meta-path P, call Greedy-Optimal-Cuboid function to compute the cuboid G_c^P and compute the aggregate value (Line 3–4). If the aggregate value isn't smaller than θ, then G_c^P is an iceberg cuboid (Line 5–6).

Theorem 3. *The time complexity of* IceCube *is* $O(\sum_{P \in S} |V_{G_P}|^2 |E_{G_P}| ComAgg(G_c^P))$, *where S is the set of meta-paths, G_P is the induced subgraph of G by P, V_{G_P} is the node set, E_{G_P} is the edge set G_c^P is the cuboid w.r.t P and $ComAgg(G_c^P)$ is the cost of computing the aggregate value of cuboid G_c^P.*

Proof. For $\forall P \in S$, it will call the *Greedy-Optimal-Cuboid* algorithm to compute the cuboid, which consumes $O(|V_{G_P}|^2 |E_{G_P}|)$ time (Theorem 2). The aggregate value of the cuboid needs to be computed in $ComAgg(G_c^P)$. Thus the time complexity of IceCube is $O(\sum_{P \in S} |V_{G_P}|^2 |E_{G_P}| ComAgg(G_c^P))$. $\square$

4 A General Approximate Algorithm

Generally, nodes rarely have exactly the same path instances. Similarities of nodes should be relaxed. To tackle the hard problem, we propose a general approximate algorithm for iceberg cube query when $\tau \in (0, 1]$. A random walk is a mathematical formalization of a path that consists of a succession of random steps. We use random walk to obtain the number of path instances p between v and u following P. To modify the previous random walk to make it applicable to heterogeneous information networks, we restrict the random walk implemented on meta-paths. The transition matrix is dependent on meta-paths. For $P = T_i - T_j - \ldots - T_k$, the transition exists from T_i to T_j, and finally arrives at T_k.

Definition 7 (Transition probability on meta-path). *Given a meta-path $P = T_i - T_j - \ldots - T_k$, for $\forall v$, $\phi_V(v) = T_i$, the transition probability from nodes of T_i to nodes of T_j is: $M_{ij}(v) = \frac{1}{d_{T_j}}$, where d_{T_j} is the number of v_i's neighbors whose type is T_{i+1}.*

The length of each walk is related to the length of meta-paths. For example, the length of random walk following meta-path $A - M - D$ is 2. We use random walk to compute the similarities of nodes. For a meta-path $P = T_1 - T_2 - \ldots - T_l$, random walk starts from a node v whose type is T_1, the transition probability of the random walk following meta-path P is given in Definition 7.

Based on the random walk on meta-paths, we propose an approximate slice to guarantee the nodes in B are similar with each other within τ. Random walk is used to estimate the number of path instances $p_{v \rightsquigarrow u}$ between v and u. Then use the estimate value $p_{v \rightsquigarrow u}$ to compute $sim(v, u)$. The pseudo-code for

---

Algorithm 4. Appro-Slice

Input: A network G, similarity threshold τ, meta-path P, a set of nodes X
Output: Best slice t

```
 1: Initialize num = 0, max = 0, R
 2: for each v ∈ X do
 3:     repeat
 4:         Random walk starts from v following P
 5:         if Random walk ends with v then
 6:             p_{v⇝v}++
 7:     until Random walk times reach R
 8: for each v ∈ X do
 9:     for each u ∈ X do
10:         repeat
11:             Random walk starts from v following P
12:             if Random walk ends with u then
13:                 p_{v⇝u}++
14:         until Random walk times reach R
15:         if ∀w ∈ B, sim(u, w) ≥ τ then
16:             Insert u into B, num++
17:     if num > max then
18:         t ← s(v, X, P), max = num
19: return t
```

---

approximate slice is given in Algorithm 4. For each node v in X, start a random walk following meta-path P from v. If the random walk stops at v, the number of path instances $p_{v \rightsquigarrow v}$ increases by one. Repeat random walk R times (Line 2–7). For each node v in X, compute the similarities between v and other nodes in X by random walk (Line 8–14). If the similarities between u and the other nodes in B are not smaller than τ, insert u into B (Line 15–16). Finally, select the slice which produces the maximum size of B (Line 17–18).

Theorem 4. *The time complexity of Appro-Slice is* $O(|X|^2(|X| + R|P|))$.

Proof. For each $v \in X$, the random walks starting from v and following meta-path P cost $O(R|P|)$ time. Then for each $v \in X$ and $u \in X$, the random walks starting from v and following meta-path P execute, which cost $O(R|X||P|)$ time, for $\forall w \in B$, compute $sim(w, u)$ and decide if u can be added into B cost $O(|B|)$. The time complexity if $O(|X|^2(|X| + R|P|))$. $\qquad \square$

Based on the approximate slice, we propose an algorithm called Greedy-Appro-Cuboid that computes the cuboids approximately. To save space, the pseudo-code of the algorithm is omitted here, which is the same with Algorithm 2 except replacing the Optimal-Slice with Appro-Slice.

5 Pruning Strategy

5.1 Prune 1

As defined in cuboid, the aggregate value of G_c is $\Psi(\varphi(G_1), \varphi(G_2), \dots, \varphi(G_k))$, where $\Psi(\cdot)$ is the aggregate function of cuboid. If the aggregate functions φ and ψ are monotonic, the searching space can terminate in advance.

Theorem 5. *For a meta-path P, if the composite function $\psi(\varphi)$ is monotonically increasing with continuous calling Algorithm* Optimal-Slice *or* Appro-Slice, *the call procedure can be terminated when $\psi(G_c^P) \geq \theta$, and the cuboid w.r.t P is an iceberg cuboid. Otherwise, if $\psi(\varphi)$ is monotonically decreasing, the call procedure can be terminated when $\psi(G_c^P) < \theta$, and the cuboid w.r.t P can be pruned.*

Example 1. We set $\psi = \text{MIN}$ and $\varphi = \text{COUNT}$, then for a cuboid $G_c = \{G_1, G_2, \dots, G_k\}$, $\psi(G_c) = \text{MIN}(\text{COUNT}(V_{G_1}), \text{COUNT}(V_{G_2}), \dots, \text{COUNT}(V_{G_k}))$. The composite function $\psi(\varphi)$ is monotonically decreasing with slices. For example, when $G_c^{'} = \{G_1, G_2, G_3\}$, in next slice, we will slice the subgraph G_3 into B and $G_3 \setminus B$. $\text{COUNT}(V_{G_3}) > \text{MIN}(\text{COUNT}(B), \text{COUNT}(V_{G_3} \setminus B))$. Thus, when $\psi(G_c^{'}) < \theta$, the slices can be stopped and the cuboid is not an iceberg cuboid.□

Example 2. If we set $\psi = \text{MAX}$ and $\varphi = \text{COUNT}$, $\psi(G_c) = \text{MAX}(\text{COUNT}(G_1), \text{COUNT}(G_2), \dots, \text{COUNT}(G_k))$. The composite function $\psi(\varphi)$ is monotonically increasing with slices. when $\psi(G_c^{'}) \geq \theta$, the slices can be stopped and the cuboid is an iceberg cuboid.   □

5.2 Prune 2

For meta-paths in DBLP, $P_1 = Author - Paper - Author$ and $P_2 = Author - Paper - Venue - Paper - Author$, the similarities of authors based on P_1 is smaller than P_2. P_1 captures the coauthor relationship, whereas the P_2 represents the relationship between a pair of authors through their papers published on the common venues. More authors are aggregated by P_2 than P_1. For the meta-path pairs like P_1 and P_2, we define $P_1 \prec P_2$. We give the following pruning strategy:

Theorem 6. *If the composite function $\psi(\varphi)$ is monotonically increasing with the sizes of subgraphs in cuboids and $P_1 \prec P_2$, there exists $\psi(G_c^{P_1}) < \psi(G_c^{P_2})$. If $\psi(G_c^{P_2})$ isn't an iceberg cuboid, then $\psi(G_c^{P_1})$ can be pruned.*

Example 3. The aggregate functions are set as the same with Example 1. We have $\psi(G_c^{P_1}) < \psi(G_c^{P_2})$, since the number of authors who have published papers in the same venues is larger than the number of co-authors.   □

6 Experiments

All the experiments are implemented on a Microsoft Windows 7 machine with an Intel(R) Core i5-2400 CPU 3.1 GHz and 8 GB main memory. Programs are compiled by Microsoft Visual Studio 2010 with C language.

Fig. 3. Synthetic network structure

Table 1. Statistics of synthetic networks

Network	Node type	Edge type	Node size/type
1	50	80	10 K
2	50	80	15 K
3	50	80	20 K
4	50	80	25 K
5	50	80	30 K

6.1 Datasets

DBLP Network. The network [13] includes fours types of nodes, *Paper*, *Author*, *Conference* and *Term*. Four types of edges exist, i.e., *Write* relation between authors and papers, *Publish* relation between papers and conferences and *Belong* relation between papers and terms representing the papers have the terms as key contents. It contains 8,340 authors, 5,000 papers, 4,729 terms and 37 venues. There are 13,351 edges of *Write*, 5,000 edges of *Publish*, and 32,540 edges of *Belong*.

Synthetic Networks. To study the scalability of our solutions, we generate a collection of synthetic networks with size varying from 500 K to 1.5 M. The *Densities* of networks are set to $\frac{|E|}{|V|} = 2$. The statistics of synthetic networks are shown in Table 1. The synthetic networks are generated in two steps. First, we generate k types of nodes, and each type of nodes have m nodes. Second, we randomly select a node type T_i from Q and another node type T_j from $T \setminus Q$, then construct an edge type between T_i and T_j. Insert T_j into Q. Repeat this process until $|Q| = n$. For each type of edges, randomly select $Density \cdot \frac{km}{n}$ pairs of nodes to insert edges. The structure of networks is shown in Fig. 3, where a node represents a node type and an edge represents an edge type.

6.2 Experiments on DBLP Dataset

Parameters. We evaluate the optimal algorithm by $\tau = 1$, and the approximate algorithms by $\tau = 0.8$, 0.9. Aggregate functions are set to $\psi = MAX$ and $\varphi = COUNT(V)$. Three case studies are conducted, given (a) an iceberg threshold θ; (b) a set of meta-paths. The three queries are shown as follows:
Query 1. (a) $\theta = 15$; (b) meta-paths:
(1) $P_1 = A - P - A$; (2) $P_2 = V - P - V$; (3) $P_3 = T - P - T$;
(4) $P_4 = P - A - P$; (5) $P_5 = P - V - P$; (6) $P_6 = P - T - P$.
Query 2. (a) $\theta = 100$; (b) meta-paths:
(7) $P_7 = A - P - V - P - A$; (8) $P_8 = A - P - T - P - A$;
(9) $P_9 = V - P - A - P - V$; (10) $P_{10} = T - P - A - P - T$;
(11) $P_{11} = T - P - V - P - T$; (12) $P_{12} = V - P - T - P - V$.
Query 3. (a) $\theta = 100$; (b) meta-paths: meta-paths (1)~(12).

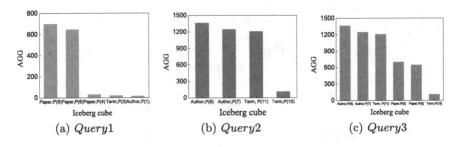

Fig. 4. Optimal iceberg cube of on DBLP network.

Effectiveness Evaluation. x-axis represents the iceberg cube including: type of aggregate nodes and meta-path. y-axis shows the *aggregate values (AGG)*.

Optimal Algorithm. Figure 4(a) shows the iceberg cube of query 1. There are five cuboids above threshold θ. Three of them are paper nodes, for the papers are the central core of the networks. The paper cuboid based on meta-path $P-V-P$ has the highest AGG 700, followed with AGG 648 by $P-T-P$. The meta-path $P-A-P$ has small AGG on papers. Papers are more similar on their venues and terms than authors. Every venue publishes hundreds of papers every year. Plenty of papers on similar topics are published.

Optimal Algorithm. Figure 4(b) shows the optimal iceberg cube of query 2. The AGG of top-3 cuboids are all beyond 1200. Followed by two cuboids with 2-length meta-path, whose AGG decrease sharply. The top 2 AGG illustrates that the number of authors who published papers with similar topics in the same venues is huge. Meanwhile, the top 3th AGG states that papers in the same venues have similar terms. The authors correspond with $T-P-A-P-T$ means that there are 113 terms at most, which are used in the papers of the same authors.

Figure 4(c) gives the optimal iceberg cube of query 3. It is obvious that the cuboids with longer meta-path have higher AGG. When the path length is 1, for example $P-A-P$, two papers are partitioned into one group when they must be written by the same authors. When the path length is more than 1, for example $A-P-V-P-A$ which represents the authors of the papers published in the same venues are similar. More path instances are found by longer meta-paths.

Approximate Algorithm. We set the round of random walk to 20. Figures 5 and 6 display the iceberg cube by approximate algorithm with $\tau = 0.9$ and 0.8, respectively. Approximate algorithm with smaller τ could get iceberg cube with higher AGG. The algorithm with larger τ could produce more accurate results.

Figures 5(a) and 6(a) give the iceberg cube of query 1. Compared with optimal algorithm, the approximate algorithm could get iceberg cube with larger AGG, for the similarities based on meta-paths are relaxed by τ. The smaller τ is, the more similar of nodes. Figures 5(b) and 6(b) exhibit the iceberg cube of query 2. When $\tau = 0.8$, cuboid $Venue, P(12)$ is output as an iceberg cuboid. This is because approximate algorithm with smaller τ makes more nodes aggregated together. Figures 5(c) and 6(c) show the approximate results of query 3.

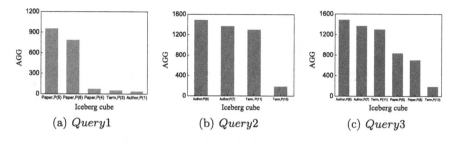

(a) *Query1*   (b) *Query2*   (c) *Query3*

Fig. 5. Approximate iceberg cube of $\tau = 0.9$ on DBLP network.

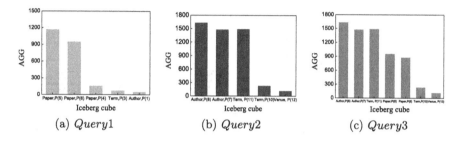

(a) *Query1*   (b) *Query2*   (c) *Query3*

Fig. 6. Approximate iceberg cube of $\tau = 0.8$ on DBLP network.

The algorithms could get more iceberg cube when the number meta-paths become larger. At the same time, the $AGGs$ of iceberg cube increase greatly, for the relaxed constrain in similarity. The sizes of groups in cuboids become larger. Figure 6(c) shows the approximate algorithm with $\tau = 0.8$ gets 1 more iceberg cube.

Recall. Recall is defined as $\frac{|G_c^{Exp-p}|}{|G_c^{Opt}|}$, where G_c^{Exp-p} is the positive iceberg cube produced by algorithms and G_c^{Opt} is the iceberg cube produce by optimal algorithm. Figure 7 gives the recall of the optimal algorithms and the approximate algorithms. In approximate algorithms, τ is set to 0.8 and 0.9. We can see the three algorithms could find the whole iceberg cube in three queries.

Accuracy. Accuracy is defined as $\frac{|G_c^{Exp-p}|}{|G_c^{Exp}|}$, where $G_c^{Exp-true}$ is the positive iceberg cube produced by algorithms and G_c^{Exp} is the iceberg cube found by algorithms. Figure 8 shows the accuracy of the optimal and approximate algorithms. We can see the optimal algorithm and the approximate algorithm τ have 100 percent accuracy. The approximate algorithm $\tau = 0.8$ produces negative iceberg cuboids in query 2 and 3.

Efficiency Evaluation. *Runtime.* The comparisons of runtime by optimal and approximate algorithms are shown in Fig. 9. Both the optimal and approximate algorithms take more time on query 3 than query 1 and 2. The approximate algorithms cost much more time than the optimal one, since the approximate algorithm computes the similarities by multiple random walks. Long meta-paths

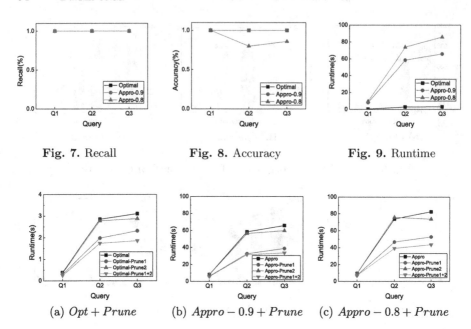

Fig. 7. Recall   Fig. 8. Accuracy   Fig. 9. Runtime

(a) *Opt + Prune*   (b) *Appro − 0.9 + Prune*   (c) *Appro − 0.8 + Prune*

Fig. 10. Efficiency of pruning strategies on DBLP network

need long random walk. The approximate algorithm with smaller τ costs more time, because more nodes are partitioned into the same groups when τ is small. This will lead to more time consumption on approximate slice function.

Prune Efficiency. We demonstrate the runtime improvement of the pruning strategy over optimal and approximate algorithms. Four partial versions of pruning algorithms are evaluated respectively: (1) Optimal/Appro ($\tau = 0.8/0.9$) without pruning; (2) Optimal/Appro ($\tau = 0.8/0.9$) with Prune 1; (3) Optimal/Appro ($\tau = 0.8/0.9$) with Prune 2; (4) Optimal/Appro ($\tau = 0.8/0.9$) with Prune $1 + 2$.

Figure 10 shows the improvement of pruning strategy on optimal and approximate algorithms. The original algorithms without any pruning cost the most time. The pruning methods outperform significantly when approximate algorithm with $\tau = 0.8$ than $\tau = 0.9$ and the optimal algorithm. By including Prune $1 + 2$, the approximate algorithm with $\tau = 0.8$ costs about 40 s less than with pruning, achieving 2 times faster. It implies the pruning methods can indeed avoid some unnecessary computation. From Fig. 10(a) and (b) we can see, Prune 2 has no effect on query 1 and 2, since there are no meta-paths overlapped in query 1 and 2. The runtime of query 1 and 2 can further be reduced by employing Prune 1. Prune 2 shows a clear advantage on runtime improvement on query 3. For example, $P_2 = V − P − V$ can be pruned after eliminating $P_9 = V − P − A − P − V$ from iceberg cube.

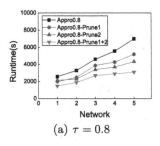
(a) $\tau = 0.8$

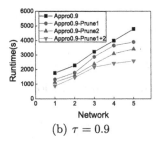
(b) $\tau = 0.9$

Fig. 11. Runtime of approximate algorithms on synthetic networks.

6.3   Scalable Experiments on Synthetic Networks

Parameters. We test the scalability of approximate algorithms. We set iceberg threshold $\theta = 100$ in the scalability tests. All the meta-paths on synthetic networks are generated automatically and as input. We repeat each query 10 times and report the average runtime. The round of random walk is $\frac{Nodesize/type}{500}$.

Experimental Results. Figure 11 compares the runtime with varying the size of network. The run time of approximate algorithms increases with the size of network enlarging. Pruning strategies perform better when τ is smaller. As shown in Fig. 11(a), the approximate algorithm runs 2 times faster with Prune $1 + 2$ than without any pruning. While in Fig. 11(b), the approximate algorithm runs about 1.8 times faster. When the size of network increases to $30\,\text{K} \times 50$, the time costs reach up to $3000\,\text{s}$ and $2500\,\text{s}$ when $\tau = 0.8$ and $\tau = 0.9$, respectively.

7   Related Work

Iceberg Cube and Graph OLAP. Data cube has been widely researched in data warehouse and OLAP [4]. Many researchers have contributed to developing efficient cube materialization and iceberg query algorithms for traditional databases [1,2,5]. Iceberg analysis on graphs has been under-explored due to the absence of dimensionality in graph data. The iceberg cube query in large graphs is first proposed by [7], which focuses on the iceberg nodes for which the aggregation of an attribute in their vicinities as above a given threshold. The first work to place graphs in a rigid multidimensional and multi-level framework is [3]. The aggregate dimension is the attributes of nodes. [19] follows [3] to study graph cube in homogeneous networks. [16] proposes parallel graph cube over homogeneous networks. With a distinct focus, we defined the concept of aggregate graphs on the heterogeneous dimensions and propose scalable solutions.

Graph Clustering. [20] proposes a method that summarizes the graphs based on different partitioned regions. SNAP operations were introduced in [15,18]. All the nodes are homogeneous in terms of both attributes and relationships. [11] proposes a model-based method for clustering heterogeneous information networks with different edge types and different attribute types. [17] displays a cluster algorithm on multiple meta-paths over heterogeneous information networks. The above two papers focus on the clustering of networks, ignoring finding the iceberg cube with higher aggregate values.

Heterogeneous Information Networks. [6] constructs the multi-type networks into star schema to rank nodes. [13] searches the top-k similar nodes by meta-paths. [12] proposes a link prediction method. [9] measures the relatedness between nodes with different types. [14] utilizes user guidance as seeds to automatically learn the best meta-path for clustering. [8] investigates the entity identification problem.

8   Conclusions

This paper introduces a novel concept, iceberg cube query in heterogeneous information networks. Random walk is used to aggregate the nodes in heterogeneous information networks for approximate computation. Efficient pruning strategies are introduced for reducing search space. Experiments on real world and synthetic data sets demonstrate the algorithms effective and efficient. Iceberg cube can help users discover interesting results from networks.

Acknowledgement. This work is supported by National Grand Fundamental Research 973 Program of China under grant 2012CB316200, National Natural Science Foundation of China under Grant 61190115, 61173023 and 61532015.

References

1. Agarwal, S., Agrawal, R., Deshpande, P.: On the computation of multidimensional aggregates. In: Proceedings of Very Large Database Conference, pp. 506–521. ACM (1996)
2. Beyer, K.S., Ramakrishnan, R.: Bottom-up computation of sparse and iceberg cubes. In: Proceedings of ACM SIGMOD International Conference on Management of Data, pp. 359–370. ACM (1999)
3. Chen, C., Yan, X., Zhu, F., Han, J., Yu, P.S.: Graph OLAP: towards online analytical processing on graphs. In: Proceedings of International Conference on Data Mining, pp. 103–112. IEEE (2008)
4. Gray, J., Bosworth, A., Reichart, D.: Data cube: a relational aggregation operator generalizing group-by, cross-tab, and sub-totals. In: Proceeding of IEEE International Conference on Data Engineering. IEEE (1996)
5. Harinarayan, V., Rajaraman, A., Ullman, J.D.: Implementing data cube efficiently. In: Proceedings of ACM SIGMOD International Conference on Management of Data, pp. 205–216. ACM (1996)
6. Ji, M., Han, J., Danilevsky, M.: Ranking-based classification of heterogeneous information networks. In: Proceedings of ACM SIGKDD International Conference on Knowledge Discovery and Data Mining, pp. 1298–1306. ACM (2011)

7. Li, N., Guan, Z., Ren, L., Wu, J., Han, J., Yan, X.: gIceberg: Towards iceberg analysis in large graphs. In: Proceedings of IEEE International Conference on Data Engineering, pp. 1021–1032. IEEE (2013)
8. Shen, W., Han, J., Wang, J.: A probabilistic model for linking named entities in web text with heterogeneous information networks. In: Proceedings of ACM SIGMOD International Conference on Management of Data. ACM (2014)
9. Shi, C., Kong, X., Yu, P.S.: Relevance search in heterogeneous networks. In: Proceedings of International Conference on Extending Database Technology, pp. 180–191. ACM (2012)
10. Silva, A., Bogdanov, P., Singh, A.K.: Hierarchical in-network attribute compression via importance sampling. In: International Conference on Data Engineering, pp. 951–962. IEEE (2014)
11. Sun, Y., Aggarwal, C.C., Han, J.: Relation strength-aware clustering of heterogeneous information networks with incomplete attributes. Proc. Very Large Data-Bases Endowment 5(5), 394–405 (2012)
12. Sun, Y., Barber, R., Gupta, M.: Co-author relationship prediction in heterogeneous bibliographic networks. In: International Conference on Advances in Social Networks Analysis and Mining, pp. 121–128. IEEE (2011)
13. Sun, Y., Han, J., Yan, X., Yu, P.S., Wu, T.: PathSim: meta path-based top-k similarity search in heterogeneous information networks. Proc. Very Large Databases Endowment 4(11), 992–1003 (2011). ACM
14. Sun, Y., Norick, B., Han, J., Yan, X., Yu, P.S., Yu, X.: Integrating meta-path selection with user-guided object clustering in heterogeneous information networks. In: Proceedings of ACM SIGKDD International Conference on Knowledge Discovery and Data Mining, pp. 1348–1356. ACM (2012)
15. Tian, Y., Hankins, R.A., Patel, J.M.: Efficient aggregation for graph summarization. In: Proceedings of ACM SIGMOD International Conference on Management of Data, pp. 567–580. ACM (2008)
16. Wang, Z., Fan, Q., Wang, H., Tan, K.-L., Agrawal, D., Abbadi, A.E., Pagrol: parallel graph olap over large-scale attributed graphs. In: Proceeding of IEEE International Conference on Data Engineering, pp. 496–507. IEEE (2014)
17. Yang, Z., Ling, L., David, B.: Integrating vertex-centric clustering with edge-centric clustering for meta path graph analysis. In: Proceedings of ACM SIGKDD Conference on Knowledge Discovery and Data Mining, pp. 1563–1572. ACM (2015)
18. Zhang, N., Tian, Y., Patel, J.M.: Discovery-driven graph summarization. In: Proceeding of IEEE International Conference on Data Engineering, pp. 880–891. IEEE, Piscataway (2010)
19. Zhao, P., Li, X., Xin, D., Han, J.: Graph cube: on warehousing and olap multidimensional networks. In: Proceedings of ACM SIGMOD International Conference on Management of Data, pp. 853–864. ACM (2011)
20. Zhou, Y., Cheng, H., Yu, J.X.: Graph clustering based on structural/attribute similarities. Proc. Very Large Database Endowment 2(1), 718–729 (2009)

Pre-computed Region Guardian Sets Based Reverse kNN Queries

Wei Song[1]([✉]), Jianbin Qin[1], Wei Wang[1], and Muhammad Aamir Cheema[2]

[1] The University of New South Wales, Sydney, Australia
{wsong,jqin,weiw}@cse.unsw.edu.au
[2] Monash University, Melbourne, Australia
aamir.cheema@monash.edu

Abstract. Given a set of objects and a query q, a point p is q's reverse k nearest neighbour (RkNN) if q is one of p's k-closest objects. RkNN queries have received significant research attention in the past few years. However, we realise that the state-of-the-art algorithm, SLICE, accesses many objects that do not contribute to its RkNN results when running the filtering phase, which deteriorates the query performance. In this paper, we propose a novel RkNN algorithm with pre-computation by partitioning the data space into disjoint rectangular regions and constructing the *guardian set* for each region R. We guarantee that, for each q that lies in R, its Rk'NN results are only affected by the objects in R's guardian set, where $k' \leq k$. The advantage of this approach is that the results of a query $q \in R$ can be computed by using SLICE on only the objects in its guardian set instead of using the whole dataset. Our comprehensive experimental study on synthetic and real datasets demonstrates the proposed approach is the most efficient algorithm for RkNN.

Keywords: RkNN · Pre-computation · Guardian Set · SLICE

1 Introduction

Reverse k nearest neighbour queries (RkNN) are classified into *bichromatic* RkNN and *monochromatic* RkNN.

Bichromatic RkNN. Given a set of facilities F, a set of users U and a query $q \in F$, the Bichromatic RkNN (denoted as $biRkNN$) returns every user $u \in U$ for which q is one of its k-closest facilities.

Example : For a given McDonald's q, the people for which q is one of their k-closest McDonald's restaurants are its $biRkNN$. These people are its potential customers and can be attracted by targeted marketing. In this paper, the objects providing some service (e.g., McDonald's, supermarkets) are called *facilities* and the objects that use the facilities (e.g., residents, customers) are called *users*.

Monochromatic RkNN. Given a set of facilities F and a query $q \in F$, the Monochromatic RkNN (denoted as $monoRkNN$) returns every facility f for which q is one of its k-closest facilities.

© Springer International Publishing Switzerland 2016
S.B. Navathe et al. (Eds.): DASFAA 2016, Part II, LNCS 9643, pp. 98–112, 2016.
DOI: 10.1007/978-3-319-32049-6_7

Example : Consider the example of hospitals. Given a hospital q, its *monoRkNN* are the ones for which q is one of their k nearest hospitals. Such hospitals may seek assistance (e.g., blood, staff) from q in case of emergencies.

Like most of the existing work on RkNN, we address the problem in a Euclidean space. Our proposed algorithm can by applied to both *monoRkNN* and *biRkNN*. For clear presentation, we focus on the *biRkNN* unless mentioned specifically.

As shown in a recent experimental study [14], SLICE is the state-of-the-art RkNN algorithm. Like most other RkNN algorithms, SLICE consists of two phases namely **filtering phase** and **verification phase**. SLICE's filtering phase dominates the total query processing cost [3]. We observe that SLICE needs to access many unnecessary facilities in its filtering phase and this adversely affects the query performance.

Motivated by the above observation, in this paper, we propose a solution based on pre-computation that divides the whole data space into a set of disjoint rectangular regions. Given a value k, for each rectangular region R, we compute a set of objects $F_g \subseteq F$ such that the results of every Rk'NN query q that lies in R (and $k' \leq k$) can be computed using only the facilities in F_g. The set of objects F_g is called the guardian set of R and a facility $f \in F_g$ is called a guardian facility of R. During the query processing time, we determine the region R that contains q and then use SLICE on its guardian set instead of the whole dataset to compute the results. Since the size of guardian set is significantly smaller than the whole dataset, this approach significantly improves the performance as demonstrated in our experimental study.

We remark that although there exists other pre-computation based approaches, our approach is unique in that its pre-computation does not depend on the set of users. For example, the technique proposed in [7] pre-computes, for each user u_i, its k-th closest facility f_k and creates a circle C_i centred at u_i with radius $dist(u_i, f_k)$. All such circles are indexed by an R-tree and a Rk'NN query q for which $k' \leq k$ is answered using the circles that contain q. A disadvantage of such approach is that any change in the set of users U requires updating or reconstructing the index. On the other hand, our guardian sets do not depend on the set of users U and do not require update with the change in U. This is a desirable property especially because in many real world applications the updates in the locations of facilities (e.g., restaurants, fuel stations) is less common as compared to the locations of users (e.g., people, cars).

Next, we summarise our contributions.

- To the best of our knowledge, we are the first to propose a pre-computation based approach that does not depend on the set of users. Our pre-computation significantly reduces the number of facilities to be accessed for SLICE and improves the query processing cost. The proposed index can be used to answer any Rk'NN query for which $k' \leq k$.
- Our comprehensive experimental study on real and synthetic datasets demonstrates that our algorithm significantly improves SLICE in query processing cost and outperforms all existing RkNN algorithms.

The rest of the paper is organized as follows. We introduce related works in Sect. 2. Section 3 presents our techniques constructing rectangular region based guardian set F_g. We also present our method partitioning universe to many small regions and prove each region's guardian set F_g can be used flexibly for any k' s.t. $k' \leq k$. Section 4 briefly recalls SLICE to do the experimental study in Sect. 5. We conclude this paper in Sect. 6.

2   Related Works

Six-Regions [2] is a region-based technique proposed for $RkNN$. It consists of two phases, namely **Filtering phase** and **Verification phase**. In filtering phase, Six-Regions centres at query q partitioning universe into six regions, each of which has a subtending angle of 60°. In each region, it computes q's k-th nearest neighbour NN_k and construct an arc by centering at q with a radius of $dist(q, NN_k)$. In verification phase, each u locates above the arc in its region as shown u in P_2 (Fig. 1(a)) cannot be a result and only users lie under the arc of its region can be returned as candidates to be verified.

Influence Zone [1] is a half-space based technique proposed for $RkNN$, denotes $InfZone$. It keeps constructing perpendicular bisector between each facility f_i and q and halving the universe to two parts. The half where f_i locates is pruned by f_i as any u in this area must have closer distance to f_i than to q. For areas pruned by at least k facilities (shaded area in Fig. 1(b)) cannot return any result. $InfZone$ guarantees every u in the unpruned area is a result. Such unpruned area is called **Influence Zone**.

SLICE [3] is the most efficient algorithm before our work for $RkNN$, which is integrated in our query processing algorithm in this paper. We introduce SLICE in detail in Sect. 4.

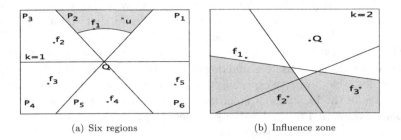

(a) Six regions   (b) Influence zone

Fig. 1. Related works

3   Techniques

As motivated by [1] that given a query q and a k, we may compute a set of facilities that q's $RkNN$ is only affected by such facilities. Similarly, given a rectangular region R and k, we compute such a set F_g of facilities that for any $q \in R$, all facilities affecting q's RkNN are contained in F_g and the rest facilities $f \notin F_g$ cannot affect any q's RkNN. As for any $q \in R$, whose RkNN results are only guarded by R's F_g. We define F_g **guardian set** and $f_g \in F_g$ **guardian facility** w.r.t. rectangular region R.

Therefore, by partitioning U into many small rectangular regions R s.t for any two $R_i, R_{j(i \neq j)} \in U$, we have $R_i \cap R_j = \emptyset$ and $\bigcup R_{i=1\ldots n} = U$, we can compute and obtain every guardian set F_{gi} of R_i.

Next, we present our techniques to compute R's guardian set F_g w.r.t.the value of k.

3.1   Computing Guardian Set of a Rectangular Region

First, we give some definitions and Lemmas.

Definition 1. *Given a set F of facility f, a rectangular region R and k, the **guardian set** F_g of R consists of a few facilities that for any $q \in R$, q's RkNN results are only affected by $f \in F_g$. For any facility $f \notin F_g$, it does not affect q's RkNN result. Such $f \in F_g$ is **guardian facility**, denote f_g.*

Definition 2. *For any f locates outside R, we draw the line segment with minimum distance from f to R joining R at v. We define f is owned by its nearest side L of R and f is in the range of L if v lies on L. If v is a vertex of R, we define f is owned by R's two sides intersecting at v and f is out of range of any R's side. As shown f_1, f_2 in Fig. 2(a).*

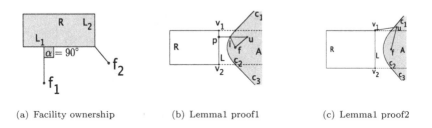

(a) Facility ownership   (b) Lemma1 proof1   (c) Lemma1 proof2

Fig. 2. Definition 2 and Lemma 1 proof

Lemma 1. *For any f locates outside R and owned by only one side L of R, we construct a combined curve C consists of: (i) **sub-curve1**: A partial parabola in the range of L constructed by f as the focus and L as the directrix; (ii) **sub-curve2**: Two radials that are parts of perpendicular bisectors between f and two vertices of L respectively toward directions that are out of L's range. C divides universe into two parts, for any user locates in the same area A with f, it has a closer distance to f than to any point in R and we define: f **prunes** A.*

Proof (Lemma 1). We prove Lemma 1 by considering two cases:

Case 1: For any p lies on L, any u locates in the same area A with f has smaller distance to f than to p:

Case 1.1: For any u lies in A within the range of L (as shown in Fig. 2(b)), we set the min distance $mindist(u, L) = |up|$, where up is the line segment passing c_2 at i and p is the intersection between up and L. According to the property of parabola, $|pi| = |if|$, then $|up| = |ui| + |if|$. By triangle inequality, $|up| > |uf|$.

Case 1.2: For u lies in A but is out of L's range (shown in Fig. 2(c)), the minimum distance from u to L is $|uv_1|$, where v_1 is the vertex of L locates at the same side with u w.r.t. f. As u is in the half dominated by f, $|uf| < |uv_1| \leq |up|$.

Case 2: For any p lies in R or on R's sides (exclude L), any u locates in A has smaller distance to f than to p:

Case 2.1: It is easy to show for any u lies in A within the range of L the triangle inequality still holds by changing $|pi| = |if|$ to $|pi| > |if|$, then $|up| > |uf|$.

Case 2.2: For u in A locates out of range of L (in Fig. 2(c)), as $mindist(u, R) = |v_1u| < |up|$, and $|uf| < |v_1u|$, we have $|up| > |uf|$.

Lemma 2. *For any f locates outside R and owned by R's two sides L_1, L_2, we construct its combined curve C consists of: (i) **Sub-curve1:** Two partial parabolas in the range of L_1 and L_2 constructed by f as the focus and L_1,L_2 as directrixes, respectively; (ii) **Sub-curve2:** Three partial perpendicular bisectors between f and three vertices of L_1,L_2 respectively toward directions that are out of range of L_1,L_2. C divides universe into two parts, for any u lies in the same area A with f, u has a closer distance to f than to any point p in R and we define: f **prunes** A (Fig. 3(a)). (note that two perpendicular bisector radials at the two ends of C do not necessary both exist due to location between R and f.)*

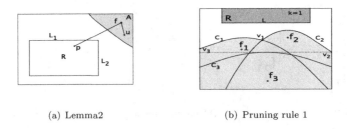

(a) Lemma2   (b) Pruning rule 1

Fig. 3. Lemma 2 and Pruning Rule 1

With Lemmas 1 and 2, we compute R's guardian set F_g by traversing all $f \in F$ and keep f_g whose pruning area is pruned by other facilities at most $k - 1$ times as a guardian facility.

Apparently, this algorithm is costly that every accessed facility f will be checked by every existed facility, costing a time complexity of $O(|N|^2)$, where $|N|$ is the total number of facilities. To avoid such problem, we propose two pruning rules to improve the efficiency.

Pruning Rule 1. *For any $f \in F$ owned by only one side L of R and locates outside R, we find the vertex v_g on the combined curve C_{fgi} of current guardian facility f_{gi} s.t. v_g*

- *is an intersection between other guardian facility's combined curve and C_{fgi} or between C_{fgi} and universe boundary;*
- *lies at the same side with f w.r.t. L;*
- *is pruned exactly k-1 times and has the farthest distance to L or L's extension if v_g is out of L's range;*
  if $dist(L, f) > 2dist(v_g, L)$, f has no influence on constructing R's guardian set and can be pruned.

Proof. As Fig. 3(b) shows, with f_1 and f_2 existed, the shaded area has been pruned by f_1 and f_2 jointly and v_2 is the vertex described in Pruning Rule 1. As $dist(L, f_3) > 2dist(v_2, L)$, it guarantees the area pruned by f_3 is pruned by f_1 and f_2 jointly, therefore, f_3 is pruned.

Pruning Rule 2. *For facility f fails meeting Pruning Rule 1, we compute its combined curve C_f and keep each intersection on C_f that is:*

- *intersection between C_f and combined curves of other guardian facilities;*
- *intersection between C_f and universe boundaries;*
- *intersection that is joint point between different sub-curves of C_f.*

If all such intersections on the C_f have been pruned by other facilities for at least k times, f has no effect on constructing R's guardian set and can be pruned.

Proof. We prove it by contradiction. Assume f and its combined curve C_f meet statement of Pruning Rule 2 but cannot be pruned, it has at least a part C_i on C_f cannot be pruned and at least existing two intersections that are two vertices of C_i pruned by other facilities for at most k-1 times, which contradicts with the statement of Pruning Rule 2. Therefore, Pruning Rule 2 holds.

Algorithm. We compute R's guardian set by indexing all facilities in a R-tree and accessing each node in an ascending order of its minimum distance to R. For each f that is not pruned by Pruning Rule 1, its combined curve C_f is created. Then we compute intersections between C_f and existed combined curves and apply Pruning Rule 2 to prune more existed guardian facilities. We update the temporary guardian set by removing facilities that meet Pruning Rule 2 and adding f if it is not pruned. The algorithm finishes when all nodes are accessed.

3.2   Partition Universe

Before the computation in Sect. 3.1, we partition the universe U to small rectangular regions R. It is not avoidable that each R contains facilities. However, with more facilities contained inside, more guardian facilities locate outside of R are likely to be pruned when q is given. Therefore, we set a threshold T_f as the

---

Algorithm 1. ConstructGuardianSet(R, F, k)

Input: Rectangular region R, a set F of facilities, value k
Output: R's Guardian set
1 initialise F_g as $\emptyset$;
2 insert root of R-tree in a min-heap h; /* we access f in an ascending order of mindist(R,f) */;
3 **while** h *is not empty* **do**
4   deheap an entry e;
5   **if** e *is not a facility* **then**
6 | Put e's child entry in h;
7   **else**
8   **if** f *cannot be pruned by pruning 1* **then**
9   Compute C_f w.r.t. R;
10   UpdateExistedGuardianSet(F_g,f,k,C_f);

11 Return F_g;

---

maximum number of facilities each R contains when partitioning to reduce the number of such guardian facilities that are possibly pruned in query process.

We partition U in a kd-tree [6] liked manner. Specifically, a big region R_u is split into four disjoint child regions R_l if R_u contains more than T_f facilities. For each region partitioned, we first find the median x-coordinate of all facilities in R_u and partition R_u into two smaller intermediate regions by the median. Then we partition two intermediate regions into four smallest ones following the same way by focusing on their y coordinates instead. Such procedure is conducted recursively until every region meets the threshold.

As we do not consider those facilities locate inside R when computing F_g of R in Sect. 3.1 and they are likely to affect q's RkNN. Therefore, we add all of them to R's guardian set after computing guardian facilities of R in case missing facilities that may affect q's RkNN results. Such facilities are also guardian facilities w.r.t.R.

Theoretical Analysis: [How many regions to be computed for guardian sets] As each time a region R_u is split into four smaller regions R_l if it contains at least $|T_f| + 1$ facilities. Therefore the whole universe has at most $\log_4(\frac{|N|}{|T_f|+1}) + 1$ level, where $|N|$ is the total number of facilities. At the lowest level that is level $\log_4(\frac{|N|}{|T_f|+1}) + 1$, there is no more region to be split. As a result, the total number of regions to be computed for guardian sets is $4^{\log_4(\frac{|N|}{|T_f|+1})+1}$, which is $4\frac{|N|}{|T_f|+1}$.

3.3 Region R's Guardian Set w.r.t k is Compatible for k' s.t. $k' \leq K$

Next, we prove R's guardian set of k can be used flexibly on any Rk'NN s.t. $k' < k$.

Algorithm 2. UpdateExistedGuardianSet(F_g, f, k, C_f)

Input: Existing Guardian Set F_g, new facility f, value k, combined curve C_f
Output: Updated Guardian Set F_g

1   **for each** $f_g \in F_g$ **do**
2     $\lfloor$ compute and keep intersections between C_f and C_{f_g};

3   **for each** $f_g \in F_g$ **do**
4   update pruned times of intersections on C_f and C_{fg};
5   **if** f_g *can be pruned by Pruning Rule 2* **then**
6       $\lfloor$ remove f_g from F_g;

7   **if** f *is not pruned by Pruning Rule 2* **then**
8     $\lfloor$ Put f in F_g;

9   Return F_g;

Lemma 3. *Given a rectangle region R, R's guardian set $F_{gk'}$ of k' is a subset of its guardian set F_{gk} of k, where $k' < k$.*

Proof. We prove Lemma 3 by contradiction. Assume there is at least one facility $f_{outside}$ excluded from F_g that is R's guardian facility of k', where $k' < k$. Therefore, the area pruned by $f_{outside}$ (denote $A_{foutside}$) holds:

$$((\bigcup A_{f_{gk}}) \cap A_{foutside}) \subset A_{foutside} \tag{1}$$

where f_{gk} is the guardian facility of R and whose partial combined curve contributes to the boundary of unpruned area of R w.r.t.k. However, as $f_{outside}$ is excluded from F_g of R, according to Lemmas 1 and 2, $((\bigcup A_{f_{gk}}) \cap A_{foutside}) = A_{foutside}$, which contradicts with Eq. 1, then Lemma 3 holds.

By Lemma 3, we mark facilities to each R's Rk'NN guardian set they are included. When given q, we retrieve R's guardian set w.r.t. k' ($k' \le k$) to avoid accessing more facilities that have no influence on R, then start processing query by SLICE.

4   Query Processing

With guardian sets, given a query q and k, we locate the region R where q locates and retrieve its guardian set w.r.t. k. Then SLICE starts processing query.

Filtering Phase. Given a query q and k, SLICE partitions the universe U into several regions, each of which has same subtending angle. In [3], the best partition number is 12 but it is 9 in our algorithm according to our preliminary experimental study.

    Figure 4 shows an example where the space is divided into 9 regions. Consider the perpendicular bisector between f_1 and q in region P, f_1 contributes two arcs in P, namely upper arc U_1(with radius r_U) and lower arc L_1 (with radius r_L).

Normally, every f constructs at least a lower arc to each region that its perpendicular bisector L_{fq} passes and it will contribute an upper arc for each region that L_{fq} passes and the max subtending angle between L_{fq} and such region is smaller than $90°$.

In filtering phase, each partitioned region maintains a k-th lower arc and a k-th upper arc dynamically, for each f whose lower arcs are lower than k-th upper arcs in corresponding regions will be kept as a significant facility of such regions to verify users, otherwise it is discarded (like f_2 is discarded in P in Fig. 4). After all facilities are accessed, the filtering phase finishes.

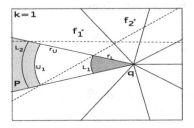

Fig. 4. SLICE filtering

Verification Phase. All users u are indexed in a R-tree and accessed in the ascending order of their distances to q. For u lies above the k-th upper arc of its region is discarded and every u lies under the k-th lower arc is returned. The remain users will be checked by significant facilities of this region.

Algorithm 3. RkNN($IndexFile, q, k$)

Input: IndexFile, query point q,value k
Output: RkNN results
1 locate q's region R;
2 retrieve R's guardian set F_gw.r.t. k;
3 invoke $SLICE$ with F_g;
4 return RkNN results;

---

5   Experimental Study

5.1   Experimental Setup

We compare our algorithm with InfZone [1] and SLICE [3] which are the latest algorithms for RkNN and SLICE is also the most efficient algorithm before our pre-computed one. All algorithms are implemented in C++ and the experiments are run on a 64-bit PC with Intel Xeon 2.66 GHz quad CPU and 4 GB memory running Linux.

We use the synthetic and real datasets in experiments. The real dataset consists of 175,812 points in North America and we randomly divide it into two sets of equal size. One of them is facility set, the other one is user set. For synthetic datasets, each dataset consists of 50,000, 100,000, 150,000, 200,000 points following either Uniform or Normal distribution and the facility sets, user sets are obtained like real dataset. The default synthetic dataset contains 100,000 points following Normal distribution unless mentioned otherwise. For k, we set it from 1 to 25 and make 10 at the default value. The page size for R-tree used in our experiment is 4096 bytes.

As big indices are kept in disk practically, I/O cost cannot be avoided when processing query. Therefore a penalty of I/O will be charged for each communication between disk and CPU. In our study, we estimate the I/O cost [15], [16] by 0.1ms which is the lowest time cost at the moment. We calculate the total time cost of each query in experiments by:

$$T = Cost_{I/O} * P_{I/O} + Cost_{CPU} \qquad (2)$$

where $Cost_{I/O}$, $Cost_{CPU}$ are I/O cost, CPU time cost and $P_{I/O}$ is the I/O penalty. Through experimental study, our algorithm runs times faster than Inf-Zone and improves SLICE performance significantly. Although [14] claims the I/O cost is highly system specific, we will prove our algorithm is the most efficient whatever I/O penalty charged and our algorithm outperforms others much more largely when higher I/O cost charged.

5.2   Evaluating Query Performance

In this section, we compare performance of three algorithms on both $monoRkNN$ and $biRkNN$. Three algorithms InfZone, SLICE and our pre-computed one are shown as INF,SLICE and PRE respectively. The number of partition for SLICE and PRE are 12 and 9 based on [3] and our preliminary experiments (see Fig. 11(a)). The experimental results shown in this sub-section is an average cost for one query in terms of total time (in millisecond) and I/O.

Effect of Data Size: In Figs. 5 and 6, we study the effect of data size for both $monoRkNN$ and $biRkNN$. For $monoRkNN$, Fig. 5(a) and (b) show the average total time and I/O cost. In $biRkNN$, Fig. 6(a) and (b) show average total time and I/O cost for both filtering and verification phases.

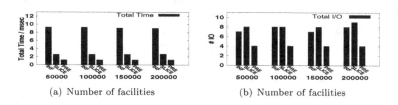

(a) Number of facilities   (b) Number of facilities

Fig. 5. Monochromatic Queries: Effect of data set size (Normal Distribution)

In Fig. 5(a), PRE performs best as only guardian facilities f_g of the region R where query locates are accessed. SLICE starts with considering all facilities in filtering phase, costing more time. $InfZone$ costs the most time as every facility to be pruned needs to compute with every vertex of Influence Zone, which is time-consuming.

Figure 5(b) demonstrates PRE costs least I/O as all guardian facilities of a region can be indexed in one disk page and can be located by a small constant number of I/Os. As expected, the I/O cost of SLICE is slightly larger than InfZone's as InfZone prunes a larger area.

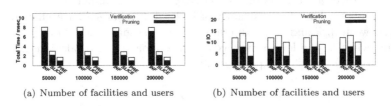

(a) Number of facilities and users   (b) Number of facilities and users

Fig. 6. Bichromatic Queries: Effect of data set size (Normal Distribution)

Figure 6 shows results for $biRkNN$. For INF and SLICE, the major cost in Fig. 6(a) is filtering phase whereas PRE which improves the filtering phase by one time faster by pre-computing guardian sets narrows the difference between two phases.

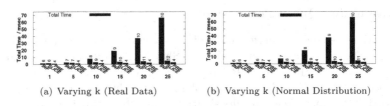

(a) Varying k (Real Data)   (b) Varying k (Normal Distribution)

Fig. 7. Monochromatic Queries: Effect of k

Due to space limitations, in the rest part, we focus on the total time cost **(I/O cost included)** of algorithms. **The numbers displayed above the bars correspond to the number of I/Os unless mentioned otherwise.**

Effect of k. In Figs. 7 and 8, we study the effect of k on $monoRkNN$ and $biRkNN$. The performance of InfZone deteriorates rapidly with the increasing k as time complexity is $O(km^2)$ and m increases as k increases, where m is the number of influence zone's vertices. PRE still performs best with least cost on both total time and I/O.

Figure 8(b) shows results for normal distribution using lines to demonstrate clearly how algorithms scale with the increasing k. Under our experimental setting, with accessing less facilities in filtering phase, PRE absolutely outperforms other algorithms.

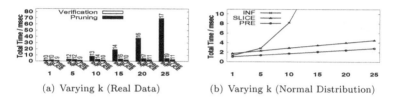

(a) Varying k (Real Data)   (b) Varying k (Normal Distribution)

Fig. 8. Bichromatic Queries: Effect of k

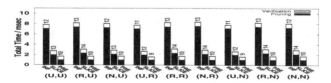

Fig. 9. Effect of Data Distribution

Effect of Data Distribution: In Fig. 9, we study the effect of data distribution on algorithms. Distribution of the facilities and users is shown as (D_f, D_u) where D_f and D_u correspond to distributions of facilities and users. **U**, **R** and **N** correspond to Uniform, Real and Normal distribution respectively. In this experiment, the synthetic datasets contain the same number of points as real dataset. Figure 9 demonstrates our algorithm outperforms others whatever combination of data distributions used.

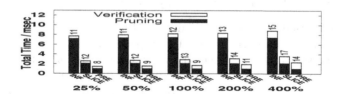

Fig. 10. users = x%facilities

Effect of Number of Users Relative to Number of Facilities: In this experiment, we fix the number of facilities to 100,000 and change the number of users to see the effect of change in the relative size of the two data sets. Figure 10 shows that all three algorithms' verification phases cost more with increasing number of users. PRE's filtering phase is much efficient making its total time cost dominated by verification phase. For the rest two, filtering phase dominates the total time cost as it is still costly. Overall, PRE cost least I/O and runs much faster than InfZone and SLICE.

Total Time Cost and I/Os. To clearly demonstrate our pre-computation algorithm change under different partitions, we process 500 queries and collect the total I/O and total time cost. Figure 11(a) shows the total I/O cost for 500 queries and the number above each bar is the total time cost in second.

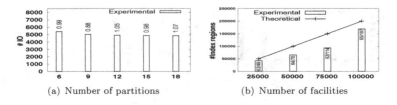

(a) Number of partitions   (b) Number of facilities

Fig. 11. Evaluation on Pre-computation

We conclude that when partition number is 9, our algorithm reaches a best performance as when partition number is smaller than 9, verification phase takes much time due to only a small area pruned by filtering phase. Although its pruning power becomes stronger with larger partition number, the filtering phase is still costly and affects the performance.

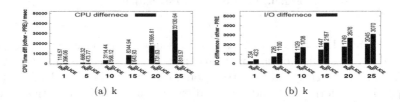

(a) k   (b) k

Fig. 12. CPU and I/O difference

5.3 Evaluating Pre-computation Algorithm

Number of Pre-computed Rectangular Regions. Figure 11(b) shows the number of regions computed practically. Experimental results are under our theoretical analysis in Sect. 3.2, demonstrating the number of regions computed practically is bounded by our theoretical analysis. The number inside each bar is the average number of guardian facilities in the guardian set followed by the size of index file in MB.

5.4 PRE is Efficient Whatever I/O Penalty Charged

Following Eq. 2, let total time of PRE : $T_{PRE} = Cost_{PRE.I/O} * P_{I/O} + Cost_{PRE.CPU}$. T' denotes total time cost of any other algorithm: $T' = Cost'_{I/O} * P_{I/O} + Cost'_{CPU}$. The total time difference between other algorithm and PRE is $\delta = T' - T_{PRE}$:

$$\delta = (Cost'_{I/O} - Cost_{PRE.I/O}) * P_{I/O} + (Cost'_{CPU} - Cost_{PRE.CPU}) \quad (3)$$

Then we prove PRE is the most efficient algorithm processing RkNN whatever I/O penalty charged.

Proof: Proving PRE is the most efficient is equivalent to prove $\delta > 0$, i.e.

$$P_{I/O} > \frac{Cost_{PRE.CPU} - Cost'_{CPU}}{Cost'_{I/O} - Cost_{PRE.I/O}} \quad (4)$$

Fig. 12(a) shows CPU time cost difference and Fig. 12(b) shows I/O difference by using corresponding cost of other algorithm minus ours. To clearly show results, we collect data by processing 500 queries.

As shown in Fig. 12(a) and (b), PRE cost least I/O and CPU time, which makes right part of inequality 4 become small than 0, as the I/O penalty is positive, the inequality always holds whatever is the I/O cost. As a result, PRE outperforms other RkNN algorithms no matter what I/O penalty charged.

Total Time Difference with Larger $P_{I/O}$. As Eq. 3 is a δ's function of $P_{I/O}$, which increases monotonically with the increasing I/O penalty. Thus, PRE will outperform other algorithm by a larger margin with larger I/O penalty charged.

According to [14]'s claim that SLICE is the most efficient RkNN algorithm before us, we conclude that PRE outperforms all existed RkNN algorithms.

6   Conclusion

In this paper, we propose a novel pre-computed RkNN algorithm by computing guardian set for each disjoint rectangular region R in universe and guarantee that for each possible query q in R, all its Rk'NN results are only affected by those facilities in guardian set, where $k' \leq k$. Through pre-computation, we reduce the number of facilities to be accessed from the whole facility set to only tens before processing query by SLICE. The extensive experimental study demonstrates our algorithm outperforms all existed algorithms on both total time and I/O cost.

Acknowledgements. Research of Wei Wang is supported by ARC DP130103401 and DP130103405. Muhammad Aamir Cheema is supported by ARC DE130101002 and DP130103405.

References

1. Cheema, M.A., Lin, X., Zhang, W., Zhang, Y.: Influence zone: efficiently processing reverse k nearest neighbors queries. In: 27th International Conference on Data Engineering (ICDE), pp. 577–588. IEEE (2011)
2. Wu, W., Yang, F., Chan, C.Y., Tan, K.L.: Finch: evaluating reverse k-nearest-neighbor queries on location data. Proc. VLDB Endow. **1**(1), 1056–1067 (2008)
3. Yang, S., Cheema, M.A., Lin, X., Zhang, Y.: Slice: reviving regions-based pruning for reverse k nearest neighbors queries. In: 30th International Conference on Data Engineering (ICDE), pp. 760–771. IEEE (2014)
4. Tao, Y., Papadias, D., Lian, X.: Reverse kNN search in arbitrary dimensionality. In: Proceedings of the Thirtieth International Conference on Very Large Data Bases. vol. 30, pp. 744–755. VLDB Endowment (2004)
5. Stanoi, I., Agrawal, D., El Abbadi, A.: Reverse nearest neighbor queries for dynamic databases. In: ACM SIGMOD Workshop on Research Issues in Data Mining and Knowledge Discovery, pp. 44–53 (2000)
6. Güting, R.H.: An introduction to spatial database systems. Int. J. Very Large Data Bases **3**(4), 357–399 (1994)

7. Korn, F., Muthukrishnan, S.: Influence sets based on reverse nearest neighbor queries. ACM SIGMOD Rec. **29**(2), 201–212 (2000). ACM
8. Yang, C., Lin, K.I.: An index structure for efficient reverse nearest neighbour queries. In: 17th International Conference on Data Engineering Proceedings, pp. 485–492. IEEE (2001)
9. Tao, Y., Papadias, D., Lian, X., Xiao, X.: Multidimensional reverse kNN search. VLDB J. **16**(3), 293–316 (2007)
10. Cao, X., Chen, L., Cong, G., Jensen, C.S., Qu, Q., Skovsgaard, A., Wu, D., Yiu, M.L.: Spatial keyword querying. In: Atzeni, P., Cheung, D., Ram, S. (eds.) ER 2012 Main Conference 2012. LNCS, vol. 7532, pp. 16–29. Springer, Heidelberg (2012)
11. Cao, X., Cong, G., Jensen, C.S., Ooi, B.C.: Collective spatial keyword querying. In: Proceedings of the 2011 ACM SIGMOD International Conference on Management of data, pp. 373–384. ACM (2011)
12. Cheema, M.A., Brankovic, L., Lin, X., Zhang, W., Wang, W.: Multi-guarded safe zone: an effective technique to monitor moving circular range queries. In: IEEE 26th International Conference on Data Engineering (ICDE), pp. 189–200. IEEE (2010)
13. Cheema, M.A., Zhang, W., Lin, X., Zhang, Y.: Efficiently processing snapshot and continuous reverse k nearest neighbors queries. VLDB J. **21**(5), 703–728 (2012)
14. Yang, S., Cheema, M.A., Lin, X., Wang, W.: Reverse k nearest neighbors query processing: experiments and analysis. Proc. VLDB Endow. **8**(5), 605–616 (2015)
15. Ruemmler, C., Wilkes, J.: UNIX disk access patterns. In: USENIX Winter, vol. 93, pp. 405–420 (1993)
16. Tsirogiannis, D., Harizopoulos, S., Shah, M.A., Wiener, J.L., Graefe, G.: Query processing techniques for solid state drives. In: Proceedings of the 2009 ACM SIG-MOD International Conference on Management of data, pp. 59–72. ACM (2009)

Hacube: Extending MapReduce for Efficient OLAP Cube Materialization and View Maintenance

Zhengkui Wang[1][(✉)], Yan Chu[2][(✉)], Kian-Lee Tan[3], Divyakant Agrawal[4], and Amr EI Abbadi[4]

[1] Singapore Institute of Technology, Singapore, Singapore
zhengkui.wang@singaporetech.edu.sg
[2] Harbin Engineering University, Harbin, China
chuyan@hrbeu.edu.cn
[3] National University of Singapore, Singapore, Singapore
tankl@comp.nus.edu.sg
[4] University of California, Santa Barbara, USA
{agrawal,amr}@cs.ucsb.edu

Abstract. Data cubes are widely used as a powerful tool to provide multi-dimensional views in data warehousing and On-Line Analytical Processing (OLAP). However, with increasing data sizes, it is becoming computationally expensive to perform data cube analysis. In this paper, we introduce `HaCube`, an extension of MapReduce, designed for efficient parallel data cube computation on large-scale data. We also provide a general data cube materialization solution which is able to facilitate the features in MapReduce-like systems towards an efficient data cube computation. Furthermore, we demonstrate how `HaCube` supports view maintenance through either incremental computation (e.g. used for SUM or COUNT) or recomputation (e.g. used for MEDIAN or CORRELATION). We implement `HaCube` by extending Hadoop and evaluate it based on the TPC-D benchmark over billions of tuples on a cluster with over 320 cores. The experimental results demonstrate the efficiency, scalability and practicality of `HaCube` for cube computation over a large amount of data in a distributed environment.

1 Introduction

In many industries, such as sales, manufacturing and finance, there is a need to make decisions based on aggregation of data over multiple dimensions. Data cubes [9] are one such critical technology that has been widely used in data warehousing and On-Line Analytical Processing (OLAP) for data analysis in support of decision making.

In OLAP, the attributes are classified into **dimensions** (the grouping attributes) and **measures** (the attributes which are aggregated) [9]. Given n dimensions, there are a total of 2^n **cuboids**, each of which captures the aggregated data over one combination of dimensions. To speed up query processing,

© Springer International Publishing Switzerland 2016
S.B. Navathe et al. (Eds.): DASFAA 2016, Part II, LNCS 9643, pp. 113–129, 2016.
DOI: 10.1007/978-3-319-32049-6_8

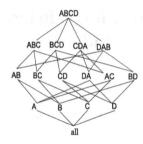

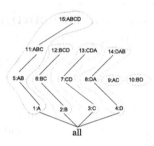

Fig. 1. A cube lattice with 4 dimensions: A, B, C and D

Fig. 2. The numbered cube lattice with execution batches

these cuboids are typically stored into a database as views. The problem of **data cube materialization** is to efficiently compute all the views ($\mathbb{V}$) based on the data ($\mathbb{D}$). Figure 1 shows all the cuboids represented as a cube lattice with 4 dimensions A, B, C and D.

In many append-only applications (no UPDATE and DELETE operations), the new data ($\Delta\mathbb{D}$) will be incrementally INSERTed or APPENDed to the data warehouse for view update. For instance, the logs in many applications (like the social media or stocks) are incrementally generated/updated. There is a need to update the views in a manner of one-batch-per-hour/day. The problem of **view maintenance** is to efficiently calculate the latest views while $\Delta\mathbb{D}$ are produced.

Both data cube materialization and view maintenance are computationally expensive, and have received considerable attention in the literature [12,16,20,23]. However, existing techniques can no longer meet the demands of today's workloads. On the one hand, the amount of data is increasing at a rate that existing techniques (developed for a single server or a small number of machines) are unable to offer acceptable performance. On the other hand, more complex aggregate functions (like complex statistical operations) are required to support complex data mining and statistical analysis tasks. Thus, this calls for new scalable systems to efficiently support data cube analysis over a large amount of data.

Meanwhile, MapReduce (MR) [7] has emerged as a powerful computation paradigm for parallel data processing on large-scale clusters. Its high scalability and append-only features have made it a potential target platform for data cube analysis in append-only applications. Therefore, exploiting MR for data cube computation has become an interesting research topic. However, deploying an efficient data cube computation using MR is non-trivial. A naive implementation of cube materialization and view maintenance over MR can result in high overheads.

Therefore, in this paper, we are motivated to explore the techniques of developing new scalable data cube analysis systems by leveraging the MR-like paradigm, as well as to develop new techniques for efficient data cube computation to broaden the application of data cubes primarily for append-only environments. Our main contributions are as follows:

1. *New system design and implementation:* We present HaCube, an extension
   of MR, for large-scale data cube computation. HaCube tries to integrate the
   good features from both MR and parallel DBMS. It extends MR to better
   support data cube computation by integrating new features, e.g. a new local
   store for data reuse among jobs, a layer with user-friendly interfaces and
   a new computation paradigm MMRR (MAP-MERGE-REDUCE-REFRESH). HaCube
   illustrates one way to develop a scalable and efficient decision making system,
   such that cube computation can be utilized in more applications.
2. *A General Cubing Algorithm:* We provide a general and efficient data cubing
   algorithm, CubeGen, which is able to complete the entire cube lattice using
   one MR job. We show how cuboids can be batched together to minimize the
   read/shuffle overhead and salvage partial work done. On the basis of batch
   processing principle, CubeGen further leverages the ordering property of the
   reducer input provided by the MR-like framework for an efficient materialization.
3. *Efficient View Maintenance Mechanisms:* We demonstrate how views can be
   efficiently updated under HaCube through either recomputation (e.g. used for
   MEDIAN or CORRELATION) or incremental computation (e.g. used for
   SUM or COUNT).
4. *Experimental Study:* We evaluate HaCube based on the TPC-D benchmark
   with more than two billions tuples. The experimental results show that
   HaCube has significant performance improvement over MR.

The rest of the paper is organized as follows. In Sect. 2, we provide an
overview of HaCube. Sections 3 and 4 present our proposed cube materialization and view maintenance approaches. We report our experimental results in
Sect. 5. In Sects. 6 and 7, we review some related works and conclude the paper.

2   HaCube: The Big Picture

2.1   Architecture

Figure 3 gives an overview of the basic architecture of HaCube. We implement
HaCube by modifying Hadoop which is an open source equivalent implementation
of MR [1]. Similar to MR, all the nodes in the cluster are divided into two
different types of function nodes, including the master and processing nodes.
The master node is the controller of the whole system and the processing nodes
are used for storage and computation.

Master Node: The master node consists of two functional layers:

1. The **cube converting layer** contains two main components: Cube Analyzer
   and Cube Planner. The cube analyzer is designed to accept the user request
   of data cube analysis, analyze the cube, such as figuring out the cube id (the
   identifier of the cube analysis application), analysis model (materialization
   or view update), measure operators (aggregation function), and input and
   output paths etc.

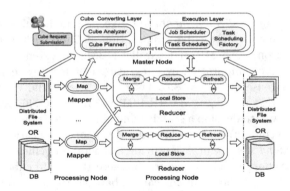

Fig. 3. HaCube architecture

The cube planner is developed to convert the cube analysis request into an execution job (either a materialization job or a view update job). The execution job is divided into multiple tasks each of which handles part of the cuboid calculation. The cube planner consists of several functional components such as the execution plan generator (combine the cuboids into batches to reduce the overhead), and load balancer (assign the right number of computation resources for each batch).

2. The **execution layer** is responsible for managing the execution of jobs passed from the cube converting layer. It has three main components: `job scheduler`, `task scheduler` and `task scheduling factory`. We use the same job scheduler as in Hadoop which is used to schedule different jobs from different users. In addition, we add a task scheduling factory which is used to record the task scheduling information of a job which can be reused in other jobs. Furthermore, we develop a new task scheduler to schedule the tasks in terms of the scheduling history stored in the task scheduling factory rather than the random scheduler used in MR.

Processing Node: A processing node is responsible for the task execution assigned from the master node. Similar to MR, each processing node contains one or more processing units each of which can either be a `mapper` or a `reducer`. Each processing node has a `TaskTracker` which is in charge of communicating with the master node through heartbeats, reporting its status, receiving the task, reporting the task execution progress and so on. Unlike MR, there is a `Local Store` built at each processing node running `reducers`. The local store is developed to cache useful data of a job in the local file system of the reducer node. It is a persistent storage in the local file system and will not be deleted after a job execution. In this way, tasks (possibly from other jobs) assigned to the same reducer node can access the local store directly from the local file system.

2.2 Computation Paradigm

HaCube inherits some features from MR, such as data read/process/write format of (key, value) pairs, sorting all the intermediate data and so on. However, it further enhances MR to support a new computational paradigm. HaCube adds two optional phases - a Merge phase and a Refresh phase before and after the Reduce phase - to support the MAP-MERGE-REDUCE-REFRESH (MMRR) paradigm as shown in Fig. 3.

The Merge phase has two functionalities. First, it is used to cache the data from the reduce input to the local store. Second, it is developed to sort and merge the partitions from mappers with the cached data in the local store. The Refresh phase is developed to perform further computations based on the reduce output data. Its functionalities include caching the reduce output data to the local store and refreshing the reduce output data with the cached data in the local store. These two additional phases are intended to fit different application requirements for efficient execution support.

As mentioned, these two phases are optional for the jobs. Users can choose to use the original MR computation or MMRR computation. More details can be found in Sect. 4 about how MMRR benefits the data cube view maintenance.

3   Cube Materialization

In this section, we provide our proposed data cubing algorithm, CubeGen, under the MR-like systems. We first present some principles of sharing computation through cuboid batching and a batch generator, then followed by the detail implementation. For simplicity, we assume that we are materializing the complete cube. Note that our techniques can be easily generalized to compute a partial cube (compute only selective cuboids). We also omit the cuboid "all" from the lattice. This special cuboid can be easily handled through an independent processing unit.

3.1   Cuboid Computation Sharing

To build the cube, computing each cuboid independently is clearly inefficient. A more efficient solution, which we advocate, is to combine cuboids into batches so that intermediate data and computation can be shared and salvaged.

We have the following observation: Let A and B be a set of dimensions such that $A \cap B = \emptyset$. In MR-like systems, given cuboids A and AB, A can be combined and processed together with AB, once AB is set of the key and is partitioned by A in one MR job. A is referred to as the **ancestor** of AB (denoted as $A \prec AB$). Meanwhile, AB is called the **descendant** of A. Note that the ancestor and descendant require them share the same prefix. This observation is the formal basis for combining and batching the cuboids computation under the MR-like systems.

The above observation can be generalized using transitivity: Since we can combine the processing of the pair of cuboids $\{A, AB\}$ and the pair $\{AB, ABC\}$, we can also combine the processing of the three cuboids $\{A, AB, ABC\}$. Thus, given one cuboid, all its ancestors can be calculated together as a batch. For instance, in Fig. 1, as $A \prec AB \prec ABC \prec ABCD$, the cuboids A, AB, ABC

can be processed with $ABCD$. Note that BC cannot be processed with $ABCD$ because $BC \not\prec ABCD$.

Given a batch, the principle to calculate this batch is to set the *sort dimensions* as the key and partition the (k,v) pairs based on the *partition dimensions* in the key in the MR-like paradigm. We formally define these two dimension classes below:

Definition 1 ***Sort Dimensions:*** *The dimensions in cuboid A are called the sort dimensions if A is the descendant of all other cuboids in one batch.*

Definition 2 ***Partition Dimensions:*** *The dimensions in cuboid A are called the partition dimensions if A is the ancestors of all other cuboids in one batch.*

For instance, given the batch $\{A, AB, ABC, ABCD\}$, $ABCD$ and A can be set as the sort and partition dimensions respectively.

The benefits of this approach are: (1) In the reduce phase, the group-by dimensions are all in sorted order for every cuboid in the batch, since MR would sort the data before supplying to the reduce function. This is an efficient way of cube computation since it obtains sorting for free and no other extra sorting is needed before aggregation. (2) All the ancestors do not need to shuffle their own intermediate data but use their descendant's. This would significantly reduce the intermediate data size, and thus remove a lot of data sort/partition/shuffle overheads.

3.2   Plan Generator

A plan generator is developed to generate the batches among the given cuboids. Intuitively, the more cuboids can be combined, the more sharing operations can be achieved. Therefore, the plan generator is responsible for generating the minimum number of batches based on the aforementioned principles. Note that each cuboid may have different permutations. For instance, the cuboid $ABCD$ can also be permutated as $ABDC$, $ACBD$, $BCDA$, $CDAB$, $DABC$ and so on. Thus, as the number of dimensions increases, it is no longer applicable to enumerate all the possible plans exhaustively. As such, some heuristic algorithm can be used to find a suboptimal execution plan.

Recall that one cuboid can be batched with all its ancestors. In this paper, we adopt a greedy algorithm to combine one cuboid with as many of its ancestors as possible. Intuitively, each batch construction starts from one unbatched cuboid with the maximum number of dimensions. The chosen cuboid then searches its different permutations with all its unbatched ancestors and the one with the largest number of ancestors is used to form this batch. The construction continues until all the cuboids are batched. In addition, we propose different optimizations to further reduce the search space, such as how to choose the right permutation and how to stop the permutation evaluation earlier. More details and proof can be found in our technical report [18]. For instance, given $2^n - 1$ cuboids (excluding "*all*") in Fig. 1, the algorithm generates 6 batches marked using the dotted lines as shown in Fig. 2.

Algorithm 1. CubeGen Algorithm

Function: Map(t)
1   # t is the tuple value from the raw data
2   Let $\mathbb{B}$ (resp. I_i) be the batch set with B_0, B_1, ..., B_{b-1} (resp. the identifier of batch B_i)
3   **for** *each B_i in $\mathbb{B}$* **do**
4   k (resp. v) $\Leftarrow$ get sort dimensions (resp. the measure m) in B_i from t
5   # If there are multiple measures (e.g. m_1, m_2), then v $\Leftarrow$ (m_1, m_2)
6   v.append(I_i); emit(k,v);

Function: Partitioning(k, v)
7   Let R_i (resp. *attr*) be the number of reducers (resp. the partition dimensions) for B_i
8   $S_i \Leftarrow \sum_{j=0}^{i-1} R_j$
9   return $S_i + hash(attr, R_i)$;

Function: Reduce/Combine $(k, \{v_1, v_2, ..., v_m\})$
10   Let $\mathbb{C}$ (resp. $\mathbb{M}$) be the cuboid set in the batch identifier (resp. the aggregate function)
11   **for** C_i *in* $\mathbb{C}$ **do**
12   **if** C_i *is ready* **then**
13   k'' (resp. v'')$\Leftarrow$ get the group-by dimensions in C_i (resp. $\mathbb{M}(v_1, ..., v_m, v_1', ..., v_k', ...)$)
14   # Perform multiple aggregate functions e.g. $(\mathbb{M}_1, \mathbb{M}_2)$ here: $v_1'' \Leftarrow \mathbb{M}_1(v_1, ..., v_m, v_1', ..., v_k', ...)$ and $v_2'' \Leftarrow \mathbb{M}_2(v_1, ..., v_m, v_1', ..., v_k', ...)$
15   emit(k'', v'');
16   **else**
17   Buffer the measure for aggregation

3.3   Implementation of CubeGen

Consider the batch plan B with b batches (B_0, B_1, ..., B_{b-1}) generated from the plan generator. There is a need to determine the number of computation resources (reducers) assigned to each batch. To achieve this, we also propose one load balancing approach based on sampling to guarantee that the computation task in each reducer can be balanced. Due to space constraint, we omit the discussion; interested readers are referred to [18]. Suppose that the number of reducers needed for each batch is R=(R_0, R_1, ..., R_{b-1}). Given B and R, the proposed CubeGen algorithm materializes the entire cube in one job and its pseudo-code is provided in Algorithm 1.

Map Phase: The base data is split into different chunks each of which is processed by one mapper. CubeGen parses each tuple and emits multiple (k,v) pairs each of which is for one batch (lines 3–6). The sort dimensions in the batch are set as the key and the measure is set as the value.

To distinguish which (k,v) pair is for which batch with which cuboids, we add a batch identifier appended after the value. The identifier is developed as one Bitmap with 2^n bits where n is the number of dimensions and each bit corresponds to one cuboid. First, we number all the 2^n cuboids from 0 to $2^n - 1$. Second, if the cuboid is included in one batch, its corresponding bit is set as 1, otherwise 0. For instance, Fig. 2 depicts an example of a numbered cube lattice. Assume that B_0 consists of cuboids $\{A, AB, ABC$ and $ABCD\}$. The identifier for B_0 is set as '10001000 00100010'.

The partitioning function partitions the pairs to the appropriate partition based on the identifier and the load balancing plan R. CubeGen first schedules the data into the right range of reducers. Recall that the batch B_i is assigned R_i reducers. Therefore, the assigned reducers for batch B_i are from $\sum_{j=0}^{i-1} R_j$ to $\sum_{j=0}^{i-1} R_j + R_i - 1$. Then the (k,v) pairs are hash partitioned among these R_i reducers according to the partition dimensions in the key (lines 7–9).

Reduce Phase: In the Reduce phase, the MR library sorts all the (k,v) pairs based on the key and passes them to the reduce function. Each reducer obtains its computation tasks (the cuboids in the batch) by parsing the batch identifier in the value. The reduce function extracts the measure and projects the group-by dimensions for each cuboid in the batch. For the descendant cuboid, the aggregation can be performed directly based on input tuple, since each input tuple is one complete group-by cell. For other cuboids, the measures of the group-by cell are buffered until the cell receives all the measures it needs for aggregation (lines 11–17). We develop multiple file emitters to write different aggregated results to different destinations.

Note that if the (k,v) pairs can be pre-aggregated in the map phase, users can specify a combine function to conduct a first round aggregation. The combine function is normally similar to the reduce function as shown in lines 10–17, but only aggregates the pairs with the same key. This pre-aggregation is able to reduce the data shuffle size between mappers to reducers.

We emphasize that if there are muliple measures (e.g. m_1, m_2, ..., m_n) and multiple aggregate functions ($\mathbb{M}_1$, $\mathbb{M}_2$, .., $\mathbb{M}_m$), they can be processed in the same MR job as shown in the line 5 and 14 in Algorithm 1. Compared to the naive solution, CubeGen minimizes the cube materialization overheads by sharing the data read/shuffle/computation to the maximum, which obtains significant performance improvement as we shall see in Sect. 5.

4   View Maintenance

There are two different manners to update the views, namely recomputation and incremental computation. Recomputation computes the latest views by reconstructing the cube based on the entire base data $\mathbb{D}$ and $\Delta\mathbb{D}$. In append-only applications, this manner is normally used for the holistic aggregate functions, e.g. STDDEV, MEDIAN, CORRELATION and REGRESSION [9].

Incremental computation, on the other hand, updates the views using only $\mathbb{V}$ and $\Delta\mathbb{D}$ in two steps: (1.) In the propagate step, a delta view $\Delta\mathbb{V}$ is calculated based on the $\Delta\mathbb{D}$. (2.) In the refresh step, the latest view is obtained by merging

---

Algorithm 2. A Refresh Job in MR

Function: Map(t)
1 # t is the tuple value from either $\mathbb{V}$ or $\Delta\mathbb{V}$
2 k (resp. v) $\Leftarrow$ get dimensions (resp. aggregate value) from t;
3 emit(k,v)
   Function: Reduce(k, $\{v_1, v_2\}$)
4 emit($k, \mathbb{M}(v_1, v_2)$)

$\mathbb{V}$ and $\Delta\mathbb{V}$ without visiting $\mathbb{D}$ [14]. In append-only applications, this manner is normally used for the distributive and algebraic aggregate functions, e.g. SUM, COUNT, MIN, MAX and AVG [9]. Note that the update for these functions can also be conducted through recomputation.

4.1   Supporting View Maintenance in MR

To support recomputation in MR, when $\Delta\mathbb{D}$ is inserted, the latest views can be calculated by issuing one MR job using our CubeGen algorithm to reconstruct the cube over $\mathbb{D} \cup \Delta\mathbb{D}$. The key problem with such an MR-based recomputation view updates is that reconstruction from scratch in MR is expensive. This is because the base data (which is large and increases in size at each update) has to be reloaded to the mappers from DFS and shuffled to the reducers for each view update, which incur significant overheads.

To support incremental computation in MR, the latest views can be calculated by issuing two MR jobs. The first propagate job generates $\Delta\mathbb{V}$ from $\Delta\mathbb{D}$ using our proposed CubeGen algorithm. The second refresh job merges $\mathbb{V}$ and $\Delta\mathbb{V}$ as shown in Algorithm 2. However, this would incur significant overheads. For instance, the materialized $\Delta\mathbb{V}$ from the propagate job has to be written back to DFS, reloaded from DFS again and shuffled from mappers to reducers in the refresh job. Likewise, $\mathbb{V}$ has to be reloaded and shuffled around in the refresh job. Therefore, it is highly expensive to support view update operations directly over the traditional MR.

4.2   HaCube Design Principles

HaCube avoids the aforementioned overheads through storing and reusing the data between different jobs. We extend MR to add a local store in the reducer node which is intended to store useful data of a job in the local file system. Thus, the task shuffled to the same reducer is able to reuse the data already stored there. In this way, the data can be read directly from the local store (and thus significantly reducing the overhead that would have been incurred to read the data from DFS and shuffle them from mappers).

We further extend MR to develop a new task scheduler to guarantee that the same task is assigned to the same reducer node and thus the cached data can be reused among different jobs. Specifically, the task scheduler records the

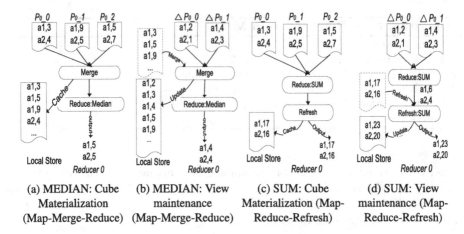

Fig. 4. Recomputation for MEDIAN and incremental computation for SUM in HaCube

scheduling information by storing a mapping between the data partition number (corresponds to the task) and the TaskTracker (corresponds to the reducer node) and puts it to the task scheduling factory from one job. When a new job is triggered to use the scheduling history from previous jobs, the task scheduler fetches and adopts the scheduling information from the factory to distribute the tasks. The scheduler automatically checks the situation of the over-loaded nodes and re-assigns the task to a nearby processing node.

In addition, two computation phases (`Merge` and `Refresh`) are added to conduct more computation with the cached data locally. The `Merge` phase is added to either cache the intermediate reduce input data in one job or preprocess the data between the newly arriving data and cached data before the `Reduce` phase. The `Refresh` phase is added to either cache the reduce output data in one job or postprocess the reduce output result with the cached data after the `Reduce` phase.

4.3   Supporting View Maintenance in HaCube

Recomputation. The recomputation view update can be efficiently supported in HaCube using a Map-Merge-Reduce (MMR) paradigm. We demonstrate this procedure through one running example by introducing the cube materialization and update jobs.

In the first cube materialization job, HaCube is triggered to cache the intermediate reduce input data to the local store in the `Merge` phase, such that this data can be reused during the view update job. For instance, Fig. 4(a) shows an example of calculating the cuboid A for MEDIAN. Assume that reducer 0 is assigned to process cuboid A. In this job, each mapper emits one sorted partition for reducer 0, such as P_0_0, P_0_1 and P_0_2. Here, each partition is a sequence of (dimension-value, measure-value) pairs, e.g., (a1, 3), (a2, 4). Recall that once these partitions are shuffled to the reducer 0, it first performs a merge-sort (the same as MR does) to sort all the partitions based on the key in the `Merge` phase. The sorted data is further supplied to the reduce function to calculate the MEDIAN for each group-by cell (e.g. $< a1, 5 >$ and $< a2, 5 >$) where this view will be written to DFS.

Different to MR (which deletes all the intermediate data after one job), since recomputation requires the base data for update, `HaCube` caches the sorted reduce input data in the `Merge` phase for subsequent reuse. This caching operation is conducted while the `Reduce` phase finishes, which guarantees the atomicity of the operation - if the reduce task fails, the data will not be written to the local store. Meanwhile, the scheduling information is recorded.

A view update job is launched when $\Delta \mathbb{D}$ is added for view updates. Intuitively, this job conducts a cube materialization job using the `CubeGen` algorithm based on $\Delta \mathbb{D}$. It differs from the first materialization job in the scheduling and the `Merge` phase. For task scheduling, instead of randomly distributing the tasks to reducer nodes, it distributes the tasks according to the scheduling history from the first materialization job to guarantee that the same tasks are processed at the same reducer. For instance, the partitions of cuboid A (ΔP_0_0 and ΔP_0_1) are scheduled to the same node running reducer 0 as shown in Fig. 4(b). In the `Merge` phase, since the base data is already cached in the local store, `HaCube` merges the delta partitions with the cached base data from the local store. Recall that the cached data is the sorted reduce input data from the previous job, and so it has the same format as the delta partition. Thus, it can be treated as a local partition and a global merge-sort is further performed. Then the sorted data will be supplied to the reduce function for recalculation in the `Reduce` phase. When the `Reduce` phase finishes, the local store is updated with both the base and delta data (becoming an updated base dataset) for further view update use.

Compared to MR, `HaCube` does not need to reload the base data from DFS and shuffle them from mappers to reducers for recomputation. This significantly reduces the data read/shuffle overheads. Another implementation optimization is proposed to minimize the data caching overhead. To cache the data to the local store, it is expensive to push the data to the local store, as this would incur much overhead of moving a large amount of data. Based on the observation that the intermediate sorted data are maintained in temporary files in the local disk in each reducer, `HaCube` simply registers the file locations to the local store rather than moving them. Note that the traditional MR would delete these temporary files once one job finishes. As we shall see, the experimental study shows that there is almost no overhead added for caching the data with this optimization.

Incremental Computation. `HaCube` adopts a `Map-Reduce-Refresh` (MRR) para- digm for incremental computation. Intuitively, different to MR in the first materialization job, it triggers to invoke a `Refresh` phase after the `Reduce` phase, to cache the view $\mathbb{V}$ to the local store for further reuse. For instance, Fig. 4(c) shows an example of calculating cuboid A for SUM in reducer 0. In this job, $\mathbb{V}$ ($< a1, 17 >$ and $< a2, 16 >$) is cached to the local store in the `Refresh` phase, and the scheduling information is also recorded.

When $\Delta\mathbb{D}$ is added for view updates, HaCube conducts both the propagate and refresh steps in one view update job, as $\mathbb{V}$ is already cached in the reducer node. This view update job in HaCube also executes in an MRR paradigm where MR (Map-Reduce) phases obtain $\Delta\mathbb{V}$ based on $\Delta\mathbb{D}$ (propagate step) and the Refresh phase merges $\Delta\mathbb{V}$ with $\mathbb{V}$ locally (refresh step). Intuitively, this can be achieved by running the CubeGen algorithm on $\Delta\mathbb{D}$ using the same scheduling plan as the previous materialization job. Meanwhile, the cached views in the local store will be updated with the latest ones. For instance, in Fig. 4(d), the Reduce phase calculates the $\Delta\mathbb{V}$ ($< a1, 6 >$ and $< a2, 4 >$) based on $\Delta\mathbb{D}$. In the Refresh phase, the updated view ($< a1, 23 >$ and $< a2, 20 >$) is obtained by merging $\Delta\mathbb{V}$ with $\mathbb{V}$ ($< a1, 17 >$ and $< a2, 16 >$) cached in the local store.

Different to MR, HaCube is able to finish the incremental computation in one job where there is no need to reload and shuffle the delta views and old views among DFS and the cluster during the propagate and refresh steps. This provides an efficient view update using the incremental computation by removing much overheads.

5   Performance Evaluation

We evaluate HaCube on the Longhorn Hadoop cluster in TACC (Texas Advanced Computing Center) [2]. Each node consists of 2 Intel Nehalem quad-core processors (8 cores) and 48 GB memory. By default, the number of nodes used is 35 (and 280 cores).

We perform our studies on the classical dataset generated by TPC-D benchmark generators [3]. The TPC-D benchmark offers a rich environment representative of many decision support systems. We study the cube views on the fact table, *lineitem* in the benchmark. The attributes *l_partkey*, *l_orderkey*, *l_suppkey* and *l_shipdate* are used as the dimensions and the *l_quantity* as the measure. We choose MEDIAN and SUM as the representative functions for evaluation.

5.1   Cube Materialization Evaluation

Baseline Algorithms: To study the benefit of the optimizations adopted in CubeGen, we design two corresponding baseline algorithms to study each of them including MulR_MulS (compute each cuboid using one MR job) and SingR_MulS (compute all the cuboids using one MR job without batching them), which are widely used for cube computations in MR. MulR_MulS (Resp. SingR_MulS) is used to study the benefit of removing multiple data read overheads (Resp. sharing the shuffle and computation through batch processing).

In the following set of experiments, we vary the data size from 600M (Million) to 2.4B (Billion) tuples. We study two versions of the CubeGen algorithm where CubeGen_Cache caches the data and CubeGen_NoCache does not. This provides insights into the overhead of caching the data to the local store.

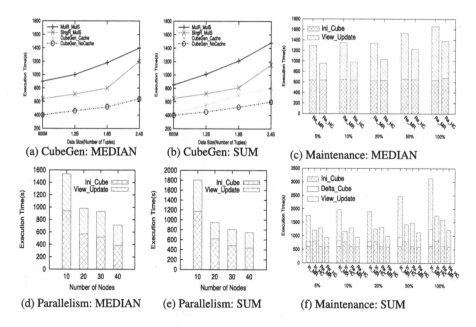

(a) CubeGen: MEDIAN   (b) CubeGen: SUM   (c) Maintenance: MEDIAN

(d) Parallelism: MEDIAN   (e) Parallelism: SUM   (f) Maintenance: SUM

Fig. 5. *CubeGen* performance evaluation for cube materialization

Efficiency Evaluation. We first evaluate the performance improvement of CubeGen for cube materialization. Figure 5(a) and (b) show the execution time of all four algorithms for MEDIAN and SUM respectively. As expected, for both MEDIAN and SUM, our CubeGen-based algorithms are 2.2X and 1.6X faster than MulR_MulS and SingR_MulS on average respectively. This indicates that computing the entire cube in one MR job reduces the overheads significantly compared to the case where multiple MR jobs were issued which requires reading data multiple times. In addition, it also demonstrates that batch processing highly reduces the size of intermediate data which can consequently minimize the overheads of data sorting, shuffling as well as computing.

Impact of Caching Data: Figure 5(a) and (b) also depict the impact of caching data. For MEDIAN, the execution time of the CubeGen_Cache is almost the same as CubeGen_NoCache as shown in Fig. 5(a). This confirms that our optimization to cache the data through file registration instead of actual data movement does not cause much overhead. For SUM, we observe that CubeGen_Cache performs worse than CubeGen_ NoCache. This is not surprising as the former needs to write an extra view to the local file system. However, even though CubeGen_Cache incurs around 16 % overhead to cache the view, as we will see later, it is superior to CubeGen_NoCache when it comes to view updates.

5.2   View Maintenance Evaluation

Efficiency Evaluation: We next study the efficiency of performing the view maintenance in HaCube compared with Hadoop. We fix $\mathbb{D}$ with 2.4B tuples in the first cube materialization job and vary the size of $\Delta\mathbb{D}$ from 5 % to 100 % of $\mathbb{D}$ for view updates.

Figure 5(c) shows the execution time for both the cube materialization (Ini_Cube) and the view updates (View_Update) for MEDIAN. In this set of experiments, we adopt recomputation for view updates of MEDIAN using MR (Re_MR) and HaCube (Re_HC). The result shows that Re_HC is 2X and 1.4X faster than Re_MR, when $\Delta\mathbb{D}$ is 5 % and 100 % respectively. The gains come from avoiding reloading and reshuffling $\mathbb{D}$ among the cluster. Thus, the larger $\mathbb{D}$ is, the bigger the benefit will be.

Figure 5(f) depicts the result for SUM. As view updates for SUM can either be done by incremental computation or recomputation, we evaluate both approaches to update the view. In Fig. 5(f), In_MR and Re_MR (resp. In_HC and Re_HC) are MR (resp. HaCube) -based methods using incremental computation and recomputation respectively.

In_MR and Re_MR are implemented in the way described in Sect. 4.1. In In_MR, Delta_Cube (in the figure) corresponds to the propagate job to generate the delta view and View_Update is the refresh job. The result shows that, for incremental computation, In_HC is 2.8X and 2.2X faster than In_MR when $\Delta\mathbb{D}$ is in 5 % and 100 % as shown in Fig. 5(f). For recomputation, Re_HC is about 2.1X and 1.4X faster than the Re_MR when the $\Delta\mathbb{D}$ is in 5 % and 100 % as shown in Fig. 5(f). This indicates that HaCube has significant performance improvement compared to MR for the view update for both recomputation and incremental computation.

We observe that incremental computation performs worse than recomputation in both MR and HaCube. While this seems counter-intuitive, our investigation reveals that DFS does not provide indexing support; as such, in incremental computation, the entire view which is much larger than the base data (in our experiments) has to be accessed. Another insight we gain is the smaller the $\Delta\mathbb{D}$ is, the more effective HaCube is. As future work, we will integrate more existing techniques (e.g. indexing) in DBMS into HaCube, which will further improve the view update performance.

Impact of Parallelism: We further analyze the impact of parallelism on HaCube for both cube materialization and view update while varying the number of nodes from 10 to 40. The experiments use $\mathbb{D}$ with 600M tuples and $\Delta\mathbb{D}$ in 20 % of $\mathbb{D}$.

Figures 5(d) and (e) report the execution time for MEDIAN and SUM. Note that, in this experiment, incremental computation is used for SUM. We observe that for both recomputation and incremental computation, HaCube scales linearly on the testing data set from 10 to 20 nodes, where the execution time almost reduces to half when the resources are doubled. From 20 nodes to 40 nodes, the benefit of parallelism decreases a little bit. This is reasonable, since the entire overheads include two parts, the setup of the framework and the cube computation; the former one may reduce the benefits of increasing the computation resources while cube computation cost is not big enough.

Due to the space limitation, interested readers are referred to our technical report [18] for more experimental evaluations (e.g. load balancing, impact of dimensions) and other issues (e.g. fault tolerance mechanism, storage analysis).

6 Related Work

Much research has been devoted to the problem of data cube analysis [9]. A lot of studies have investigated efficient cube materialization [4,20,21,23] and view maintenance [12,16]. Three classic cube computation approaches (Top-down [23], Bottom-Up [4] and Hybrid [20]) have been well studied to share computation among the lattice in a centralized system or a small cluster environment. Different to these approaches, CubeGen adopts a new strategy to partition and batch the cuboids according to their prefix order to tackle the new challenges brought by MR. It utilizes the sorting feature better in MR-like systems such that no extra sorting needed during materialization.

Existing works [17,22] have adopted MR to build closed cubes for algebraic measures. However, both of these works do not provide a generic algorithm that can balance the load to materialize the cube for different measures. Nandi et al. [15] provided a solution to a special case during the cube computation under MR where one reducer gets the "hot spot" group-by cell with a large number of tuples. This complements our work and can be employed to handle such a case in HaCube. We note that HaCube is able to support all these existing cube materialization algorithms. More importantly, none of these aforementioned works have developed any techniques for view maintenance. In addition, [13] provided one OLAP system by extending HBase for real-time analysis and [19] provides pagrol system for graph OLAP computation.

Our work is also related to the problem of incremental computations. Existing works [5,10,11] have studied some techniques for incremental computations for single operators in MR. HaLoop [6] is designed to support iterative operations through a similar caching mechanism which is used for different purposes under a different application context. Restore [8] also shares the similar spirit to keep the intermediate results (either the output of one MR job or the data operated within one job) to DFS in a workflow and reuse them in the future. For data cube computation, as the size of intermediate results is large, HaCube adopts a different data caching mechanism to guarantee the data locality that the cached data can be directly used from local store. This avoids the overhead incurred by Restore in reloading and reshuffling data from DFS. Furthermore, none of these existing works provide explicit support and techniques for data cube analysis under OLAP and data warehousing semantics.

7 Conclusion

It is of critical importance to develop new scalable and efficient data cube computation systems on a big cluster with low-cost commodity machines to tackle the challenges brought by the large-scale of data, to provide a better query response and decision making support. In this paper, we made one step towards developing such a system, HaCube an extension of MapReduce, by integrating the good features from both MapReduce (e.g. Scalability) and parallel DBMS (e.g. Local Store). We showed how to batch and share the computations to salvage partial work done by facilitating the features in MapReduce-like systems towards an efficient cube materialization. We also demonstrated how HaCube supports an efficient view maintenance by facilitating the extension leveraging a new computation paradigm. The experimental results showed that our proposed cube materialization approach is at least 1.6X to 2.2X faster than the naive algorithms and HaCube performs at least 2.2X to 2.8X faster than Hadoop for view maintenance. We expect HaCube to further improve the performance by integrating more techniques from DBMS, such as indexing techniques.

Acknowledgements. Kian-Lee Tan is partially supported by the MOE/NUS grant R-252-000-500-112. This work used the Extreme Science and Engineering Discovery Environment (XSEDE), which is supported by National Science Foundation grant number OCI-1053575.

References

1. Hadoop. http://hadoop.apache.org/
2. Tacc longhorn cluster. https://www.tacc.utexas.edu/
3. Tpc-h, ad-hoc, decision support benchmark. www.tpc.org/tpch/
4. Beyer, K.S., Ramakrishnan, R.: Bottom-up computation of sparse and iceberg cubes. In: SIGMOD, pp. 359–370 (1999)
5. Bhatotia, P., Wieder, A., Rodrigues, R., Acar, U.A., Pasquini, R.: Incoop: mapreduce for incremental computations. In: SOCC (2011)
6. Yingyi, B., Howe, B., Balazinska, M., Ernst, M.D.: Haloop: efficient iterative data processing on large clusters. PVLDB **3**(1), 285–296 (2010)
7. Dean, J., Ghemawat, S.: Mapreduce: simplified data processing on large clusters. In: OSDI, pp. 137–150 (2004)
8. Elghandour, I., Aboulnaga, A.: Restore: reusing results of mapreduce jobs. PVLDB **5**(6), 586–597 (2012)
9. Gray, J., Bosworth, A., Layman, A., Reichart, D.: Data cube: a relational aggregation operator generalizing group-by cross-tab and sub-totals. In: ICDE, pp. 152–159 (1996)
10. Jörg, T., Parvizi, R., Yong, H., Dessloch, S.: Incremental recomputations in mapreduce. In: CloudDB, pp. 7–14 (2011)
11. Lämmel, R., Saile, D.: Mapreduce with deltas. In PDPTA, (2011)
12. Lee, K.Y., Kim, M.H.: Efficient incremental maintenance of data cubes. In: VLDB, pp. 823–833 (2006)
13. Feng Li, M., Ozsu, T., Chen, G., Ooi, B.C.: R-store: a scalable distributed system for supporting real-time analytics. In: ICDE, pp. 40–51 (2014)
14. Mumick, I.S., Quass, D., Mumick, B.S.: Maintenace of data cubes and summary tables in a warehouse. In: SIGMOD, pp. 100–111 (1997)
15. Nandi, A., Cong, Y., Bohannon, P., Ramakrishnan, R.: Distributed cube materialization on holistic measures. In: ICDE, pp. 183–194 (2011)
16. Palpanas, T., Sidle, R., Cochrane, R., Pirahesh, H.: Incremental maintenance for non-distributive aggregate functions. In: VLDB, pp. 802–813 (2002)
17. Sergey, K., Yury, K.: Applying map-reduce paradigm for parallel closed cube computation. In: DBKDA, pp. 62–67 (2009)
18. Wang, Z., Chu, Y., Tan, K.-L., Agrawal, D., Abbadi, A.E., Xiaolong, X.: Scalable data cube analysis over big data. In: CORR (2013). arxiv:1311.5663

19. Wang, Z., Fan, Q., Wang, H., Tan, K.-L., Agrawal, D., El Abbadi, A.: Pagrol: parallel graph olap over large-scale attributed graphs. In: ICDE, pp. 496–507 (2014)
20. Xin, D., Han, J., Li, X., Wah, B.W.: Computing iceberg cubes by top-down and bottom-up integration: the starcubing approach. TKDE **19**(1), 111–126 (2007)
21. Xin, D., Han, J., Wah, B.W.: Star-cubing: Computing iceberg cubes by top-down and bottom-up integration. In VLDB, pp. 476–487 (2003)
22. You, J., Xi, J., Zhang, P., Chen, H.: A parallel algorithm for closed cube computation. In ACIS-ICIS, pp. 95–99, (2008)
23. Zhao, Y., Deshpande, P.M., Naughton, J.F.: An array-based algorithm for simultaneous multidimensional aggregates. In: SIGMOD, pp. 159–170 (1997)

Similarity Computing

TSCluWin: Trajectory Stream Clustering over Sliding Window

Jiali Mao, Qiuge Song, Cheqing Jin$^{(\boxtimes)}$, Zhigang Zhang, and Aoying Zhou

Institute for Data Science and Engineering,
School of Computer Science and Software Engineering,
East China Normal University, Shanghai, China
maojl1231@163.com, {sugar_song,zzg22936}@sina.com,
{cqjin,ayzhou}@sei.ecnu.edu.cn

Abstract. The popularity of GPS-embedded devices facilitates online monitoring of moving objects and analyzing movement behaviors in a real-time manner. Trajectory clustering acts as one of the most important trajectory analysis tasks, and the researches in this area have been studied extensively in the recent decade. Due to the rapid arrival rate and evolving feature of stream data, little effort has been devoted to online clustering trajectory data streams. In this paper, we propose a framework that consists of two phases, including a *micro-clustering* phase where a number of micro-clusters represented by compact synopsis data structures are incrementally maintained, and a *macro-clustering* phase where a small number of macro-clusters are generated based on micro-clusters. Experimental results show that our proposal is both effective and efficient to handle streaming trajectories without compromising the quality.

1 Introduction

With the vigorous development of modern mobile devices and location acquisition technologies, the positions of moving objects are collected continuously in a streaming fashion. For instance, the taxis embedded with GPS sensors transmit their current location information to a data center frequently, so that the taxi-company is capable of processing taxi-hailing requests efficiently. Effective analyzing trajectory data stream fosters a broad range of critical applications, such as location-based social network, route recommendation [16], intelligent transportation management [5], road infrastructure optimization, etc.

As an important trajectory analysis task, clustering attempts to group a large amount of trajectories into a few comparatively homogeneous clusters to find the representative paths or common moving trend shared by different objects [6–8, 12,14,15,21]. While trajectory data keep updating rapidly in stream scenarios, it is intrinsically quite difficult to track the cluster changing. For example, Fig. 1(a) illustrates a small example of five taxis' trajectories from 9:00 to 9:30 a.m. There exist two clusters, the left three taxis have the similar driving itinerary, while the rest two behave similarly. But in the next period (from 9:30 to 10:00 a.m.), as shown in Fig. 1(b), the left three trajectories are no longer in one cluster.

© Springer International Publishing Switzerland 2016
S.B. Navathe et al. (Eds.): DASFAA 2016, Part II, LNCS 9643, pp. 133–148, 2016.
DOI: 10.1007/978-3-319-32049-6_9

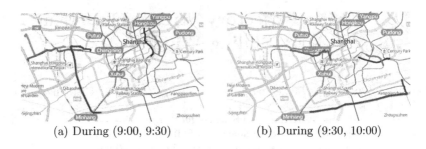

(a) During (9:00, 9:30)   (b) During (9:30, 10:00)

Fig. 1. Movement distribution

Hence, it is imperative to consider a model that focuses on the evolving feature
of stream data. The sliding-window model that eliminates the obsolete data and
only keeps most recent W tuples in the stream can satisfy this demand.

However, it is challenging to cluster trajectory data streams due to the following reasons. First, a trajectory in the stream may evolve as time progresses.
Second, a suitable processing algorithm must be with strict time- and space- com-
plexities since the stream volume is huge and the arrival rate of data is rapid.
Finally, as the old tuples progressively expire, the algorithm must be capable of
updating the synopsis data structure dynamically. So far, although there exists
abundant work on incremental clustering for trajectories, or on data stream clustering over the sliding-window model, to the best of our knowledge, the issue
of online clustering trajectory data stream over the sliding-window model has
not been addressed yet. This issue is non-trivial since the existing work cannot
be adopted to tackle it in a straightforward way. For the work on trajectory
incremental clustering [8,14,15], the expired records cannot be discarded swiftly
upon the arbitrary arrival of new ones, which influence the quality of clustering
result. The work on the data stream clustering over the sliding-window model
is upon the scenario where each tuple must be a *"full"* entry [1], whereas each
tuple in the trajectory data stream is only a *"part"* of an entry.

We propose a two-phase framework to tackle this issue, including the micro-
clustering phase and the macro-clustering phase. During the micro-clustering
phase, a number of micro-clusters, represented by a novel synopsis data structure,
are maintained incrementally. During the macro-clustering phase, a handful of
macro-clusters are built upon the micro-clusters according to a clustering request
over the given time horizon. One major novelty of this paper is the construction
of two synopsis data structures, called Temporal Trajectory Cluster Feature
(TF) and Exponential Histogram of Temporal Trajectory Cluster Feature (EF).
The contributions of this paper are summarized below.

– We study the issue of online trajectory data stream clustering over the sliding-
  window model. To the best of our knowledge, there exists no prior work on
  this topic. Moreover, neither trajectory clustering algorithm nor data stream
  clustering algorithm upon the sliding-window model can be adopted to tackle
  this issue directly.

- We propose a two-phase scheme to tackle this issue, with the usage of two novel synopsis data structures (TF and EF) to summarize a cluster of trajectories, and to process the expired tuples.
- We conduct a comprehensive series of experiments on a real dataset to manifest the efficiency and the effectiveness of our proposal, as well as the superiority to other congeneric approach.

The remainder of this paper is organized as follows. In Sect. 2, we review the latest work related to our research. In Sect. 3, the problem is defined formally. In Sect. 4, we outline and analytically study TSCluWin scheme. In Sect. 5, a series of experiments are conducted on a real dataset to evaluate our proposal. Finally, we conclude this article in brief in Sect. 6.

2   Related Work

In this section, we briefly conduct a systematic review over the related work in two relevant areas: data stream clustering and trajectory stream clustering.

Data Stream Clustering. The existing work on data stream clustering are classified into two kinds: one-pass approach and evolving approach. The former constructs clusters by scanning the data stream only once, while the latter can also view the evolving process over time [24]. The work upon the sliding-window model belongs to the latter.

Babcock et al. studied the clustering issue over the sliding-window model with the focus on theoretical bound analysis [3]. Aggarwal et al. developed CluStream algorithm to cluster large evolving data streams based on the pyramid model [1]. Aggarwal et al. further proposed UMicro algorithm to deal with uncertain data streams [2]. Although [1,2] cannot deal with the sliding-window model directly, they can view the evolving process of the data stream. Zhou et al. [24] presented SWClustering algorithm to track the evolution of clusters over the sliding-window model by using a novel synopsis data structure, called EHCF, which combines the exponential histogram with the temporal cluster features to handle the in-cluster evolution and capture the distribution of recent records. Jin et al. proposed the cluUS algorithm to incrementally maintain uncertain tuples based on Uncertain Feature (UF) structure with the combination of the sliding-window model [10]. However, none of the above methods can deal with trajectory data straightforwardly due to different scenarios. In such scenario, each tuple in the data stream is an entry, while each tuple in the trajectory data stream is only a part of an entry.

Trajectory Stream Clustering. A few research work has been conducted for incremental trajectory clustering [8,14,15] in the static trajectory data set. Li et al. proposed the concept of Moving Micro-Cluster to catch the regularities of moving objects, which can accommodate the updates of moving objects and be well suited for handling large datasets [14]. Jensen et al. proposed a scheme for continuously clustering of moving objects, which employs an object dissimilarity across a period of time, and exploits an incrementally maintained clustering feature CF [8]. Li et al. proposed an incremental trajectory clustering framework, called TCMM, which includes micro-clusters maintenance based on the process of simplifying new trajectories into directed line segments, and macro-clusters generating by clustering the micro-clusters [15]. However, during the course of micro-clustering, TCMM has to accumulate a certain amount of new trajectory point data to obtain the simplified sub-trajectory by using MDL method. In addition, due to the effect of obsolete data, after continuously absorbing more records and merging the most similar pair of micro-clusters, the centers of micro-clusters will shift gradually, which more or less degrades the effectiveness of resulting clusters. More specifically, incremental clustering approaches barely consider the temporal aspects of the trajectories and cannot scale up to mine massive trajectory streams. Yu et al. [22] proposed CTraStream for clustering trajectory data stream to extract some patterns similar to convoy pattern [9]. It includes online line segment stream clustering and the update process of closed trajectory clusters based on TC-Tree index.

There also exists some approaches that are to some degree orthogonal to our work but deserve to be mentioned. Various techniques about trajectory simplification and compression have been studied in real-time trajectory tracking [11], but they refer to the problem of minimizing the amount of the position data that are communicated and stored. Due to high computational overhead to attain optimal line simplification, they are not fit to online cluster the rapidly changing stream data in the limited memory. More recent achievements have been reported for continuous query processing over trajectory stream [13,17,19,20,23], but they are unsuitable for identifying the clusters in streaming trajectories. Different from the aforementioned approaches, we focus on online capturing the clustering change in a certain temporal window, and eliminating the effect of obsolete data, which fit the feature of continuous trajectory stream.

3   Problem Statement

We define some notations in this section. Let S denote a stream that contains a totally ordered infinite records of the moving objects in the form of $(o^{(i)}, p^{(i)})$, where $p^{(i)}$ is the location of an object $o^{(i)}$ at the time stamp i in 2-D space (i.e., $p^{(i)} = (x_i, y_i)$). In general, the stream S contains multiple objects. The trajectory of one object is defined below.

Definition 1 (Trajectory). *The trajectory for an object o, denoted as Tr_o, is a sub-sequence of S affiliated to o, denoted as $\{(p_1, t_1), (p_2, t_2), \ldots\}$. Such records arrive in chronological order, i.e., $\forall i < j$, $t_i < t_j$. A line segment L_i refers to a line connecting two temporal adjacent points, i.e., L_i is denoted as (p_i, p_{i+1}). Correspondingly, the trajectory is also denoted as $\{(L_1, t_1), (L_2, t_2), \ldots\}$.*

The goal of trajectory clustering is to divide all trajectories into clusters in terms of their pair-wise distances. Although the Euclidian distance is commonly used in the spacial data management field, it is inappropriate to measure spatial proximity in the scenario with bi-directional property. Hence, we employ the distance measure based on adapted Hausdorff distance [18]. The distance between two trajectories is regarded as the longest path from each line segment to the closest line segment of another trajectory, i.e., $dist(Tr_a, Tr_b) = \max(D(Tr_a, Tr_b), D(Tr_b, Tr_a))$, where $D(Tr_a, Tr_b) = \max_{L' \in Tr_a} \min_{L'' \in Tr_b} (DL(L', L''))$. Note that $DL(L', L'')$ returns the maximal distance between two line segments after alignment, i.e., $DL(L', L'') = \max(dl(L', L''), dl(L'', L'))$, where $dl(L', L'') = \max(\min(||p'_s, p''_s||, ||p'_s, p''_e||), \min(||p'_e, p''_s||, ||p'_e, p''_e||))$. Here, p'_s and p'_e denote the starting and ending positions of L', p''_s and p''_e denote the starting and ending positions of L'', and $||p'_s, p''_s||$ denotes the length of the shortest path between p'_s and p''_s.

We study the sliding-window model in this paper, where only the most recent W records in S are considered. Due to the infeasibility to keep all trajectories in memory, it is necessary to summarize original data in memory by using a compact synopsis data structure. Our first synopsis data structure, Temporal Trajectory Cluster Feature (TF), summarizes the features of sets of incoming line segments at different intervals.

Definition 2 (Temporal trajectory cluster Feature, (TF)). *Given a set of consecutive line segments $\{L_1, L_2, \ldots, L_n\}$, which is a sub-sequence of a stream S. TF is of the form (SC, SA, BL, TR, n, t).*

- *SC: the linear sum of the line segments' center points;*
- *SA: the linear sum of the line segments' angles;*
- *BL: the bottom left corner of the MBR (Minimal Bounding Rectangle);*
- *TR: the top right corner of the MBR;*
- *n: the number of line segments;*
- *t: the timestamp of the most recent line segment;*

Note that MBR is the minimal bounding rectangle of all line segments contained in a TF. Figure 2(a) illustrates an example of the MBR for two black polyline, we can draw a line segment to represent the moving pattern of all the line segments in TF. Firstly, we obtain the central point (denoted as $TF.cen$) and the angle (denoted as θ) of that line segment by calculating them with $\frac{SC}{n}$ and $\frac{SA}{n}$ respectively. Then, we plot a line across the central point, along that angle, and finally extend it to reach the borders of MBR. The intersection points are treated as the starting and ending points of the representative line segment. For example, the blue line segment with the starting point (denoted as $TF.rp_s$) and the ending point (denoted as $TF.rp_e$) is regarded as the representative line segment in Fig. 2(a).

Property 1 (Additive property). Let $TF(C_1)$ and $TF(C_2)$ denote two TF structures for two sets C_1 and C_2 separately, $C_1 \cap C_2 = \emptyset$. We can construct $TF(C_1 \cup C_2)$ based on $TF(C_1)$ and $TF(C_2)$. The new entries SC, SA and n are equal to the sum of the corresponding entries in $TF(C_1)$ and $TF(C_2)$.

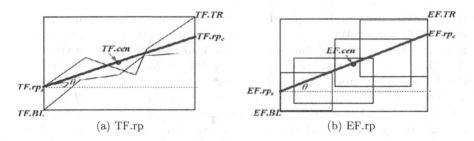

(a) TF.rp   (b) EF.rp

Fig. 2. Representative line segment

The new entry t is computed as $\max(TF(C_1).t, TF(C_2).t)$. Moreover, the corners of the new TF can be computed based on two original corners straightforwardly.

As a TF may consist of multiple line segments, and they will go out of the window one by one in the future, which necessitates a structure to deal with the expired line segments. Exponential Histogram (EH) is well-known to deal with the sliding-window model, where all the tuples in the data stream are divided into a number of buckets according to the arrival time [4]. Inspired by EH, we devise a novel structure, called Exponential Histogram of Temporal Trajectory Cluster Feature (EF) below.

Definition 3 (Exponential Histogram of Temporal Trajectory Cluster Feature (EF)). *Given a user-defined parameter ϵ, EF is a collection of TFs on some sets of line segments $C_1, C_2, \ldots$ with the following constraints:*

1. $\forall i, j, i < j$, any line segment in C_i arrives earlier than that in C_j;
2. $|C_1| = 1$. $\forall i > 1$, $|C_i| = |C_{i-1}|$ or $|C_i| = 2 \cdot |C_{i-1}|$;
3. At most $\lceil \frac{1}{\epsilon} \rceil + 1$ TFs are placed in each level.

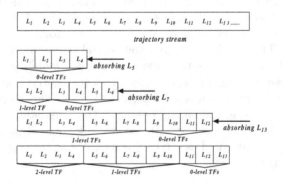

Fig. 3. Process of incorporating line segments into an EF

Theorem 1. *Given an EF that contains n_i tuples, and a user-defined parameter ϵ, the amount of obsolete tuples is within $[0, \epsilon n_i]$, and the number of TFs is at most $(\frac{1}{\epsilon} + 1)(\log(\epsilon n_i + 1) + 1)$.*

Proof. Given an EF of n_i tuples, the parameter ϵ, and the oldest TF of n_s tuples, then $\frac{1}{\epsilon}(1 + 2 + 4 + \ldots + n_s) \le n_i$. Hence, $n_s \le \epsilon n_i$ holds. In addition, an EH structure with the window size n_i and the parameter ϵ can be constructed. Each TF maps to a bucket in EH structure. Since EH Structure computes an ϵ-deficient synopsis using at most $(\frac{1}{\epsilon} + 1)(\log(\epsilon W + 1) + 1)$ buckets [4], where W represents the window size, there are at most $(\frac{1}{\epsilon} + 1)(\log(\epsilon n_i + 1) + 1)$ TFs in an EF.

EF is maintained in the following way. When a new line segment is incorporated into the existing EF, a new 0-level TF will be generated for it at first. Once the number of 0-level TFs in EF exceeds the threshold ($\lceil \frac{1}{\epsilon} \rceil + 1$), the two oldest 0-level TFs are merged to generate a 1-level TF. Note that such merging operation may repeat several times for higher levels. Figure 3 gives an example about how to maintain an EF with $\epsilon = \frac{1}{3}$. It means at most four TFs are kept at each level. When L_5 arrives, a new 0-level TF is generated, which adds to five 0-level TFs. Then, a 1-level TF is generated by merging TF($\{L_1\}$) and TF($\{L_2\}$). Similar operation occurs when L_7 arrives. Moreover, the arrival of L_{13} triggers the merging of TF($\{L_9\}$) and TF($\{L_{10}\}$), and further triggers the merging of TF($\{L_1, L_2\}$) and TF($\{L_3, L_4\}$).

Likewise, we obtain a representative line segment for an EF. The central point (denoted as $EF.cen$) and the angle (denoted as θ) of the representative line segment are the weighted mean of all TFs respectively, i.e., $(\sum_i TF_i.cen \times TF_i.n)/(\sum_i TF_i.n)$ and $(\sum_i TF_i.\theta \times TF_i.n)/(\sum_i TF_i.n)$. The corners of EF can also be computed based on the corners of all TFs involved directly. Figure 2(b) illustrates an example about the MBRs for a group of TFs contained in an EF, and the generated blue representative line segment $(EF.rp_s, EF.rp_e)$.

TF and EF synopsis structure allows us to accurately extract and incrementally maintain the feature of micro-clusters at the different intervals. Meanwhile, such synopsis structures with the combination of sliding-window model can safely eliminate the expired records.

4   General Framework of TSCluWin

In this section, we propose a scheme to cluster trajectory streams over the sliding-window model, called Trajectory Streams Clustering based on sliding Window (TSCluWin). TSCluWin is comprised of two components, including a micro-clustering phase and a macro-clustering phase. During the first phase, appropriate statistical information of the micro-clusters are extracted and maintained incrementally, as shown in Algorithm 1. Note that each micro-cluster is represented by an EF structure (Definition 3), and each bucket in an EF is a TF (Definition 2) that represents the summary statistics of a set of trajectory segments at each interval. During the second phase, a small number of macro-clusters are

---

Algorithm 1. EFs Generating and Maintenance (abbr.EFGM) $(\epsilon, \gamma, \delta, k, S, W)$

1: $Z \leftarrow \emptyset$;
2: Initialize Z;
3: **for each** line segment L_x in S **do**
4:     Find the most similar EF h for L_x;
5:     Let d_1 and d_2 denote the length of $h.rp$ and L_x respectively;
6:     **if** $(\exists h, dist(L_x, h.rp)/(d_1 + d_2) \leq \gamma \wedge (h.t - L_x.t < \delta W))$ **then**
7:         $Insert(L_x, h, \epsilon)$;
8:     **else**
9:         **if** $(|Z| = k)$ **then**
10:             Let τ denote the average participating records of all EFs in current window;
11:             **if** $(\exists h_e, (h_e.t \text{ expires}) \vee ((|t_{current} - h_e.t| \geq \frac{W}{2}) \wedge (h_e.n \leq \tau)))$ **then**
12:                 $Z \leftarrow Z - \{h_e\}$;
13:             **else**
14:                 **if** $((\text{find the most similar EF pair } (h_i, h_j)) \wedge (h_i.t - h_j.t < \delta W))$ **then**
15:                     Merge(h_1, h_2, ϵ);
16:                 **end if**
17:             **end if**
18:         **end if**
19:         Create a new EF h_n for L_x;
20:         $Z \leftarrow Z \cup \{h_n\}$;
21:     **end if**
22: **end for**
23: **return** Z;

---

generated based on EFs maintained in memory by invoking traditional weighted clustering techniques.

4.1 Maintenance of EFs

The goal of micro-clustering phase is to handle new tuples in the stream and discard expired ones. Given the window size W, at any time stamp t_c, only the records in $[t_c - W + 1, t_c]$ are active, while the records arriving before $t_c - W + 1$ are expired. Algorithm 1 shows the main framework to generate and maintain EFs. Let Z represent the set of all generated EFs. Initially, Z is emptied, and subsequently k EFs are generated one after another when continuously receiving k line segments, i.e., we create an EF respectively for each line segment through an initialization process, and regard the line segment itself as the representative line segment of such EF (line 2 in Algorithm 1). At most k EFs can be kept in memory at any time, not the trajectory data per se.

When a new line segment L_x arrives, we attempt to find its nearest EF h in terms of the spatial proximity and temporal closeness. Hence, we set a time tolerance threshold δ $(0 < \delta \leq \frac{1}{2})$ to assess the temporal closeness between L_x and the existing EFs. Only EF with the greatest spatial proximity over recent time period (time span within δW) is regarded as the appropriate EF to absorb L_x.

Algorithm 2. Insert(L_x, h, ϵ)

1: Generate $TF_0(\{L_x\})$ for h;
2: $h.t \leftarrow L_x.t$;
3: **for** $l = 0$ to L **do**
4:     **if** $(|TF_l| < \lceil \frac{1}{\epsilon} \rceil + 2)$ **then**
5:         **break**;
6:     **else**
7:         Merge two oldest l-level TFs into one $(l+1)$-level TF;
8:     **end if**
9: **end for**
10: **if** (the oldest L-level TF(TF_{lst}) of h expires) **then**
11:     Drop TF_{lst} from h;
12: **end if**
13: **return** h;

Also, we use a distance threshold γ ($\gamma \leq 1$) to assess the spatial proximity between L_x and it's nearest EF. Let $dist(L_x, h.rp)$ denote the distance between L_x and h's representative line segment, d_1 and d_2 denote the length of $h.rp$ and L_x separately. If $dist(L_x, h.rp)/(d_1 + d_2) \leq \gamma$, L_x is absorbed by h and the entries of h are adjusted accordingly based on L_x, the detail procedure as shown in Algorithm 2. Otherwise, a new EF h_n that only contain single line segment L_x will be created on condition that the number of EFs is less than k. When the number of EFs exceeds k, we need to take into account eliminating the expired EFs or merging EFs to make room for the new created EF.

4.2   Elimination of Expired Records and Merging of EFs

To eliminate the adverse effect of expired records, when a line segment L_x is incorporated into its nearest EF h, h must be checked to discard obsolete TFs (line 10 in Algorithm 2). Moreover, when the number of EFs exceeds the given threshold k, we not only remove the expired EFs, but also filter out EFs with the earlier updated time and fewer participating trajectory line segments, which no longer contribute to subsequent clustering. We set τ equal to the average participating records of all EFs in current window, and the least recent updated EFs that participating records are smaller than τ will be discarded (line 11 in Algorithm 1).

If none of the aforementioned eliminating criteria are met, we try to find the most similar EFs pair to merge until the number of EFs meets the space constraints, as shown in Algorithm 3. Similarly, only the nearest EF pair within the time span δW will be merged into a new EF. This is rational since two EFs with the greatest spatial proximity in the most recent period, are more similar in actual situation than that only take spatial proximity into account. The merging process is akin to the process of incorporating line segments into an EF. Once the number of corresponding $l - level$ TFs in two EFs exceeds $\lceil \frac{1}{\epsilon} \rceil + 1$, the oldest $l - level$ TFs need to be merged into a $(l+1) - level$ TF. Such operation will cascade to level $l = 0, 1, 2, \ldots$, until all TFs of two EFs are handled.

Algorithm 3. Merge(h_i, h_j, ϵ)

1:    **for** $l = 0$ to L **do**
2:        **if** ($|h_i.TF_l| + |h_j.TF_l| > \lceil \frac{1}{\epsilon} \rceil + 1$) **then**
3:            Merge two oldest $l - level$ TFs of h_i and h_j into a new $(l+1) - level$ TF of h_{new};
4:        **else**
5:            $h_{new}.TF_l$ is comprised of $h_i.TF_l$ and $h_j.TF_l$;
6:        **end if**
7:    **end for**
8: $Z \leftarrow Z - \{\{h_i\} \cup \{h_j\}\} \cup \{h_{new}\}$;
9: **return** Z;

However, the computation overhead of finding the closest EF pair is costly, as a nested loop to calculate and compare the distance between all pairs of EFs is inevitable. The cost of such step is $O(k^2)$. Due to the evolving feature of data and the high update cost, Tree-based indexes cannot be adapted to trajectory streams. We opt for a strategy to speed up this process. For each EF h_i, we maintain its closest EF h_{c_i} and the minimal distance d_s between h_i and h_{c_i}. When receiving a new line segment L_x, we attempt to search the closest EF h_{n_1} and the second closest EF h_{n_2} for it. Meanwhile, the distance between h_{n_1} and h_{n_2} (denoted as $dist(h_{n_1}.rp, h_{n_2}.rp)$) is computed and compared with d_s of h_{n_1}, if $dist(h_{n_1}.rp, h_{n_2}.rp) \leq d_s$, h_{n_1}'s original closest EF is replaced with h_{n_2}. If the closest EF (h_{n_1}) of L_x cannot satisfy the defined spatial proximity and temporal closeness, a new EF h_n that only containing L_x will be created, and h_{n_1} is intuitively regarded as the closest EF for h_n. Only when the closest EF h_{c_i} of an EF h_i is eliminated or merged into the other EF, we need to search the nearest EF for h_i by the distance calculation between EFs. In this way, when searching the closest EF pair to merge, we simply need to traverse the closest EF h_{c_i} of all EF h_i, to find the EF pair with the shortest distance. As a result, the cost of searching the closest EF pair to merge is reduced to $O(k)$.

4.3 Macro-Cluster Creation

Given a time horizon of length len and the current time t_c, we can draw curves to reveal the relation between the amount of trajectory segments that have been absorbed in micro-clusters and the interval from current time t_c to $t_c - len$. Through curve graph, the analysts can better differentiate the micro-clusters to further discover the meaningful characteristic of micro-cluster.

Additionally, given the clustering request over specific time horizon, we can further explore macro-clusters based on the previously generated micro-clusters. As mentioned above, the micro-clusters are represented by the representative trajectory line segments of EFs. All EFs in the user-specific time interval $[t_c - len, t_c]$ are treated as pseudo line segments, and re-clustered to produce macro-clusters by using a variant of DBScan algorithm [12,15]. If the user-defined time horizon exceeds the current window size, we will find the most recent EFs that

Fig. 4. Movement distri-   **Fig. 5.** Micro-clustering   **Fig. 6.**   Macro-clustering
bution of taxis (Color   results   (Color   figure   results   (Color   figure
figure online)   online)   online)

maintained in main memory, and the historical EFs that stored on disk within
the given time horizon. Subsequently, the representative trajectory line segments
of such EFs can be offline re-clustered to generate the macro-clusters by using
DBScan algorithm.

4.4   Performance Analysis

The space complexity of TSCluWin is as follows. Given the error parameter ϵ, the
maximal number of EF k, the window size W, and the number of line segments
absorbed in the i-th micro-cluster n_i, then $\sum_{i=1}^{k} n_i = W$, and the number of
obsolete records is within $[0, \epsilon W]$. There are at most $(\frac{1}{\epsilon}+1)(\log(\epsilon n_i +1)+1)$ TFs
in an EF. The total number of TFs in k EFs is at most $k(\frac{1}{\epsilon}+1)(\log(\epsilon n_i +1)+1)$.
In addition, the number of TFs required by merging two EFs is $(\frac{1}{\epsilon}+1)(\log(\epsilon(n_i + n_j))+1)$. As a consequence, the total required memory (the total number of TFs)
is $O(\frac{k}{\epsilon}\log(\epsilon\lceil\frac{W}{k}\rceil))$.

Concerning time complexity, the cost of incorporating a new line segment L_x
into the nearest EF involves lines 4, 11 and 14 in Algorithm 1. The cost of line
4 (finding the closest EF for L_x) is simply $O(k)$. At line 11, when the number
of EF exceeds k, the cost of removing obsolete EF is $O(k)$. At line 14, the cost
of computing the shortest distance between EFs and merging a pair of EFs is
$O(k)$ (as illustrated in Sect. 4.2). Consequently, the per-record processing cost is
$O(k)$, the total processing cost of dealing with W records is $O(kW)$.

5   Experiments

In this section, we conduct extensive experiments to evaluate the clustering
performance and efficiency of our proposed method by comparing against TCMM
algorithm (the work most similar to our proposal) on a real life dataset. Though
TCMM employs the micro- and macro-clustering framework to cluster trajectory
data incrementally, it doesn't take the temporal aspects of the trajectories into
account and cannot fit for clustering trajectory stream. All codes written in Java,
are conducted on a PC with Intel Core CPU 3.1 GHz and 8.00 GB RAM. The
operating system is Windows 8.1. In the experiments, TSCluWin maintains the

same number of micro-clusters as that of TCMM. Unless mentioned otherwise, the parameters are set below, $\epsilon = 0.5$, $\gamma = 0.75$, $\delta = \frac{1}{2}$, and $d_{max} = 800$.

5.1 Dataset

We use a real-life dataset about the trajectories of the taxis in Shanghai, China. This dataset contains the GPS logs of about 30000 taxis during three months (October, November, December) in 2013, covering 93 % main road network of Shanghai. Figure 4 shows the movement distribution of partial taxis (in green) during the interval (from 10:30 to 11:00 a.m.) on October 7th. After micro-clustering phase, 33 micro-clusters are extracted (in blue). As shown in Fig. 5, they capture most traffic flows in Fig. 4. Given the interval (from 10:40 to 11:00 a.m.) on October 7th, 33 micro-clusters are grouped into 3 macro-clusters (in red, green, blue respectively) after clustering by DBScan algorithm, as shown in Fig. 6.

5.2 Effectiveness

We quantify the clustering effect according to the sum of square distance (SSQ). For each cluster C_i, we compute the sum of square distance between any two line segments' center points (denoted as $L_j^i.cen$, $0 < j \leq n_i$) in C_i and the centroid (denoted as Cen_i) of C_i. Therefore, SSQ is calculated with $\sum_{i=1}^{K} dist^2(L_j^i.cen, Cen_i)$, where K denotes the number of clusters (micro-clusters or macro-clusters). Generally, a small SSQ value means the high-quality clustering result.

At first, we report the performance upon different window sizes. To improve credibility, both algorithms are executed for ten times and the average SSQs are reported. Figure 7 shows the average SSQ obtained by both algorithms as time progresses. We observe that TSCluWin (the left bar) always behaves better than TCMM (the right bar) when the window size is set to 160,000 and 330,000 respectively. Since the obsolete records are promptly eliminated by TSCluWin, the micro-clusters are maintained relatively compact with fewer records whenever the cluster center drifts. Conversely, since TCMM does not consider eliminating the influence of the expired records, a micro-cluster may continuously

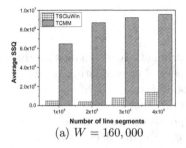
(a) $W = 160,000$

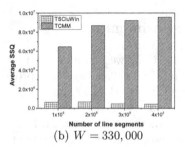
(b) $W = 330,000$

Fig. 7. Quality comparison

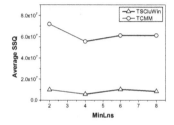

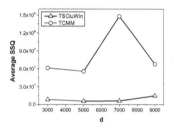

Fig. 8. Average SSQ versus $MinLns$ **Fig. 9.** Average SSQ versus d

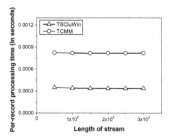

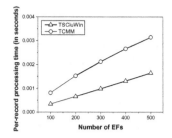

Fig. 10. Execution time versus length of stream

Fig. 11. Execution time versus number of EFs

increase on the boundary rather than be split into multiple small micro-clusters. Additionally, the gap of SSQ between two algorithms varies since the underlying distribution of the positional stream data is always changing.

In order to verify the robustness of TSCluWin in the presence of uncertainty, we proceed to study the effect of input parameters ($MinLns$, d) on the macro-clustering results. Figures 8 and 9 show the average SSQ obtained by TSCluWin and TCMM when varying the values of $MinLns$ and d respectively. Note that the same parameters $MinLns$ and d are used by TSCluWin and TCMM. According to Figs. 8 and 9, TSCluWin is superior to TCMM in all situations, and both algorithms attain the best quality when $MinLns = 4$ and $d = 5,000$.

5.3 Execution Time

Figure 10 shows the execution time of TSCluWin and TCMM. The per-record processing time of TSCluWin fluctuates smoothly, and keeps superior to that processed by TCMM with the progression of trajectory stream. The quicker implementation of TSCluWin is due to that micro-clustering is executed on the original trajectory data (without disregard any trajectory point). While TCMM needs to partition trajectories by using MDL method before micro-clustering phase, which consumes additional waiting time of partitioning accumulated trajectory data into sub-trajectories. Figure 11 shows the per-record processing time when varying the amount of micro-clusters. Both approaches scale linearly with the amount of micro-clusters, since the distance computation cost of finding

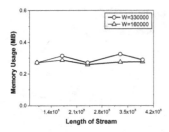
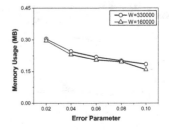

Fig. 12. Memory usage versus length of stream

Fig. 13. Memory usage versus parameter ϵ

the most similar micro-cluster for incoming line segment keeps growing as the number of micro-clusters increases.

5.4 Memory Usage

Figure 12 shows the memory footprint (in MBytes) of TSCluWin when the window size W is set to 160,000 and 330,000 respectively. We can see that the memory usage fluctuates with the progression of the trajectory stream. For $W = 330,000$, when the number of incoming records exceed 160,000, the memory usage of TSCluWin approach decreases at first and then gradually rises with the new records. The same change trend happens when the number of incoming records exceed 330,000. The main reason is that the number of TFs drops along with eliminating of the expired records in terms of two erasure criteria (as illustrated in Sect. 4.2). Similarly, when $W = 160,000$, the same finding can be observed when the amount of incoming records exceeds 160,000.

Figure 13 shows the memory usage of TSCluWin by tuning the value of parameter ϵ. We can observe that a larger window size enables more TFs stored in memory and thus leads to the larger memory footprint. In addition, when the value of ϵ increases from 0.02 to 0.1, the memory usage decreases significantly. It is due to that ϵ decides the amount of expired records within the sliding window. In the current window, with the increase of ϵ, more obsolete records are eliminated, and fewer TFs are stored in memory.

6 Conclusion

In this paper, we propose an efficient method to cluster evolving trajectory streams over the sliding-window model, called TSCluWin. It consists of two components, a micro-clustering component that extracts the summary of trajectory stream in the window, and a macro-clustering component that re-clusters the previously extracted summaries according to user's request. Specifically, We define two synopsis data structures (EF and TF) to maintain the most recent cluster changes of trajectory stream in memory. We validate our proposal against

TCMM algorithm for effectiveness and efficiency by conducting extensive experiments on a real dataset, and show that our proposal is efficient in coping with trajectory stream and outperforms the baseline approach.

Acknowledgement. Our research is supported by the 973 program of China (No. 2012CB316203), NSFC (U1501252, U1401256, 61370101 and 61402180), Shanghai Knowledge Service Platform Project (No. ZF1213), Innovation Program of Shanghai Municipal Education Commission(14ZZ045), and Natural Science Foundation of ShanghaiNo. 14ZR1412600).

References

1. Aggarwal, C.C., Han, J., Wang, J., Yu, P.S.: A framework for clustering evolving data streams. In: VLDB, pp. 81–92 (2003)
2. Aggarwal, C.C., Yu, P.S.: A framework for clustering uncertain data streams. In: ICDE, pp. 150–159 (2008)
3. Babcock, B., Datar, M., Motwani, R., Callaghan, L.: Maintaining variance and k-medians over data stream windows. In: PODS, pp. 234–243 (2003)
4. Datar, M., Gionis, A., Indyk, P., Motwani, R.: Maintaining stream statistics over sliding windows. In: SODA, pp. 635–644 (2002)
5. Duan, X., Jin, C., Wang, X., Zhou, A., Yue, K.: Real-time personalized taxi-sharing. In: DASFAA (2016)
6. Ester, M., Kriegel, H., Sander, J., Xu, X.: A density-based algorithm for discovering clusters in large spatial databases with noise. KDD **96**, 226–231 (1996)
7. Gaffney, S., Smyth, P.: Trajectory clustering with mixtures of regression models. In: ACM SIGKDD, pp. 63–72. ACM (1999)
8. Jensen, C.S., Lin, D., Ooi, B.C.: Continuous clustering of moving objects. IEEE TKDE **19**(9), 1161–1174 (2007)
9. Jeung, H., Yiu, M.L., Zhou, X., Jensen, C.S., Shen, H.T.: Discovery of convoys in trajectory databases. PVLDB **1**(1), 1068–1080 (2008)
10. Jin, C., Yu, J.X., Zhou, A., Cao, F.: Efficient clustering of uncertain data streams. Knowl. Inf. Syst. **40**(3), 509–539 (2014)
11. Lange, R., Dürr, F., Rothermel, K.: Efficient real-time trajectory tracking. VLDB J. **20**(5), 671–694 (2011)
12. Lee, J., Han, J., Whang, K.: Trajectory clustering: a partition-and-group framework. In: ACM SIGMOD, pp. 593–604. ACM (2007)
13. Li, X., Ceikute, V., Jensen, C.S., Tan, K.: Effective online group discovery in trajectory databases. IEEE TKDE **25**(12), 2752–2766 (2013)
14. Li, Y., Han, J., Yang, J.: Clustering moving objects. In: ACM SIGKDD, pp. 617–622 (2004)
15. Li, Z., Lee, J.-G., Li, X., Han, J.: Incremental clustering for trajectories. In: Kitagawa, H., Ishikawa, Y., Li, Q., Watanabe, C. (eds.) DASFAA 2010. LNCS, vol. 5982, pp. 32–46. Springer, Heidelberg (2010)
16. Liu, H., Jin, C., Zhou, A.: Popular route planning with travel cost estimation. In: DASFAA (2016)
17. Nehme, R.V., Rundensteiner, E.A.: SCUBA: scalable cluster-based algorithm for evaluating continuous spatio-temporal queries on moving objects. In: Ioannidis, Y., Scholl, M.H., Schmidt, J.W., Matthes, F., Hatzopoulos, M., Böhm, K., Kemper, A., Grust, T., Böhm, C. (eds.) EDBT 2006. LNCS, vol. 3896, pp. 1001–1019. Springer, Heidelberg (2006)

18. Roh, G.-P., Hwang, S.: NNCluster: an efficient clustering algorithm for road network trajectories. In: Kitagawa, H., Ishikawa, Y., Li, Q., Watanabe, C. (eds.) DASFAA 2010. LNCS, vol. 5982, pp. 47–61. Springer, Heidelberg (2010)
19. Sacharidis, D., Patroumpas, K., Terrovitis, M., Kantere, V., Potamias, M., Mouratidis, K., Sellis, T.K.: On-line discovery of hot motion paths. In: EDBT, pp. 392–403(2008)
20. Tang, L.A., Zheng, Y., Yuan, J., Han, J., Leung, A., Hung, C., Peng, W.: On discovery of traveling companions from streaming trajectories. In: ICDE, pp. 186–197 (2012)
21. Wang, W., Yang, J., Muntz, R.R.: STING: a statistical information grid approach to spatial data mining. VLDB **97**, 186–195 (1997)
22. Yu, Y., Wang, Q., Wang, X., Wang, H., He, J.: Online clustering for trajectory data stream of moving objects. Comput. Sci. Inf. Syst. **10**(3), 1293–1317 (2013)
23. Zheng, K., Zheng, Y., Yuan, N.J., Shang, S.: On discovery of gathering patterns from trajectories. In: ICDE, pp. 242–253 (2013)
24. Zhou, A., Cao, F., Qian, W., Jin, C.: Tracking clusters in evolving data streams over sliding windows. Knowl. Inf. Syst. **15**(2), 181–214 (2008)

On Efficient Spatial Keyword Querying with Semantics

Zhihu Qian[1], Jiajie Xu[1,2(✉)], Kai Zheng[1,2], Wei Sun[1], Zhixu Li[1,2], and Haoming Guo[3]

[1] Department of Computer Science and Technology, Soochow University, Suzhou, People's Republic of China
{xujj,kevinz,zhixuli}@suda.edu.cn
[2] Collaborative Innovation Center of Novel Software Technology and Industrialization, Nanjing, People's Republic of China
[3] Institute of Software, Chinese Academy of Sciences, Beijing, People's Republic of China
haoming@nfs.iscas.ac.cn

Abstract. The fast development of GPS equipped devices has aroused widespread use of spatial keyword querying in location based services nowadays. Existing spatial keyword indexing and querying methodologies mainly focus on the spatial and textual similarities, while leaving the semantic understanding of keywords in spatial web objects and queries to be ignored. To address this issue, this paper studies the problem of semantic based spatial keyword querying. It seeks to return the k objects most similar to the query, subject to not only their spatial and textual properties, but also the coherence of their semantic meanings. To achieve that, we propose a novel indexing structure called NIQ-tree, which integrates spatial, textual and semantic information in a hierarchical manner, so as to prune the search space effectively in query processing. Extensive experiments are carried out to evaluate and compare it with other two baseline algorithms.

Keywords: Spatial keyword query · Query optimization · Probabilistic topic model · Semantic similarity

1 Introduction

Location based services (LBS) is widely used nowadays [3,20,23,25], and spatial keyword query is known as an important technique for LBS systems. Extensive efforts have been made so far to support effective spatial keyword indexing and querying. Some pioneer work [5,6] mainly focuses on the Spatial Keyword Boolean Query (SKBQ) that requires exact keywords match, and apparently, they may lead too few or no results to be returned because of the diversified textual expressions. To overcome this issue, researchers proposed some novel indexing structures to support Spatial Keyword Approximate Query (SKAQ) more

© Springer International Publishing Switzerland 2016
S.B. Navathe et al. (Eds.): DASFAA 2016, Part II, LNCS 9643, pp. 149–164, 2016.
DOI: 10.1007/978-3-319-32049-6_10

recently in [16,18,21], which are able to handle the spelling errors and conventional spelling differences (e.g., 'theater' vs. 'theatre') that frequently appear in real applications. But still, they cannot retrieve the objects that are synonym but literally different to the keywords in query, such as 'theater' and 'cinema', due to the lack of understanding of the semantics in objects and queries. This gap motivates us to investigate other *semantic-aware* approaches that are able to capture the semantic meanings of spatial keywords.

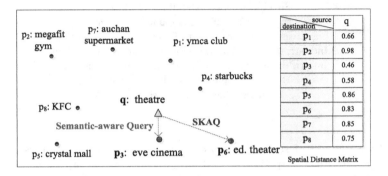

Fig. 1. An example of spatial keyword query

Example 1. Considering the example with eight spatial web objects in Fig. 1, where each object can be seen as a place of interest that has a spatial location and attached keywords. A user issues a query q to find a theatre close to the query location. If the SKBQ approaches [5,6] are applied in the search engine, no result can be returned to the user because of the none precise match to the keyword 'theatre' in query. Alternatively, by using the SKAQ techniques [16,18,21], the search engine can return the object p_6, which seems to be a relatively reasonable object to recommend in terms of spatial and textual similarities. However if checking those objects more carefully, we can easily observe that p_3 is the best match object should be returned, because it is not only closer to q in spatial, but also more relevant to q in semantics, meaning that the user intention can be satisfied as well. In order to make a more reasonable recommendation such as p_3, the key problem is how to interpret and represent the semantics of keywords, and then take the semantic meanings into consideration of query processing.

To fulfill the gap mentioned above, we apply the probabilistic topic model (e.g., LDA [1]), known as a powerful tool in the field of machine learning, to convert the textual descriptions of objects into semantic representations (i.e. distribution over topics, or called topic distribution). By applying the LDA model on p_1 in Fig. 1, we can obtain five latent topics, and its topic distribution (over the five topics) is (0.72, 0.07, 0.07, 0.07, 0.07) (in Table 1). The topic distribution of p_1 indicates the semantic relevance between its textual description and each topic, e.g., 0.72 for topic 'exercise', implying that p_1 is very relevant to the this topic. Similarly, we can compute the topic distributions of the query and all other objects (in Table 1). Note that, each topic distribution is a high dimensional vector in essential. The semantic similarity between textual descriptions of a query and a spatial object can be measured by their topic distributions (e.g. cosine distance of the two vectors). From Table 1, we can thus infer that 'theatre' and 'eve cinema' have close semantic similarity.

Once the topic model is incorporated, spatial keyword querying becomes challenging and time-consuming in despite of more meaningful feedbacks can be found. The reason can be summarized in three main aspects. Firstly, the topic distribution based indexing method has much higher dimensions associating to each object, which severely deteriorates the pruning efficiency (known as the 'curse of dimensionality' [11]) of most multi-dimensional search algorithms. Secondly, compared to the conventional SKBQ and SKAQ, it incurs more memory and I/O cost because additional space is required to store the topic distribution based object information, and I/O cost increases accordingly. Last but not the least, it is necessary to integrate information of multiple dimensions in the indexing and query processing, which makes the hybrid representation more difficult.

To address all above difficulties, we define a new type of spatial keyword query that incorporates spatial, textual and semantic similarities into account. To prune the search space effectively in query processing, we carefully design a hierarchical indexing structure called NIQ-tree, which can integrate spatial, textual and semantic information seamlessly in a hierarchical manner. Since iDistance [14] is one of the best known high dimensional indexing methods, which coincides to our topic distribution based representation of spatial web objects, we incorporate iDistance into the NIQ-tree to avoid the large dead space when indexing all objects in high dimensional space. To make efficient retrival, a novel query processing mechanism on top of the NIQ-tree is proposed to prune the search space effectively based on some theoretical bounds. To sum up, our main contributions of this paper can be briefly summarized as follows:

– We introduce and formalize a new type of probabilistic topic model based similarity measure between a query and a object.
– We propose a novel hierarchical indexing structure, namely NIQ-tree, to integrate the spatial, semantic and textual information of the objects seamlessly while avoiding large dead space.
– We further design an efficient search algorithm, which can greatly prune the high dimensional search space in query processing based on some theoretical bounds.
– We conduct an extensive experiment analysis based on spatial databases and make the comparisons with two baseline algorithms, and then demonstrate the efficiency of our method.

2   Preliminaries and Problem Definition

In this section, we introduce some preliminaries and formalize the problem of this paper.

2.1 Probabilistic Topic Model

In order to recommend spatial web objects that can fulfill user's intention, it is necessary not only to understand the semantic meanings of the textual descriptions embedded in objects and queries [12], but also to measure their semantic relevance accurately. Probabilistic topic model is a mature nature language processing technique that has been proven to be successful on theme interpretation and document classification. Therefore in this paper, we apply the Latent Dirichlet Allocation (LDA) model, i.e. one of the most frequently used probabilistic topic models, to understand the semantic meanings of textual description formed by words with respect to topics. Here, topics can be understood as possible semantic meaning of textual data defined as follows:

Definition 1 (Topic). A topic z represents a type of intended activity that a user may be interested in, such as 'Chinese restaurant', 'coffee shop', 'supermarket' and so on. Z is a preprocessed topic set, which is the union of all topics used to describe the meaningful semantics of textual descriptions.

By carrying out statistical analysis on the large amount of textual descriptions, the LDA model derives the semantic relevance of a topic to all relevant words, known as words distribution defined as follows:

Definition 2 (Words Distribution). Given the topic set Z and the set of all possible words V, the matrix $M = Z \times V_z$ ($V_z \subseteq V$) is used to represent the words distributions of all topics in Z, where V is the collection of all keywords that may appear in textual descriptions, and V_z is the keywords collection belonging to the topic z, apparently, $V_z \subseteq V$. Each M_z represents a distribution of a single topic over all words which belong to this topic and $M_z[w]$ is the topic proportion satisfies $\sum_{w \in V_z} M_z[w] = 1$, where $z \in Z$.

Definition 3 (Topic Distribution). Given a textual description W, the topic distribution of W, denoted as TD_W, is the statistical proportion for each keyword in W, where the topic proportion $TD_W[z]$ from W to topic z is calculated as

$$TD_W[z] = \frac{N_{w \in (W \cap W_z)} + \alpha}{|W| + |Z| \times \alpha} \tag{1}$$

where $N_{w \in (W \cap W_z)}$ is the number of keywords belongs to the given textual description of W in W_z; α is the symmetric Dirichlet prior and generally set to 0.1. $|W|$ and $|Z|$ are the number of keywords in W and topics in Z respectively.

A topic distribution TD_W is a $|Z|$-dimensional vector, which can be regarded as a point in a high dimensional topic space. Therefore, the topic distance of two textual descriptions can be calculated as the following definition.

Definition 4 (Topic Distance). Given two textual descriptions W and W', their topic distance can be quantified by several similarity measures (e.g., Euclidean

Table 1. Topic distributions of textual descriptions

topics textual descriptions	exercise	movie	drink	shop	food
ymca club (in p_1)	0.72	0.07	0.07	0.07	0.07
megafit gym (in p_2)	0.88	0.03	0.03	0.03	0.03
eve cinema (in p_3)	0.04	0.84	0.04	0.04	0.04
starbucks (in p_4)	0.07	0.07	0.72	0.07	0.07
crystal mall (in p_5)	0.07	0.07	0.07	0.72	0.07
ed. theater (in p_6)	0.07	0.72	0.07	0.07	0.07
auchan supermarket (in p_7)	0.07	0.07	0.07	0.72	0.07
KFC (in p_8)	0.04	0.04	0.04	0.04	0.84
theatre (in q)	0.07	0.72	0.07	0.07	0.07

and Cosine Distance). Here, we adapt the cosine distance to measure their distance in high dimensional topic space. The topic distance $\mathcal{D}_T(TD_W, TD_{W'})$ is defined as

$$\mathcal{D}_T(TD_W, TD_{W'}) = \frac{\sum_{z \in Z}(TD_W[z] \times TD_{W'}[z])}{||TD_W|| \times ||TD_{W'}||} \tag{2}$$

where $||TD_W||$ is the modulus of TD_W in $|Z|$ dimensions. It is obvious that the less topic distance of two arbitrary textual descriptions is, the more relevant they are in semantics according to the LDA interpretation.

Example 2. Table 1 shows the LDA interpretation on all the objects in Fig. 1. After running LDA, we derive the distribution of topic (e.g., 'exercise') over relevant words (e.g., 'club', 'gym') based on statistical concurrence. Then we derive the topic distribution of each textual description, where each number in the table is a topic proportion (e.g., $M'_{gym'}[exercise] = 0.88$) that indicates their semantic relevance. Therefore, the topic distance between the textual description W of query point (e.g., $q.W$ = 'theatre') and textual description W' of point in the database (e.g., $p_3.W'$ = 'eve cinema') can be further quantified as $\mathcal{D}_T(W, W') = 0.09$.

2.2 Problem Definition

In this subsection, we give some basic definitions and then formalize the problem of this paper.

Definition 5 (Spatial Web Object). A spatial web object can be a shop, a restaurant or other place of interest whose location and textual descriptive information can be accessed through Internet. It is formalized as $o = (o.\lambda, o.\varphi)$, where $o.\lambda$ is the position of o and $o.\varphi$ is the textual description for describing o. We use the term *spatial object* to represent it in short in the rest of this paper.

Definition 6 (Spatial Keyword Query). Consider a query $q = (q.\lambda, q.\varphi, \tau)$, where $q.\lambda$ is the query location represented by a longitude and a latitude in the two dimensional geographical space; $q.\varphi$ is a group of words that describe user's intention, such as 'Chinese restaurant'; τ is a user-specified threshold of textual distance in case that strict textual similarity is required. The textual distance between the query q and spatial object o is denoted as $TD(q,o)$, which is measured by the Edit Distance [15] of their keywords.

Definition 7 (Candidate Object). Given a query q, a spatial object o is said to be a candidate object, if and only if its textual distance to q is no more than the threshold of q, i.e. $TD(q,o) \leq q.\tau$.

Note that, a spatial object can be returned to user only if it is a candidate object. Among all candidate objects, we rank them by the distance function subject to their spatial proximity and semantic coherence.

Definition 8 (Distance Function). Given a query q, a set of spatial object o, their spatial distance are calculated by the Euclidean distance of their geographical locations. We normalize it to range [0,1] by using the sigmoid function as shown in Eq. 3.

$$\mathcal{D}_S(q,o) = \frac{2}{1 + e^{-dist(q.\lambda, o.\lambda)}} - 1 \tag{3}$$

By combining the spatial distance and the topic distance, we further define a distance function of q and o, denoted as $\mathcal{D}(q,o)$ in Eq. 4.

$$\mathcal{D}(q,o) = \lambda \times \mathcal{D}_S(q,o) + (1 - \lambda) \times \mathcal{D}_T(q,o) \tag{4}$$

where λ is a user-specified parameter to balance the weight of the spatial and semantic distance.

Problem Statement. Given a query q, a set of spatial objects D, and user-specified integer k, this paper returns the k candidate objects that have the minimum distance to q.

3   Baseline Algorithms

In this section, we propose two baseline algorithms which explore the possibility of using existing techniques to solve the problem in this paper.

3.1   Quadtree Based Algorithm

The first baseline algorithm uses the Quadtree [7] to prune the search space in spatial dimension. In the method, the Quadtree, which only utilizes the spatial coordinates of the points, is used to index these points in two-dimensional space directly. Given a query q, the first baseline traverses the index structure to find the spatial nearest object incrementally in terms of the *spatial best match distance* which is computed as follows:

$$\mathcal{D}_{sbm}(q,o) = \lambda \times \mathcal{D}_S(q,o) \qquad (5)$$

It is easy to see that the spatial best match distance is always the lower bound of the distance between q and o.

In the processing of the query, we keep finding the next nearest point o' (based on spatial best match distance) and computing its distance $\mathcal{D}(q,o')$ to q. During the process, we keep track of the k-th minimum distance as the upper bound of final results $\mathcal{D}_{ub}$ based on a priority queue. If the spatial best match distance of next obtained object exceeds $\mathcal{D}_{ub}$ already, the search algorithm terminates since all remaining objects have no chance to be better than the current top-k results.

3.2   MHR-tree Based Algorithm

The basic idea of the second baseline algorithm is mainly motivated by some early work for approximate string search in spatial database [5,6,21]. We use a hybrid indexing structure called Min-wise signature with linear Hashing R-tree (MHR-tree) [21], which combines R-tree [10] with signatures embedded in the nodes, to prune the search space in both spatial and textual dimension.

The indexing structure of MHR-tree embeds the min-wise signature in a R-tree node. For every leaf node u in the MHR-tree, we compute the n-grams G_p and the corresponding min-wise signature $S(G_p)$ of every point p in this node, then store all $(p, S(G_p))$ pairs in node u. For every non-leaf node u in the MHR-tree, with its child node entries $c_1, ..., c_f$ and every child node w_i pointed by c_i, we store the min-wise signature of the node pointed to by c_i, i.e., $s(G_{w_i})$. Then the signature of a non-leaf node u can be computed using $s(G_{w_i})$ as

$$s(G_u)[i] = min(s(G_{w_i})[i], ..., s(G_{w_f})[i]) \qquad (6)$$

where $s(G_u)[i]$ is proportion of $s(G_u)$. Finally, we store $s(G_u)$ of the non-leaf node in the index entry that points to u in u's parent.

In processing a query q, we use a min-priority queue that orders objects in the queue according to their distance to the query point. We search from the root of the MHR-tree and prune the search space by spatial best match distance similar to the search in the first baseline. Additionally, here we use some strategies to avoid checking all points in the node of MHR-tree according to their textual information. When we reach a leaf node, we traverse every point p in the node and insert it into the queue according to Lemma 1 [8,19] if it satisfies:

$$|G_p \cap G_q| \geq max(|p|, |q|) - 1 - (\tau - 1) \times n \qquad (7)$$

where G_p and $|p|$ are the set of n-grams and string length in p respectively, τ is the user-specified threshold of textual distance (e.g., Jaccard distance [13] and edit distance [15]) and n is 3 if we choose 3-gram.

Lemma 1 *(From [8]). For strings σ_1 and σ_2, if their edit distance is τ, then*
$|G_{\sigma_1} \cap G_{\sigma_2}| \geq max(|\sigma_1|, |\sigma_2|) - 1 - (\tau - 1) \times n.$

However, when we reach a non-leaf node, its child w_i will be added to the queue according to Lemma 2 if it satisfies:

$$|\widehat{G_{w_i} \cap G_q}| \geq |q| - 1 - (\tau - 1) \times n \tag{8}$$

where $|\widehat{G_{w_i} \cap G_q}|$ is the estimation of $|G_{w_i} \cap G_q|$. Whenever a point is removed from the head of the queue, it is added to the result set. The search terminates when there are k points in the result or the priority queue becomes empty.

Lemma 2. *Let G_u be the set for the union of n-grams of strings in the subtree of node u in a MHR-tree. Given a query q, if $|G_u \cap G_q| \geq |q| - 1 - (\tau - 1) \times n$, then the subtree of node u does not contain any point in the result.*

Proof: G_u is a set, which contains distinct n-grams. The proof follows by the definition of G_u and Lemma 1.   □

4   NIQ-tree Based Algorithm

In this section, we propose an improved hybrid indexing structure NIQ-tree based on iDistance [14]. The iDistance is a well-known index scheme for high-dimensional similarity search, with a basic idea to group all objects by clustering (e.g., by k-means, k-medoids, etc.), which enables us to achieve superior pruning effect in query processing. By utilizing iDistance to sketch the topic distributions, the NIQ-tree is expected to support effective pruning on spatial, textual and semantic dimensions simultaneously.

Indexing Structure. The NIQ-tree is a three layered hybrid indexing structure shown as Fig. 2, where the spatial, semantic and textual layers are integrated in a vertical way. In designing NIQ-tree, we adopt a spatial first method because of the better pruning effect in spatial domains, which can be explained by its two dimensional nature (v.s. high dimensions of topic and textual domains). The basic form of a NIQ-tree node is $N = (p, rect, o, r)$, where p is the pointer(s) to its child node(s); $rect$ is the minimum bounding rectangle (MBR) in spatial of all objects contained by N; o and r are the center point and radius of a topic space hyper-sphere that covers the topic distributions of all objects contained by N respectively. On top of all spatial objects, we use Quadtree to index them in the spatial domain according to their spatial closeness first. For each leaf node of Quadtree, all objects are further organized by iDistance index in the topic layer, such that objects are grouped and managed by their topic coherence. For each leaf nodes N of the topic layer, it is referenced to a set of n-gram based inverted lists in the textual layer, and similar to MHR-tree, the n-gram inverted lists functionally sketch out the textual descriptions of objects contained in N. In this way, it is possible to filter irrelevant objects according to the $q.\varphi$ and $q.\tau$ specified in query.

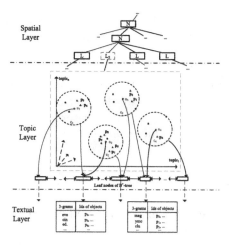

Fig. 2. NIQ-tree

Example 3. A three-layered NIQ-tree is shown in Fig. 2. Assuming that all POIs in Fig. 1 are divided into the same leaf node L_2 in spatial layer, these points are partitioned into four clusters in topic layer, where C_1 contains $\{p_3, p_6\}$, C_2 contains $\{p_5, p_7, p_8, p_9\}$, C_3 contains $\{p_1, p_2\}$ and C_4 contains $\{p_4\}$. It is clear that all points in the same cluster have high semantic similarity. At last, in textual layer, we construct a n-gram based inverted list.

In constructing the NIQ-tree, we build up a Quadtree for all points in spatial database first like the first baseline algorithm. Then for the points in every leaf node of the Quadtree, we use iDistance to cluster these points based on their topic distributions of all contained objects, and construct a B^+-tree to organize the nodes (each node represents a cluster) according to the *key* value computed as follows:

$$key = i \times c + \mathcal{D}_T(p, o_i) \tag{9}$$

where i is the identifier of the partition P_i, c is a constant to partition the single dimension space into regions so that all points in P_i will be mapped to the range $[i \times c, (i+1) \times c)$, o_i is a reference point to P_i and p is the point in this partition. In this way, the high dimensional topic space is expressed by a transformed point *key* in single dimension space, and B^+-tree can thus be applied directly. Next, we set the $N.o$ and $N.r$ for all spatial layer nodes of the NIQ-tree in an bottom up fashion, such that they can cover the center point and radius of N's child nodes in minimum topic space cost. Finally, we build the inverted lists for every leaf node of the B^+-tree by the n-gram method.

Query Processing. Algorithm 1 illustrates the query processing mechanism over the NIQ-tree. Given a query q, the objects retrieval is carried out on the spatial, topic and textual domains of the index alternately. Starting from the root of index, we traverse the spatial layer nodes in the ascending order of the *best match distance* $D_{bm}(q, N)$ with respect to q defined as the following formula:

$$D_{bm}(q, N) = \lambda \times min_{p \in N.mbr} \mathcal{D}_S(q, p) + (1 - \lambda) \times min \mathcal{D}_T(q, N) \qquad (10)$$

where $min_{p \in N.mbr} \mathcal{D}_S(q, p)$ and $min \mathcal{D}_T(q, N)$ denote the minimum possible spatial and topic distance from q to any object contained in the node N. Let $\|TD_{q.\varphi}, N.o\|$ be the cosine distance between textual description $q.\varphi$ and reference point o in topic layer, the minimum possible topic distance $\mathcal{D}_T(q, N)$ can be computed as follows:

$$min \mathcal{D}_T(q, N) = \begin{cases} 0 & \|TD_{q.\varphi}, N.o\| \leq N.r \\ \|TD_{q.\varphi}, N.o\| - N.r & \|TD_{q.\varphi}, N.o\| > N.r \end{cases} \qquad (11)$$

It is noted that $D_{bm}(q, N)$ is the lower bound distance D_{lb} to q for all unvisited points according to its definition.

In the query processing, the node we fetch from the priority queue is a non-leaf node, we add all its child nodes to the queue; otherwise, we access the topic and textual layer indices subject to this node to access the candidate objects they covers. During the search in topic layer, the leaf node in the B$^+$-tree can be identified according to key value. Similar to iDistanceKNN search [14], we browse the space by expanding the radius R of the hyper-sphere centered at the query point q. At each time, R is increased by ΔR (i.e., $R = R + \Delta R$). If the leaf node of this layer intersects with the searching sphere, we traverse the points in this node according to its key value in the range of $[i \times c + dis(q, o_i) - R, i \times c + dist(q, o_i) + R]$, where $dist(q, o_i)$ is the distance between q and reference point o_i. Then, by checking its inverted list in textual layer, we find the spatial objects whose textual distance to q is no more than $q.\tau$. Especially, we dynamically maintain the top-k minimum distance for all scanned points and keep the k-th minimum distance as an upper bound $\mathcal{D}_{ub}$. The radius of search sphere R stops increasing when the following condition holds for all unvisited topic layer leaf nodes:

$$\lambda \times min_{p \in N.mbr} \mathcal{D}_S(q, p) + (1 - \lambda) \times R \geq \mathcal{D}_{ub} \qquad (12)$$

where $N.mbr$ is the MBR of a spatial layer leaf node N. Obviously, ($\lambda \times min_{p \in N.mbr} \mathcal{D}_S(q, p) + (1 - \lambda) \times R$) is a lower bound distance from q to a topic layer leaf node rooted in N according to Lemma 3. The whole search algorithm terminates when $\mathcal{D}_{lb}$ is no less than $\mathcal{D}_{ub}$ because the remaining unvisited points have no opportunity to be better than the current top-k results.

Lemma 3. *Given a query q and a NIQ-tree, if N is a spatial layer leaf node of the NIQ-tree, then the search in topic layer with respect to N terminates when* ($\lambda \times min_{p \in N.mbr} \mathcal{D}_S(q, p) + (1 - \lambda) \times R$) $\geq \mathcal{D}_{ub}$.

Proof: $min_{p \in N.mbr} \mathcal{D}_S(q, p)$ is the minimum spatial distance from q to a spatial layer node N, which is the lower bound of spatial distance from q to any unvisited point in this node. R is the minimum topic distance from q to any unvisited point in the cluster, which is the lower bound of topic distance. So ($\lambda \times min_{p \in N.mbr} \mathcal{D}_S(q, p) + (1 - \lambda) \times R$) is a lower bound distance from q to a topic layer leaf node. Lemma 3 can be proven.   □

Algorithm 1. NIQ-tree based Search Algorithm

    Input: dataset D, query q and user-specified k and λ
    Output: top-k result set V
1 Upper Bound $\mathcal{D}_{ub} = +\infty$;
2 Lower Bound $\mathcal{D}_{lb} = 0$;
3 Search radius $R = 0.1$;
4 Put NIQ-tree root into a priority queue U;
5 **while** $U \neq \emptyset$ **do**
6   Pop an element N from U;
7   **if** N *is a non-leaf node* **then**
8   Add its children to U;
9   **else if** N *is a leaf node* **then**
10   **for** *every iDistance node N' in N intersecting with searching sphere* **do**
11   **for** *every object o in N'* **do**
12   Check the n-gram inverted lists of N';
13   **if** o *is a candidate object* **then**
14   Compute $\mathcal{D}(q, o)$ using Eq. 4;
15   **if** $\mathcal{D}(q, o) < \mathcal{D}_{ub}$ **then**
16   Update V with o included;
17   Update $\mathcal{D}_{lb}$ and $\mathcal{D}_{ub}$;
18   **if** $(\lambda \times min_{p \in N.mbr} \mathcal{D}_S(q, p) + (1 - \lambda) \times R) \geq \mathcal{D}_{ub}$ **then**
19   break;
20 $R = R + \Delta R$;
21   **if** $\mathcal{D}_{lb} \geq \mathcal{D}_{ub}$ **then**
22   break;
23 return V;

---

5 Experiment Study

In this section, we conduct extensive experiments on real datasets to evaluate the performance of our proposed index and search algorithms.

5.1 Experiment Settings

We create the real object datasets by using the online check-in records of Foursquare within the area of New York City. Each check-in record of Foursquare contains the user ID, venue with geo-location (place of interest), time of check-in and the tips written in plain English. We put the records belonging to the same object to form textual descriptions of the objects. The topic distributions over words are obtained by the textual descriptions in the tips associated with the location, and then the textual descriptions for each place are interpreted into a probabilistic topic distribution by LDA model. The number of objects in the whole dataset is 422030 in sum.

Table 2. default values of parameters

Parameter	Default value	Description
k	10	Top-k results
λ	0.5	Weight factor
τ	3	Threshold of edit distance
D	200 K	No. of objects
c	4 K	Capacity of quadtree leaf node
m	20	Number of clusters

We compare the query time cost and number of visited objects during the search processing of the proposed method (NIQ-tree) with the two baseline algorithms proposed in Sect. 3. The default values for parameters are given in Table 2. In the experiments, we vary one parameter and keep the others constant to investigate the effect of this parameter. All algorithms are implemented in Java and run on a PC with 2-core CPUs at 3.2 GHz and 8 GB memory.

5.2   Performance Evaluation

In this part, we vary the values of parameters in Table 2 to compare the NIQ-tree with two baseline algorithms and investigate the effect of each parameter.

Effect of k**.** In the first set of experiments, we study the effect of the intended number of results k by plotting the query time and visited objects (denoting I/O) on the dataset. As shown in Fig. 3, our proposed indexing structure, NIQ-tree, significantly outperforms all other two baseline indexing methods on the same dataset. Particularly, the NIQ-tree based method is almost 2–3 times faster than the MHR-tree with respect to query time. All algorithms incur high cost in both number of visited objects and query time as k increases, which is not beyond our expectation because the k-th match distance becomes greater, which leads more candidates to be retrieved.

Effect of λ**.** Next we study the effect of different weight factors λ. As shown in Fig. 4, all algorithms including Quadtree, MHR-tree and NIQ-tree based methods have ascending tendency of I/O and query time when the value of λ goes up. In contrast, the NIQ-tree is superior than two other approaches because of its superior pruning effect in spatial, semantic and textual domains.

Effect of τ**.** Then we investigate the query performance of these algorithms when the threshold τ of the edit distance between object and query is varying. Figure 5 shows the results of our experiment. With the increase of τ, all algorithms incur more time cost and more visited objects because more candidate objects are retrieved since their edit distance to query are less than the threshold. Especially, the NIQ-tree still outperforms the other two baseline algorithms with respect to query time and visited objects.

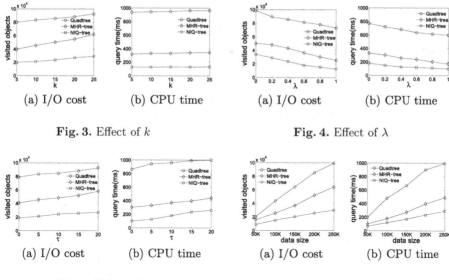

(a) I/O cost   (b) CPU time

Fig. 3. Effect of k

(a) I/O cost   (b) CPU time

Fig. 4. Effect of λ

(a) I/O cost   (b) CPU time

Fig. 5. Effect of τ

(a) I/O cost   (b) CPU time

Fig. 6. Effect of D

Effect of D. In order to evaluate the scalability of all algorithms, we sample the dataset of New York City to generate datasets with different number of objects varying from 50 K to 250 K, and report the query time and visited points in Fig. 6. Within our expectation, the query time and the number of visited objects of all three methods increase linearly with respect to the size of dataset. But it is worth to note that the NIQ-tree based method performs much more efficient than the others.

In NIQ-tree, there are several parameters of iDistance index which may have effects on the performance, including the size c of leaf node in spatial layer and the number m of clusters in iDistance.

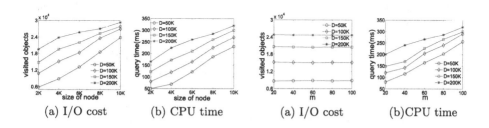

(a) I/O cost   (b) CPU time

Fig. 7. Effect of c

(a) I/O cost   (b)CPU time

Fig. 8. Effect of m

Effect of c. As shown in Fig. 7, the capacity of leaf node in Quadtree affects the performance of our proposed indexing structure. With the increase of node size c, both visited objects and query time increase. That is to say that the query time and I/O increase with the size of c. From Fig. 7, it is noted that the increase of data size D also makes more I/O and query time when c remains the same.

Effect of m. It is shown in Fig. 8 that the performance of our proposed index are affected when the number of clusters m changes. On one hand, we can observe that the visited objects remain almost constant with respect to m but increase with the data size D from the left figure of Fig. 8. On the other hand, the query time has a nearly linear increase with m and it also increases when the data size D becomes larger.

6   Related Work

The related work to out study mainly include probabilistic topic model and spatial keyword query.

Probabilistic Topic Model. Probabilistic topic model is a statistical method to analyze the words in documents and to discover the themes that run through the words, how those themes are connected to each other, with no prior annotations or labeling of documents been required. Based on the topic models, it is possible to measure the semantic relevance between a text to a theme, as well as that between different texts (in semantic level). There are several classical topic models including LDA [1], Dynamic Topic Model, Dynamic HDP, Sequential Topic Models, etc. Above techniques have been widely used in applications like document classification, user behavior understanding, functional region discovery, etc. In this paper, we tend to integrate topic model and spatial objects for efficient spatial keyword querying with semantics.

Spatial Keyword Query. With the prevalence of spatial objects associated with textual information on the Internet, spatial keyword queries that exploit both location and textual description are gaining in prominence. Some efforts are made to support the SKBQ [5,6,17,22] that requires exact keywords match, which may lead few or no results to be found. To overcome this problem, some efforts are further made to support the SKAQ [16,18,21], so that the query results are no longer sensitive to spelling errors and conventional spelling differences. Many novel indexing structures are proposed to support efficient processing on SKBQ and SKAQ, such as IR-tree [5], IR^2-tree [6], MHR-tree [21], S2I [18], etc. Numerous work studies the problem of spatial keyword query on collective querying [2], fuzzy querying [26], confidentiality support [4], continuous querying [9], interactive querying [24], etc. But as far as we know, none of those existing approaches can retrieve spatial objects that are semantically relevant but morphologically different. Therefore, in this paper, we investigate the topic model based spatial keyword querying to recommend users spatial objects that have both high spatial and semantic similarities to query.

To the best of our knowledge, this is the only work to consider the fusion of topic model and spatial keyword query, so that spatial objects can be recommended more rationally by the interpretation of textual descriptions for objects and user intentions.

7 Conclusion

This paper studies the problem of searching spatial objects more effectively by converting keywords matching to semantic interpretation. The probabilistic topic model is utilized to interpret the textual descriptions attached to spatial objects and user queries into topic distributions. To support the efficient top-k spatial keyword query in spatial, topic and textual dimension, we propose a novel hybrid index structure called NIQ-tree, and effective searching algorithm to prune the high dimensional search space regarding to spatial, semantic and textual similarities. Extensive experimental results on real datasets demonstrate the efficiency of our proposed method.

Acknowledgement. This work was partially supported by Chinese NSFC project under grant numbers 61402312, 61402313, 61572335, 61232006, the Key Research Program of the Chinese Academy of Sciences under grant number KGZD-EW-102-3-3, and Collaborative Innovation Center of Novel Software Technology and Industrialization.

References

1. Blei, D.M., Ng, A.Y., Jordan, M.I.: Latent dirichlet allocation. J. Mach. Learn. Res. **3**, 993–1022 (2003)
2. Cao, X., Cong, G., Jensen, C.S.: Collective spatial keyword querying. In: SIGMOD (2011)
3. Chen, L., Cong, G., Jensen, C.S.: Spatial keyword query processing: An experimental evaluation. PVLDB **6**(3), 217–228 (2013)
4. Chen, Q., Hu, H., Xu, J.: Authenticating top-k queries in location-based services with confidentiality. PVLDB **7**(1), 49–60 (2013)
5. Cong, G., Jensen, C.S., Wu, D.: Efficient retrieval of the top-k most relevant spatial web objects. PVLDB **2**(1), 337–348 (2009)
6. De Felipe, I., Hristidis, V., Rishe, N.: Keyword search on spatial databases. In: ICDE (2008)
7. Finkel, R.A., Bentley, J.L.: Quad trees a data structure for retrieval on composite keys. Acta informatica **4**(1), 1–9 (1974)
8. Gravano, L., Ipeirotis, P.G.: Approximate string joins in a database (almost) for free. In: ICDE (2001)
9. Guo, L., Shao, J., Aung, H.H., Tan, K.-L.: Efficient continuous top-k spatial keyword queries on road networks. Geoinformatica **19**(1), 29–60 (2015)
10. Guttman, A.: R-trees: A dynamic index structure for spatial searching. In: SIGMOD (1984)
11. Har-Peled, S., Indyk, P., Motwani, R.: Approximate nearest neighbors: Towards removing the curse of dimensionality. In: ACM symposium on Theory of computing (1998)
12. Hua, W., Wang, Z., Wang, H., Zheng, K., Zhou, X.: Short text understanding through lexical-semantic analysis. In: ICDE (2015)
13. Jaccard, P.: Etude comparative de la distribution florale dans une portion des alpes et du jura. Impr. Corbaz (1901)
14. Jagadish, H.V., Ooi, B.C., Tan, K.-L.: idistance: An adaptive b+-tree based indexing method for nearest neighbor search. ACM TODS **30**(2), 364–397 (2005)

15. Levenshtein, V.I.: Binary codes with correction for deletions and insertions of the symbol 1. Problemy Peredachi Informatsii **1**(1), 12–25 (1965)
16. Li, F., Yao, B., Tang, M.: Spatial approximate string search. TKDE **25**(6), 1394–1409 (2013)
17. Li, G., Feng, J., Xu, J.: Desks: Direction-aware spatial keyword query. In: ICDE (2012)
18. Rocha-Junior, J.B., Gkorgkas, O., Jonassen, S., Nørvåg, K.: Efficient processing of top-k spatial keyword queries. In: Pfoser, D., Tao, Y., Mouratidis, K., Nascimento, M.A., Mokbel, M., Shekhar, S., Huang, Y. (eds.) SSTD 2011. LNCS, vol. 6849, pp. 205–222. Springer, Heidelberg (2011)
19. Ukkonen, E.: Approximate string-matching with q-grams and maximal matches. Theor. Comput. Sci. **92**, 191–211 (1992)
20. Wang, H., Zheng, K.: Sharkdb: An in-memory column-oriented trajectory storage. In: CIKM (2014)
21. Yao, B., Li, F., Hadjieleftheriou, M., Hou, K.: Approximate string search in spatial databases. In: ICDE (2010)
22. Zhang, C., Zhang, Y., Zhang, W., Lin, X.: Inverted linear quadtree: Efficient top k spatial keyword search. In: ICDE (2013)
23. Zheng, K., Huang, Z., Zhou, A.: Discovering the most influential sites over uncertain data: A rank based approach. TKDE **24**(12), 2156–2169 (2012)
24. Zheng, K., Su, H.: Interactive top-k spatial keyword queries. In: ICDE (2015)
25. Zheng, K., Zheng, Y.: Online discovery of gathering patterns over trajectories. TKDE **26**(8), 1974–1988 (2014)
26. Zheng, K., Zhou, X.: Spatial query processing for fuzzy objects. VLDB **21**(5), 729–751 (2012)

Approximation-Based Efficient Query Processing with the Earth Mover's Distance

Merih Seran Uysal[1]($^{(\boxtimes)}$), Daniel Sabinasz[1], and Thomas Seidl[2]

[1] Data Management and Exploration Group,
RWTH Aachen University, Aachen, Germany
{uysal,sabinasz}@cs.rwth-aachen.de
[2] Database Systems Group,
Ludwig-Maximilians-Universität (LMU) Munich, Munich, Germany
seidl@dbs.ifi.lmu.de

Abstract. The Earth Mover's Distance (EMD) is an effective distance-based similarity measure which determines the dissimilarity between data objects by the minimum amount of work required to transform one signature into another one. Although the EMD has been proven to reflect the human perceptual similarity very well in prevalent applications and domains, its high computational time complexity hinders its application to large-scale datasets where the user is rather interested in receiving an answer from the underlying application within a short period of time than requesting an exact and complete query result set. To this end, we propose to improve the efficiency of the query processing with the EMD on signature databases by utilizing signature compression approximations. We introduce an efficient signature compression algorithm to alleviate query computation cost. Furthermore, we theoretically explicate and analyze the approximation-based EMD and the relationship between the proposal and the original EMD. Moreover, our extensive experiments on 4 real world datasets point out the accuracy and efficiency of our approach.

Keywords: Earth mover's distance · Similarity search · Approximate query processing · Signature databases

1 Introduction

The rapid generation and dissemination of data in numerous applications and the high usage of data-sharing web sites on the Internet have resulted in an explosion in collection of various kinds of data recently. One of the most crucial tasks in large data collections is the similarity search where the user specifies a query object to which the most similar objects are output within a result set.

Initially introduced in the computer vision domain as an effective distance-based similarity measure [14], the Earth Mover's Distance (EMD) determines the dissimilarity between any two data objects by the minimum amount of work required to transform one feature distribution (*signature*) into another one.

© Springer International Publishing Switzerland 2016
S.B. Navathe et al. (Eds.): DASFAA 2016, Part II, LNCS 9643, pp. 165–180, 2016.
DOI: 10.1007/978-3-319-32049-6_11

The EMD is known to be robust against outliers, which allows for effective similarity search, when compared to other approaches [14]. Furthermore, the EMD can be utilized for partial similarity search and can be applied to signatures with arbitrary number of representatives. For these reasons, it has attracted many researchers from a wide range of domains, such as multimedia retrieval [2,20,24], computer vision [17], and data analysis [17].

The incorporation of signatures (also referred to as *adaptive binning*) into the similarity search process has been witnessed in prevalent domains and applications. As will be presented in Sect. 2, signatures are utilized to model various kinds of data, such as probabilistic data [25], multimedia data [14,19–21], and text documents [4]. The key advantage of the signatures is the high quality of content approximation utilized for similarity search and query processing in various types of databases.

The EMD can be solved by using transportation simplex algorithms [6] with some appropriate initial basic feasible solutions. Hence, although the theoretical time complexity of the EMD is exponential regarding the number of representatives (features), it can be empirically computed in at least super-cubic time with respect to the number of representatives. Therefore, the database community has attempted to develop EMD-based efficiency improvement techniques [3,14,15,20,21,24]. All such approaches mainly focus on answering the queries with a complete k-nearest-neighbor (k-nn) result set without any false drops. However, in numerous applications dealing with large-scale databases today, it is appropriate for the user to receive approximate answers where an exact answer is not required. This comes up in various applications where, for instance, the query processing time cost is higher than that the user expects. Interestingly, there is little previous work dealing with approximate EMD-based query processing in signature databases. To this end, we propose to improve the query processing with the EMD in signature databases by utilizing signature compression approximations. Our main contributions are listed as follows:

- We introduce an efficient signature compression algorithm to alleviate query computation cost.
- We theoretically explicate the approximation-based EMD and the relationship between the proposal and the EMD.
- The extensive experiments on 4 real world databases point out the accuracy and efficiency of our approach.

The paper is structured as follows: Section 2 presents data representation and the EMD. Then, Sect. 3 gives the related work, followed by Sect. 4 introducing the approximation-based EMD. After Sect. 5 presents our extensive experimental evaluation, the paper is concluded by Sect. 6.

2 Preliminaries

Data Representation. In order to represent a data object by a signature, first the features are extracted from that corresponding object and then aggregated by

(a) image

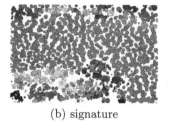

(b) signature

Fig. 1. An image and visualization of its signature with 1000 representatives [23].

utilizing various methods, such as k-means clustering algorithm directly applied to the features. A feature space can be denoted by $(\mathbb{F}, \delta)$, including a set of features, where a ground distance function $\delta : \mathbb{F} \times \mathbb{F} \rightarrow \mathbb{R}$ is utilized to determine the distance between any two features. Hence, any data object can be represented via using *features* $x_1, \ldots, x_n \in \mathbb{F}$ each of which is assigned a real number denoting the number of features assigned to it. In the literature these features are also found as *representatives*.

A *signature* $X : \mathbb{F} \rightarrow \mathbb{R}$ includes a finite set of representatives, where each representative denotes a real number corresponding to the weight of that representative in the feature space. The remainder of the features in the feature space exhibit a zero weight, i.e. they do not play a role with respect to the determination of the signature of the corresponding data object.

An example image and its corresponding signature are depicted in Fig. 1. The signature comprises 1000 representatives, which are visualized by circles, and are based on position, color, and texture information. As illustrated on this example, signatures are able to visually approximate the contents of the images by utilizing individual representatives of contributing features. In the remainder of the paper, for the sake of simplicity we assume that signatures belong to a class of positive signatures $\mathbb{S}^+$ comprising signatures which consist of representatives with only positive weights, and the notation R_X refers to the set of representatives of the signature X. Furthermore, the definitions are based on a feature space $(\mathbb{F}, \delta)$ and a ground distance function δ.

Earth Mover's Distance. The Earth Mover's Distance is a well-known transformation-based similarity measure which determines the cost of transforming one signature into another one. The formal definition is given below.

Definition 1. *Given two signatures* $X, Y \in \mathbb{S}^+$, *the* Earth Mover's Distance $\mathrm{EMD} : \mathbb{S}^+ \times \mathbb{S}^+ \rightarrow \mathbb{R}$ *between* X *and* Y *is defined as a minimization over all possible flows* $F = \{f | f : \mathbb{F} \times \mathbb{F} \rightarrow \mathbb{R}\} = \mathbb{R}^{\mathbb{F} \times \mathbb{F}}$ *as follows:*

$$\mathrm{EMD}(X, Y) = \min_{F} \left\{ \frac{\sum_{x \in \mathbb{F}} \sum_{y \in \mathbb{F}} f(x, y) \cdot \delta(x, y)}{\min\{\sum_{x \in \mathbb{F}} X(x), \sum_{y \in \mathbb{F}} Y(y)\}} \right\},$$

subject to the following constraints:

- *Non-negativity:* $\forall x, y \in \mathbb{F} : f(x,y) \geq 0$
- *Source:* $\forall x \in \mathbb{F} : \sum_{y \in \mathbb{F}} f(x,y) \leq X(x)$
- *Target:* $\forall y \in \mathbb{F} : \sum_{x \in \mathbb{F}} f(x,y) \leq Y(y)$
- *Total Flow:* $\sum_{x \in \mathbb{F}} \sum_{y \in \mathbb{F}} f(x,y) = \min\{\sum_{x \in \mathbb{F}} X(x), \sum_{y \in \mathbb{F}} Y(y)\}$

The EMD is defined as a linear optimization problem where the constraints guarantee a feasible solution, i.e. all flows are positive and do not exceed the corresponding limitations given by the weights of the representatives of both signatures. As mentioned before, the EMD can be solved by the utilization of simplex algorithms [6] where the empirical computational time complexity is super-cubic with respect to the number of the representatives of the corresponding signatures, denoting a real bottleneck.

In order to alleviate the computational cost, the EMD has been utilized within multi-step filter-and-refine architectures where k-nearest-neighbor (k-nn) queries can be performed more efficiently [1,5,7–9,16]. As Fig. 2 depicts, k-nearest-neighbor queries are processed in multiple steps generating candidate objects which are subsequently refined to gather the final results. This approach yields complete result sets, since the exact distance function d is approximated by a lower-bounding distance function LB_d which is utilized in the filter step. In other words, for all objects x, y it holds that $\mathrm{LB}_d(x,y) \leq d(x,y)$.

In this paper, for our approximation-based approach we utilize the multi-step algorithm proposed in [16] which is proven to be optimal in the number of candidates, i.e. the algorithm is optimal in the number of distance computations for exact k-nn queries. The algorithm first generates a ranking by means of the lower bound, and then the ranking is processed as long as the lower bound does not exceed the distance of the kth-nearest neighbor. The algorithm updates the result set as long as data objects with smaller distances have been found.

After presenting the EMD and the efficient query processing, we below give the related work with respect to the filter distance functions utilized to improve the efficiency of the EMD-based query processing.

3   Related Work

There have been numerous attempts to speed up the EMD-based query processing towards similarity search on *fixed-binned signatures*. The authors in [3] aimed at proposing to lower-bound the EMD via L_p-based distances and constraint relaxation. Furthermore, dimensionality reduction techniques were developed for the EMD where reduced cost matrices are utilized relying on the original cost matrix [24]. The approach in [25] derived a lower bound of the EMD by utilizing the primal-dual theory in linear programming on top of B^+-trees. Another further method proposed to lower-bound the EMD by projecting histograms on a vector and approximating their distance by a normal distribution [15].

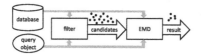

Fig. 2. Multi-step filter-and-refine architecture

In [18], it is proposed to optimize the refinement phase of EMD-based similarity search by presenting a dynamic distance bound. The limitation of the aforementioned approaches is that they are mainly applicable to fixed-binned signatures, i.e. signatures sharing the same representatives.

In contrast, approaches applicable to signatures (irrespective of shared representatives) expose a high flexibility and comprehensive solution. Although adaptive-binned signatures generate more effective results than those with fixed-binned signatures [14], there is unfortunately insufficient research into this field: Rubner and IM-Sig filter distances are currently the state of the art.

Given two signatures X, Y with an equal total weight of m over a feature space $(\mathbb{F}, \delta)$ with a norm-based distance function δ, the Rubner lower bound between X and Y is defined as $\delta\big((\sum_{x \in R_X} X(x) \cdot x)/m, (\sum_{y \in R_Y} Y(y) \cdot y)/m\big)$, where the filter distance simply computes the ground distance between the mean features of both signatures, and it holds $\text{Rubner}(X, Y) \leq \text{EMD}(X, Y)$ [14]. The computation time complexity of Rubner lies in $\mathcal{O}(|R_X|)$ provided that $|R_X| \geq |R_Y|$, however, according to [20] and the results in [22], Rubner filter shows a relatively low selectivity and very high query time on image and video databases.

The authors in [20] devised research towards developing new efficient query processing and indexing techniques for the EMD on signatures. The aim is to reduce the computational time complexity of the k-nn query processing by utilizing constraint relaxation at the level of linear programming examined on signatures. The Independent Minimization for Signatures (IM-Sig) is a lower-bounding technique which is based on the relaxation of the target constraint of the EMD via local examination of each feature independently. Regarding the IM-Sig, the target constraint of the EMD is replaced by another constraint stating that each flow from a source feature may not exceed the destination feature capacity. The basic notion lies in the restriction of the number of earth flows locally for each representative related with outgoing flows. To this end, for each representative $x \in R_X$ the smallest set S can be defined which comprises the nearest neighbors of x in R_Y whose total capacity is sufficient to receive all amount of earth from x. Below, the formal definition of the IM-Sig is presented.

Definition 2 (IM-Sig Filter). *Given signatures X, Y with $m = \sum\limits_{x \in \mathbb{F}} X(x) = \sum\limits_{y \in \mathbb{F}} Y(y)$, IM-Sig between X and Y is a minimization over all possible flows i.e.*

$$IM - Sig(X, Y) = \min_{f \in F}\left\{\sum_{x \in \mathbb{F}}\sum_{y \in \mathbb{F}} \frac{\delta(x,y)}{m} f(x,y)\right\}, \text{ subject to non-negativity con-}$$

straint $\forall x, y \in \mathbb{F} : f(x,y) \geq 0$, source constraint $\forall x \in \mathbb{F} : \sum\limits_{y \in \mathbb{F}} f(x,y) \leq X(x)$,

IM-Sig target constraint $\forall x, y \in \mathbb{F} : f(x,y) \leq Y(y)$, *and flow constr.* $\sum_{x \in \mathbb{F}} \sum_{y \in \mathbb{F}} f(x,y) = m$.

The IM-Sig target constraint states that for any feature no incoming flow may exceed the weight of that corresponding feature in the feature space, while the other three constraints remain the same as for the EMD. In other words, each flow $f(x,y) \in F$ from any feature $x \in \mathbb{F}$ to $y \in \mathbb{F}$ may possess the value of at most $Y(y)$, which provides a greater search space than the original space including the result set of the EMD. It is worth noting that the authors in [20,22] report that Rubner filter yields much lower selectivity and significantly higher query time on image and video databases than for the IM-Sig within a filter-and-refine algorithm.

4 Approximations for the EMD-based Query Processing

As mentioned in Sect. 1, in numerous applications dealing with large-scale databases today, it is appropriate for the user to receive approximate answers where the precise answer is not required. For this reason, we propose to improve the query processing with the EMD on signature databases by utilizing signature compression approximations. First, we introduce an efficient signature compression algorithm to alleviate query computation cost. Then, we theoretically explicate the approximation-based EMD and the relationship between the proposal and the EMD.

The time complexity of the query processing increases with the number of representatives of the signatures. The time complexity of the distance computation between any two signatures can be reduced by decreasing the number of representatives. A simple and straightforward way to achieve it would be to arbitrarily leave out some representatives so that the representative set cardinalities are smaller. However, such a method would lead to low accurate approximate distance results due to the arbitrarily chosen representatives which are neglected for the signature representation. Thus, below we propose to approximate the signatures by merging representatives.

Definition 3. *Given a signature* X *and two representatives* $x, y \in R_X$, *the merged representative of* x *and* y *is defined as:*

$$M(x,y) = \frac{X(x) \cdot x + X(y) \cdot y}{X(x) + X(y)}.$$

Definition 4. *Given a signature* X *and two representatives* $x, y \in R_X$, *the merged weight of* x, y *is defined as:*

$$MW(x,y) = X(x) + X(y).$$

For any two clusters C_x and C_y of features whose centroids (i.e. representatives) are denoted by x and y (i.e. $x = \frac{1}{|C_x|} \sum_{z \in C_x} z$, $y = \frac{1}{|C_y|} \sum_{z \in C_y} z$), $M(x,y)$ denotes the centroid of $C_x \cup C_y$: $M(x,y) = \frac{1}{|C_x|+|C_y|} \sum_{z \in C_x \cup C_y} z$.

Given a set $\mathcal{C}$ of representative pairs, the compressed signature of the underlying signature is defined by merging the representative pairs as follows.

Definition 5. *Given a signature X and a set $\mathcal{C} \subset \{\{x,y\} \mid x,y \in R_X\}$ of representative pairs such that for all $\{x,y\}, \{x',y'\} \in \mathcal{C}$, the representatives x,y,x',y' are pairwise distinct, the compressed signature $Comp_{\mathcal{C}}(X) : \mathbb{F} \to \mathbb{R}$ with respect to X and the set $\mathcal{C}$ of merged pairs of representatives is defined as:*

$$Comp_{\mathcal{C}}(X)(x) = \begin{cases} 0 & \text{if } \exists x', x'' \in R_X. \{x',x''\} \in \mathcal{C} \wedge (x = x' \vee x = x'') \\ MW(x',x'') & \text{if } \exists x', x'' \in R_X. \{x',x''\} \in \mathcal{C} \wedge x = M(x',x'') \\ X(x) & \text{else} \end{cases}$$

Note that the following statement denotes that x,y,x',y' are pairwise distinct: $\forall x,y \in R_X. \{x,y\} \in \mathcal{C} \Rightarrow (x \neq y \wedge \forall \{x',y'\} \in \mathcal{C} - \{x,y\}. x \neq x' \wedge x \neq y' \wedge y \neq x' \wedge y \neq y')$. Definition 5 states that every representative may be merged with at most one representative other than itself. The compression of any signature involves in some information loss which directly depends on the set $\mathcal{C}$ of the pairs of merged representatives. A threshold distance value ϵ is utilized which affects the compression process: Using a small value for ϵ leads to less information loss than using a greater value. We first compute the distance values among all representatives, and then sort the pairs $\{x,y\}$ in ascending order of the distances. If the distance of the considered pair does not exceed ϵ and both representatives of that pair are not utilized in $\mathcal{C}$ yet, then they are merged and their corresponding pair is added to the set. The algorithm is presented in Algorithm 1.

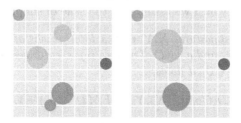

Fig. 3. An example signature (left) and its compressed signature (right) (Color figure online)

An example signature and its compressed signature are presented in Fig. 3 which depicts a 5-dimensional feature space with 2 positional and 3 color dimensions. Each representative is denoted by a circle whose diameter corresponds to the weight of that representative. The green and the orange representatives are merged, respectively, since the distance between them does not exceed the predefined threshold value. The blue and the rose-colored representatives are not

---

Algorithm 1. Signature Compression Algorithm

---

    input : signature $X : \mathbb{F} \rightarrow \mathbb{R}^{\geq 0}$ with $R_X = \{x_1, ..., x_n\}$,
              ground distance function $\delta : \mathbb{F} \times \mathbb{F} \rightarrow \mathbb{R}^{\geq 0}$,
              distance threshold $\epsilon \in \mathbb{R}^{\geq 0}$
    output: the set $\mathcal{C}$ of merged representatives

1 Create an empty distance matrix $D \in \mathbb{R}^{n \times n}$
2 **for** $1 \leq i \leq n$ **do**
3   **for** $1 \leq j < i$ **do**
4        $D_{i,j} \leftarrow \delta(x_i, x_j)$
5   **end**
6 **end**
7 Sort the values of D in ascending order, and let $\{i_k, j_k\}$ be the pair of representatives with the k-th distance value in this order
8 $\mathcal{C} \leftarrow \emptyset$
9 **for** $1 \leq k \leq \frac{n^2 - n}{2}$ **do**
10   **if** $D_{i_k, j_k} > \epsilon$ **then**
11   **break**
12   **end**
13    $\mathcal{C} \leftarrow \mathcal{C} \cup \{\{x_{i_k}, x_{j_k}\}\}$
14 **end**
15 **return** $\mathcal{C}$

---

merged with any other ones due to their high distance values. The computational time complexity of Algorithm 1 lies in $\mathcal{O}(n^2 \ log \ n)$ due to the sorting phase (line 8). Below, we analyze the relationship between the approximate EMD and the original EMD, and show that the approximate EMD value has an upper bound of $EMD(X, Y) + 1.5\,\epsilon$ for any 1-normalized signatures X, Y, i.e. signatures with the total weight of 1, which we assume for the sake of simplicity:

Theorem 1. *Given signatures X, Y with a total weight of 1, a metric ground distance $\delta : \mathbb{F} \times \mathbb{F} \rightarrow \mathbb{R}$, and a set $\mathcal{C}$ of merged representatives satisfying $\delta(x, y) \leq \epsilon \in \mathbb{R}^{\geq 0}$ for all $\{x, y\} \in \mathcal{C}$, it holds:*

$$EMD(Comp_{\mathcal{C}}(X), Y) \leq EMD(X, Y) + 1.5\,\epsilon$$

Proof. Let f be a minimum-cost flow from X to Y. Let us define a flow f' from $Comp_{\mathcal{C}}(X)$ to Y and show the feasibility of this flow:

$$f'(x, y) = \begin{cases} f(x', y) + f(x'', y) \text{ if } \exists x', x'' \in R_X. \{x', x''\} \in \mathcal{C} \wedge x = M(x', x'') \\ \qquad f(x, y) \qquad \text{ else} \end{cases}$$

– Non-negativity constraint: It follows by the non-negativity of f.

– Source constraint: For any given $x \in R_{Comp_C(X)}$, if $x = M(x', x'')$ holds for some $\{x', x''\} \in C$, then the following statement holds:

$$\sum_{y \in R_Y} f'(x, y) = \sum_{y \in R_Y} (f(x', y) + f(x'', y)) \leq X(x') + X(x'')$$

$$= MW(x', x'') = Comp_C(X)(x)$$

For any other case it holds:

$$\sum_{y \in R_Y} f'(x, y) = \sum_{y \in R_Y} f(x, y) \leq X(x) = Comp_C(X)(x)$$

– Target constraint: For any $y \in R_Y$, it holds:

$$\sum_{x \in R_{Comp_C(X)}} f'(x, y)$$

$$\stackrel{(1)}{=} \sum_{x \in R_X \cap R_{Comp_C(X)}} f(x, y) + \sum_{\{x', x''\} \in C} \left[(f'(M(x', x''), y) \right]$$

$$\stackrel{(2)}{=} \sum_{x \in R_X \cap R_{Comp_C(X)}} f(x, y) + \sum_{\{x', x''\} \in C} \left[f(x', y) + f(x'', y) \right]$$

$$\stackrel{(3)}{=} \sum_{x \in R_X} f(x, y) \stackrel{(4)}{\leq} Y(y)$$

(1) follows by the fact that any representative $x \in R_{Comp_C(X)}$ is either a representative which is not merged with any representative in R_X or it is an already merged representative. (3) holds by the following statement: $R_X = (R_X \cap R_{Comp_C(X)}) \cup \{x', x'' \in \mathbb{F} \mid \{x', x''\} \in C\}$.

– Total flow constraint:

$$\sum_{x \in R_{Comp_C(X)}} \sum_{y \in R_Y} f'(x, y)$$

$$\stackrel{(5)}{=} \sum_{x \in R_{Comp_C(X)} \cap R_X} \sum_{y \in R_Y} f(x, y) + \sum_{\{x', x''\} \in C} \sum_{y \in R_Y} \left[f'(M(x', x''), y) \right]$$

$$\stackrel{(6)}{=} \sum_{x \in R_{Comp_C(X)} \cap R_X} \sum_{y \in R_Y} f(x, y) + \sum_{\{x', x''\} \in C} \sum_{y \in R_Y} \left[f(x', y) + f(x'', y) \right]$$

$$\stackrel{(7)}{=} \sum_{x \in R_X} \sum_{y \in R_Y} f(x, y) \stackrel{(8)}{=} \min\{ \sum_{x \in R_X} X(x), \sum_{y \in R_Y} Y(y) \}$$

$$\stackrel{(9)}{=} \min\{ \sum_{x \in R_{Comp_C(X)}} Comp_C(X)(x), \sum_{y \in R_Y} Y(y) \}$$

The equations of (5) and (7) hold by the same reasons given for (1) and (3), respectively. In addition, the Eq. (8) follows from the total flow constraint of f, and (9) follows directly by Definition 5.

So far, we have seen that f' is a feasible flow between the compressed signature $Comp_{\mathcal{C}}(X)$ and Y. Now, we can derive an upper bound to the overall cost which is gathered by the utilization of the flow f':

$$\sum_{x \in R_{Comp_{\mathcal{C}}(X)}} \sum_{y \in R_Y} f'(x,y) \cdot \delta(x,y)$$

$$\overset{(10)}{=} \sum_{x \in R_X} \sum_{y \in R_Y} f(x,y) \cdot \delta(x,y) + \sum_{\{x',x''\} \in \mathcal{C}} \sum_{y \in R_Y} \Big[(f'(M(x',x''),y)$$
$$\cdot \delta(M(x',x''),y) - f(x',y) \cdot \delta(x',y) - f(x'',y) \cdot \delta(x'',y) \Big]$$

$$\overset{(11)}{\leq} \sum_{x \in R_X} \sum_{y \in R_Y} f(x,y) \cdot \delta(x,y) + \sum_{\{x',x''\} \in \mathcal{C}} \sum_{y \in R_Y} \Big[(f'(M(x',x''),y) \cdot (\delta(x',y)$$
$$+ \delta(x',x'')) - f(x',y) \cdot \delta(x',y) - f(x'',y) \cdot (\delta(x',y) - \delta(x',x'')) \Big]$$

$$\overset{(12)}{=} \sum_{x \in R_X} \sum_{y \in R_Y} f(x,y) \cdot \delta(x,y) + \sum_{\{x',x''\} \in \mathcal{C}} \sum_{y \in R_Y} \Big[(f'(M(x',x''),y) + f(x'',y))$$
$$\cdot \delta(x',x'') + (f'(M(x',x''),y) - f(x',y) - f(x'',y)) \cdot \delta(x',y) \Big]$$

$$\overset{(13)}{=} \sum_{x \in R_X} \sum_{y \in R_Y} f(x,y) \cdot \delta(x,y) + \sum_{\{x',x''\} \in \mathcal{C}} \sum_{y \in R_Y} (f(x',y) + 2 \cdot f(x'',y)) \cdot \delta(x',x'')$$

$$\overset{(14)}{\leq} \sum_{x \in R_X} \sum_{y \in R_Y} f(x,y) \cdot \delta(x,y) + \sum_{\{x',x''\} \in \mathcal{C}} (X(x') + 2 \cdot X(x'')) \cdot \delta(x',x'')$$

$$\overset{(15)}{\leq} \sum_{x \in R_X} \sum_{y \in R_Y} f(x,y) \cdot \delta(x,y) + \sum_{\{x',x''\} \in \mathcal{C}} (X(x') + 2 \cdot X(x'')) \cdot \epsilon$$

$$\overset{(16)}{\leq} \sum_{x \in R_X} \sum_{y \in R_Y} f(x,y) \cdot \delta(x,y) + 1.5 \epsilon \overset{(17)}{=} EMD(X,Y) + 1.5 \epsilon$$

The Eq. (10) holds for the same reason given for (1), and (11) holds by the triangle inequality. After reordering terms, (12) is acquired and (13) holds by the definition of f'. (14) holds by the source constraint of f, and (15) is gathered by applying $\delta(x',x'') \leq \epsilon$. The normalization of X, and w.l.o.g. the assumption $X(x'') \leq X(x')$ yield (16). As f' is a feasible flow from $Comp_{\mathcal{C}}(X)$ to Y, the min-cost flow results in an overall cost which does not exceed the cost determined by f', hence $EMD(Comp_{\mathcal{C}}(X),Y) \leq EMD(X,Y) + 1.5 \epsilon$ holds.   □

It is worth noting that the approximation error coming up in real world data experiments may differ from the approximation error given in Theorem 1. The empirical approximation error is observed to be much smaller, which we omit due to space limitations. By using the information of merging representatives in $\mathcal{C}$, it is possible to estimate an a-posteriori error bound more accurately by utilizing the following term in (14): $\sum_{\{x',x''\} \in \mathcal{C}} (X(x') + 2 \cdot X(x'')) \cdot \delta(x',x'')$. Furthermore, it is of particular importance to point out that the distance threshold ϵ is a trade-off parameter between accuracy and efficiency: The higher the parameter ϵ, the smaller the cardinality of the compressed signature, resulting in speeded up distance computation at the expense of a lower accuracy.

So far, we have seen how to define and perform approximate k-nn search utilizing compressed signatures. In the following section we present our extensive experimental results on 4 real world databases.

5 Experiments

In this section, we conduct experimental evaluation of approximate k-nn retrieval using signature compression with respect to accuracy and efficiency.

Experimental Setup. We perform k-nn queries on 4 real world image and video signature databases. We generated a video database named *PUBVID* consisting of 250,000 videos downloaded from vine.co, while *NDVINE* consists of 350,000 near-duplicate videos generated out of 3636 original videos from [13] by altering brightness, contrast, playback speed, resolution, frame order, adding overlay text, borders, and modifying content by frame deletion. The third database is the well-known medical image database *IRMA* [10,11] consisting of 12,675 anonymous radiographs. *UKBench* [12] is another database comprising 10,200 images, which we use as the last database in our experiments.

Each signature is generated by extracting features by sampling of 10,000 pixels of an image (or video) and then clustering them with k-means algorithm. Each representative comprises the dimensions of the relative spatial information of the corresponding pixel (x and y positional values), CIELAB color values, coarseness and contrast values. Thus, each feature is represented by a 7-dimensional feature vector for the image databases, and 8-dimensional feature vector for the video databases, which additionally includes the temporal dimension.

In order to achieve reliable experimental evaluation, we present the results denoting the averages over a query workload of 100. The experiments were conducted on a single-core 2.2 GHz machine equipped with Windows Server 2008 OS and 6 GB of main memory. Note that no parallelization of the implementations was performed. As ground distance between the features, we utilize the Manhattan distance, but our approach can be coupled with other distance functions, too. In addition, due to space limitation we choose k as 100 and we utilize our approach within the multi-step k-nn filter and refinement algorithm.

We evaluate the percentage of reduced representatives (centroids), the accuracy and the query processing time for different values of the distance threshold ϵ. The accuracy of the approximate k-nn retrieval for a query q is defined as:

$$accuracy(q) = \frac{|knn_{approx}(q) \cap knn_{exact}(q)|}{|knn_{exact}(q)|}$$

As mentioned before and as reported in [20,22], it is worth noting that Rubner filter yields much lower selectivity and significantly higher computational query time on signature databases than for the IM-Sig within a filter-and-refine

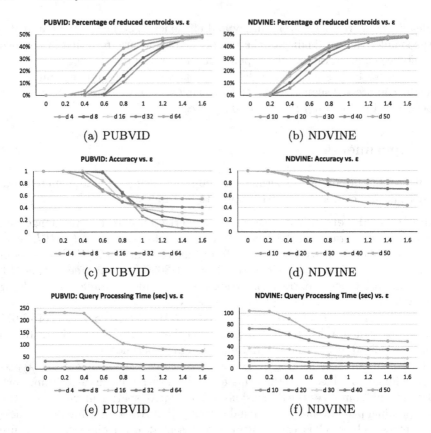

Fig. 4. Percentage of reduced centroids, average accuracy, and query processing time vs. ϵ for video databases PUBVID and NDVINE.

algorithm. Considering these verified results in the literature, we below investigate the efficiency and accuracy performance of our proposal by comparing with the results of the IM-Sig utilized in a filter-and-refine algorithm. The results corresponding to $\epsilon = 0$ denote the *complete* results which are gathered by applying the IM-Sig in filter step and the EMD in the refinement step of the optimal multi-step algorithm, i.e. no signature compression occurs. For the case $\epsilon > 0$, the database signatures are compressed by the corresponding percentage of the centroids given in the figures within a precomputation step.

Experimental Results. Figure 4 depicts the results for the video databases PUBVID and NDVINE, and Fig. 5 shows the results for the image databases IRMA and UKBench. For $\epsilon = 0$, no compression occurs, yielding an accurate retrieval with no false drops. Hence, the number of reduced centroids remains as 0 and the accuracy is observed as 1. Obviously, the higher we choose ϵ, the higher the number of reduced centroids are. For high values of ϵ, the percentage of reduced centroids approaches 50%, since each centroid is merged with another centroid in its ϵ-neighborhood, halving the number of centroids.

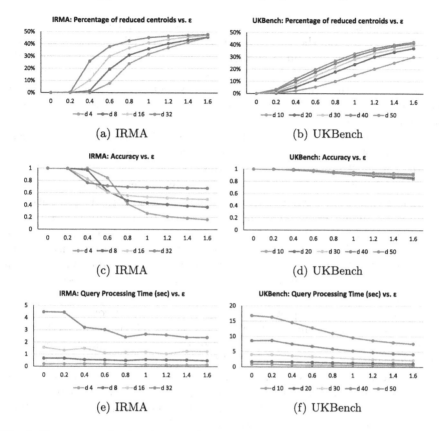

Fig. 5. Percentage of reduced centroids, average accuracy, and query processing time vs. ϵ for image databases IRMA and UKBench.

The number of reduced centroids for the same values of ϵ differ between the databases, suggesting that the databases have a different distribution of the centroids and different inter-centroid distances. The percentage of reduced centroids is reflected in the accuracy: The less centroids we have, the worse the accuracy becomes, since a smaller number of centroids yields a less expressive representation of the underlying video, impairing the discriminability between the videos. For smaller signature sizes, the accuracy suffers more than for higher signature sizes, since the information loss induced by a reduction of the centroids through compression is higher when the signature size is small.

The accuracy for NDVINE is observed to remain above the accuracy of 0.7 for the signature dimensions 20-50. This can be explained by the fact that PUB-VID consists of a random collection of videos where for a given query the most similar videos are not significantly more similar than the other videos. Hence, the distances to the nearest neighbors do not differ substantially from the distances to other videos. Thus, the set of nearest neighbors is unrobust to the approximation error in the EMD resulting from the compression, causing the nearest neighbors with respect to the EMD to be quite different from the nearest neighbors regarding the approximate EMD. In NDVINE, on the other hand, the nearest neighbors of a query mostly comprise near-duplicate versions of the query. Here, the distances to the nearest neighbors are significantly smaller than the distances to other videos. As a result, the set of nearest neighbors is more robust to the approximation error induced by the compression, resulting in a comparatively high accuracy.

A similar observation can be made for IRMA and UKBench: IRMA consists of a random collection of images, resulting in the same kind of unrobustness to compression as PUBVID, deteriorating the accuracy. However, UKBench contains categories of images, each category comprising an image of the same object from four different perspectives. Hence, the nearest neighbors are most likely to contain these similar images, causing the distances to the nearest neighbors to be significantly smaller than the distances to other images, again making the set of nearest neighbors more robust to the approximation error induced by the compression, resulting in high accuracy lying above 0.8. In addition, UKBench is a small database, which also improves the robustness of the nearest neighbors to approximation errors.

For all databases, the query processing time for $\epsilon = 0$ corresponds to the elapsed query time for accurate retrieval. As expected, a higher ϵ causes the query time to decrease, which can be attributed to the reduced number of centroids, which alleviates the query time required for a single distance computation. For higher signature sizes, this effect is stronger than for smaller signature sizes, since the relative number of reduced centroids corresponds to a higher absolute number of reduced centroids, which has an amplified effect on the query processing time due to the super-cubic time complexity of the EMD.

Summary of Experiments. The results yielded by our experiments show that approximate k-nn search through signature compression is a reasonable method for speeding up the query processing time while ensuring accurate results, especially for high signature sizes. The choice of the parameter ϵ enables us to control the trade-off between accuracy and efficiency, allowing us to ensure a certain expected accuracy while reducing the query processing time cost.

6   Conclusion

The high computational time complexity of the well-known Earth Mover's Distance (EMD) hinders its application in large-scale datasets where the user is rather interested in receiving an answer from the underlying application within a short period of time than requesting an exact and complete query result set. To overcome this limitation, in this paper, we propose to improve the efficiency of the query processing with the EMD on signature databases by utilizing signature compression approximations. To this end, we introduce an efficient signature compression algorithm to alleviate query computation cost, and theoretically analyze the approximation-based EMD. On top of this, the comprehensive experiments show that our approach can be successfully applied to 4 different kinds of real world datasets, pointing out the accuracy and efficiency of the proposal and sheding light on the applicability of our approach to signature databases.

Acknowledgments. This work is funded by DFG grant SE 1039/7-1.

References

1. Agrawal, R., Faloutsos, C., Swami, A.: Efficient similarity search in sequence databases. In: Lomet, David B. (ed.) FODO 1993. LNCS, vol. 730. Springer, Heidelberg (1993)
2. Assent, I., Kremer, H., Seidl, T.: Speeding up complex video copy detection queries. In: Kitagawa, H., Ishikawa, Y., Li, Q., Watanabe, C. (eds.) DASFAA 2010. LNCS, vol. 5981, pp. 307–321. Springer, Heidelberg (2010)
3. Assent, I., Wenning, A., Seidl, T.: Approximation techniques for indexing the earth mover's distance in multimedia databases. In: ICDE, p. 11 (2006)
4. Barrio, P., Gravano, L., Develder, C.: Ranking deep web text collections for scalable information extraction. In: CIKM, pp. 153–162 (2015)
5. Faloutsos, C., Ranganathan, M., Manolopoulos, Y.: Fast subsequence matching in time-series databases. In: SIGMOD, vol. 23, no. 2, pp. 419–429 (1994)
6. Hillier, F., Lieberman, G.: Introduction to Linear Programming. McGraw-Hill, New York (1990)
7. Houle, M.E., Ma, X., Nett, M., Oria, V.: Dimensional testing for multi-step similarity search. In: ICDM, pp. 299–308 (2012)
8. Korn, F., Sidiropoulos, N., Faloutsos, C., Siegel, E.L., Protopapas, Z.: Fast nearest neighbor search in medical image databases. In: VLDB, pp. 215–226 (1996)
9. Kriegel, H.-P., Kröger, P., Kunath, P., Renz, M.: Generalizing the optimality of multi-step k-nearest neighbor query processing. In: Papadias, D., Zhang, D., Kollios, G. (eds.) SSTD 2007. LNCS, vol. 4605, pp. 75–92. Springer, Heidelberg (2007)
10. Lehmann, T., et al.: Content-based image retrieval in medical applications. Methods Inf. Med. **43**(4), 354–361 (2004)
11. Lehmann, T., et al.: IRMA project site (2009). http://www.irma-project.org/datasets
12. Nister, D., Stewenius, H.: Scalable recognition with a vocabulary tree. In: CVPR, pp. 2161–2168 (2006)
13. Redi, M., OHare, N., Schifanella, R., Trevisiol, M., Jaimes, A.: 6 seconds of sound and vision: Creativity in micro-videos. In: CVPR, pp. 4272–4279 (2014)
14. Rubner, Y., Tomasi, C., Guibas, L.: A metric for distributions with applications to image databases. In: ICCV, pp. 59–66 (1998)
15. Ruttenberg, B.E., Singh, A.K.: Indexing the earth mover's distance using normal distributions. PVLDB **5**(3), 205–216 (2011)
16. Seidl, T., Kriegel, H.: Optimal multi-step k-nearest neighbor search. In: SIGMOD, pp. 154–165 (1998)
17. Solomon, J., Rustamov, R., Guibas, L., Butscher, A.: Earth mover's distances on discrete surfaces. ACM Trans. Graph. **33**(4), 67:1–67:12 (2014)
18. Tang, Y., Cai, L.H., Mamoulis, N., Cheng, R.: Earth mover's distance based similarity search at scale. PVLDB **7**(4), 313–324 (2013)

19. Uysal, M.S., Beecks, C., Sabinasz, D., Seidl, T.: FELICITY: A flexible video similarity search framework using the earth mover's distance. In: Amato, G., Connor, R., Falchi, F., Gennaro, C. (eds.) Similarity Search and Applications. LNCS, vol. 9371, pp. 347–350. Springer, Heidelberg (2015)

20. Uysal, M.S., Beecks, C., Schmücking, J., Seidl, T.: Efficient filter approximation using the Earth Mover's Distance in very large multimedia databases with feature signatures. In: CIKM, pp. 979–988 (2014)

21. Uysal, M.S., Beecks, C., Schmücking, J., Seidl, T.: Efficient similarity search in scientific databases with feature signatures. In: SSDBM, pp. 30:1–30:12 (2015)

22. Uysal, M.S., Beecks, C., Seidl, T.: On efficient content-based near-duplicate video detection. In: CBMI, pp. 1–6 (2015)

23. Uysal, M.S., et al.: Large-scale efficient and effective video similarity search. In: LSDS-IR@CIKM, pp. 3–8 (2015)

24. Wichterich, M., Assent, I., et al.: Efficient emd-based similarity search in multimedia databases via flexible dimensionality reduction. In: SIGMOD, pp. 199–212 (2008)

25. Xu, J., Zhang, Z., et al.: Efficient and effective similarity search over probabilistic data based on earth mover's distance. PVLDB 3(1), 758–769 (2010)

Effective Similarity Search on Indoor Moving-Object Trajectories

Peiquan Jin[1]([⊠]), Tong Cui[1], Qian Wang[1], and Christian S. Jensen[2]

[1] University of Science and Technology of China, Hefei, China
jpq@ustc.edu.cn
[2] Department of Computer Science, Aalborg University, Aalborg, Denmark

Abstract. In this paper, we propose a new approach to measuring the similarity among indoor moving-object trajectories. Particularly, we propose to measure indoor trajectory similarity based on spatial similarity and semantic pattern similarity. For spatial similarity, we propose to detect the critical points in trajectories and then use them to determine spatial similarity. This approach can lower the computational costs of similarity search. Moreover, it helps achieve a more effective measure of spatial similarity because it removes noisy points. For semantic pattern similarity, we propose to construct a hierarchical semantic pattern to capture the semantics of trajectories. This method makes it possible to capture the implicit semantic similarity among different semantic labels of locations, and enables more meaningful measures of semantic similarity among indoor trajectories. We conduct experiments on indoor trajectories, comparing our proposal with several popular methods. The results suggest that our proposal is effective and represents an improvement over existing methods.

Keywords: Indoor space · Similarity search · Trajectory

1 Introduction

Most people spend most of their time in indoor spaces, such as in homes, office buildings, shopping malls, and airports. The increasingly deployment of indoor positioning technologies offers the possibility to obtain the locations and trajectories of people in indoor spaces [4]. Therefore, there are increasing needs for analyzing the trajectories of indoor moving objects. For instance, it is helpful to find potential shopping patterns by identifying similar paths of customers in a shopping mall [15]. This development calls for the research on similarity search in indoor spaces.

Similarity search is an important issue in many applications. A key objective is to define efficient methods to measure the trajectory similarity for indoor moving objects. Previous efforts on trajectory similarity search focus on outdoor spaces, e.g., Euclidean and road-network spaces [7–10]. However, indoor spaces have unique features compared with outdoor spaces. First, in indoor spaces, we cannot use GPS for positioning. Instead, indoor locations have to be determined through a deployment graph of readers like RFID readers or Wi-Fi adapters [4]. Due to the small-area property of indoor spaces, indoor positions reported by readers are much closer than outdoor locations, yielding a polyline with dense points and introducing many noisy points in trajectory

S.B. Navathe et al. (Eds.): DASFAA 2016, Part II, LNCS 9643, pp. 181–197, 2016.
DOI: 10.1007/978-3-319-32049-6_12

similarity measurement. Second, a typical indoor space consists of different floors and many indoor elements like rooms, doors, hallways, stairs, and elevators. Thus, indoor spaces are actually constrained three-dimensional spaces. Therefore, we need new techniques to compute indoor distances when measuring trajectory similarity.

In this paper, we propose a new approach to measuring trajectory similarity for indoor moving objects. We consider both spatial and semantic similarity between trajectories, and we make the following contributions:

(1) We propose to detect *critical points* in a trajectory and then define spatial similarity based on critical points (Sect. 3.1).
(2) We propose to use a hierarchical categorization of indoor locations to capture the semantic patterns of trajectories, and we define hierarchical semantic similarity for indoor moving objects (Sect. 3.2).
(3) We compare our proposal with the *LCSS*-based and *Edit-Distance*-based methods using synthetic indoor trajectory data. The results suggest that our proposal is effective and improves on existing proposals (Sect. 4).

2 Problem Statement

2.1 Indoor Space

Many models of indoor space have been proposed [1–4]. In this paper, we define indoor space by the following symbolic model

Definition 1 (Indoor Space). An indoor space is represented as a triple:

$$IndoorSpace = (Cell, Sensor, Deployment)$$

Here, *Cell* is a set of cells in the indoor space. According to previous researches, an indoor space can be partitioned into cells [3]. *Sensor* is a set of positioning sensors deployed in the indoor space. Typical sensors are RFID readers, Bluetooth detectors, and Wi-Fi signal receivers. *Deployment* records the placement information of the sensors in the indoor space.   □

In real applications, rooms can be regarded as cells and sensors are usually used to identify cells in indoor space, e.g., to identify shops in a shopping mall. Thus, the *Deployment* information of sensors can be pre-determined and maintained in a database.

In order to introduce semantics into the model, we assign semantic labels to each sensor. As a result, the set *Sensor* can be represented as follows.

$$Sensor = \{s | s = (sensorID, location, label)\}$$

The *label* of a sensor provides descriptions on thematic attributes of the location identified by *sensorID*. For instance, we can use labels like "*elevator*" and "*stair*" to indicate the functions of the cell identified by a sensor. We can also use other labels like "*Starbucks*" and "*Burger King*" to annotate semantic features of the cell. The location of a sensor is a three-dimensional coordinate (x, y, z), which reflects the relative position of the sensor inside the indoor space where the sensor is deployed. In real applications, indoor maps are usually designed by AutoCAD [22]. Thus, if we import an indoor map into a database system, we can simply use the coordinates and floors in the map to represent the locations of sensors.

2.2 Similarity Search in Indoor Spaces

In outdoor spaces, a trajectory is a series of GPS locations, while a trajectory in indoor spaces is typically a series of sensor identifiers. Thus, we first define the indoor location of a moving object as follows:

Definition 2 (Indoor Location). An indoor location LOC of a moving object mo is defined as LOC_{mo}:

$$LOC_{mo} = \{p | p = (s, [t_s, t_e]) \wedge (t_s < t_e) \wedge s \in Sensor\},$$

Here, s is a sensor, mo is the object identifier, t_s is the instant that the object enters the sensor's range and t_e is the instant that the object leaves the sensor range. $\square$

Due to the special properties of some kinds of sensors, an indoor moving object may be detected by two or more sensors simultaneously. Thus, we need to find out the most exact location from a set of sensors' signals. This is an important issue in indoor location-based applications, which has been studied in many previous works on indoor localization [23, 24]. We assume that an indoor moving object has only one indoor location at each time instant. Then, we define the indoor trajectory for a moving object.

Definition 3 (Indoor Trajectory). An indoor trajectory TR of a moving object mo is defined as a sequence of indoor locations of mo:

$$TR_{mo} = \{\langle p_1, p_2, \ldots p_n, \rangle \mid (\forall p_i)(p_i \in LOC_{mo} \wedge (p_i.t_e < p_{i+1}.t_s))\}$$

$\square$

Definition 4 (Indoor Trajectory Similarity Search). Given a set of indoor trajectories T, a query trajectory q, and an integer k, an *indoor trajectory similarity search* retrieves a set $S \subseteq T$ that consists of k trajectories, such that:

$$\forall x \in S \wedge \forall y \in T - S, SimTra(x, q) \geq SimTra(y, q)$$

$\square$

Here, $SimTra(x, q)$ returns the similarity between x and q.

3   Indoor Trajectory Similarity

Given two indoor trajectories x and y, their similarity is defined as follows.

$$Sim(x, y) = a \cdot spaSim(x, y) + (1 - a) \cdot semSim(x, y) \qquad (3.1)$$

Here, a is the weight reflecting the importance of the spatial similarity. Next, we give the details of measuring spatial similarity and semantic similarity.

3.1   Critical Point Based Spatial Similarity

We propose a critical-point-based solution for the measurement of spatial similarity for indoor trajectories. The idea of critical points can be illustrated by Fig. 1. As different points in a trajectory usually have different impacts in describing the most important features of the trajectory, e.g., direction, we can detect those important points, called *critical points*, and remove other non-critical points to make the original trajectory more simple and clear. As shown in Fig. 1(b), after constructing the critical-point-based trajectory for the original one in Fig. 1(a), the trajectory contains fewer points (original 23 points, only 8 critical points after transformation).

The benefits of critical points are two-fold. First, the computation time of similarity search can be reduced. Second, this design can be more effective than traditional methods like *Longest Common* Sub-sequence (*LCSS*) [15] and *Edit Distance* [9]. For example, the two trajectories shown in Fig. 1 will be regarded as dissimilar according to either the *LCSS* or *Edit Distance* approach. Note that the *LCSS* approach uses the longest common points to measure similarity (in this example, there are only 4 longest common points between the trajectories.), and the *Edit Distance* approach uses the count of edits transforming Fig. 1(a) to (b) to measure the similarity (in this example, fifteen edits have to be performed.). However, we can see from Fig. 1 that these two trajectories have very similar shapes and moving patterns (e.g., crossing through the hallway on a floor).

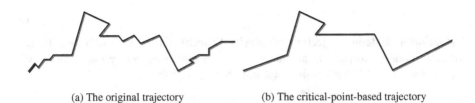

(a) The original trajectory   (b) The critical-point-based trajectory

Fig. 1.   The idea of critical points

One previous approach similar to the critical-point-based trajectory reduction is the Douglas-Peucker (DP) algorithm [25]. However, the DP algorithm is not effective in reducing complex indoor trajectories. For example, in Fig. 2, the trajectory from A to H will be reduced into A-E-H according to the DP algorithm, which omits some critical

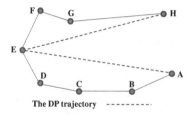

Fig. 2. The ineffectiveness of the DP algorithm in reducing indoor trajectories

points such as D and F. On the other hand, the trajectory of Fig. 2 is very common in indoor space, where each point can represent one room in a floor.

3.1.1 Critical Point

Definition 5 (Position Distance). Let p, p_c, and p_1 be indoor locations. The Euclidean distance from p to the line segment $\overrightarrow{p_c p_1}$ is called the *position distance* from p to p_c, denoted as $PD_{p \rightarrow p_c}$.   □

Definition 6 (Position Angle). Let p_1, p_2, and p_3 be indoor locations. The angle between the line segment $\overrightarrow{p_1 p_2}$ and $\overrightarrow{p_2 p_3}$ is called the *position angle* of p_2, denoted as PA_{p_2} □

Figure 3 shows an example of *position distance (PD)* and *position angle (PA)*. Generally, we can find that a large position distance or a large position angle implies a substantial change on the moving direction (in Fig. 3 the referenced direction is $\overrightarrow{p_c p_1}$). Thus we can utilize the *position distance (PD)* and *position angle (PA)* to detect critical points in an indoor trajectory.

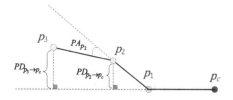

Fig. 3. An example of *position distance (PD)* and *position angle (PA)*

Definition 7 (Critical Point). Given an indoor trajectory x, an angle threshold θ and a distance threshold d, an indoor location p in x is called a *critical point* of x such that p satisfies one of the following criteria:

(1) p is the start or end point of x.
(2) p is an inter-connection location, e.g., an "*elevator*" room or a "*stair*" room, which can be identified through the labels in p.

(3) $PA_p > \theta$

(4) $PD_{p \to p_c} > d$, where p_c is the most-recently critical point preceding p in x.

□

The third condition is introduced to detect those points that lead to a substantial change on the movement direction. The last condition is used to find those points that gradually change the moving direction. We also identify inter-connection locations as critical points, e.g., elevator rooms and stair rooms, because those locations typically indicate dramatic changes on the moving direction. The initial critical point needed for computing *position distance (PD)* is the starting point of the trajectory.

Figure 4 shows the algorithm for detecting critical points of indoor trajectories.

Algorithm *CriticalPoint_Detection (x, d, θ)*

Input: $x=\{p_1, p_2, ..., p_n\}$ is an indoor trajectory, d is the distance threshold, and θ is the angle threshold.

Output: *CP*: the set of critical points.

1: $CP \leftarrow \{p_1\}$;
2: $c \leftarrow p_1$; // *c is the newest critical point, which is used for computing $OD_{p_i \to c}$*
3: **for** $i = 2$ to $n-1$ **do**
4:      **if** (*p_i is an interconnection location*) or *(OA_{p_i} > θ)* or *(OD_{p_i \to c} > d)* **then**
5:          $CP \leftarrow CP \cup \{p_i\}$;
6:          $c \leftarrow p_i$; // *p_i becomes the newest critical point for computing $OD_{p_i \to c}$*
7:      **end if**
8:      $CP \leftarrow CP \cup \{p_n\}$;
9: **end for**
10: **return** *CP*;
End *CriticalPoint_Detection*

Fig. 4. The algorithm to detect critical points

3.1.2   Indoor Distance

We now define the indoor distance. For each trajectory x, we can acquire its critical points using the algorithm in Fig. 4. Each two adjacent points form a line segment; then the trajectory can be represented as a sequence of three-dimensional line segments composed with these critical points.

Given two line segment L_i and L_j, in an indoor space (x, y, z), we suppose that L_j is on the (x, y) plane and L_i' denote the projection of L_i on the (x, y) plane. Then, we adapt the distance function defined in [18], which is proposed for two-dimensional Euclidean space, for three-dimensional indoor spaces. As illustrated in Fig. 5, the indoor distance between two line segments L_i and L_j is composed of four distances, namely the *perpendicular distance, horizontal distance, projection distance,* and *shifting distance,* which are defined as follows and based on the symbols shown in Fig. 5.

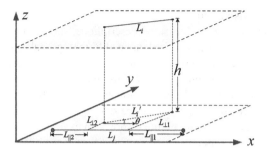

Fig. 5. Illustration of the indoor distance between two line segments

(1)   The *perpendicular distance*: $d_\perp = \frac{L_{\perp 1}^2 + L_{\perp 2}^2}{L_{\perp 1} + L_{\perp 2}}$

(2)   The *horizontal distance*: $d_\| = \min\left(L_{\|1}, L_{\|2}\right)$

(3)   The *shifting distance*: $d_\theta = |\,L_j\,| \times \sin(\theta)$

(4)   The *projection distance*: $d_h = h$

Then we can compute the distance between L_i and L_j by (3.2):

$$dist(L_i, L_j) = w_\perp \times d_\perp + w_\| \times d_\| + w_\theta \times d_\theta + w_h \times d_h \qquad (3.2)$$

This distance function can accurately reflect the differences between the two segments, because they both consider the lengths and the directions of the segments. We can adjust the four weighting parameters, namely $w_\perp, w_\|$, w_θ, and w_h, according to real applications. In our experiment, we simply set these parameters to the same value.

Then, given two indoor trajectories x and y, whose number of locations are m and n, respectively, we can compute the indoor distance between x and y by (3.3).

$$distTR(x, y) = \begin{array}{c} dist(x.L_1, y.L_1) + \min\{dist(x - \{x.L_1\}, y - \{y.L_1\}), \\ dist(x - \{x.L_1\}, y), dist(x, y - \{y.L_1\}) \end{array} \qquad (3.3)$$

Finally, we define the spatial similarity between two trajectories x and y by (3.4). The $\frac{1}{1 + distTR(x,y)}$ part is used to reflect the effect of indoor distance, where a farther distance means less similarity. The $\frac{\min(|x|,|y|)}{\max(|x|,|y|)}$ part is used to reflect the influence of the length of trajectories, which is based on the assumption that if the lengths of two trajectories are much different they are less similar.

$$spaSim(x, y) = \frac{1}{1 + distTR(x, y)} \times \frac{\min(|x|, |y|)}{\max(|x|, |y|)} \qquad (3.4)$$

3.2   Hierarchical Semantic Similarity

We present the semantic classification tree, and then propose a new method to compute the hierarchical semantic similarity between trajectories.

3.2.1 Semantic Classification Tree

Generally, each indoor location has some descriptions, such as restaurant, movie theater, KTV, etc. These descriptions can have different granularities and levels. For example, the coarse-grained description by the label *"restaurant"* can be further explained using some fine-grained descriptions like *"Chinese restaurant"*, *"Western restaurant"*, etc. Therefore, we build a semantic classification tree to represent the hierarchical semantic relationship between the descriptions with different granularities. We denote such a tree as *SC_Tree*. Figure 6 shows an example of *SC_Tree*. In this tree, location classifications of different levels means different granularity of partition. Each non-leaf node in *SC_Tree* represents a category of locations, and each leaf node represents a semantic label of a location.

Fig. 6. An example of *SC_Tree*

Given a trajectory, we can find its corresponding leaf nodes in *SC_Tree*. For example, suppose that we have the following indoor trajectory (here each p_i represents a *sensorID* and a geographical coordinate, and for simplicity we omit the time information), the corresponding leaf nodes in *SC_Tree* in Fig. 6 are $p1, p2, ..., p6$.

$$TRS = \langle (p_1, Nike), (p_2, Adidas), (p_3, LI - NING), (p_4, KFC), (p_5, Starbucks), (p_6, Lancome) \rangle$$

3.2.2 Hierarchical Semantic Patterns

Definition 8 (Hierarchical Semantic Pattern). Assume that *labelS* represents the labels of the leaf nodes in *SC_Tree*, *classS* is the set of labels of the non-leaf nodes, and H is the height of *SC_Tree*, the *hierarchical semantic pattern MP* of an indoor trajectory x is defined as $MP(x)$, which is a set of $MP_k(x)$, where $0 \leq k \leq H$. $MP_k(x)$ represents the semantic pattern of x at the k-th level in *SC_Tree*, which is defined as follows:

$$MP_k(x) = \begin{cases} \{\langle a_1, a_2, \ldots a_{|x|} \rangle | \forall i \in [1, |x|], a_i = x.p_i.label\}, & k = H \\ \{\langle a_1, a_2, \ldots a_{|MP_k(x)|} \rangle | \forall i \in [1, |MP_k(x)|], a_i \in (labelS \cup classS)\}, & k < H \end{cases}$$

□

Fig. 7. An example for hierarchical moving patterns

For example, consider our trajectory *TRS*, its hierarchical semantic pattern is shown in Fig. 7. The lowest level of semantic pattern, i.e., $MP_3(TRS)$, is the sequence of semantic labels in *TRS*, while $MP_0(TRS), MP_1(TRS), MP_2(TRS)$ are sequences of non-leaf level labels.

3.2.3   Semantic Similarity

Given two indoor trajectory *x* and *y*, we first construct their hierarchical semantic patterns. Next, for the semantic pattern at each level of *SC_Tree*, we use the *LCSS* method [8] to calculate the similarity between the semantic patterns.

$$simMP(MP_i(x), MP_i(y)) = \frac{|LCSS(MP_i(x), MP_i(y))|}{min(|MP_i(x)|, |MP_i(y)|)} \qquad (3.5)$$

Then, the semantic similarity of two trajectories *T* and *R* is defined.

$$semSim = \sum_{i=0}^{H} simMP(MP_i(x), MP_i(y)) \times w_i \qquad (3.6)$$

Here, w_i is the importance of the similarity of *i*th semantic pattern to the total semantic similarity. This allows us to decrease the users' interests gradually from the leaf nodes up to the root, so that upper patterns contribute less to the users' interests. The following definition of w_i reflects this observation.

$$w_i = \frac{1}{\sum_{k=0}^{H} 2^k} \times 2^i \qquad (3.7)$$

4   Experiment

We explain the experimental settings and then discuss the detailed results, including comparisons with several existing methods with respect to time performance and effectiveness.

4.1 Experimental Setup

Data Set. We simulate a shopping mall with two floors and generate indoor trajectory data using the indoor data generator *IndoorSTG* [20]. *IndoorSTG* can simulate different indoor spaces consisting of elements such as rooms, doors, corridors, stairs, elevators, and virtual positioning devices like RFID and Bluetooth readers. Besides, it can generate semantic-based trajectories for indoor moving objects. The simulated shopping mall has 94 indoor locations that represent different types of indoor elements such as rooms, elevators, corridors, and stairs. We simulate moving objects in such an indoor space during a time period of 20 days to generate different sets of trajectory data. The number of moving objects is set to 250, 500, 1,000, and 2,000, respectively, which yields 5,000, 10,000, 20,000, and 40,000 trajectories when the number of moving objects is varied from 250 to 2,000. We manually add the semantic label to each of the 94 indoor locations to represent the semantics. In the experiments, we use 20,000 trajectories as the default data set and randomly select 100 trajectories to serve as queries.

Metrics. We focus mainly on the time performance and effectiveness of similarity search. For time performance, we record the overall run time for processing queries when varying the number of trajectories. In order to evaluate the effectiveness, we use two metrics. The first is the *average distance* between the query trajectories and the results. The second is the *precision* of similarity search. Since we are using simulated data, it is not feasible to evaluate the precision from users' perspective. Instead, we utilize the idea of *Cumulative Gain (CG)* [21] to define precision. *CG* is well-known in information retrieval as part of the commonly-used metric *Normalized Discounted Cumulative Gain (NDCG)*, which is used to evaluate the relevance of search results.

Let q be a query trajectory and let R be the set of returned trajectories for q. We define the semantic relevance between q and a returned trajectory $r \in R$ as follows.

$$rel(r, q) = \sum_{i=1}^{\min(|r|,|q|)} \frac{1}{d(r_i, q_i)} \qquad (4.1)$$

Here, $\frac{1}{d(r_i,q_i)}$ returns the relevance between two locations from q and r. And these two locations have the same sequence in q and r. When computing $rel(r, q)$, we first get the sum of the relevance between each location pair from q and r, and then map the sum into a value in the unit interval [0, 1]. We denote this normalized $rel(r, q)$ as *normalized_rel(r, q)*. Considering that q and r may contain different number of locations, we use the shorter trajectory between q and r as the basic referential trajectory.

Given two locations r_1 and q_1, $d(r_1, q_1)$ returns the distance between the labels of the two locations in *SC_Tree*. As *SC_Tree* is constructed according to the semantics of locations, the distance between two nodes in *SC_Tree* is able to represent the semantic difference between the two locations.

If r_1 and q_1 have the same label then $d(r_1, q_1) = 1$. Otherwise, $d(r_1, q_1)$ is the number of nodes in the path from r_1 to q_1 in *SC_Tree* (including r_1 and q_1). For instance, in the *SC_Tree* shown in Fig. 6, $d(p_1, p_3) = 3$, while $d(p_1, p_4) = 7$.

As *SC_Tree* is organized in terms of the hierarchy of semantic classification, the closer two nodes are in *SC_Tree* the more similar their semantics are.

As a result, assume that q is a query trajectory and $R = \{r_1, r_2, \cdots, r_n\}$ is the set of returned trajectories for q, we give the computation of average distance and precision as follows:

$$average\_distance(q, R) = \frac{\sum_{0 \leq i \leq |R|} distTR(r_i, q)}{|R|} \qquad (4.2)$$

$$precision(q, R) = \frac{\sum_{0 \leq i \leq |R|} normalized\_rel(r_i, q)}{|R|} \qquad (4.3)$$

Here, $distTR(r_i, q)$ is defined in Formula (3.3), and $normalizde\_rel(r_i, q)$ is the normalized value of Formula (4.1) in the unit interval [0, 1]. In the experiments, we compute the mean average distances and precisions of 100 query trajectories as the final results.

Comparative Methods. We implement two classical algorithms as the baseline methods for measuring spatial similarity, *LCSS* (*Longest Common Subsequence*) [15] and *Edit Distance* [9]. They are denoted *LCSS_Indoor* and *ED_Indoor* in the following. Regarding semantic similarity, we use the *Cosine Similarity* [21] as the baseline method, which is one of the most popular methods of evaluating document similarity in information retrieval.

We call our method **SIT** (*Similarity of Indoor Trajectories*), and we consider four variations of *SIT* in the experiments:

(1) *SIT_S* only considers spatial similarity and without introducing critical points.
(2) *SIT_SCP* only considers spatial similarity, but uses critical points.
(3) *SIT_SS* considers both spatial and hierarchical semantic similarities, but does not use critical points.
(4) *SIT_SSCP* considers both spatial and semantic similarities and uses critical points. For semantic similarity, it considers all the points in the trajectories.

4.2 Results

In the following, we present the results of the evaluation. For each experiment, we choose 100 trajectories as query trajectories from the dataset trajectories generated by *IndoorSTG*, and the results shown in the following are average results.

4.2.1 Precision

Figure 8 shows the precision on semantic similarity of all the methods. In this experiment, we set the parameter a to 0.5 and vary k from 5 to 30. Obviously, our methods *SIT_SS* and *SIT_SSCP* have much higher precision than *LCSS_indoor* and *ED_indoor*,

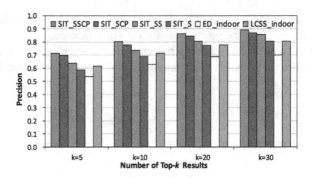

Fig. 8. Precision

which suggest that our proposal is effective. *SIT_S* and *SIT_SCP* perform poorly because they do not consider the influence of semantics.

In summary, *SIT_SSCP* and *SIT_SS* are effective for indoor trajectory similarity search as they consider both spatial and semantic similarities.

4.2.2 Average Distance
In this section, we evaluate the effectiveness of our proposal for spatial similarity search by calculating the average distance between the query trajectory and the trajectories returned.

As shown in Fig. 9, *SIT_S* has the smallest average distance, while *LCSS_indoor* and *ED_indoor* get the worst performance. This is mainly because *SIT_S* considers the features of indoor spaces and uses indoor distance. When considering critical points, *SIT_SCP* slightly larger the average distance compared with *SIT_S*. In fact, *SIT_SCP* can be regarded as an approximation of *SIT_S*, because it only considers the critical points of trajectories.

We can adjust the parameter *a* in Formula (3.1) to adapt special similarity-search needs of applications. For example, for user behavior analysis in shopping malls, semantic similarity may be important; thus, we can use a small value for parameter *a*. On the other hand, for public emergency monitoring in metro stations, spatial similarity could be more important, and we can use a high value for parameter *a*.

We can also see in Fig. 9 that *SIT_SSCP* and *SIT_SS* have higher average distances than *SIT_S* and *SIT_SCP*, which do not consider semantic similarity. Note that the returned set of top-*k* trajectories is influenced when we add semantic similarity into the computation of relevance. Therefore, the semantic similarity of the results will increase, but the spatial similarity (average distance) will decrease.

4.2.3 Time Performance
Figure 10 compares the running times of all the methods when executing 100 similarity searches on various number of trajectories. We vary the number of trajectories from 5,000 to 40,000, and calculate the time between issuing the queries and returning the ranked results.

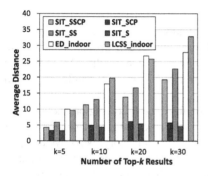

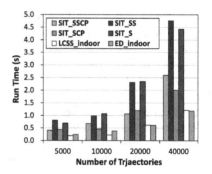

Fig. 9. Average distance   **Fig. 10.** Run Time

As Fig. 10 shows, the critical-point-based methods, including *SIT_SSCP* and *SIT_SCP*, performs faster than the methods that consider all points, namely *SIT_SS* and *SIT_S*. In particular, when increasing the number of trajectories, the benefit of critical points becomes more notable. On average, the critical-point-based methods are able to reduce the running times to about 50 % percent of run time compared with the methods without using critical points.

Both *LCSS_indoor* and *ED_indoor* get good time performance in the experiment, owing to their simple computation on distances and similarity measurement. However, as we have shown, they are not suitable for practical applications because of their poor effectiveness on both spatial similarity and semantic similarity.

4.2.4   Impact of Parameter a

Figure 11 shows the influence of parameter *a* on similarity measurement. Here, the fundamental algorithm is *SIT_SSCP*, and *k* is set to 20. This parameter is used to balance the impact of spatial and semantic similarity in the similarity evaluation. As shown in the figure, with the increase of *a*, the average spatial similarity decreases while the precision increases. This is simply because a large *a* means we give spatial similarity more weights in the computation of similarity. In real applications, we can tune this parameter to make it suit for the needs.

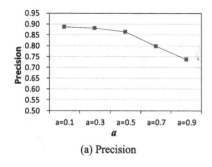

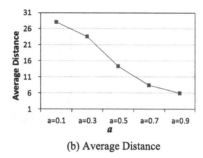

(a) Precision   (b) Average Distance

Fig. 11. Effect of parameter *a*

4.2.5   Comparison with Cosine Similarity

Cosine Similarity [21] is commonly used in information retrieval to evaluate document similarity. In this section, we aim to compare the performance of *Cosine Similarity* and our hierarchical semantic similarity. For this purpose, we modify *SIT_SSCP* by replacing the part of semantic similarity with *Cosine Similarity*. We denote this *Cosine Similarity* based method as *Cosine_SS*.

In order to computer *Cosine Similarity*, we perform the following procedure. First, each trajectory as well as the query trajectory is transformed into a vector representing the *Term Frequencies* (*TF*) of each semantic label in the trajectory. Next, we compute the *Cosine* value of the angle θ between vector A and vector B to measure the similarity between A and B. Here, A and B are vectors representing the term frequencies of the semantic labels in the trajectories.

For example, given the following two trajectories:

$$x = \langle (p_1, Nike), (p_2, LI - NING), (p_3, Adidas), (p_4, KFC) \rangle$$

$$y = \langle (p_1, Nike), (p_2, Adidas), (p_3, KFC), (p_4, Adidas), (p_5, Starbucks) \rangle$$

We first get the vectors of labels.

$$x' = \langle Nike, LI - NING, Adidas, KFC \rangle, \quad y' = \langle Nike, Adidas, KFC, Adidas, Starbucks \rangle$$

Then, we compute the term frequency for each semantic label, and get the vectors of term frequencies.

$$A = \langle 1, 1, 1, 1, 0 \rangle, \quad B = \langle 1, 0, 2, 1, 1 \rangle$$

Figures 12 and 13 compare precision and average distance, where parameter a is set to 0.5. Both figures show that *SIT_SSCP* performs better than *Cosine_SS*. Particularly, the precision of *SIT_SSCP* is about 1.1 times higher than that of *Cosine_SS*, and the average distance of *Cosine_SS* is over 4 times that of *SIT_SSCP*.

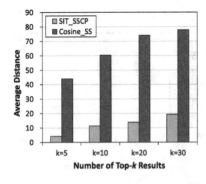 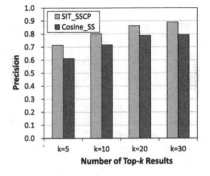

Fig. 12. Average distances of *SIT_SSCP* and *Cosine Similarity (Cosine_SS)*

Fig. 13. Precisions of *SIT_SSCP* and *Cosine Similarity (Cosine_SS)*

5   Related Work

Previous efforts on moving-object trajectories mainly focus on outdoor space such as Euclidean space and road network space. Many approaches for measuring outdoor trajectory similarity have been proposed, including *Dynamic Time Warping (DTW)* [5], *One Way Distance* [6], *Longest Common Subsequence (LCSS)* [7, 8], *Edit Distance* based approaches [9, 10]. Although these approaches can also be used for indoor trajectory similarity search, they are not effective in indoor spaces because of the major difference between the computation of indoor distance and that of outdoor distance. Recently, some researchers begin to study the semantic similarity among trajectories [12–14].

However, trajectories in indoor spaces and outdoor spaces are different and most outdoor similarity measures have to be re-considered for indoor scenarios. Currently, there have been few studies for indoor trajectory similarity analysis. The only one that exactly focuses on indoor settings is called *CVTI (Common Visit Time Interval)* [15], which is actually based on the *LCSS* approach. As the *LCSS* approach has been proposed to analyze the similarity between strings or sequences, they make each character in strings corresponds to a cell id. It aims to find common time interval at the same location between two trajectories, and then use the common time intervals to define the similarity. If two trajectories stay at the same cell during the same time interval, they will be considered more similar than the case where there is no common time interval. However, this approach only concerns the common time intervals among trajectories, but neglects many other important factors such as closeness between trajectories as well as the semantics of trajectories.

Semantics of trajectories have attracted much attention in trajectory analysis [16–19]. Josh et al. [16] take into account the semantics of trajectories and propose a novel approach for recommending potential friends based on users' labels on trajectories in location-based social networks. They mine users' similarity from GPS trajectory data by considering semantic meanings of trajectories. Since the semantic labels of trajectories can reflect the preference and interests of users, many researchers propose to integrate semantics into trajectory analysis and further provide personalized location services or recommendations [17, 18]. For example, in [18], Haibo et al. propose to take users' preferences into consideration to provide personalized location searching.

In a trajectory, there are some points that can describe the spatial characteristics of the trajectory, such as turning points or others. These points are similar to the critical points proposed in this paper. Our proposal of critical points is inspired by [19], where they find that critical points are useful for region partitioning and location clustering. The major differences of our proposal and the work in [19] are two folds. First, we develop new algorithms suitable for indoor spaces to detect critical points in indoor moving trajectories. Second, we first use critical points in similarity search on indoor moving trajectories.

6 Conclusions

Similarity trajectory search is mostly based on Euclidean space or road network space before. They are not suitable for indoor spaces. In this paper, we present a new similarity measure considering both spatial and semantic similarities between indoor trajectories. We propose a critical-point-based method for spatial similarity as well as a hierarchical semantic pattern based method for semantic similarity. Comparative experiments suggest that our proposal is effective for indoor trajectory similarity search.

Our future work will focus on taking into account the time dimension [11] into indoor trajectory similarity search. Specifically, we will concentrate on users' stay times in indoor locations.

Acknowledgement. This work is supported by the National Science Foundation of China under the grant number 61379037.

References

1. Dudas, P., Ghafourian, M., Karimi, H.: ONALIN: ontology and algorithm for indoor routing. In: Proceedings of MDM, pp. 720–725 (2009)
2. Jin, P., Zhang, L., Zhao, J., Zhao, L., Yue, L.: Semantics and modeling of indoor moving objects. Int. J. Multimedia Ubiquit. Eng. 7(2), 153–158 (2012)
3. Li, D., Lee, D.: A topology-based semantic location model for indoor applications. In: Proceedings of ACM GIS, pp. 1–10 (2008)
4. Jensen, C.S., Lu, H., Yang, B.: Graph model based indoor tracking. Mobile data management. In: Proceedings of MDM, pp. 17–24 (2008)
5. Berndt, D.J., Clifford, J.: Finding patterns in time series: a dynamic programming approach. In: Advances in Knowledge Discovery and Data Mining, pp. 229–248. AAAI/MIT Press (1996)
6. Lin, B., Su, J.: One way distance: for shape based similarity search of moving object trajectories. GeoInformatica 12(2), 117–142 (2008)
7. Boreczky, J.S., Rowe, L.A.: Comparison of video shot boundary detection techniques. J. Electron. Imaging 5(2), 122–128 (1996)
8. Vlachos, M., Kollios, G., Gunopulos, D.: Discovering similar multidimensional trajectories. In: Proceedings of ICDE, pp. 673–684 (2002)
9. Chen, L., Ozsu, M.T., Oria, V.: Robust and efficient similarity search for moving object trajectories. In: Proceedings of SIGMOD, pp. 491–502 (2005)
10. Wang, Y., Yu, G., Gu, Y., Yue, D., Zhang, T.: Efficient similarity query in RFID trajectory databases. In: Chen, L., Tang, C., Yang, J., Gao, Y. (eds.) WAIM 2010. LNCS, vol. 6184, pp. 620–631. Springer, Heidelberg (2010)
11. Yuan, Y., Raubal, M.: Measuring similarity of mobile phone user trajectories- a Spatio-temporal Edit Distance method. Int. J. Geogr. Inf. Sci. 28(3), 496–520 (2014)
12. Pelekis, N., Kopanakis, I., Marketos, G.: Similarity search in trajectory databases. In: Proceedings of TIME, pp. 129–140 (2007)
13. Frentzos, E., Gratsias, K., Theodoridis, Y.: Index-based most similar trajectory search. In: Proceedings of ICDE, pp. 816–825 (2007)

14. Dodge, S., Weibel, R., Laube, P.: Exploring movement-similarity analysis of moving objects. SIGSPATIAL Special (SIGSPATIAL) **1**(3), 11–16 (2009)
15. Kang, H.-Y., Kim, J.-S., Li, K.-J., Hwang, J.-R.: Similarity measures for trajectory of moving objects in cellular space. In: Proceedings of ACM SAC, pp. 1325–1330 (2009)
16. Ying, J.J., Lu, E.H., Lee, W.-C., Weng, T.-C., Tseng, V.S.: Mining user similarity from semantic trajectories. In: Proc. of GIS-LBSN, pp. 19–26 (2010)
17. Ma, C., Lu, H., Shou, L., Chen, G.: KSQ: Top-k similarity query on uncertain trajectories. IEEE Trans. Knowl. Data Eng. **25**(9), 2049–2062 (2013)
18. Wang, H., Liu, K.: User oriented trajectory similarity search. In: Proceedings of UrbComp, pp. 103–110 (2012)
19. Lee, J.-G., Han, J., Whang, K.-Y.: Trajectory clustering: a partition-and-group framework. In: Proceedings of SIGMOD, pp. 593–604 (2007)
20. Huang, C., Jin, P., Wang, H., Wang, N., Wan, S., Yue, L.: IndoorSTG: a flexible tool to generate trajectory data for indoor moving objects. In: Proceedings of MDM, pp. 341–343 (2013)
21. Manning, C.D., Raghavanm, P., Schütze, H.: An Introduction to Information Retrieval. Cambridge University Press, Cambridge (2008)
22. Schafer, M., Knapp, C., Chakraborty, S.: Automatic generation of topological indoor maps for real-time map-based localization and tracking. In: Proceedings of International Conference on Indoor Positioning and Indoor Navigation (IPIN), pp. 1–8. IEEE CS (2011)
23. Zhang, D., Yang, L.T., Chen, M., Zhao, S., Guo, M., Zhang, Y.: Real-time locating systems using active RFID for internet of things. IEEE Syst. J. **PP**(99), 1–10 (2014)
24. Stojanović, D., Stojanović, N.: Indoor localization and tracking: methods, technologies and research challenges. Autom. Control Robot. **13**(1), 57–72 (2014)
25. Douglas, D., Peucker, T.: Algorithms for the reduction of the number of points required to represent a line or its caricature. Can. Cartographer **10**(2), 112–122 (1973)

Graph Databases

Towards Neighborhood Window Analytics over Large-Scale Graphs

Qi Fan[1(✉)], Zhengkui Wang[2], Chee-Yong Chan[3], and Kian-Lee Tan[1,3]

[1] NUS Graduate School for Integrative Science and Engineering,
Singapore, Singapore
fan.qi@nus.edu.sg
[2] Singapore Institute of Technology, Singapore, Singapore
zhengkui.wang@singaporetech.edu.sg
[3] School of Computing, National University of Singapore, Singapore, Singapore
{chancy,tankl}@comp.nus.edu.sg

Abstract. Information networks are often modeled as graphs, where the vertices are associated with attributes. In this paper, we study neighborhood window analytics, namely k-hop window query, that aims to capture the properties of a local community involving the k-hop neighbors (defined on the graph structures) of each vertex. We develop a novel index, *Dense Block Index (DBIndex)*, to facilitate efficient processing of k-hop window queries. Extensive experimental studies conducted over both real and synthetic datasets with hundreds of millions of vertices and edges show that our proposed solutions are four orders of magnitude faster in query performance than the non-index algorithm, and are superior over the state-of-the-art solution in terms of both scalability and efficiency.

Keywords: Graph analytics · Graph window · Neighborhood aggregation

1 Introduction

Information networks such as social networks, biological networks and phone-call networks are typically modeled as graphs [4] where the vertices correspond to objects and the edges capture the relationships between these objects. For instance, in social networks, every user is represented by a vertex and the friendship between two users is reflected by an edge between the vertices. In addition, a user's profile can be maintained as the vertex's attributes. Such graphs contain a wealth of valuable information which can be analyzed to discover interesting patterns. For example, we can find the top-k influential users who can reach the most number of friends within 2 hops. With increasingly larger network sizes, there is an urgent need to develop effective and efficient mechanisms over large-scale graph data.

Recent research on graph analytics focuses on discovering the global graph properties and characteristics. To name a few, graph summarization [14] aims

© Springer International Publishing Switzerland 2016
S.B. Navathe et al. (Eds.): DASFAA 2016, Part II, LNCS 9643, pp. 201–217, 2016.
DOI: 10.1007/978-3-319-32049-6_13

to provide a compressed representation of a given graph based on its structure and vertex/edge attributes, while graph aggregation [4,17,19] focuses on aggregating the graph based on its vertex/edge attributes to discover the underlying characteristics of large graphs.

In this paper, we study a new type of query that analyzes each vertex's local community (e.g., neighborhoods) in a graph. To each vertex, these local communities (also referred to as windows in this paper) carry the most important information that captures the vertex's social influence and relations in the graph. Unlike graph summarization and aggregation that discover the entire graph's property, graph window queries (GWQs) explore the underlying characteristics of a small window related to each individual vertex. We identify one instantiation of graph "windows", namely *k-hop* window. We first demonstrate the k-hop window semantics with the following example.

Example 1 (K-hop window). *In social network scenario, it is of great interest to summarize the most relevant connections to each user such as the neighbors within 2-hops. Some analytic queries such as summarizing the related connections' distribution among different companies, and computing age distribution of the related friends can be useful. In order to answer these queries, collecting data from every user's neighborhoods within 2-hop is necessary.*

A k-hop window forms a window for one vertex by using its k-hop neighbors. In the Example 1, every user needs to gather data from his/her friends and friends-of-friends. k-hop neighbors are important to one vertex, as these are the vertices showing structural closeness as in Example 1.

To the best of our knowledge, existing graph databases or graph query languages do not directly support our proposed GWQ. There are two major challenges in processing GWQ. First, we need an efficient scheme to calculate the neighborhood window of each vertex. Second, we need efficient solutions to process the aggregation over a large number of windows that may overlap. However, it is nontrivial to address these two challenges. The state-of-the-art algorithm for k-hop like query is EAGR [12]. EAGR builds an overlay graph to leverage the shared components of different windows through multiple iterations. However, EAGR requires all vertices' k-hop neighbors to be pre-computed and resides in memory during every iteration. This heavily limits the efficiency and scalability of EAGR.For instance, a LiveJournal social network graph[1] (4.8 M vertices, 69M edges) generates over 100GB neighborhood information for k = 2 in adjacency list representation. In addition, the overlay graph construction is not a one-time task, but is periodically performed after a certain number of structural updates in order to maintain the overlay quality. The high memory requirement renders EAGR impractical when k and the graph size increase.

In this paper, we propose Dense Block Index (DBIndex), which enables an efficient query processing by integrating the optimized query execution plan for shared aggregation computation. Additionally, for index construction, we apply a hash-based technique to cluster the vertices based on the window similarity,

---

[1] Available at http://snap.stanford.edu/data/index.html, which is used in [12].

which ensures memory efficiency. On the basis of the clusters, we further develop different optimizations to extract the shared components efficiently.

Our contributions are summarized as follows:

– We introduce a new type of graph analytic query, *Graph Window Query* and formally define the *k-hop* window. We illustrate how these window queries would help users better query and understand the graphs under different semantics.
– We propose *Dense Block Index (DBIndex)* to support the proposed k-hop window queries. The index integrates the window aggregation sharing techniques to salvage partial work done to enable efficient query processing over large-scale graphs.
– We perform extensive experiments over both real and synthetic datasets with hundreds of millions of vertices and edges on a single machine. Our experiments indicate that our proposed index-based algorithms outperform the naive non-index algorithm by up to four orders of magnitude. In addition, our experiments also show that DBIndex is superior over EAGR in terms of both scalability and efficiency.

The rest of the paper is organized as follows: Sect. 2 formulates the GWQ. In Sect. 3, we introduce the DBIndex for k-hop window query. Section 4 presents the experimental evaluations. In Sects. 5 and 6, we provide the related works and the conclusion respectively.

2   Problem Formulation

In this section, we provide the formal definition of graph window query. We use $G = (V, E)$ to denote a directed/undirected data graph, where V is its vertex set and E is its edge set. Each vertex/edge is associated with a (possibly empty) set of attributes.

Figure 1 shows an undirected graph representing a social network that we will use as our running example. The table shows the values of the five attributes (User, Age, Gender, Industry, and Number of posts) associated with each vertex. For convenience, each vertex is labeled with its user attribute value; and there is one edge between a user X and another user Y if X and Y are connected in the social network.

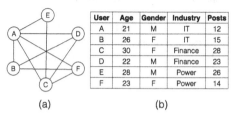

(a)   (b)

Fig. 1. Example of Social Graph. (a) Graph structure; (b) Attributes associated with the vertices in (a).

Given a data graph $G = (V, E)$, a *Graph Window Function (GWF)* over G can be expressed as a quadruple (G, W, Σ, A), where $W(v)$ denotes a *window specification* (or *window* for short) for a vertex $v \in V$ that determines the set of vertices (refer to as *nodes*) in some subgraph of G, Σ denotes an *aggregate function*, and A denotes a *vertex attribute*. The evaluation of a GWF (G, W, Σ, A) on G

computes for each vertex v in G, the aggregation Σ on the values of attribute A over all the vertices in $W(v)$, which is denoted by $\Sigma_{v' \in W(v)} v'.A$. We consider all common aggregate functions (e.g., sum, count, average, max, min etc.) in this paper.

Definition 1 (k-hop Window). *Given a vertex v in a graph G, the k-hop window of v, denoted by $W_{kh}(v)$ (or $W(v)$ when there is no ambiguity), is the set of neighbors of v in G which can be reached within k hops. For an undirected graph G, a vertex u is in $W_{kh}(v)$ iff there is a α-hop path between u and v where $\alpha \leqslant k$. For a directed graph G, a vertex u is in $W_{kh}(v)$ iff there is a α-hop directed path from v to u^2 where $\alpha \leqslant k$.*

Intuitively, a k-hop window selects the neighboring vertices of a vertex within a k-hop distance. These neighboring vertices typically represent the most important vertices to a vertex wrt their structural relationship in a graph. Thus, k-hop windows provide meaningful specifications for many applications, such as customer behavior analysis [1,6] and digital marketing [10]. As an example, in Fig. 1, the 1-hop window of vertex E is $\{A, C, E\}$ and the 2-hop window of vertex E is $\{A, B, C, D, E, F\}$.

We emphasize that there are different types of windows which can be formalized under different application scenarios. For instance, we have identified another useful window, namely the *topological window*, which captures the set of ancestor vertices of each vertex in a directed acyclic graph (DAG). There are many DAGs in real-world applications (such as biological networks, citation networks and dependency networks) where topological windows represent meaningful relationships that are of interest. For example, in a citation network where (X,Y) is an edge iff paper X cites paper Y, the topological window of a paper represents the citation impact of that paper [3,11]. Based on the topological window, we have proposed another index, *inheritance index* as in our technical report [8] to facilitate an efficient topological window query processing and systematically evaluated the index. In general, a graph window query is defined as:

Definition 2 (Graph Window Query). *A graph window query on a data graph G is of the form $GWQ(G, W_1, \Sigma_1, A_1, \cdots, W_m, \Sigma_m, A_m)$, where $m \geq 1$ and each quadruple (G, W_i, Σ_i, A_i) is a graph window function on G.*

In this paper, we focus on efficiently processing k-hop window queries with indexes. Due to space constraint, we only present the static solution for illustration, and the strategy for handling updates are described in our technical report [8].

---

² Other variants of k-hop window for directed graphs are possible; e.g., a vertex u is in $W_{kh}(v)$ iff there is a α-hop directed path from u to v where $\alpha \leqslant k$.

3   Dense Block Index

A straightforward approach to process a graph window query $Q = (G, W, \Sigma, A)$, where $G = (V, E)$, is to dynamically compute the window $W(v)$ for each vertex $v \in V$ and its aggregation $\Sigma_{v' \in W(v)} v'.A$ independently from other vertices. We refer to this approach as *Non-Indexed* method. Given that many of the windows would share many common nodes (e.g., the k-hop windows of two adjacent vertices), such a simple approach would be very inefficient due to the lack of sharing of the aggregation computations.

To efficiently evaluate graph window queries, we propose an indexing technique named *Dense Block Index* (*DBIndex*), which is both space and query efficient. The main idea of DBIndex is to try to reduce the aggregation computation cost by identifying subsets of nodes that are shared by more than one window so that the aggregation for the shared nodes could be computed only once instead of multiple times.

For example, consider a graph window query on the social graph in Fig. 1 using the 1-hop window. We have $W(B) = \{A, B, D, F\}$ and $W(C) = \{A, C, D, E, F\}$ sharing three common nodes A, D, and F. By identifying the set of common nodes $S = \{A, D, F\}$, its aggregation $\Sigma_{v \in S} v.A$ can be computed only once and then reuse to compute $\Sigma_{v \in W(B)} v.A$ and $\Sigma_{v \in W(C)} v.A$.

Given a window W and a graph $G = (V, E)$, we refer to a non-empty subset $B \subseteq V$ as a *block*. Moreover, if B contains at least two nodes and B is contained by at least two different windows (i.e., if $|B| \geq 2$, and $\exists v_1 \neq v_2 \in V, B \subseteq W(v_1)$, and $B \subseteq W(v_2)$), then B is a *dense block*. Thus, in the last example, $\{A, D, F\}$ is a dense block.

We say that a window $W(X)$ is *covered* by a collection of disjoint blocks $\{B_1, \cdots, B_n\}$ if the set of nodes in the window $W(X)$ equals to the union of all nodes in the collection of disjoint blocks; i.e., $W(X) = \bigcup_{i=1}^{n} B_i$ and $B_i \cap B_j = \emptyset$ if $i \neq j$.

To maximize the sharing of aggregation computations for a graph window query, the objective of DBIndex is to identify a small set of blocks $\mathcal{B}$ such that for each $v \in V$, $W(v)$ is covered by a small subset of disjoint blocks in $\mathcal{B}$. Clearly, the cardinality of $\mathcal{B}$ is minimized if $\mathcal{B}$ contains a few large dense blocks.

Thus, given a window W and a graph $G = (V, E)$, a DBIndex to evaluate W on G consists of three components in the form of a bipartite graph. The first component is a collection of vertex (i.e., V); the second component is a collection of blocks $\mathcal{B} = \{B_1, \cdots, B_n\}$ where each $B_i \subseteq V$; and the third component is a collection of links from blocks to vertices such that if a set of blocks $B(v) \subseteq \mathcal{B}$ is linked to a vertex $v \in V$, then $W(v)$ is covered by $B(v)$. Note that a DBIndex is independent of both the aggregate function (i.e., Σ) and the attribute to be aggregated (i.e., A). Figure 2(d) shows an example of a DBIndex wrt the social graph in Fig. 1 and the 1-hop window. There are three dense blocks detected which are $\{A, F, D\}, \{C, E\}$, and $\{A, C\}$.

3.1   Query Processing Using DBIndex

Given a DBIndex wrt a graph G and a window W, a graph window query $Q = (G, W, \Sigma, A)$ is processed by the following two steps. First, for each block B_i in the index, we compute the aggregation (denoted by T_i) over all the nodes in B_i; i.e., $T_i = \Sigma_{v \in B_i} v.A$. Thus, each T_i is a partial aggregate value. Next, for each window $W(v)$, $v \in V$, the aggregation for the window is computed by aggregating over all the partial aggregates associated with the blocks linked to $W(v)$; i.e., if $B(v)$ is the collection of blocks linked to $W(v)$, then the aggregation for $W(v)$ is given by $\Sigma_{B_i \in B(v)} T_i$.

3.2   DBIndex Construction

In this section, we discuss the construction of the DBIndex (wrt a graph $G = (V, E)$ and window W) which has two key challenges.

The first challenge is the time complexity of the index construction. From our discussion of query processing using DBIndex, we note that the number of aggregation computations is determined by both the number of blocks as well as the number of links in the index; the former determines the number of partial aggregates to compute while the latter determines the number of aggregations of the partial aggregate values. Thus, to maximize the shared aggregation computations using DBIndex, both the number of blocks in the index as well as the number of blocks covering each window should be minimized. However, finding the optimal DBIndex to minimize this objective is NP-hard[3]. Therefore, efficient heuristics are needed to construct the DBIndex.

The second challenge is the space complexity of the index construction. In order to identify large dense blocks to optimize for query efficiency, a straightforward approach is to first derive the window $W(v)$ for each vertex $v \in V$ and then use this derived information to identify large dense blocks. However, this direct approach incurs a high space complexity of $O(|V|^2)$. Therefore, a more space-efficient approach is needed in order to scale to large graphs.

MinHash-based Index Construction (MC). To reduce both the time and space complexities for the index construction, instead of trying to identify large dense blocks among a large collection of windows, MC first partitions all the windows into a number of smaller clusters of similar windows and then identifies large dense blocks from each of the smaller clusters. Intuitively, two windows are considered to be highly similar if they share a larger subset of nodes. We apply the well-known *MinHash based Clustering* algorithm [2] to partition the windows into clusters of similar windows. The MinHash clustering algorithm uses *Jaccard Coefficient* to measure the similarity of two sets. Given two windows $W(v)$ and $W(u)$, $u, v \in V$, their *Jaccard Coefficient* is given by $J(u, v) = \frac{|W(u) \cap W(v)|}{|W(u) \cup W(v)|}$. The *Jaccard Coefficient* ranges from 0 to 1, where a larger value means that the windows are more similar.

---

[3] Note that a simpler variation of our optimization problem has been proven to be NP-hard [16].

Our heuristic approach to construct DBIndex I operates as in Algorithms 1 and 2. Let $vertices(I)$, $blocks(I)$, and $links(I)$ denote the collection of vertices, blocks, and links in I. Initially, we have $vertices(I) = V$, $blocks(I) = \emptyset$, and $links(I) = \emptyset$.

Algorithm 1. CreateDBIndex

Require: Graph $G = (V, E)$, window W

Ensure: DBIndex I

1: Initialize DBIndex I: $vertices(I) = V$, $blocks(I) = \emptyset$, $links(I) = \emptyset$
2: **for all** $v \in V$ **do**
3:    Traverse G to determine $W(v)$
4:    Compute the hash signature $H(v)$ for $W(v)$
5: **end for**
6: Partition V into clusters $\mathcal{C} = \{C_1, C_2, \cdots\}$ based on hash signatures $H(v)$
7: **for all** $C_i \in \mathcal{C}$ **do**
8:    **for all** $v \in C_i$ **do**
9:       Traverse G to determine $W(v)$
10:    **end for**
11:    IdentifyDenseBlocks (I, W, C_i)
12: **end for**
13: **return** I
14:

Algorithm 2. IdentifyDenseBlocks

Require: DBIndex I, window W, a cluster $C_i \subseteq V$

1: Return if C_i is empty.
2: Partition V into blocks wrt to C_i, $DenseNodes = \emptyset$
3: **for all** dense block B **do**
4:    Insert B into $blocks(I)$ if $B \notin blocks(I)$
5:    Insert (B, v) into $links(I)$ for each $v \in C_i$ where $B \subseteq W(v)$
6:    $DenseNodes = DenseNodes \cup B$
7: **end for**
8: $C_n \leftarrow \emptyset$, $W_n \leftarrow \emptyset$
9: **for all** $v_i \in C_i$ **do**
10:    **if** $(W(v_i) - DenseNodes \neq \emptyset)$ **then**
11:       Insert v_i to C_n
12:       Insert $(W(v_i) - DenseNodes)$ to W_n
13:    **end if**
14: **end for**
15: IdentifyDenseBlocks (I, W_n, C_n)
16: **return**

The first step (Lines 1–6 Algorithm 1) is to partition the vertices in V into clusters using MinHash algorithm such that vertices with similar windows belong to the same cluster. For each vertex $v \in V$, we first derive its window $W(v)$ by an appropriate traversal (e.g., k-hop BFS) of the graph G. Next, we compute hash signatures (denoted by $H(v)$) for each v by applying MinHash on $W(v)$. Vertices with identical hash signatures are considered to have highly similar windows and are grouped into the same cluster. To ensure that our approach is scalable, we do not retain $W(v)$ in memory after its hash signature $H(v)$ has been computed and used to cluster v; i.e., our approach does not materialize all the windows in the memory to avoid high space complexity. Let $\mathcal{C} = \{C_1, C_2, \cdots\}$ denotes the collection of clusters obtained from the first step, where each C_i is a subset of vertices.

The second step (Lines 7–12 Algorithm 1) is to identify dense blocks from each of the clusters computed in the first step. The identification of dense blocks in each cluster C_i is based on the notion of node equivalence defined as follows. Two distinct nodes $u, v \in V$ are defined to be equivalent (denoted by $u \equiv v$) wrt C_i iff u and v are both contained in the same set of windows wrt C_i; i.e., for every window $W(x), x \in C_i$, $u \in W(x)$ iff $v \in W(x)$. Based on this notion of node equivalence, V is partitioned into blocks of equivalent nodes. To perform this partitioning, we need to again traverse the graph for each vertex $v \in C_i$ to determine its window $W(v)^4$.

However, since C_i is now a smaller cluster of vertices, we can now materialize all the windows for the vertices in C_i in memory without exceeding the memory space. In the event that a cluster C_i is still too large for all its windows to be materialized in main memory, we can further partition C_i into equal sized sub-clusters. This re-partition process can be recursively performed until the sub-clusters created are small enough such that the windows for all vertices in the sub-cluster fit in memory.

Recall that a block B is a dense block if B contains at least two nodes and B is contained in at least two windows. Thus, we can classify nodes in V as either dense or non-dense nodes: a node $v \in V$ is classified as a *dense node* if v is contained in a dense block; otherwise, v is a *non-dense node*.

For each dense block B in C_i, we update the blocks and links in the DBIndex I recursively as follows: If the current cluster or window only contains one element, then algorithm stops. Otherwise, we insert dense block B into $block(I)$; and we insert (B, v) into $links(I)$ for each $v \in C_i$ where $B \subseteq W(v)$ (Lines 3–7 Algorithm 2). For each vertex v in C_i, we remove dense nodes from its window $W(v)$. This forms the refined window $W_n(v)$. If $W_n(v)$ is not empty, we then add v to a refined cluster C_n. C_n and W_n are then processed recursively (Lines 8–15 Algorithm 2).

Figure 2 illustrates the construction of the DBIndex wrt the social graph in Fig. 1(a) and 1-hop window using the MC algorithm. First, the set of graph vertices are partitioned into clusters using MinHash clustering; Fig. 2(a) shows that the set of vertices $V = \{A, B, C, D, E, F\}$ are partitioned into two clusters $C_1 = \{A, B, C\}$ and $C_2 = \{D, E, F\}$. Table 1 in Fig. 2(b) shows the node-vertex mapping in C_1, i.e. for each node $u \in V$, the corresponding row is the set $\{v \in C_1 | u \in W(v)\}$. Similarly, Table 2 in Fig. 2(b) shows the node-vertex mapping in C_2.

Consider the identification of dense blocks in cluster C_1. As shown in Fig. 2(c), based on the notion of equivalence nodes, cluster C_1 is partitioned into three blocks of equivalent nodes: $B_1 = \{A, D, F\}$, $B_2 = \{B\}$, and $B_3\{C, E\}$. Among these three blocks, only B_1 and B_3 are dense blocks. The MC algorithm then tries to repartition the window A, B, C using non-dense nodes in C_1, (i.e., B) as next window. Since B is the only node, it directly outputs. At the end of processing cluster C_1, the DBIndex I is updated as follows: $blocks(I) = \{B_1, B_2, B_3\}$ and $links(I) = \{(B_1, \{A, B, C\}), (B_2, \{A, B\}), (B_3, \{A, C\})\}$. The identification of dense blocks in cluster C_2 is of similar process.

---

[4] Note that although we could have avoided deriving $W(v)$ a second time if we had materialized all the derived windows the first time, our approach is designed to avoid the space complexity of materializing all the windows in memory at the cost of computing each $W(v)$ twice. We present an optimization later in this section to avoid the recomputation cost on k-hop window query.

Assume that, the average neighborhood size of each vertex is $\overline{w}$. The MinHash cost is thus $\overline{w}|V|$. The cost of traversal for all vertex is $\overline{w}|E|$. In Algorithm 1, Lines 1–10 have the cost of $\overline{w}(|V|+2|E|)$; In Algorithm 2, since we can simply partition nodes using hashing, the time cost is thus $\overline{w}|C_i|$. The recursive procedure runs at most $log(|C_i|)$ times, and the total cost for Algorithm 2 is $\Sigma(\overline{w}|C_i|log(|C_i|))$. Since $\Sigma(|C_i|) = |V|$, the total cost for Algorithm 2 is less than $\overline{w}|V|log(|V|)$. Therefore, the total cost for Algorithm 1 and

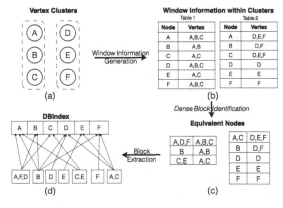

Fig. 2. DBIndex Construction over Social Graph in Fig. 1. (a) Two clusters after MinHash clustering; (b) Window information of involved vertices within each cluster; (c) Dense blocks within each cluster; (d) Final DBIndex.

Algorithm 2 is $\overline{w}(|V| + 2|E| + |V|log(|V|))$, thus the complexity is $O(\overline{w}(|E| + |V|log(|V|)))$.

Next, we provide two specific optimizations for constructing DB-Index for k-hop window queries.

Estimation Optimization. For k-hop window query with a large value of k, the cost of graph traversals to compute the k-hop windows is high. Moreover, the cost of initial MinHash in MC approach equals to the initial number of vertex-window mappings, which is of the same order as graph traversal.

To address the high computation issue, we make an observation that if the m-hop windows for two vertices are similar, the n-hop windows for them are also similar, where $m < n$. The intuition is that the shared component becomes larger via hop expansion. This observation is formally described as follows:

Theorem 1. *Let $\langle u, v \rangle$ be a randomly chosen vertex pair from a graph, let $J_k(u, v)$ be their Jaccard similarity wrt k-hop window. Then with high probability $J_m(u, v) \leq J_n(u, v)$, where $m < n$.*

We omit the theoretic proof here. Interested readers are referred to [8] for details. Based on Theorem 1, one optimization to improve the efficiency of Algorithm 1 with the tradeoff of a possible lower "quality" dense blocks (in terms of their sizes) is to use the m-hop window as an estimation for a n-hop index construction during the clustering, where $m < n$. In particular, for the first round of window computation (Lines 3–4 in Algorithm 1), we can use the hash signatures of the lower hop windows cluster the vertices in V to approximate k-hop windows. This approximation has the advantage of improved time-efficiency as traversal and MinHash clustering on lower-hop window is significantly faster. In particular, if the average number of neighbors for each vertex in a n-hop window is denoted by $\overline{w}_n$, then the optimization reduces the index construction cost by $(|V| + |E|)(\overline{w}_n - \overline{w}_m)$ from $(|V| + 2|E| + |V|log(|V|))\overline{w}_n$ to $(|V| + |E|)\overline{w}_m + (|E| + |V|(log|V|))\overline{w}_n$. This improvement is significant as $\overline{w}_n$ is exponentially greater than $\overline{w}_m$, where $m < n$, in k-hop windows. Our experimental results show with this optimization, the reduction in the quality of dense blocks is actually only marginal which makes this optimization a good tradeoff.

Batching Optimization. When multiple k-hop indexes are required, one applicable optimization is to batch the index constructions to share the graph traversal and clustering. Suppose there is a need to compute the DBIndex for 1-hop, 2-hop, ..., k-hop windows. Constructing each index independently would incur high overhead on both clustering and graph traversal which can be alleviated by batching their computations. The overall idea of the batching construction is to utilize the lower hop (e.g., 1-hop) traversal information to build the clustering and reuse it for all the h-higher hops. In addition, the second time graph traversal after obtaining the clustering can also be shared. Intuitively, while we expand the k-hop window, we can calculate the i-hop window as well. This can be achieved as the BFS is adopted where the $(i + 1)$-hop window can be directly derived based on the i-hop windows, thus the traversal overhead can be shared.

4   Experimental Evaluation

In this section, we present the results of our experiments on both real-world networks and synthetic graphs. Due to space limitations, we can only present partial experimental results here and more results can be found in our technical report [8].

All experiments are conducted on an Amazon EC2 r3.2xlarge machine[5], with an 8-core 2.5 GHz CPU, 60 GB memory and 320 GB hard drive running with 64-bit Ubuntu 12.04. We implement EAGR algorithm as a reference in our comparative study. All algorithms are implemented in Java and run under JRE 1.6.

Name	Type	# of Vertices	# of Edges
LiveJournal	undirected	3,997,962	34,681,189
Pokec	directed	1,632,803	30,622,564
Orkut	undirected	3,072,441	117,185,083
DBLP	undirected	317,080	1,049,866
YouTube	undirected	1,134,890	2,987,624
Google	directed	875,713	5,105,039
Amazon	undirected	334,863	925,872
Stanford-web	directed	281,903	2,312,497

Fig. 3. Large-scale Real Datasets

Datasets. For real datasets, we use 8 information networks which are available at the Stanford *SNAP* website[6]: The detail description of these datasets is provided in Fig. 3. For synthetic datasets, we use *SNAP* graph generator to create graphs with various sizes.

---

[5] http://aws.amazon.com/ec2/pricing/.
[6] http://snap.stanford.edu/snap/index.html.

Query. In all the experiments, the window query is conducted by using SUM() as the aggregate function.

4.1 Index Construction Optimization

To study the performance of index construction, we compare two indexing methods, namely MC and MC++. MC method uses the MinHash clustering as described in Algorithm 1 while MC++ adapted the estimation optimization as in Theorem 1. We then present the results on the Amazon and Stanford-web graphs for a series of k-hop queries.

Index Construction. Figure 4(a) and (c) compare the index construction time between MC and MC++ when we vary the windows from 1-hop to 4-hop under Amazon and Stanford-web datasets. To better understand the time difference, the construction time is split into two parts: the MinHash cost (MC++-hash or MC-hash) and the BFS traversal (to compute the k-hop window) cost (MC++-bfs or MC-bfs). The results show the same trend for the two datasets. We made several observations. First, as the number of hops increases, the indexing time increases as well. This is expected as a larger hop count results in a larger window size and the BFS and MinHash computation time increase correspondingly. Second, as the hop count increases, the difference between the index time of MC++ and that of MC widens. For instance, as shown in Fig. 4(a), for the 4-hop window queries, compared to MC, MC++ can save 62 % construction time. MC++ benefits from both the low MinHash cost and low BFS cost. From Fig. 4(a), we can see that the MinHash cost of MC increases as the number of hops increases, while that for MC++ remains almost the same as the 1-hop case. The similar pattern can be found in Fig. 4(c) as well. These show that the cost of MinHash becomes more significant for larger windows. Thus, using 1-hop clustering for larger hop counts reduces the MinHash cost in MC++. Similarly, as MC++ saves on BFS cost for k-hop queries where $k > 1$, the BFS cost of MC++ is much smaller than that of MC as well.

Query Performance. Figure 4(b) and (d) present the query time of MC and MC++ on Amazon and Stanford-web datasets as we vary the number of hops from 1 to 4. To appreciate the benefits of an index-based scheme, we also implemented a *Non-indexed* algorithm which computes window aggregate by performing k-hop breadth first search for each vertex individually in real time. In Fig. 4(b), the execution time shown on the y-axis is in log scale. The results show that the index-based schemes outperform the non-index approach by four orders of magnitude. For instance, for the 4-hop query, our algorithm is 13,000 times faster than the non-index approach. This confirms that it is necessary to have well-designed index support for efficient window query processing. By utilizing DBIndex, for these graphs with millions of edges, every aggregation query can be processed in just between 30 ms to 100 ms. In addition, we can see that as the number of hops increases, the query time decreases. This is the case because a larger hop count eventually results in a larger number of dense blocks where more (shared) computation can be salvaged. Furthermore, we can see that the

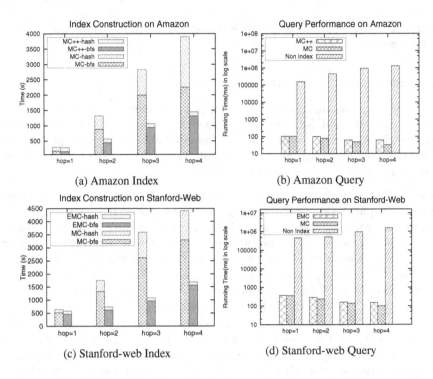

Fig. 4. The evaluation of the index construction optimization.

query time of MC++ is slightly longer than that of MC when the number of hops is large. This is expected as MC++ does not cluster based on the complete window information; instead, it uses only partial information derived from the 1-hop windows. However, the performance difference is quite small even for 4-hop queries - the difference is only 20 ms. For small number of hops, the time difference is even smaller. This performance penalty is acceptable as tens of milliseconds time difference will not affect user's experience. As MC++ is significantly more efficient than MC in index construction, MC++ may still be a promising solution for many applications. In addition, we also observe the same pattern in Fig. 4(d). As such, in the following sections, we adopt MC++ for DBIndex in our experimental evaluations.

4.2 Comparison Between DBIndex and EAGR

We then compare DBIndex and EAGR [12][7] using both real and synthetic datasets.

Real Datasets. We first study the index construction and query time performance of DBIndex and EAGR for 1-hop and 2-hop windows using 6 real

---

[7] As in [12], for each dataset, EAGR is run for 10 iterations in the index construction.

datasets: DBLP, YouTube, Livejournal, Google, Pokec and Orkut. The results for 1-hop window and 2-hop window are presented in Fig. 5(a)-(d). As shown in Fig. 5(a) and (c) both DBIndex and EAGR can build the index for all the real datasets for 1-hop but EAGR ran out of the memory for 2-hop window queries on LiveJournal and Orkut datasets. This further confirms that EAGR incurs high memory usage as it needs to maintain the vertex-window mapping information. We also observe that DBIndex is significantly faster than EAGR in index creation. For instance, for Orkut dataset, EAGR takes 4 hours to build the index while DBIndex only takes 33 min.

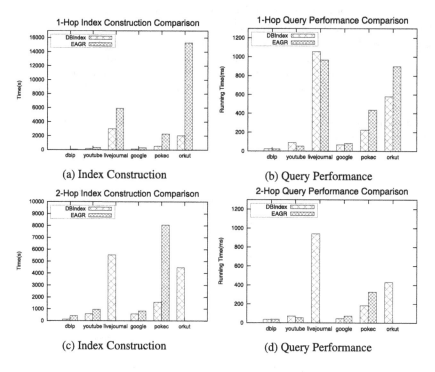

(a) Index Construction   (b) Query Performance

(c) Index Construction   (d) Query Performance

Fig. 5. DBIndex VS. EAGR (a)(b) are for 1-hop queries;(c)(d) are for 2-hop queries

Figure 5(b) and (d) show the query performance for 1-hop and 2-hop queries respectively. The results indicate that the query performance is comparable. For most of the datasets, DBIndex is faster than EAGR. In some datasets (e.g., Orkut and Pokec), DBIndex performs 30 % faster than EAGR. We see that, for 1-hop queries on YouTube and LiveJournal datasets and 2-hop queries on YouTube dataset, DBIndex is slightly slower than EAGR. We observe that these datasets are very sparse graphs where the intersections among windows are naturally small. For very sparse graphs, both DBIndex and EAGR are unable to find much computation sharing. In this case, the performance of DBIndex and EAGR is very close. For instance, in the worst case, as in the livejournal dataset,

DBIndex is 9 % slower than EAGR where the actual time difference remains tens of milliseconds.

We also observe that, both algorithms process 2-hop queries faster than 1-hop queries. This is because there is more computation sharing for 2-hop window query. In summary, DBIndex takes much shorter time to build but offers comparable, if not much faster, query performance than EAGR.

Synthetic Datasets. We generated synthetic datasets using the SNAP generator to study the scalability of DBIndex.

Impact of Number of Vertices. First, we study how the performance changes when we fix the degree [8] at 10 and vary the number of vertices from 2 M to 10 M. Figure 6(a) and (b) show the execution time for index construction and query performance respectively. From the results, we can see that DBIndex outperforms EAGR in both index construction and query performance. For the graph with 10 M vertices and 100 M edges, the DBIndex query time is less than 450 milliseconds. Moreover, when the number of vertices changes from 2M to 10M, the query performance only increases 3 times. This shows that DBIndex is not only scalable, but offers acceptable performance.

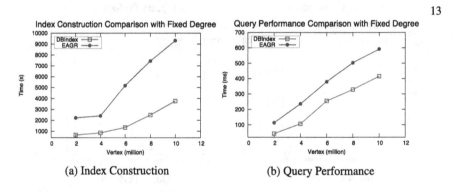

(a) Index Construction   (b) Query Performance

Fig. 6. Impact of number of vertices

Impact of Degree. Our proposed DBIndex is effective when there are significant overlaps between windows of neighboring vertices. As such, it is interesting to study how it performs under various graph degrees. Here, we report the results from dense graphs. More results of sparse graphs can be found in the technical report [8]. We fix the number of vertices in a graph to be 200k. and then vary the degree from 80 to 200. Figure 7(a) shows the index construction and query time for 1-hop query. We can see that DBIndex outperforms EAGR significantly. As the degree increases, EAGR's performance degrades much faster than DBIndex.

---

[8] Degree means average degree of the graph. The generated graph is of Erdos-Renyi model .

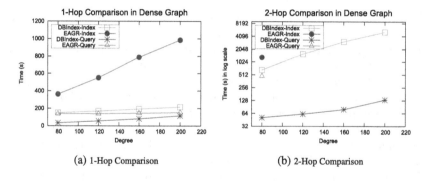

Fig. 7. Impact of Degree on Dense Graphs

It is notable that DBIndex indexing time almost matches EAGR's query time. Figure 7(b) shows the comparison under 2-hop queries. EAGR is only able to work on the dataset with degree 80 due to the memory issue. This is because the number of edges is large (e.g., 40 M edges for a graph of degree 200).

In summary, the insight we obtain is that the scalability of EAGR is highly limited by two factors: the graph size and the number of hops. DBIndex achieves better scalability as it does not need to create a large amount of intermediate data in memory.

5   Related Work

GWFs are different from graph aggregation [4,17,19] in graph OLAP. In graph OLAP, information in a graph are summarized by partitioning the graph's vertices/edges (based on some attribute values) and computing aggregate values for each partition. GWFs, on the other hand, compute aggregate values for each graph vertex wrt the subgraph associated with the vertex. Indeed, such differences also arise in the relational context, where different techniques are developed to evaluate OLAP and window function queries.

In [18], the authors investigated the problem of finding the vertices that have top-k highest aggregate values over their h-hop neighbors. They proposed mechanisms to prune the computation by using two properties: First, the locality between vertices is used to propagate the upper bound of aggregation; Second, the upper bound of aggregates can be estimated from the distribution of attribute values. However, all these pruning techniques are not applicable in our work, as we need to compute the aggregation for every vertex. In such a scenario, techniques in [18] degrade to the non-indexed approach as described in Sect. 4.

Indexing techniques have been proposed to efficiently determine whether a pair of vertices is within a distance of k-hops (e.g., k-reach index [5]). However, such techniques are not suitable for k-hop window queries due to the time complexity of $O(n^2)$ in determining each vertex's window. Moreover, such techniques do not leverage shared components among windows to boost query processing.

In distributed databases community, some works considered utilizing partial aggregates to facilitate efficient aggregate computation (e.g., [9,15]). However, their primary goal is to optimize the communication cost between sites, hence the optimization problem is fundamentally different from ours.

In network science community, *egocentric* analysis is emerging in recent years. However, their main focus is on structural analyses of a vertex's k-hop neighborhood. For example, Everett et al. [7] looked at finding the betweenness centrality among vertices' k-hop neighbors; and Moustafa et al. [13] developed techniques for matching specialized patterns among k-hop neighborhoods. These works are different from ours as they do not consider attribute aggregation.

The work that is most related to ours is [12] - referred to as EAGR which examined the evaluation of egocentric (similar to our k-hop window) aggregation queries. EAGR and our DBIndex share the similar spirit in terms of discovering the shared components among different windows to speed up the query processing. However, as elaborated in Sect. 1, our techniques are more memory-efficient, as well as more scalable than those in EAGR.

6   Conclusion

In this paper, we have proposed *Graph Window Query* to facilitate analytics over a local community of each graph vertex, and studied one instantiations, namely k-hop window in detail. We proposed the Dense Block Index (DBIndex) to facilitate efficient processing of k-hop window query. DBIndex integrates window aggregation sharing techniques to salvage partial work done, which is both space and query efficient. Results of an extensive experimental study on both large-scale real and synthetic datasets showed the efficiency and scalability of our proposed index.

Acknowledgment. Qi Fan is supported by NGS Scholarship. This work is supported by the MOE/NUS grant R-252-000-500-112 and AWS in Education Grant award.

References

1. Briscoe, E.J., Appling, D.S., Mappus IV, R.L., Hayes, H.: Determining credibility from social network structure. In: ICASNAM 2013, pp. 1418–1424. ACM (2013)
2. Broder, A.Z., Glassman, S.C., Manasse, M.S., Zweig, G.: Syntactic clustering of the web. Comput. Netw. ISDN Syst. **29**(8), 1157–1166 (1997)
3. Campanario, J.M.: Empirical study of journal impact factors obtained using the classical two-year citation window versus a five-year citation window. Scientometrics **87**(1), 189–204 (2011)
4. Chen, C., Yan, X., Zhu, F., Han, J., Yu, P.S.: Graph OLAP: towards online analytical processing on graphs. In: ICDM 2008, pp. 103–112 (2008)
5. Cheng, J., Shang, Z., Cheng, H., Wang, H., Yu, J.X.: K-reach: who is in your small world. VLDB **5**(11), 1292–1303 (2012)
6. Dai, L., Luo, J.-D., Fu, X., Li, Z.: Predicting offline behaviors from online features: an ego-centric dynamical network approach. In: HotSocial 2012, pp. 17–24 (2012)

7. Everett, M., Borgatti, S.P.: Ego network betweenness. Soc. Netw. **27**(1), 31–38 (2005)
8. Fan, Q., Wang, Z., Chan, C.Y., Tan, K.L.: Supporting window analytics over large-scale dynamic graphs, CORR (2015). arxiv:1510.07104
9. Huebsch, R., Garofalakis, M., Hellerstein, J.M., Stoica, I.: Sharing aggregate computation for distributed queries. In: SIGMOD 2007, pp. 485–496 (2007)
10. Ma, H.H., Gustafson, S., Moitra, A., Bracewell, D.: Ego-centric network sampling in viral marketing applications. In: Ting, I.-H., Wu, H.-J., Ho, T.-H. (eds.) Mining and Analyzing Social Networks. SCI, vol. 288, pp. 35–51. Springer, Heidelberg (2010)
11. Ma, N., Guan, J., Zhao, Y.: Bringing pagerank to the citation analysis. Inf. Process. Manage. **44**(2), 800–810 (2008)
12. Mondal, J., Deshpande, A.: Eagr: supporting continuous ego-centric aggregate queries over large dynamic graphs. In: SIGMOD 2014, pp. 1335–1346 (2014)
13. Moustafa, W.E., Deshpande, A., Getoor, L.: Ego-centric graph pattern census. In: ICDE 2012, pp. 234–245 (2012)
14. Navlakha, S., Rastogi, R., Shrivastava, N.: Graph summarization with bounded error. In: SIGMOD 2008, pp. 419–432 (2008)
15. Trigoni, N., Yao, Y., Demers, A., Gehrke, J., Rajaraman, R.: Multi-query optimization for sensor networks. In: Prasanna, V.K., Iyengar, S.S., Spirakis, P.G., Welsh, M. (eds.) DCOSS 2005. LNCS, vol. 3560, pp. 307–321. Springer, Heidelberg (2005)
16. Vassilevska, V., Pinar, A.: Finding nonoverlapping dense blocks of a sparse matrix. Lawrence Berkeley National Laboratory, Livermore (2004)
17. Wang, Z., Fan, Q., Wang, H., Tan, K.-L., Agrawal, D., El Abbadi, A.: Pagrol: parallel graph OLAP over large-scale attributed graphs. In: ICDE 2014, pp. 496–507 (2014)
18. Yan, X., He, B., Zhu, F., Han, J.: Top-k aggregation queries over large networks. In: ICDE 2010, pp. 377–380 (2010)
19. Zhao, P., Li, X., Xin, D., Han, J.: Graph cube: on warehousing and OLAP multi-dimensional networks. In: SIGMOD 2011, pp. 853–864 (2011)

Bitruss Decomposition of Bipartite Graphs

Zhaonian Zou$^{(\boxtimes)}$

School of Computer Science and Technology,
Harbin Institute of Technology, Harbin, China
znzou@hit.edu.cn

Abstract. In this paper, we propose *bitruss*, a new notion of a dense subgraph of a bipartite graph. Specifically, the *k-bitruss* of a bipartite graph is the largest edge-induced subgraph H such that every edge of H is contained in at least k rectangles within H. The *bitruss decomposition* of a bipartite graph is the set of all nonempty k-bitrusses of the bipartite graph for $k \geq 0$. In this paper, we show that the bitruss decomposition of a bipartite graph have three important properties. First, the bitruss decomposition is unique. Second, the bitruss decomposition is hierarchical, that is, the $(k+1)$-bitruss is a subgraph of the k-bitruss for all $k \geq 0$. Third, the bitruss decomposition can be computed in polynomial time. These three interesting properties make bitruss a promising notion of dense bipartite subgraphs.

Keywords: Bipartite graph · Dense subgraph · Bitruss · Bitruss decomposition

1 Introduction

A graph is a *bipartite graph* if its vertices can be partitioned into two disjoint sets L and R such that every edge connects a vertex in L to a vertex in R. A bipartite graph is a general data structure for representing complicated relationships between two disjoint sets of entities. For example, in an online recommender system, the ratings on items by users can be represented by a bipartite graph, where users and items are two disjoint vertex sets, and there is an edge connecting a user to an item if the user posts a rating on that item. So far, vast amounts of data represented by bipartite graphs has been accumulated in a variety of applications, including online recommender systems, online social networks, and bookmarking systems [13].

Discovering dense subgraphs of a bipartite graph is of great significance in graph mining that encompasses many diverse applications. Examples include identifying similar users and similar items based on the ratings on items posted by the users of an online recommender system, and building taxonomy of bookmarks based on tags assigned to bookmarked URL's.

Z. Zou—was partially supported by the NSF of China (No. 61532015 and No. 61173023) and the 973 Program of China (No. 2011CB036202).

S.B. Navathe et al. (Eds.): DASFAA 2016, Part II, LNCS 9643, pp. 218–233, 2016.
DOI: 10.1007/978-3-319-32049-6_14

Various notions of dense bipartite subgraphs have been used in the literature, including *maximum bicliques* [1,8,16], *maximum average-degree subgraphs* [6,7], *maximum density subgraphs* [11], (i,j)-*cores* [12], and *dense k-subgraphs* [5,10]. Section 2 provides a brief description of these notions. Although each of these notions has its own strengths, it has one or more limitations described below.

- The dense subgraphs in terms of these definitions are generally very small, compared with the size of the underlying bipartite graph. Hence, they cannot capture the global density characteristics of the underlying bipartite graph.
- The dense subgraphs in terms of these definitions are generally scattered in the underlying bipartite graph and may be overlapped or disjoint.
- It is generally computationally expensive to find the optimal answers under most of these definitions.

To overcome these drawbacks, we propose *"bitruss"*, a new notion of a dense bipartite subgraph, in this paper. This notion is inspired by the concept of *trusses* [3] defined exclusively for non-bipartite graphs. The definition of a truss is based on *triangles*, that is, cycles with three vertices. Specifically, the k-*truss* in a non-bipartite graph is the maximal edge-induced subgraph H such that every edge of H is contained in at least k triangles within H. The *truss decomposition* of a non-bipartite graph is composed by all nontrivial k-trusses in the graph for $k \geq 0$. The truss decomposition of a non-bipartite graph possesses two important properties. First, the truss decomposition is unique. Second, the $(k + 1)$-truss is a subgraph of the k-truss for all $k \geq 0$. Therefore, the truss decomposition provides a hierarchical organization of subgraphs of different granularities and densities. Furthermore, the truss decomposition of a non-bipartite graph can be computed in $O(m^{1.5})$ time [17], where m is the number of edges of the graph.

Unfortunately, the notion of a truss cannot be applied to a bipartite graph because there is definitely *no* triangles in a bipartite graph. Li et al. [14] attempt to apply the concept of a truss to a bipartite graph by augmenting the bipartite graph to a general non-bipartite graph. In particular, they add an auxiliary edge between a pair of vertices if they share a common neighbor vertex in the bipartite graph. Then, they compute the truss decomposition of the augmented non-bipartite graph and remove the auxiliary edges from the discovered trusses. Although simple, this approach has two main drawbacks.

- This approach cannot distinguish bipartite graphs with significantly different densities. To illustrate, consider the complete bipartite graphs G_1, G_2 and G_3 shown in Fig. 1. All of them consist of 8 vertices, while the numbers of left vertices are 1, 2 and 4, respectively. Intuitively, G_2 is denser than G_1, and G_3 is even denser than G_2. However, after augmenting all these graphs into non-bipartite graphs, we obtain the same one—the complete graph with 8 vertices, which is in turn a 6-truss. After removing the auxiliary edges, we obtain G_1, G_2 and G_3 again. Hence, this approach regards G_1, G_2 and G_3 to be at the same level of cohesiveness, which contradicts with our intuition.
- Since a large number of auxiliary edges are added to the bipartite graph, computing the truss decomposition of the augmented graph tends to be more

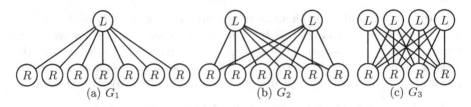

Fig. 1. Three bipartite graphs with the same number of vertices but different densities. Vertices labeled by L are left vertices, and vertices labeled by R are right vertices.

expensive. In the worst case, $\binom{l}{2} + \binom{r}{2}$ auxiliary edges can be added to a bipartite graph with l left vertices and r right vertices.

In this paper, we propose a new notion of a dense bipartite graph, called *bitruss*. The definition of a bitruss is based on *rectangles*, where a rectangle refers to a complete bipartite graph with two left vertices and two right vertices [18]. In particular, the k-bitruss in a bipartite graph is the maximal edge-induced subgraph H such that every edge of H is contained in at least k rectangles within H. The *bitruss decomposition* of a bipartite graph is the set of all nontrivial k-bitrusses in the bipartite graph for all $k \geq 0$.

The rationale behind the definition of a bitruss is that a rectangle is a commonly observed *motif*, that is, a subgraph that repeats itself more frequently in a real graph than in a randomly linked graph [15]. A rectangle represents the homogeneity between its left vertices as well as the homogeneity between its right vertices. For example, in a user-item bipartite graph, two users and two items such that both users post high ratings on both items constitute a rectangle, which indicates that the users are likely to have common interests and the items are likely to be similar.

The bitruss decomposition of a bipartite graph provides a hierarchical organization of bipartite subgraphs with increasing sizes and decreasing densities. To better illustrate this, we investigate the properties of the bitruss decomposition of a bipartite graph, which are described below.

- *Uniqueness Property:* The bitruss decomposition of a bipartite graph is unique.
- *Nestedness Property:* The $(k+1)$-bitruss is a subgraph of the k-bitruss for all $k \geq 0$.

Because of the nestedness property, the larger k is, the smaller and denser the k-bitruss is. Hence, the hierarchical organization of bitrusses enables us to explore subgraphs of different density levels by adjusting the parameter k.

Another advantage of the bitruss notion is that the bitruss decomposition of a bipartite graph can be computed in polynomial time. Specifically, we propose an algorithm, called Peel, that computes the bitruss decomposition of a bipartite graph in $O(m^2)$ time and in $O(l + r + m)$ space, where l, r and m are the number of left vertices, the number of right vertices and the number of edges in the bipartite graph, respectively.

We performed an extensive set of experiments to evaluate the effectiveness of the bitruss notion as well as the efficiency of the `Peel` algorithm. The experimental results verify the theoretical results given above. Furthermore, the density of the k-bitruss of a bipartite graph generally increases as k becomes larger, and hence, the bitruss notion is generally consistent with the classical notions of dense bipartite subgraphs.

2   Related Work

A bipartite graph is said to be a *biclique* or a *complete bipartite graph* if there is an edge between every left vertex and every right vertex. The *maximum biclique* is a classic notion of a dense bipartite subgraph, which has two specific definitions. The *maximum vertex biclique* is the biclique with the maximum number of vertices, and the *maximum edge biclique* is the biclique with the maximum number of edges [1,16]. The maximum vertex biclique can be found in polynomial time, while finding the maximum edge biclique is a NP-complete problem [16].

In [7], the density of a graph is defined by the ratio of the number of edges to the number of vertices, which is half of the average degree of all vertices. This definition is applicable to both bipartite graphs and non-bipartite graphs. Goldberg [7] proposed a max-flow-based algorithm to find a subgraph of the maximum average degree. Gallo et al. [6] improves the running time to $O(nm \log(n^2/m))$ by using the parametric max-flow algorithms, where n and m are the number of vertices and the number of edges of the input graph, respectively.

Khuller and Saha [11] defines the density of a bipartite graph by $|E|/\sqrt{|L||R|}$, where E is the set of edges, L is the set of left vertices, and R is the set of right vertices. Without any size constraints, a subgraph of the maximum density can be found in polynomial time. When a constraint on the minimum size is specified, the problem is NP-complete, and fast algorithms have been proposed to find subgraphs within a factor 2 of the maximum density [11].

The (i, j)-*core* of a bipartite graph is the maximal subgraph such that every vertex on the left side has degree at least i and that every vertex on the right side has degree at least j [12]. In order to enumerate all nontrivial (i, j)-cores, Kumar et al. [12] proposed an algorithm based on frequent itemset mining.

Although each of the notions above has its own strengths, it has one or more limitations as described in Sect. 1. The notion of a *bitruss* proposed in this paper is inspired by the concept of a *truss* [3] defined for a non-bipartite graph. Cohen [3] proposed the first algorithm for computing the truss decomposition of a non-bipartite graph. Wang and Cheng [17] improved Cohen's algorithm by reducing random access to adjacency lists of vertices of high degrees. The improved algorithm runs in $O(m^{1.5})$ time, where m is the number of edges of the graph. Cohen [2] proposed the first parallel algorithm. Wang and Cheng [17] argued that the iterative triangle listing process in the MapReduce algorithm hinders parallelization and proposed two I/O-efficient algorithms [17] for handling large non-bipartite graphs that cannot fit in main memory. Huang et al. [9] proposed the first distributed algorithm for this problem. Unfortunately, the concept of a truss cannot be applied to a bipartite graphs because there is *no* triangles in a bipartite graph. Moreover, as we discussed in Sect. 1, the attempt to apply the notion of trusses to bipartite graphs [14] has several drawbacks.

3   Bitruss Decomposition Problem

Preliminaries. In this paper we consider *undirected graphs*, that is, (u, v) and (v, u) refer to the same edge. A *bipartite graph* is a graph whose vertices can be partitioned into two disjoint sets L and R such that each edge connects a vertex in L to one in R. The sets L and R are called *partite sets*. One often uses (L, R, E) to denote a bipartite graph with partite sets L and R and edges E. A *complete bipartite graph* is a bipartite graph (L, R, E) such that each vertex in L is adjacent to all vertices in R. One often denote a complete bipartite graph (L, R, E) with $|L| = l$ and $|R| = r$ by $K_{l,r}$. Clearly, $K_{2,2}$ is a cycle of length 4, which is called a *rectangle* figuratively.

In the paper we denote the partite sets of a bipartite graph G by $L(G)$ and $R(G)$, respectively, and the edge set of G by $E(G)$. Let $N_G(v)$ denote the set of vertices adjacent to a vertex v in G. The cardinality of $N_G(v)$ is called the *degree* of v, denoted by $d_G(v)$. In the paper, when G is explicit in context, we simplify $N_G(v)$ and $d_G(v)$ into $N(v)$ and $d(v)$, respectively.

Problem Statement. We now define some new concepts and give a formal statement of the *bitruss decomposition problem*.

Definition 1. *Given a bipartite graph G and an integer $k \geq 0$, the k-bitruss of G, denoted by $T_k(G)$ is the largest edge-induced subgraph H of G such that every edge of H is contained in at least k rectangles within H.*

Definition 2. *The* bitruss decomposition *of a bipartite graph G is a set $\{T_k(G)|\ 0 \leq k \leq K\}$, where K is the largest integer such that $T_K(G)$ is nontrivial.*

Definition 3. *The* bitruss number *of an edge e of a bipartite graph G, denoted by $t_G(e)$, is the largest integer k such that $T_k(G)$ contained e. Let $\tau(G)$ denote the maximum bitruss number of edges of G.*

The *bitruss decomposition problem* is therefore computing the bitruss decomposition of a bipartite graph. Figure 2 illustrates a bipartite graph G. The 0-bitruss of G is G itself. The k-bitrusses of G for $k = 1, 2, 3$ are shown in Fig. 2(b)–(d), respectively. The maximum bitruss number $\tau(G)$ is 3.

Next, we study the properties of a bitruss decomposition.

Theorem 1. *The k-bitruss of a bipartite graph is unique for all $k \geq 0$.*

Proof. Let H and H' be two different k-bitrusses. Let $H+H'$ denote the bipartite graph obtained by merging H and H', that is, $L(H + H') = L(H) \cup L(H')$, $R(H + H') = R(H) \cup R(H')$, and $E(H + H') = E(H) \cup E(H')$. Clearly, each edge of H is contained in at k rectangles in H, which are certainly rectangles in

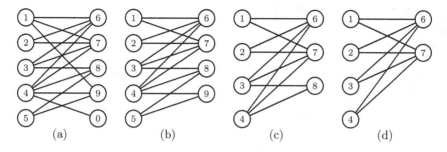

Fig. 2. An example of bitruss decomposition of a bipartite graph. (a) a bipartite graph G; (b) the 1-bitruss of G; (c) the 2-bitruss of G; (d) the 3-bitruss of G.

$H + H'$. Similarly, each edge of H' is also contained in at k rectangles in $H + H'$. Thus, every edge of $H + H'$ is contained in at least k rectangles in $H + H'$. By Definition 1, neither H nor H' is a k-bitruss of G, which is a contradiction. Thus, the theorem holds.   □

Corollary 1. *The bitruss decomposition of a bipartite graph is unique.*

The next theorem shows the hierarchy property of a bitruss decomposition.

Theorem 2. *For all bipartite graphs G and all integers $k \geq 0$, $T_{k+1}(G)$ is a subgraph of $T_k(G)$.*

Proof. Suppose $T_{k+1}(G)$ is not a subgraph of $T_k(G)$. Let $H = T_{k+1}(G) + T_k(G)$. Each edge of $T_k(G)$ is contained in at least k rectangles in $T_k(G)$, which are certainly rectangles in H. Similarly, each edge of $T_{k+1}(G)$ is contained in at least $k + 1$ rectangles in H. By Definition 1, $T_k(G)$ cannot be the k-bitruss of G, which is a contradiction. This completes the proof.   □

4   Bitruss Decomposition Algorithm

In this section we propose a simple yet effective algorithm for the Bitruss Decomposition Problem. We are given as input a bipartite graph G with $|L(G)| = l$, $|R(G)| = r$, and $|E(G)| = m$. Rather than outputting the bitruss decomposition of G, the algorithm outputs the bitruss numbers of all edges in G since the bitruss decomposition of G can be reconstructed by sorting all edges $E(G)$ in decreasing order of bitruss numbers in $O(m \log m)$ time and then by scanning the sorted $E(G)$ in $O(m)$ time. The algorithm removes (peels off) k-bitrusses in increasing order of k, which runs in the following steps.

1. Set a variable k to 0. Initialize a variable $t[e]$ to keep the bitruss number of each edge e of G. Count the number of rectangles in G containing each edge e and keep the rectangle count in $c[e]$.
2. If there is no edge in G, output t and terminate; otherwise goto next step.

Algorithm. Peel

Input: a bipartite graph G
Output: the bitruss numbers of all edges of G
1: $k \leftarrow 0$
2: $c \leftarrow$ CountRectangles(G) //$c[e]$ keeps the number of rectangles in G containing e
3: **while** $E(G) \neq \emptyset$ **do**
4:     **if** there is an edge e of G such that $c[e] \leq k$ **then**
5:         **for all** $v' \in N(u)$ **do**
6:             **for all** $u' \in N(v)$ **do**
7:                 **if** $(u', v') \in E(G)$ **then**
8:                     $c[(u, v')] \leftarrow c[(u, v')] - 1;\ c[(u', v)] \leftarrow c[(u', v)] - 1$
                        $c[(u', v')] \leftarrow c[(u', v')] - 1$
9:         $t[e] \leftarrow k$
10:        $E(G) \leftarrow E(G) \setminus \{e\}$
11:    **else**
12:        $k \leftarrow k + 1$
13: **return** t

---

3. If there is no edge e such that $c[e] \leq k$, increase k by 1 and goto Step 2; otherwise carry out the following steps.

   3a. For each rectangle containing e, decrease $c[e']$ by 1 for all edges e' in the rectangle other than e.

   3b. Set $t[e]$ to k, remove e from G, and goto Step 2.

The pseudocode of the algorithm is given in Algorithm Peel[1]. At line 2, we count the number of rectangles containing each edge of G. The main loop at lines 3–12 repeats Step 3 (lines 4–12) until there is no edge in G. In each iteration of the main loop, if there exists an edge $e \in E(G)$ such that $c[e] \leq k$, we carry out Step 3a at lines 5–8, followed by Step 3b at lines 9–10. In particular, for each vertex v' in $N(u)$ and each u' in $N(v)$, if u' is adjacent to v', vertices u, v, u', v' induce a rectangle. The removal of e must eliminate the rectangle containing u, v, u', v', so we decrease all of $c[(u, v')]$, $c[(u', v)]$ and $c[(u', v')]$ by 1. In each iteration of the main loop, if there is no edge $e \in E(G)$ such that $c[e] \leq k$, we increase k by 1 at line 12. Once G becomes empty, the bitruss numbers of all edges stored in t are outputted at line 13, then the algorithm terminates.

Next, we discuss some issues related to the implementation of Peel.

– We can use the in-memory algorithm proposed in [18] to count the rectangles containing each edge of G. The algorithm runs in $O(\sum_{u \in L(G)} d^2(u))$ time and $O(l+r+m)$ space [18]. Since $\sum_{u \in L(G)} d^2(u) = O((\sum_{u \in L(G)} d(u))^2) = O(m^2)$, the algorithm runs in $O(m^2)$ time.

– To enable fast adjacency test, we use a hash table to store $N(v)$ for all vertices v. Thus, the condition at line 7 can be tested by accessing the hash table of $N(u')$ (or $N(v')$) with key v' (or u') in $O(1)$ time. If $|N(u')| < |N(v')|$, we access $N(u')$ with key v'; otherwise, we access $N(v')$ with key u'.

---

[1] The Peel algorithm mimics the process of peeling an onion, so is named Peel.

Correctness. We now prove the correctness of `Peel`. Note that the input graph G is constantly changed during the execution of `Peel`, so we use G^* to denote G when inputted to `Peel`. The following lemma gives a loop invariance of `Peel`.

Lemma 1. *Every time when line 4 is executed, $T_{k+1}(G^*) \subseteq G \subseteq T_k(G^*)$.*

Proof. We prove the lemma by induction. *Base Case:* When line 4 is executed for the first time, we have $k = 0$ and $G = G^*$. Clearly, $T_1(G^*) \subseteq G^* = G = T_0(G^*)$.

Induction: Assume that $T_{k+1}(G^*) \subseteq G \subseteq T_k(G^*)$ when line 4 is executed for the ith time. If `Peel` continues to execute line 5 after executing line 4, we have that e is contained in at most k rectangles in G. By the assumption that $T_{k+1}(G^*) \subseteq G$, e is contained in at most k rectangles in $T_{k+1}(G^*)+e$, so e cannot be an edge of $T_{k+1}(G^*)$. Thus, we have $T_{k+1}(G^*) \subseteq G - e \subseteq T_k(G^*)$. At line 10, `Peel` removes e from G and goes to line 3. Therefore, $T_{k+1}(G^*) \subseteq G \subseteq T_k(G^*)$ when line 4 is executed for the $(i+1)$th time.

If `Peel` continues to execute line 12 after executing line 4, all edges of G are contained in at least $k+1$ rectangles, so $G \subseteq T_{k+1}(G^*)$. By the assumption that $T_{k+1}(G^*) \subseteq G$, we have $T_{k+1}(G^*) = G$. Moreover, it follows from Theorem 2 that $T_{k+2}(G^*) \subseteq T_{k+1}(G^*) = G$. At line 12, `Peel` increases k by 1 and goes to line 3. Hence, $T_{k+1}(G^*) \subseteq G \subseteq T_k(G^*)$ when line 4 is executed for the $(i+1)$th time. By induction the lemma holds. □

Based on the loop invariance, we have the following sufficient and necessary condition of bitruss numbers.

Theorem 3. *The bitruss number of an edge e is l if and only if $k = l$ when e is deleted from G.*

Proof. We first prove the sufficiency. If $k = l$ when e is deleted from G, it follows from Lemma 1 that $T_{l+1}(G^*) \subseteq G \subseteq T_l(G^*)$ just before e is removed from G. According to the proof of Lemma 1, e is not an edge of $T_{l+1}(G^*)$. Thus, the bitruss number of e is l.

Next, we prove the necessity. We first prove that $k \leq l$ when e is deleted from G. Suppose $k > l$. By Theorem 2, we have $T_k(G^*) \subseteq T_{l+1}(G^*)$. By Lemma 1, we have $G \subseteq T_k(G^*)$ just before e is deleted from G. Thus, $G \subseteq T_{l+1}(G^*)$, so the bitruss number of e is greater than l, which is a contradiction. Next, we prove that $k \geq l$ when e is deleted from G. Suppose $k < l$, so edge e is contained in less than l rectangles in G. By Theorem 2, we have $T_l(G^*) \subseteq T_{k+1}(G^*)$. By Lemma 1, we have $T_{k+1}(G^*) \subseteq G$ just before e is deleted from G. Thus, $T_l(G^*) \subseteq G$, so e is contained in at least l rectangles in G, which is a contradiction. Hence, $k = l$ when e is deleted from G. The theorem thus holds. □

By Theorem 3, the `Peel` algorithm is correct.

Complexity. We now analyze the complexity of `Peel`. For each edge $e = (u, v)$ selected at line 4, the loop at lines 5–8 runs in $O(d(u)d(v))$ time. Thus, the loop totally runs in $O(\sum_{(u,v)\in E(G)} d(u)d(v))$ time. Note that

$$\sum_{(u,v)\in E(G)} d(u)d(v) \leq \sum_{u\in L(G)} \sum_{v\in R(G)} d(u)d(v) \leq \sum_{u\in L(G)} d(u) \sum_{v\in R(G)} d(v) = m^2 \ ,$$

where the last equality follows from the fact that every edge connects a vertex in $L(G)$ to one in $R(G)$, so $\sum_{u \in R(G)} d(u) = \sum_{v \in R(G)} d(v) = m$. Thus, Peel runs in $O(m^2)$ time, which is optimal since counting all rectangles in G requires $\Theta(m^2)$ in the worst case. The Peel algorithm requires $O(l + r + m)$ space to store the input graph G and $O(m)$ space to store $c[e]$ and $t[e]$ for all $e \in E(G)$. Hence, the space complexity of Peel is $O(l + r + m)$.

5   Experimental Evaluation

5.1   Experimental Setting

We implemented the Peel algorithm in C++ and compiled it using g++ with option -O3. All experiments were performed on a computer powered by a 2.8GHz Intel Core i7 CPU and 16GB of RAM, running Mac OS X 10.10.1.

Evaluation Metrics. We evaluated the Peel algorithm by the metrics below.

- *Size.* The size of the k-bitruss of a bipartite graph is evaluated by the number of vertices and the number of edges of the k-bitruss.
- *Density.* For a sufficiently large integer k, the k-bitruss of a bipartite graph is expected to be a dense subgraph. Moreover, the k-bitruss is expected to be denser as k becomes larger. The *density* of a bipartite graph (L, R, E) is evaluated by $|E|/(|L||R|)$. In this sense, a bipartite graph is dense if its density is close to 1.
- *Execution time.* The time efficiency of the Peel algorithm is measured by its execution time.

Datasets. Our experiments were performed on the following bipartite graphs. The structural statistics of these bipartite graphs are summarized in Fig. 3.

Bipartite graph	L	R	E	d_L	d_R	d	$d_L^{\max}$	$d_R^{\max}$
Delicious	68755	14346	487131	7.1	34.0	5.9	10	9435
IMDB	303617	896302	3782463	12.5	4.2	3.2	1334	1590
MovieLens-1M-4-5	6038	3533	575281	95.3	162.8	60.0	1435	2853
Netflix	998	280406	1134613	1136.9	4.0	4.0	57537	457
NotreDame	383640	127823	1470404	3.8	11.5	2.9	646	294
Sandi	314	360	613	2.0	1.7	0.9	26	6
WikiElec	2043	5504	81862	40.1	14.9	10.9	323	766

Fig. 3. Summary of bipartite graphs used in experiments. L: the number of left vertices; R: the number of right vertices; E: the number of edges; d_L: the average degree of left vertices; d_R: the average degree of right vertices; d: the average degree of all vertices; $d_L^{\max}$: the maximum degree of left vertices; $d_R^{\max}$: the maximum degree of right vertices.

- **Delicious**. This bipartite graph contains the bookmarking and tagging information from 2000 users of the social bookmarking system Delicious.com. This graph was released in HetRec'11 (http://ir.ii.uam.es/hetrec2011). In this graph, a left vertex represents a user, a right vertex represents a bookmark, and an edge connects a user vertex to a bookmark vertex if the user assigns a tag to the bookmark.
- **IMDB**. This bipartite graph was obtained from the University of Florida sparse matrix collection [4]. It represents the relationships between movies and actors. In particular, there is an edge connecting a movie vertex to a actor vertex if the actor played in the movie.
- **MovieLens**. This dataset contains 10000054 ratings applied to 10681 movies by 71567 users of the online movie recommender service MovieLens. It is publicly available at the GroupLens Datasets. We extracted a set of bipartite graphs from this dataset, where left vertices represent users, right vertices represent movies, and an edge connects a user vertex to a movie vertex if the user's rating on the movie is within a specified interval $[l, u]$. In our experiments, we use **MovieLens**-m-l-u to denote the bipartite graph constructed based on the ratings within score interval $[l, u]$ among the first m ratings, where $l, u \in \{1, 2, \ldots, 5\}$ and $m \leq 10000054$.
- **Netflix**. The Netflix dataset contains over 100 million ratings from 480 thousand randomly-chosen, anonymous Netflix customers over 17 thousand movie titles. We extracted a bipartite graph that contains 998 movie vertices, 280406 user vertices and 1470404 edges.
- **NotreDame**. This bipartite graph was obtained from the University of Florida sparse matrix collection [4]. It consists of 1470404 edges connecting 383640 vertices on the left partite to 127823 vertices on the right partite.
- **Sandi**. This is a small bipartite graph obtained from the University of Florida sparse matrix collection [4]. This bipartite graph is used to show the characteristics of bitruss decomposition in details.
- **WikiElect**. This bipartite graph is the voting network of Wikipedia [13], where a link indicates a positive vote by one user on the promotion to admin status of a candidate.

5.2 Experimental Results

Bitruss Sizes and the Hierarchy Property. In this experiment, we examine the sizes of k-bitrusses with respect to k and verify the hierarchy property of a bitruss decomposition stated in Theorem 2.

Figure 4 shows the experimental results obtained on bipartite graph **Sandi**. Specifically, for every k-bitruss of **Sandi**, Fig. 4 reports the number L of left vertices, the number R of right vertices and the number E of edges, respectively. We observe that all of L, R and E monotonically decrease with respect to k. Our experimental results also show that the $(k+1)$-bitruss of **Sandi** is a proper subgraph of the k-bitruss of **Sandi** for $k = 0, 1, \ldots, 5$. Thus, the bitruss decomposition of **Sandi** satisfies the hierarchy property stated in Theorem 2. Because of this, all of L, R and E decrease as k becomes larger.

k	L	R	E	Avg. deg.	Density	k	L	R	E	Avg. deg.	Density
0	314	360	613	0.909	0.005	4	5	18	39	1.696	0.433
1	55	80	184	1.363	0.042	5	4	13	26	1.529	0.500
2	33	50	118	1.422	0.072	6	2	7	14	1.556	1.000
3	9	26	55	1.571	0.235						

Fig. 4. Structural properties of the k-bitruss of Sandi. L: the number L of left vertices, R: the number of right vertices, E: the number of edges.

Figure 5 illustrates the sizes of the k-bitrusses of bipartite graphs Delicious, IMDB, MovieLens, Netflix, NotreDame and WikiElec, particularly, the number of vertices and edges. For each bipartite graph, the number of vertices and edges of the k-bitruss all monotonically decrease as k becomes larger. Furthermore, the experimental results verify that, for all these bipartite graphs, the $(k+1)$-bitruss is a subgraph of the k-bitruss for all $k \geq 0$.

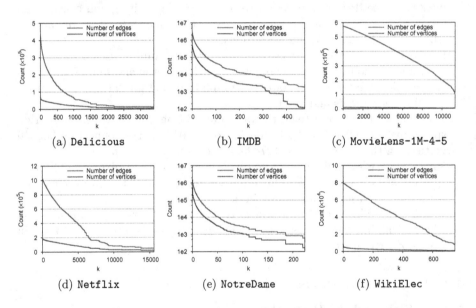

(a) Delicious   (b) IMDB   (c) MovieLens-1M-4-5

(d) Netflix   (e) NotreDame   (f) WikiElec

Fig. 5. The number of vertices and the number of edges of the k-bitrusses of bipartite graphs Delicious, IMDB, MovieLens, Netflix, NotreDame and WikiElec.

Density of Bitrusses. In this experiment, we examine the density of bitrusses of a bipartite graph to show that the notion of a bitruss is consistent with the notion of a dense subgraph.

Figure 4 shows the experimental results obtained on bipartite graph Sandi. Specifically, for every k-bitruss of Sandi, Fig. 4 reports the average degree of vertices evaluated by $E/(L + R)$ and the density of the k-bitruss evaluated by $E/(LR)$, where L, R and E refer to the number of left vertices, right vertices and

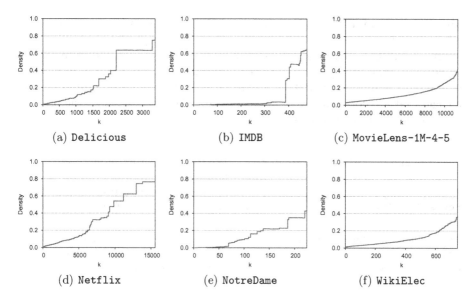

Fig. 6. Density of k-bitrusses of bipartite graphs Delicious, IMDB, MovieLens, Netflix, NotreDame and WikiElec.

edges of the k-bitruss, respectively. The maximum bitruss number of the edges in Sandi is 6. The 6-bitruss of Sandi is a complete bipartite graph with 2 left vertices and 7 right vertices, so the average degree of its vertices is $14/(2+7) = 1.556$, and its density is $14/(2 \times 7) = 1$. Interestingly, the 6-bitruss of Sandi is identical to the maximum edge biclique in Sandi. From Fig. 4, we also observe that the k-bitruss of Sandi becomes denser as k increases. Specifically, the 6-bitruss is the densest (density $= 1$). This observation verifies that the notion of a bitruss is consistent with the notion of a dense subgraph, that is, the k-bitruss for larger k captures denser region of a bipartite graph.

Figure 4 also shows that the average degree of vertices in the k-bitruss of Sandi generally becomes larger as k increases except that the 3-bitruss and the 4-bitruss have slightly higher average degrees than the 6-bitruss. This is because average degree is not a stable density metric, which may yield counterintuitive results. For instance, consider two bipartite graphs $G_1 = (L_1, R_1, E_1)$ and $G_2 = (L_2, R_2, E_2)$ with $|L_1| = 2|L_2|$, $|R_1| = 2|R_2|$, and $|E_1| = 2|E_2|$. The average degrees of G_1 and G_2 are equal; however, the density of G_1 is only half of the density of G_2. Hence, we do not use average degree as a density metric.

Figure 6 shows the density of the k-bitrusses of bipartite graphs Delicious, IMDB, MovieLens, Netflix, NotreDame and WikiElec. For each bipartite graph, the density of the k-bitruss is generally increasing with respect to k except very few values of k in IMDB. Note that, for larger k, the number of rectangles containing each edge of the k-bitruss becomes larger. Therefore, the notion of a bitruss is consistent with the notion of a dense subgraph.

Distributions of Bitruss Numbers. In this experiment, we study the distribution of bitruss numbers of edges in a bipartite graph. Figure 7 shows the minimum and the maximum bitruss numbers of the edges in bipartite graphs `Delicious`, `IMDB`, `MovieLens`, `Netflix`, `NotreDame` and `WikiElec`. The minimum bitruss numbers are all 0, indicating that all these bipartite graphs consist of edges that are not involved in any rectangles. Figure 7 also shows the proportion of edges with bitruss number of 0.

Bipartite graph	Min (proportion)	Max	Bipartite graph	Min (proportion)	Max
Delicious	0 (3.6%)	3388	Netflix	0 (8.3%)	15584
IMDB	0 (31.8%)	477	NotreDame	0 (40.0%)	224
MovieLens-1M-4-5	0 (0.02%)	11321	WikiElec	0 (3.0%)	749

Fig. 7. The minimum and the maximum bitruss numbers of edges of bipartite graphs `Delicious`, `IMDB`, `MovieLens`, `Netflix`, `NotreDame` and `WikiElec`. The numbers in brackets indicate the proportion of edges that have bitruss number of 0.

Notably, the maximum bitruss number varies significantly from hundreds to tens of thousands for different bipartite graphs. Although there is no closed-form equations for the maximum bitruss number, the maximum bitruss number of a bipartite graph is greatly affected by the size and the density of the bipartite graph. As shown in Fig. 3, `Netflix` consists of more than one million edges, and the average degree of left vertices is as high as 1136.9, so the maximum bitruss

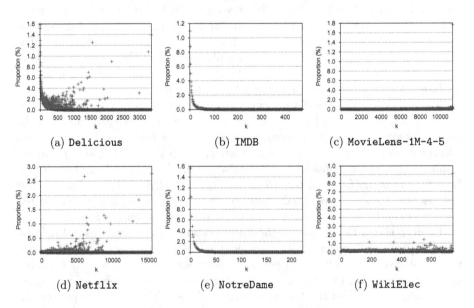

Fig. 8. Distribution of nonzero bitruss numbers of bipartite graphs `Delicious`, `IMDB`, `MovieLens`, `Netflix`, `NotreDame`, and `WikiElec`.

number of Netflix is also very large (15584). Similarly, MovieLens-1M-4-5 is also very dense, and its maximum bitruss number is 11321. Delicious contains hundreds of thousands of edges, and the average degree of the right vertices of Delicious is 34.0, so its maximum bitruss number is also large (3388).

The distributions of nonzero bitruss numbers of bipartite graphs Delicious, IMDB, MovieLens, Netflix, NotreDame and WikiElec are illustrated in Fig. 8. We observe that the distributions depends on specific bipartite graphs. In particular, the distributions for IMDB and NotreDame have long tails, that is, there are a large number of edges with very low bitruss numbers; while the edges with large bitruss numbers are very few. For other bipartite graphs, the distributions do not exhibit long tails.

Execution Time. In this experiment, we evaluate the execution time of Peel. As shown in Fig. 9, the execution time of Peel on bipartite graphs Delicious, IMDB, MovieLens, Netflix, NotreDame and WikiElec vary significantly.

Bipartite graph	Time	Memory	Bipartite graph	Time	Memory
Delicious	464	17.29	Netflix	576	53.02
IMDB	2455	107.50	NotreDame	4534	55.15
MovieLens-1M-4-5	6496	30.23	WikiElec	20	3.20

Fig. 9. Execution time and memory usage of Peel on bipartite graphs Delicious, IMDB, MovieLens, Netflix, NotreDame and WikiElec. The unit of execution time is second. The unit of memory usage is MB.

To study the effect of the input graph size on the execution time of Peel, we executed Peel on a collection of bipartite graphs generated based on the bipartite graph MovieLens, namely, MovieLens-100K-1-5, MovieLens-100K-2-5,

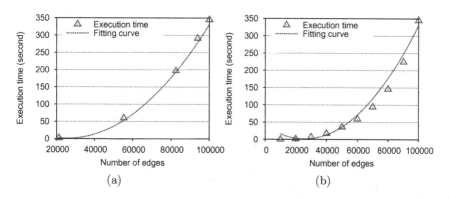

Fig. 10. Execution time of Peel with respect to the number of edges. (a) execution time of Peel on MovieLens-100K-l-5 for $l = 1, 2, \ldots, 5$. The fitting curve is $y = 6 \times 10^{-8}x^2 - 0.0031x + 41.471$. (b) execution time of Peel on MovieLens-m-1-5 for $m = 10K, 20K, \ldots, 100K$. The fitting curve is $y = 6 \times 10^{-8}x^2 - 0.0031x + 41.471$.

MovieLens-100K-3-5, MovieLens-100K-4-5 and MovieLens-100K-5-5. For $l = 1, 2, \ldots, 5$, the bipartite graphs MovieLens-100K-l-5 consist of 100000, 93890, 82520, 55357 and 21201 edges, respectively. As shown in Fig. 10(a), the execution time of Peel increases quadratically with respect to the number of edges of the input bipartite graph.

Furthermore, we produced a series of bipartite graphs MovieLens-m-1-5 for $m = 10K, 20K, \ldots, 100K$. Clearly, the number of edges of these bipartite graphs are $10000, 20000, \ldots, 100000$, respectively. We executed Peel on these bipartite graphs. Figure 10(b) shows that the execution time of Peel also increases quadratically with respect to the number of edges.

6   Conclusions

In this paper, we propose the concept of bitruss, a new notion of dense subgraphs of bipartite graphs. The bitruss decomposition of a bipartite graph satisfies two important properties. First, the bitruss decomposition of a bipartite graph is unique. Second, the $(k+1)$-bitruss is a subgraph of the k-bitruss for all $k \geq 0$. We also show that the bitruss decomposition of a bipartite graph can be computed in $O(m^2)$ time and $O(l + r + m)$ space, where l, m and n are the number of left vertices, right vertices and edges of the bipartite graph, respectively.

References

1. Ambühl, C., Mastrolilli, M., Svensson, O.: Inapproximability results for maximum edge biclique, minimum linear arrangement, and sparsest cut. SIAM J. Comput. **40**(2), 567–596 (2011)
2. Cohen, J.: Graph twiddling in a MapReduce world. Comput. Sci. Eng. **11**(4), 29–41 (2009)
3. Cohen, J.: Trusses: cohesive subgraphs for social network analysis. Technical report, National Security Agency Technical Report (2008)
4. Davis, T.A., Hu, Y.: The University of Florida sparse matrix collection. ACM Trans. Math. Softw. **38**(1), 1 (2011)
5. Feige, U., Kortsarz, G., Peleg, D.: The dense k-subgraph problem. Algorithmica **29**(3), 410–421 (2001)
6. Gallo, G., Grigoriadis, M.D., Tarjan, R.E.: A fast parametric maximum flow algorithm and applications. SIAM J. Comput. **18**(1), 30–55 (1989)
7. Goldberg, A.V.: Finding a maximum density subgraph. University of California Berkeley, CA (1984)
8. Hochbaum, D.S.: Approximating clique and biclique problems. J. Algorithms **29**(1), 174–200 (1998)
9. Huang, X., Cheng, H., Qin, L., Tian, W., Yu, J.X.: Querying k-truss community in large and dynamic graphs. In: SIGMOD, pp. 1311–1322 (2014)
10. Khot, S.: Ruling out PTAS for graph min-bisection, dense k-subgraph, and bipartite clique. SIAM J. Comput. **36**(4), 1025–1071 (2006)
11. Khuller, S., Saha, B.: On finding dense subgraphs. In: Albers, S., Marchetti-Spaccamela, A., Matias, Y., Nikoletseas, S., Thomas, W. (eds.) ICALP 2009, Part I. LNCS, vol. 5555, pp. 597–608. Springer, Heidelberg (2009)

12. Kumar, R., Raghavan, P., Rajagopalan, S., Tomkins, A.: Trawling the web for emerging cyber-communities. Comput. Netw. **31**(11–16), 1481–1493 (1999)
13. Leskovec, J., Krevl, A.: SNAP datasets: Stanford large network datasetcollection (2014). http://snap.stanford.edu/data
14. Li, Y., Kuboyama, T., Sakamoto, H.: Truss decomposition for extracting communities in bipartite graph. In: IMMM, pp. 76–80 (2013)
15. Milo, R., Shen-Orr, S., Itzkovitz, S., Kashtan, N., Chklovskii, D., Alon, U.: Network motifs: simple building blocks of complex networks. Science **298**(5594), 824–827 (2002)
16. Peeters, R.: The maximum edge biclique problem is NP-complete. Discrete Appl. Math. **131**(3), 651–654 (2003)
17. Wang, J., Cheng, J.: Truss decomposition in massive networks. PVLDB **5**(9), 812–823 (2012)
18. Wang, J., Fu, A.W., Cheng, J.: Rectangle counting in large bipartite graphs. In: BigData, pp. 17–24 (2014)

An I/O-Efficient Buffer Batch Replacement Policy for Update-Intensive Graph Databases

Ningnan Zhou[1,2], Xuan Zhou[1,2(✉)], Xiao Zhang[1,2],
Shan Wang[1,2], and Ling Liu[3]

[1] MOE Key Laboratory of DEKE, Renmin University of China, Beijing, China
zhou.xuan@outlook.com
[2] School of Information, Renmin University of China, Beijing 100872, China
[3] College of Computing, Georgia Institute of Technology, Atlanta, China

Abstract. With the proliferation of graph based applications, such as social network management and Web structure mining, update-intensive graph databases have become an important component of today's data management platforms. Several techniques have been recently proposed to exploit locality on both data organization and computational model in graph databases. However, little investigation has been conducted on buffer management of graph databases. To the best of our knowledge, current buffer managers of graph databases suffer performance loss caused by unnecessary random I/O access. To solve this problem, we develop a novel batch replacement policy for buffer management. This policy enables us to maximally exploit sequential I/O to improve the performance of graph database. To enable the policy, we devise a segment tree based buffer manager to efficiently maintains optimal replacement plan. Extensive experiments on real-world and synthetic datasets demonstrate the superiority of our method.

Keywords: Batch replacement · Buffer manager · Graph database · Data manipulation · Graph algorithm

1 Introduction

The rapid growth of graph data fosters a market of specialized graph databases such as Neo4j [9], Titan [10] and DEX [19]. To meet the needs of various graph based applications [11,12,14,26–28,34], these disk-based graph databases offer both database functionality such as insert/delete/update and analytical graph algorithms such as PageRank computation [6]. The evolving social network and the nature of some graph algorithms require graph databases to be update-friendly and update-efficient. For instance, to maintain a social network, each time a new friendship/connection establishes, a link connecting the pair of users should be inserted into the graph to reflect the change. In PageRank computation, the ranking score of every vertex needs to be updated in each iteration. This paper focuses on such update-intensive applications.

© Springer International Publishing Switzerland 2016
S.B. Navathe et al. (Eds.): DASFAA 2016, Part II, LNCS 9643, pp. 234–248, 2016.
DOI: 10.1007/978-3-319-32049-6_15

To support large scale graph databases, existing research work has mainly investigated the data organization and computational models. To achieve efficient data organization, the associated edges of each vertex are normal stored together. For example, in social networks, the friends of a user are usually stored in continuous data pages [9]. As a result, frequent requests such as "return the friends of a specific user" in Facebook or Twitter [15] can benefit from low latency of sequential I/O. As to computational model, the dominant vertex-centric [18] or edge-centric [24] processing models partition a graph based on vertices or edges, and treat each partition as a unit of computation. They can also benefit from sequential I/O.

Although existing graph databases widely adopt I/O efficient data organization and computational models, they rarely consider buffer replacement policies. In fact, they still adopt variants of Least Recently Used (LRU) or Least Frequently Used (LFU) policies [7,20], which evict one buffer page at a time and thus to some degree cancel out the effects of the specialized data organization and computational models. Figure 1 illustrates such a scenario. After the insertion of some new friends of user u, the data pages containing u's information, b_{u_1}, b_{u_2} and b_{u_3}, will be cached in the buffer. Note that b_{u_1}, b_{u_2} and b_{u_3} should be continuously located on disk. When a query such as "return the friend list of user v" is issued, the buffer manager requires to read in a new set of continuously located data pages, v_1, v_2 and v_3, which contain the friends of the user v. As the buffer is currently full, the buffer manager decides to evict b_{u_1}, b_{u_2} and b_{u_3} to make room for the incoming data pages. Following the existing replacement policy, the system will first seek to the position of u_1 to evict b_{u_1} and then seek to the position of v_1 to read in a new page. Iteratively, the system will perform 6 random I/Os according to the order marked by the arrows in Fig. 1. This is inefficient. If we can evict b_{u_1}, b_{u_2} and b_{u_3} in a batch, and read in v_1, v_2 and v_3 in a batch, we only need to perform two random disk seeks, and the other I/Os can be performed sequentially. Thus, such batch replacement can save 4 out of 6 random I/Os.

In this paper, we propose a batch replacement buffer manager for update-intensive graph databases. To the best of our knowledge, it is the first buffer replacement policy that exploits sequential I/O to speed up graph databases. Our design considers the following aspects: (1) the buffer manager should provide an unchanged interface to other layers of the graph database; (2) it should figure out the optimal replacement plan each time it needs to replace buffered pages; (3) it should minimize computational and memory overhead. To address these challenges, we first define the optimal replacement plan as the criteria to evict pages via sequential I/O. Then, we propose a segment tree based structure to organize buffered pages and to efficiently generate the optimal replacement plan. To evaluate the performance of our batch replacement buffer manager, we tried it on both real-world and synthetic datasets using typical workloads of database manipulation and graph algorithms. The experiment results show that (1) the batch replacement policy is able to achieve significant performance improvement by exploiting sequential I/O and (2) it is practical for graph databases.

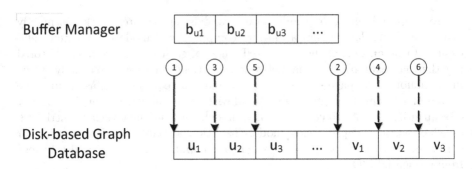

Fig. 1. An illustrative example for the effect of existing buffer manager and batch replacement in terms of random access, where the dashed arrow indicates the additional random access performed by existing buffer managers.

The contributions of this paper are threefold:

- We show the importance of exploiting sequential I/O in buffer management of graph databases.
- We propose a batch buffer replacement policy. Based on it, we define the optimal replacement plan and devise a segment tree based structure to manage buffered data pages and efficiently maintain the optimal plan.
- We conduct extensive experiments on real-world and synthetic datasets to verify the effectiveness of the batch replacement policy.

2   Related Work

Our work builds upon the existing techniques of graph databases, especially their data organization and computational models.

2.1   Data Organization

Conventionally, graph organization is built on top of the relational (*a.k.a.*, SQL) storage and graphs are stored as triplets [5,25]. In other words, each edge e directed from a vertex u to a vertex v in the graph is transformed into a triplet $\langle u, e, v \rangle$. However, it is known that RDBMS organization is not good at answering traversal types of graph queries [30]. Considering the locality of data manipulation, such as queries like "return the friends of a specific user", it is more efficient to pack in-edges and out-edges of the same vertex in two lists and store them together [22,32]. This has been adopted by most disk-based graph databases such as Neo4j. Therefore, we also assume such graph specific data organization.

2.2   Computational Model

Recently, a general iterative framework is adopted to process various graph algorithms such as PageRank and Shortest Path Computation. In the framework, every vertex and edge in the graph is associated with a value and at each iteration, the value on a vertex or an edge is updated in vertex-centric or edge-centric model.

Vertex-Centric Model. Vertex-centric model is explored by initial works such as GraphLab [16] and Pregel [18]. In vertex-centric model, each vertex and its associated edges are regarded as a unit of computation so that if the main memory can hold any single vertex and its associated edges, only sequential I/O for loading data and updating results is required for each computation unit. To improve scalability, MOCgraph further reduces the memory footprint using message online computing [33].

Edge-Centric Model. Because a single vertex in real-world graph data, such as a celebrity, may be associated with so many edges that they cannot fit in main memory, edge-centric model is proposed [14,24]. Edge-centric model partitions edges into disjoint sets and each set and its associated vertices form the unit of computation. In this way, each set can be hold in main memory to avoid random I/O access [13,34,35].

There is a significant body of work on distributed graph databases [8,23,29]. As our work focuses on speeding up a disk-based graph database on a single machine, our research is orthogonal and complementary to them.

2.3   Buffer Manager on Database

Existing buffer managers in graph databases usually adopt the variants of the LRU/LFU policy to reduce disk I/O. Neo4j adopts the LRU policy [9] while TurboGraph [13] maintains frequently used pages in memory. These works follow the same paradigm – when the buffer manager requires to read in a new page and the buffer gets overflow, only one buffered data page is evicted at a time. As a result, it introduces unnecessary random I/Os. To deal with this drawback, one recent work has proposed to remove buffer managers [17]. Besides, there are also alternative approaches which utilize index structures such as log structured merge tree [21] or fractal tree [3] to handle update-intensive workload. Both index structures process updates in a key range in a batch. However, as the physical pages of a key range may not be located consecutively on disk, random I/O still cannot be avoided completely.

In this paper, we aim to leverage sequential I/O by evicting buffered pages in a batch way rather following the existing paradigm which repeats evicting and reading one page at a time. Thus, our approach can benefit from the data organization and computational models for graph databases.

3   Batch Replacement Buffer Manager

In this section, we first present the problem definition for our batch replacement buffer manager. Then, we present the structure and algorithms of the proposed buffer manager.

3.1   Problem Formulation

As we have shown in Fig. 1 in Sect. 1, it is inefficient to follow the existing paradigm of buffer manager, which evicts only one buffered data page at a time. In this paper, we extend the single page based replacement plan to the one that considers a set of pages. Thus, the new definition of replacement plan subsumes that of the existing buffer managers.

Definition 1. *Replacement Plan. When the buffer manager gets overflow, a replacement plan is a set of buffered data pages that will be evicted before the buffer manager performs any subsequent read operation.*

For example, the ideal replacement plan in Fig. 1 is $\{b_{u_1}, b_{u_2}, b_{u_3}\}$.

Observing that evicting continuous buffered dirty data pages can maximize sequential I/O, the ideal batch replacement plan is to evict the longest sequence of such data pages.

Definition 2. *Optimal Batch Replacement Plan. Given a set of buffered pages with positions on the disk as $S = \{p_1, p_2, ..., p_n\}$, the optimal batch replacement plan is a subset $\mathcal{P} \subseteq S$ satisfying the following two conditions:*

(1) pages in $\mathcal{P}$ are continuous in disk, namely, there are $n - 1$ pairs of p_i and p_j in $\mathcal{P}$, such that $p_i \rightarrow p_j$ or $p_j \rightarrow p_i$, where $p_i \rightarrow p_j$ means that p_j is the successor data block in disk to p_i.

(2) any other subset $\mathcal{P}' \subseteq S$ satisfying Condition 1 contains less data pages than $\mathcal{P}$, namely, $|\mathcal{P}'| < |\mathcal{P}|$.

For example, in Fig. 1, the optimal batch replacement plan is $\{p_{b_1}, p_{b_2}, p_{b_3}\}$. Although its subset such as $\{p_{b_1}, p_{b_2}\}$ satisfy the first condition, they violate the second condition and are not the optimal batch replacement plan.

3.2   Overview

We would like a buffer manager to change its replacement policy to the optimal batch replacement plan. However, we also prefer the change is transparent to other components of a graph database. We identify three properties the batch replacement buffer manager should possess: (1) *transparency* requires to export the same interface to other layers in a graph database; (2) *effectiveness* requires to identify the exact optimal replacement plan and (3) *efficiency* requires to minimize the computation and space cost of buffer manager.

When a data page is being updated, if it is surrounded by a number of continuous buffered dirty pages, batch replacement may evict such an active page and cause thrashing. Therefore, we use a "using" component to keep track of such active data pages to avoid them from being evicted. Although our batch replacement buffer manager is designed for update-intensive applications, we also need to ensure transparency for mixed workloads of read and write. Therefore, we use a "clear" component to keep track of unchanged data pages.

Besides the above-mentioned two components, the core component for our batch replacement buffer manager store all dirty data pages that can be evicted.

Figure 2 shows the transitions of a data page among the three components. Whenever the buffer manager reads a data page, it is inserted into the "using" component and only when the data page is unpinned and all queries referring to it terminate, it will be moved to the "clear" component or the core component, depending on if it has been updated. When the buffer overflows, the buffered data pages in the "clear" component will be evicted first. When the "clear" component is empty, the batch replacement plans will be used.

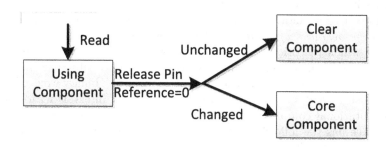

Fig. 2. The three components for Batch Replacement Buffer Manager.

To obtain an optimal replacement plan, the most straightforward approach is to sort all buffered data pages based on their positions in disk and then scan the sorted page list to find the longest continuous sequence. As shown in Algorithm 1, once we meet a continuous data page, we increase the length of the continuous page list (Lines 7–10) and once the continuous data pages terminate, we update the replacement plan (Line 11–13). Although simple, this baseline algorithm is expensive, as it needs to sort and scan all buffered data pages.

Algorithm 1. Trivial Algorithm

Require: $\mathcal{S} = \{p_1, p_2, ..., p_n\}$, the set of all buffered pages free to evict
Ensure: $\mathcal{P}$, the optimal replacement plan
 1: Compute the list $\mathcal{L}$ by sorting pages in $\mathcal{S}$ in increasing order of positions in disk
 2: $\mathcal{P} = \emptyset$
 3: $len_{\mathcal{P}} = 0$
 4: $\mathcal{P}' = \{\mathcal{L}[0]\}$
 5: $len_{\mathcal{P}'} = 1$
 6: **for** $i = 1$ to $n - 1$ **do**
 7:     **if** $\mathcal{L}[i - 1] \rightarrow \mathcal{L}[i]$ **then**
 8:         $len_{\mathcal{P}'} + +$
 9:         $\mathcal{P}' = \mathcal{P}' \cup \{\mathcal{L}[i]\}$
10:     **else**
11:         **if** $len_{\mathcal{P}'} > len_{\mathcal{P}}$ **then**
12:             $\mathcal{P} = \mathcal{P}'$
13:             $len_{\mathcal{P}} = len_{\mathcal{P}'}$
14: Return $\mathcal{P}'$

3.3 Segment Tree Based Buffer Manager

To avoid sorting and scanning, we adopt a segment tree based structure that maintains the buffered data pages that are continuous in disk[1]. In this way, each insertion routine actually amortizes the time for sorting and scanning.

To amortize the overhead of sorting, we represent each set of continuous data pages as an interval $[a, b]$, which indicates that these data pages start at the position a and end at the position b on disk. Note that such an interval represents individual data pages and continuous data pages in a unified way – the interval of an individual data page at position a on disk will be $[a, a]$. *To avoid the overhead of scanning*, we associate each interval with its interval length, on which the priority of eviction is based. In other words, the interval with the largest interval length will be chosen as the optimal replacement plan.

As Fig. 3 illustrates, a segment tree is a balanced binary tree of height $O(\log n)$, using $O(n)$ space. It can support indexing of intervals with logarithmic computational complexity for insertion, deletion and querying [4]. Such a segment tree has the following 2 properties: (1) a key value is associated with each internal node. The intervals in its left branch end with positions no more than the key value and the intervals in its right branch start with positions larger than the key value; (2) an interval is associated with each internal node; it records the longest interval among all the intervals of its descendants.

For example, given the root node associated with the key value 14 and the interval $[5, 11]$, we know that: the interval $[17, 19]$ must be in its right branch because it starts at 17 which is larger than 14 (Property 1); the associated interval $[5, 11]$ is the longest interval in the buffer and its length is 7 (Property 2). In the figure, the interval $[14, 14]$ actually represents an individual data page at the position 14 on disk.

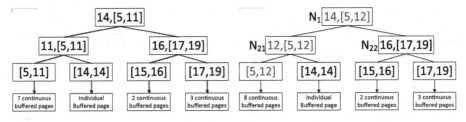

Fig. 3. An example segment tree, where leaf node represents intervals and internal node is associated with a key value and the longest interval among its descendants.

Fig. 4. The example segment tree after the page with position 12 at disk is inserted, where the updated nodes are marked in red (Color figure online).

The original segment tree is unable to maintain continuous data pages or the longest interval. It is our proposed insertion algorithm that utilizes the segment tree to maintain continuous data pages and the optimal replacement plan.

---

[1] For continence, the term "buffer manager" refers to the core component in the rest of the paper.

Algorithm 2. Buffer Insert Algorithm

Require: d, the page to be inserted into the buffer
       $tree$, the segment tree organizing buffered pages in the batch replacement
      buffer manager
 1: New Interval $new = [d.pos, d.pos]$
 2: Predecessor interval $p = tree.search(d.pos - 1)$
 3: Successor interval $s = tree.search(d.pos + 1)$
 4: **if** p exists **then**
 5:    $new = [p.start, d.pos]$
 6:    $tree.$delete(p)
 7:    update longest intervals along the path from root to p
 8: **if** s exists **then**
 9:    $new = [new.start, s.end]$
10:    $tree.$delete(s)
11:    update longest intervals along the path from root to s
12: $tree.$insert(new)
13: update longest intervals along the path from root to new

The main idea is twofold: (1) whenever a buffered data page is inserted into the buffer manager, if its predecessor interval or successor interval exists, the inserted data page will extend the interval to a new longer interval and (2) whenever an interval is updated, the longest intervals on the path percolated from the root down to the interval itself will be updated. As Algorithm 2 illustrates, if the inserted data page d is at position $d.pos$ on disk, its predecessor interval should end with $d.pos - 1$ and its successor interval should start with $d.pos + 1$ (Lines 2 and 3). If any one of the two intervals is found, it will be removed from the segment tree, and the intervals maintained by each internal node on the path from the root percolating to the interval will be updated (Lines 7 and 11). Then, a new interval combining the predecessor/successor interval and the inserted data page will be inserted into the segment tree, and the longest intervals on the path from the root to the new interval will also be updated (Lines 12 and 13). In this way, an insertion involves at most two queries, two deletions and one insertion on the segment tree. Thus its time complexity is $O(\log n)$, where n denotes the number of intervals and is normally less than the number of buffered data pages.

For example, given the segment tree in Fig. 3, if we want to insert a page with position 12, we first find its predecessor interval $[5, 11]$, and combine it with the inserted page to form the new interval $[5, 12]$. Since no successor interval starting with $12 + 1 = 13$ is found in the segment tree, only the interval $[5, 11]$ is removed from the tree and the new interval is inserted. The longest intervals are updated correspondingly as marked in red in Fig. 4.

Since the segment tree maintains the longest interval at the root node, whenever the buffer overflows, we simply pick up the data pages corresponding to the longest interval as the optimal replacement plan. After the eviction, we can remove the corresponding interval and update the segment tree with amortized and worst case time complexity of $O(\log n)$. This procedure is efficient.

Table 1. Statistics of our datasets.

Dataset	# Vertex	# Edges	Raw size
Live Journal	$4,847,571$	$68,993,773$	2.3 GB
Friendster	$65,608,366$	$1,806,067,135$	150 GB
LinkBench	$10^6 \sim 10^7$	$10^8 \sim 10^9$	$5 \sim 60$ GB

4 Experiment

In this section, we report experiment results on real-world and synthetic datasets. We demonstrate the effectiveness of our method on both database manipulation and graph algorithm execution. We also analyze the properties of the proposed batch replacement method.

4.1 Experimental Setting

Dataset. Two public real-world graph datasets were used, namely *Live Journal* [2] and *Friendster* [31]. Both datasets follow power-law distribution with parameter $\alpha \approx 1.4$, while the Friendster dataset is much larger than the Live Journal dataset. The parameter α controls the skewness of the power-law distribution, that is, with a small α such as 0.5, all vertices have similar number of edges, while with a large α such as 1.5, a small number of vertices have much more edges than others. The synthetic dataset is generated by *LinkBench*, the graph database benchmark published by Facebook [1]. It is able to generate graphs with power-law distribution under varying α. The detailed statistics are shown in Table 1.

Workload. The workloads included typical graph algorithms and database manipulation. Following [14,17,23,35], we ran typical graph algorithms including PageRank (PR), Single-Source Shortest Paths (SSSP), Weakly Connected Components (WCC) and Sparse Matrix Multiplication (SMM). LinkBench also provides a mix of insert/delete/update operations on vertices and edges as basic graph database manipulation.

All experiments were conducted on a machine with 2.5 Ghz Intel Core 2 CPU, 8 GB of RAM and 10 TB, 15,000 rpm hard drive. We implemented the proposed batch replacement buffer manager on Neo4j[2] (Neo4j-BR) and GraphChi-DB[3] (ChiDB-BR). Neo4j is a leading industry-standard graph database that adopts LRU-based buffer manager and vertex-centric programming model, while GraphChi-DB (ChiDB) is a research prototype that discards buffer manager and adopts edge-centric programming model. For database manipulation, we also report the performance of a relational database MySQL, only for the purpose of reference. ChiDB also has an option to adopt log-structured merge tree

---

[2] http://neo4j.com/.

[3] https://github.com/graphchi/graphchiDB-scala.

(ChiDB-LSM) for write-optimized database manipulation. We explicitly created appropriate indexes for all databases during the experimental study.

4.2   Performance Comparison

In this section, we first show the effectiveness of our batch replacement buffer manager for data manipulation and graph algorithms. Then, we show that our approach is robust for various buffer sizes and workloads.

Figure 5 shows the average execution time for the typical graph algorithms. The buffer size BS is set to 5 % of the dataset size. We have three observations: (1) for all graph algorithms on all datasets, the batch replacement variants of the two graph databases outperform their original versions. This shows that our batch replacement policy is superior to the LRU-based policy and the approach that does not use buffer manager; (2) on both real-world datasets, ChiDB-BR and ChiDB outperforms Neo4j-BR and Neo4j. This shows edge-centric programming model is more suitable for graph algorithms on real-world datasets. The high value of $\alpha \approx 1.4$ indicates that a few vertices may contain a huge number of edges so that data pages involved in these vertices are read and evicted repeatedly in Neo4j and Neo4j-BR. Even though, our batch replacement policy exhibits better performance than the LRU-based policy; (3) on the synthetic dataset, Neo4j-BR outperforms ChiDB. This is because under $\alpha = 0.5$ edges are distributed more uniformly on vertices and thus Neo4j-BR benefit from less buffered page eviction.

Table 2 shows the average execution time for various manipulation workload on a small dataset (5 GB) and a large dataset (50 GB) respectively. We have the following observations: (1) on both datasets, both Neo4j-BR and ChiDB-BR outperform the original databases equipped with LRU-based buffer manager or log structure merge tree or no buffer manager; This indicates that batch replacement buffer manager is more suitable for graph databases; (2) Neo4j-BR and ChiDB-BR outperform MySQL, which shows the superiority of specialized graph database; (3) Neo4j outperforms ChiDB on small dataset while ChiDB outperforms Neo4j on large dataset, revealing that LRU-based buffer management is sensitive to the scale of dataset, while batch replacement buffer management is more robust.

Both batch replacement buffer manager and log structured merge tree are designed for update-intensive applications by leveraging sequential I/O. However, ChiDB-BR outperforms ChiDB-LSM in most cases. This is because LSM-tree does not consider the optimal replacement plan. Sometimes LSM-tree's data accesses will be scattered across a wide range on disk, which incurs numerous random I/Os.

Figure 6 validates the robustness of our approach on various ratios of buffer size to data size. On Live Journal dataset, we continuously increased the buffer size until the whole dataset was hold in main memory. The execution time of the PageRank algorithm keeps dropping. We can see: (1) until the buffer holds half the dataset, graph databases employing the batch replacement policy

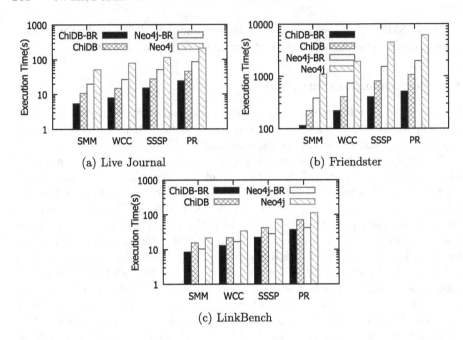

Fig. 5. Execution Time for Graph Algorithms on Three Datasets, where the synthetic dataset contains 10^6 vertices and 10^8 edges with $\alpha = 0.5$. $BS = 5\%$ of dataset size.

always outperform their counterparts; therefore, our approach can exploit available main memory efficiently; (2) when the buffer holds the whole dataset and buffer replacement is no longer needed, our approach consumes 1 % less execution time than their counterparts; this shows that our method for identifying optimal replacement plans is efficient.

Figure 7 shows the query performance on the Friendster dataset for typical read-only workloads, including retrieval of a specific vertex/edge and a traversal-heavy Friends-of-Friends (FoF) query. The FoF query is defined to find all vertices which can reach a specific vertex via any proxy vertex. We can see that although maintaining intervals of continuous buffered pages is of no use since there is no replacement for dirty pages, the overhead is still low. Therefore, although our batch replacement buffer manager is designed for update-intensive applications, its performance is acceptable for read-only applications as well.

4.3   Property of Batch Replacement

In this section, we evaluate the effectiveness of our batch replacement policy in terms of I/O and the computational overhead.

Figure 8 plots the ratios of random I/O to all disk I/O for the workloads of PageRank, node insertion and FoF query respectively, which represent typical workloads of graph algorithm, database manipulation and read-only query. We can observe that both Neo4j-BR and ChiDB-BR used the least random I/O

Table 2. Execution Time (ms) for Graph Database Manipulation on Synthetic Dataset with $\alpha = 1.5$ and $BS = 5\,GB$.

Data size	Operation	ChiDB-BR	ChiDB	ChiDB-LSM	Neo4j-BR	Neo4j	MySQL
10^6 vertices, 10^8 edges	node_insert	0.09	12.9	0.10	**0.08**	0.13	0.11
	node_delete	0.10	16.7	0.14	**0.07**	0.12	0.17
	node_update	0.12	19.1	0.16	**0.09**	0.13	0.21
	edge_insert	0.15	24.6	0.17	**0.09**	0.19	0.25
	edge_delete	0.15	26.3	0.19	**0.12**	0.19	0.34
	edge_update	0.19	29.5	0.22	**0.14**	0.22	0.41
10^7 vertices, 10^9 edges	node_insert	**31**	94	37	36	259	42
	node_delete	**33**	105	41	39	268	45
	node_update	**34**	116	46	41	280	49
	edge_insert	**42**	136	55	47	295	64
	edge_delete	**48**	152	63	57	323	69
	edge_update	**51**	159	67	62	344	73

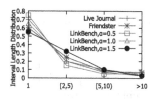
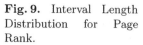

Fig. 6. Effect of RAM size on Live Journal.

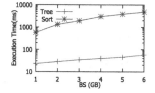

Fig. 7. Query Time on Friendst., $BS=2\,GB$.

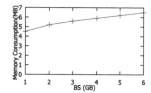

Fig. 8. Ratio of Random I/O access on Friendster dataset.

access. Therefore, it is not surprising their execution time is the shortest in aforementioned experiments. Figure 9 depicts the distribution of buffered interval lengths when running the PageRank Algorithm on the Friendster dataset. We can see that on most datasets there are sufficient segments of continuous buffered data pages. Therefore, it is always possible for our batch replacement buffer manager to exploit sequential I/Os. The distribution of random I/O and interval lengths for other graph algorithms and data manipulation are similar to Figs. 8 and 9.

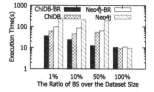

Fig. 9. Interval Length Distribution for Page Rank.

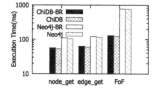

Fig. 10. CPU Time for Replacement Plan.

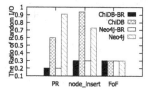

Fig. 11. Memory Overhead.

Figure 10 shows the average execution time for each batch replacement using our segment tree based solution (Tree) and the trivial sort-based algorithm (Sort, Algorithm 1) on the Friendster dataset for the PageRank Algorithm. We can see that as the buffer size increases, our segment tree based solution outperforms the trivial sort-based solution significantly. Figure 11 shows the additional memory consumption for maintaining the segment tree of continuous pages on the Friendster dataset for the PageRank Algorithm. We can see that the segment tree only consumes less than 1 % of the buffer size. Note that the computational and memory overhead are normally only influenced by buffer size, rather than the variation of workloads and datasets.

5   Conclusion

In this paper, we propose a novel approach to batch replacement buffer management for graph databases. Taking the specific data organization and vertex-centric or edge-centric programming models into consideration, the proposed method enables graph databases to make the best of sequential I/O. In addition to a sort-based trivial solution to find optimal replacement plan, we propose a segment tree based buffer structure to efficiently maintain optimal replacement plans. Extensive experiments on real-world and synthetic datasets show that our approach significantly improve the performance of existing graph databases and outperforms the LRU-based approaches and a recently proposed no-buffer approach. The experiment results also show that our approach incurs minimum computational and memory overhead and therefore is practical for real-world applications.

Acknowledgement. This work is partially funded by China Scholarship Council. Xuan Zhou's research is supported by the National High-tech R&D Program (863 Program) (2015AA015307) and the NSFC Porject (No. 61272138). Ling Liu's research is partially supported by the National Science Foundation under Grants IIS-0905493, CNS-1115375, IIP-1230740 and a grant from Intel ISTC on Cloud Computing.

References

1. Armstrong, T.G., Ponnekanti, V., Borthakur, D., Callaghan, M.: Linkbench: a database benchmark based on the facebook social graph. In: SIGMOD 2013, pp. 1185–1196
2. Backstrom, L., Huttenlocher, D., Kleinberg, J., Lan, X.: Group formation in large social networks: membership, growth, and evolution. In: KDD 2006, pp. 44–54
3. Bender, M.A., Demaine, E.D., Farach-Colton, M.: Cache-oblivious b-trees. SIAM J. Comput. **35**(2), 341–358 (2005)
4. de Berg, M., Cheong, O., van Kreveld, M., Overmars, M.: Computational Geometry: Algorithms and Applications, 3rd edn. Springer-Verlag TELOS, Heidelberg (2008)

5. Bornea, M.A., Dolby, J., Kementsietsidis, A., Srinivas, K., Dantressangle, P., Udrea, O., Bhattacharjee, V.: Buildingan efficient RDF store over a relational database. In: SIGMOD 2013, pp. 121–132
6. Brin, S., Page, L.: The anatomy of a large-scale hypertextual web search engine. Comput. Netw. **30**(1–7), 107–117 (1998)
7. Effelsberg, W., Haerder, T.: Principles of database buffer management. ACM Trans. Database Syst. **9**(4), 560–595 (1984)
8. Gonzalez, J.E., Xin, R.S., Dave, A., Crankshaw, D., Franklin, M.J., Stoica, I.: Graphx: graph processing in a distributed dataflow framework. In: OSDI 2014, pp. 599–613
9. Neo4j graph database. http://neo4j.com/
10. Titan graph database. http://thinkaurelius.github.io/titan/
11. Han, J., Wen, J.-R.: Mining frequent neighborhood patterns in a large labeled graph. In: CIKM 2013, pp. 259–268
12. Han, J., Wen, J.-R., Pei, J.: Within-network classification using radius-constrained neighborhood patterns. In: CIKM 2014, pp. 1539–1548
13. Han, W.-S., Lee, S., Park, K., Lee, J.-H., Kim, M.-S., Kim, J., Yu, V.: Turbograph: a fast parallel graph engine handlingbillion-scale graphs in a single PC. In: KDD 2013, pp. 77–85
14. Kyrola, A., Blelloch, G., Guestrin, C.: Graphchi: large-scale graphcomputation on just a PC. In: Proceedings of the 10th USENIX Conference on Operating Systems Design and Implementation, OSDI 2012, pp. 31–46
15. Twitter Developer: Get Friends List. https://dev.twitter.com/rest/reference/get/friends/list
16. Low, Y., Bickson, D., Gonzalez, J., Guestrin, C., Kyrola, A., Hellerstein, J.M.: Distributed graphlab: a framework for machine learning and data mining in the cloud. In: PVLDB 2012
17. Macko, P., Marathe, V.J., Margo, D.W., Seltzer, M.I.: LLAMA: efficient graph analytics using large multiversioned arrays. In: ICDE 2015, pp. 363–374
18. Malewicz, G., Austern, M.H., Bik, A.J.C., Dehnert, J.C., Horn, I., Leiser, N., Czajkowski, G.: Pregel: a system for large-scalegraph processing. In: SIGMOD 2010, pp. 135–146
19. Martínez-Bazan, N., Muntés-Mulero, V., Gómez-Villamor, S., Nin, J., Sánchez-Martínez, M.-A., Larriba-Pey, J.-L.: Dex: high-performance exploration on large graphs for information retrieval. In: CIKM 2007, pp. 573–582
20. O'Neil, E.J., O'Neil, P.E., Weikum, G.: An optimality proof of the LRU-K page replacement algorithm. J. ACM **46**(1), 92–112 (1999)
21. O'Neil, P., Cheng, E., Gawlick, D., O'Neil, E.: The log-structured merge-tree (LSM-tree). Acta Inf. **33**(4), 351–385 (1996)
22. Robinson, I., Webber, J., Eifrem, E.: Graph Databases. O'Reilly Media Inc., Sebastopol (2013)
23. Roy, A., Bindschaedler, L., Malicevic, J., Zwaenepoel, W.: Chaos: scale-out graph processing from secondary storage. In: SOSP 2015, pp. 472–488
24. Roy, A., Mihailovic, I., Zwaenepoel, W.: X-stream: edge-centricgraph processing using streaming partitions. In: SOSP 2013, pp. 472–488
25. Rudolf, M., Paradies, M., Bornhövd, C., Lehner, W.: The graph story of the SAP HANA database. In: BTW 2013, pp. 403–420
26. Shang, S., Ding, R., Yuan, B., Xie, K., Zheng, K., Kalnis, P.: User oriented trajectory search for trip recommendation. In: EDBT 2012, pp. 156–167
27. Shang, S., Ding, R., Zheng, K., Jensen, C.S., Kalnis, P., Zhou, X.: Personalized trajectory matching in spatial networks. VLDB J. **23**(3), 449–468 (2014)

28. Shang, S., Yuan, B., Deng, K., Xie, K., Zheng, K., Zhou, X.: Pnn query processing on compressed trajectories. Geoinformatica **16**(3), 467–496 (2012)
29. Shao, B., Wang, H., Xiao, Y.: Managing and mining large graphs: systems and implementations. In: SIGMOD 2012, pp. 589–592
30. Xia, Y., Tanase, I.G., Nai, L., Tan, W., Liu, Y., Crawford, J., Lin, C.-Y.: Graph analytics and storage. In: IEEE Big Data 2014, pp. 942–951
31. Peters, J.F.: In: Peters, J.F. (ed.). ISRL, vol. 63, pp. 1–76. Springer, Heidelberg (2014)
32. Zeng, K., Yang, J., Wang, H., Shao, B., Wang, Z.: A distributed graph engine for web scale RDF data. In: PVLDB 2013, pp. 265–276
33. Zhou, C., Gao, J., Sun, B., Yu, J.X.: MOCgraph: scalable distributed graph processing using message online computing, pp. 377–388
34. Zhou, Y., Liu, L., Lee, K., Zhang, Q.: GraphTwist: fast iterative graph computation with two-tier optimizations. In: PVLDB 2015, pp. 1262–1273
35. Zhu, X., Han, W., Chen, W.: Gridgraph: large-scale graph processing on a single machine using 2-level hierarchical partitioning. In: USENIXATC 2015, pp. 375–386

Parallelizing Maximal Clique Enumeration Over Graph Data

Qun Chen[1]([⊠]), Chao Fang[1], Zhuo Wang[1], Bo Suo[1],
Zhanhuai Li[1], and Zachary G. Ives[2]

[1] School of Computing, Northwestern Polytechnical University, Xi'an, China
{chenbenben,cfang.mail,zuow.mail,bsuo.mail,lizhh}@nwpu.edu.cn
[2] Department of Computer and Information Systems, University of Pennsylvania,
Philadelphia, PA, USA
zives@cis.upenn.edu

Abstract. In a wide variety of emerging data-intensive applications, such as social network analysis, Web document clustering, entity resolution, and detection of consistently co-expressed genes in systems biology, the detection of *dense subgraphs* (cliques) is an essential component. Unfortunately, this problem is NP-Complete and thus computationally intensive at scale — hence there is a need to come up with techniques for distributing the computation across multiple machines such that the computation, which is too time-consuming on a single machine, can be efficiently performed on a machine cluster given that it is large enough.

In this paper, we first propose a new approach for maximal clique enumeration, which identifies cliques by recursive graph partitioning. Given a connected graph $G = (V, E)$, it has a space complexity of $O(|E|)$ and a time complexity of $O(|E|\mu(G))$, where $\mu(G)$ represents the number of different cliques existing in G. It recursively divides a graph until each task is sufficiently small to be processed in parallel. We then develop parallel solutions and demonstrate how graph partitioning can enable effective load balancing. Finally, we evaluate the performance of the proposed approach on real and synthetic graph data and show that it performs considerably better than existing approaches in both centralized and parallel settings. Our parallel algorithms are implemented and evaluated on MapReduce, a popular shared-nothing parallel framework, but can easily generalize to other shared-nothing or shared-memory parallel frameworks.

Keywords: Maximal clique enumeration · Parallel graph processing · Mapreduce

1 Introduction

A variety of emerging applications are focused on computations over data modeled as a graph: examples include finding groups of actors or communities in social networks [21,30], Web mining [24], entity resolution [4], graph mining [19,35], and detection of consistently co-expressed gene groups in systems

© Springer International Publishing Switzerland 2016
S.B. Navathe et al. (Eds.): DASFAA 2016, Part II, LNCS 9643, pp. 249–264, 2016.
DOI: 10.1007/978-3-319-32049-6_16

biology [13]. For the problems just cited, as well as a number of others, a critical component of the analysis is the detection of cliques (fully connected components) in the structure of the network graph. For instance, for entity resolution, each clique may represent a block of entities that might be merged.

Maximal clique enumeration is NP-Complete. Hence a great deal of effort has been spent on efficient search algorithms [5,9–12,32]. Most of existing algorithms for maximal clique enumeration are based on the classical algorithm proposed by Bron and Kerbosch (BK) [5], which uses a backtracking technique to explore search space and limits the size of its search space by remembering the search paths it has already visited. A variant [11] of the BK algorithm also provides a worst-case-optimal solution. In practice, the BK algorithm has been widely reported as being faster than its alternatives [8,13].

Data-intensive applications usually require clique detection to be operated over large graphs, hence there is a need to parallelize it on a sufficiently large machine cluster. There have been a variety of proposals that divide the graph into smaller subcomponents and exploit parallelism to improve performance [15,22,23,33,34]. They have been empirically shown to speed computation in massive networks. However, built on classical sequential algorithms, the performance of existing parallel approaches is limited by how evenly the graph is partitioned. (In fact, as we show in Sect. 5.2, their performance is quite sensitive to particular graph characteristics.)

This paper presents a new approach for maximal clique enumeration. It computes maximal cliques by recursive graph partitioning. Versus prior work in this area, its key insight is to exploit iterative decomposition during the computation. It recursively divides a graph until each task is sufficiently small to be processed in parallel. As a result, its computation can be effectively parallelized across a machine cluster such that the computation, which may be too time-consuming on a single machine, can be efficiently performed in parallel.

Two common parallel frameworks for graph data processing are the MapReduce model [1] and the Bulk Synchronous Parallel (BSP) model [14]. The underlying computation models of MapReduce and BSP are essentially isomorphic. Our proposed approach is based on iterative data processing and can work with both MapReduce and BSP platforms. In this paper, we choose MapReduce for parallel evaluation due to the maturity and wide availability of its implementations. However, the implementation can easily generalize to other shared-nothing or shared-memory parallel architectures, such as BSP and MPI. The major contributions of this paper are summarized as follows:

1. We propose a novel and efficient approach for maximal clique enumeration. Given a connected graph $G = (V, E)$, it has a space complexity of $O(|E|)$ and a time complexity of $O(|E|\mu(G))$, where $\mu(G)$ represents the number of different cliques existing in G.

2. We develop a parallel solution to maximal clique enumeration by parallelizing the proposed algorithms and implementing the corresponding parallel algorithms based on MapReduce. By using graph partitioning to divide the tasks, the proposed solution can effectively parallelize maximal clique computation with improved load balancing.

3. We experimentally evaluate the performance of our proposed approach over a wide variety of graph data available in open-source. Our extensive experiments demonstrate that it performs considerably better than existing techniques in both centralized and parallel settings.

The rest of this paper is organized as follows: Sect. 2 provides the background information and the description of the existing techniques. Section 3 presents our new sequential algorithm for maximal clique enumeration. Section 4 presents our parallel solution to maximal clique enumeration and its MapReduce implementation. Section 5 empirically evaluates the performance of our approach on real and synthetic datasets. Section 6 discusses related work. Finally, Sect. 7 concludes this paper.

2 Preliminaries

2.1 Definition: Clique and Maximal Clique

A clique is a subgraph in which every pair of vertices is connected by an edge. The definition of a *maximal* clique is as follows:

Definition 1. *A maximal clique in a graph G is a clique not contained by any other clique in G.*

The problem of maximal clique enumeration refers to identifying all the maximal cliques in a given graph G. Since each connected component in G can be processed independently, we assume that G is a connected graph in this paper.

2.2 Background: MapReduce

The MapReduce model processes distributed data across many nodes via three basic phases. In the *Map* phase, it takes an input and produces a list of intermediate key/value pairs without communication between nodes. Next, the *Shuffle* phase repartitions these intermediate pairs according to their keys across nodes. Finally, the *Reduce* phase aggregates the intermediate pairs it receives to produce final results. This process can be repeated by invoking an arbitrary number of additional *Map-Shuffle-Reduce* cycles as necessary.

In this paper, we use Hadoop for parallel evaluation and develop a MapReduce implementation for our approach, in which recursive graph partitioning is programmed in a *Reduce* phase. If implemented on BSP platforms, it can be similarly programmed in a *superstep*. Detailed implementation of our approach on BSP platforms is however beyond the scope of this paper.

2.3 Existing Parallel Solutions

The typical parallel approaches [22,23,33] enumerate maximal cliques for different vertices in a graph in parallel. In this subsection, we describe the idea behind the typical parallel approach for maximal clique enumeration based on MapReduce.

Given a graph G and a vertex v in G, the maximal cliques of the vertex v refer to the maximal cliques containing v in G. Note that a vertex v's maximal cliques are the induced subgraphs consisting of v and its neighboring vertices in G. The parallel search consists of two steps. In the first one, the parallel approach retrieves each vertex's neighboring information relevant to its clique computation. In the second step, it searches for each vertex's maximal cliques in parallel. For the computation on an individual vertex, it simply adopts the classical sequential algorithms (e.g., the BK algorithm).

In the typical approach, enumerating the maximal cliques of a vertex is supposed to be performed on a single machine. In case that the computation on an individual vertex is extremely time-consuming due to the large number of maximal cliques (as we will show in Sect. 5), it may become a parallel performance bottleneck. The method proposed in [23] can parallelize maximal clique enumeration on an individual vertex. It uses candidate path data structures to record the search progress such that any search subtree can be traversed independently. It achieves better load balancing by allowing a computing node to steal some tasks from others when becoming almost idle. The proposed load balancing technique was implemented by MPI, but can easily generalize to other shared-nothing parallel frameworks such as MapReduce. However, as we will show in Sect. 5, its parallel efficiency may still be limited by unevenness of search subtree sizes.

3   Sequential Solution

3.1   Idea: Graph Partitioning

We illustrate the idea behind the new sequential algorithm by an example as shown in Fig. 1. As usual, we search for maximal cliques in a graph G by iteratively computing v's maximal cliques for every vertex v in G. Therefore, we focus on the general problem of identifying the maximal cliques containing a specific vertex v in G.

Suppose that G_v represents the induced subgraph of G consisting of v and its neighboring vertices. Obviously, the maximal cliques of the vertex v are contained by G_v. The challenge is how to identify maximal cliques in G_v if G_v is not a clique. We illustrate the underlying idea by the induced example subgraph G_4 shown in Fig. 1(a), which consists of Vertex 4 and its neighboring vertices (i.e.,

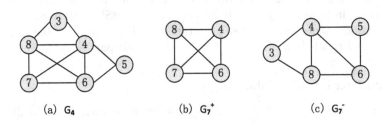

(a) G_4 (b) G_7^+ (c) G_7^-

Fig. 1. A Graph Partitioning Example

the vertices $\{4,3,5,6,7,8\}$). We randomly choose another vertex in G_4, e.g., Vertex 7, as the partitioning anchor and partition G_4 into two parts G_7^+ and G_7^-. G_7^+ denotes the induced subgraph consisting of Vertex 7 and its neighboring vertices in G_4, $\{4,6,7,8\}$. G_7^- denotes the induced subgraph of G_4 consisting of all the vertices not in G_7^+, $\{3,5\}$, and their neighboring vertices in G_4, $\{4,6,8\}$). The subgraphs G_7^+ and G_7^- are shown in Fig. 1(b) and (c) respectively. We observe that any maximal clique of G_4 is an induced subgraph of either G_7^+ or G_7^-.

Generally, we have the following theorem:

Theorem 1. *Given a graph G_v consisting of vertex v and its neighboring vertices and a vertex u in G_v, we partition G_v into two subgraphs, G_u^+ and G_u^-, in which G_u^+ is the induced subgraph consisting of vertex u and its neighboring vertices and G_u^- is the induced subgraph consisting of all the vertices not in G_u and their neighboring vertices. Then, any maximal clique of G_v is an induced subgraph of either G_u^+ or G_u^-.*

Proof. If a maximal clique contains the vertex u, it should be an induced subgraph of G_u^+. Otherwise, it should contain at least one vertex not in G_u^+. Suppose that it is the vertex w. As a result, the maximal clique is an induced subgraph of G_w, which consists of vertex w and its neighboring vertices. According to the definition of G_u^-, G_w is obviously an induced subgraph of G_u^-. Therefore, the maximal clique is an induced subgraph of G_u^-.

According to Theorem 1, maximal clique detection in G_v can be performed by searching for the maximal cliques in G_u^+ and G_u^- independently. The partitioning operation can be recursively invoked until all the resulting subgraphs become cliques. Obviously, all the maximal cliques in the original graph G_v are contained in the set of the resulting cliques. Unfortunately, a resulting clique generated by the above process cannot be guaranteed to be maximal. Therefore, enumeration algorithms should filter out those which are not maximal.

3.2   Sequential Algorithm

The algorithm, as shown in Algorithm 1, enumerates maximal cliques by recursively partitioning a graph. The function employs three sets of vertices to record the partitioning progress and prune the subtrees that can not generate maximal cliques:

- (anchor set) A set of vertices that have been selected as partitioning anchors and should be contained by the resulting cliques;
- (cand set) A set of candidate vertices that can serve as partitioning anchors in the following operations;
- (not set) A set of vertices that are connected to every vertex in the anchor set but could not produce new maximal cliques if combined with the vertices in the anchor set.

Algorithm 1. enumerateClique(anchor,cand,not)

1: **if** (G(cand) is a clique) **then**
2:     Output the clique G(anchor∪cand);
3: **else**
4:     **while** (G(cand) is NOT a clique) **do**
5:         Choose a vertex v with the smallest degree in G(cand);
6:         anchor$^+$=anchor ∪ $\{v\}$;
7:         cand$^+$=cand ∩ $N(v)$;
8:         not$^+$=not ∩ $N(v)$;
9:         **if** ($\nexists u \in$ not$^+$:u is connected to all the vertices in cand$^+$) **then**
10:             enumerateClique(anchor$^+$,cand$^+$,not$^+$);
11:         **end if**
12:         cand=cand − $\{v\}$;
13:         not=not ∪ $\{v\}$;
14:     **end while**
15:     **if** ($\nexists u \in$ not:u is connected to all the vertices in cand) **then**
16:         Output the clique G(anchor∪cand);
17:     **end if**
18: **end if**

Given a connected graph $G=(V,E)$, initially, anchor=∅, cand=V, and not=∅. We denote the induced subgraph consisting of a set of vertices V_i in G by $G(V_i)$. The recursive function first checks whether the resulting subgraph is a clique (Line 1). If yes, it simply outputs the subgraph. Otherwise, it chooses a vertex v with the smallest degree in cand as the partitioning anchor and partitions G(cand) into G(cand$^+$) and G(cand$^-$). G(cand$^+$) consists of v and its neighboring vertices in G(cand) (Line 6–8). G(cand$^-$) consists of all the vertices in G(cand) except v (Line 12–13). The algorithm recursively processes the subgraph G(cand$^+$) (Line 9–11). Note that before the recursive function is invoked, the algorithm prunes the search space by inspecting whether there exists a vertex in the current not$^+$ set that is connected to all the vertices in the current cand$^+$ set (Line 9). Updating G(cand) with G(cand$^-$) (Line 12–13), it then iteratively invokes the partition operation to search for the maximal cliques in G(cand$^-$) until G(cand$^-$) becomes a clique (Lines 4–14). After G(cand$^-$) becomes a clique, the algorithm checks whether it is maximal (Line 15–17).

Note that Algorithm 1 always chooses the vertex v with the smallest degree in G(cand) as the partitioning anchor. It can be observed that this strategy would result in a relatively small graph G(cand$^+$) and a large one G(cand$^-$). Generally, G(cand$^+$) would be partitioned into cliques after only a few iterations because of its small size. At the same time, the size of the other graph G(cand$^-$) would be effectively reduced as a result of iterative partitioning.

We have Theorems 2 and 3, whose proof can be found in our technical report [28] due to space limit. Note that in Theorem 3, *different cliques* include both the maximal cliques and the non-maximal cliques contained by the maximal cliques.

Theorem 2. *Algorithm 1 exactly returns all the maximal cliques in G.*

Theorem 3. *Given a connected graph $G=(V, E)$, Algorithm 1 has the space complexity of $O(|E|)$ and the time complexity of $O(|E|\mu(G))$, in which $\mu(G)$ represents the number of different cliques in G.*

4   Parallel Solution

In this section, we first present the parallel solution for maximal clique enumeration based on recursive graph partitioning and then briefly describe its MapReduce implementation.

4.1   Parallel Algorithm

The parallel algorithm consists of two phases. In the first phase, for every vertex v in the graph G, it retrieves an induced subgraph of G whose vertices are relevant to the computation of v's maximal cliques. In the second phase, it performs recursive graph partitioning on every vertex. Both subgraph retrieval and clique computation on individual vertices are distributed across multiple computing nodes.

The algorithm performs the computation on every vertex by Algorithm 1. Unfortunately, computational cost on individual vertices may be unbalanced. The computations on some vertices may be more expensive than on others because they have larger partitioning traversal trees. In case that the computation on a vertex is too time-consuming, it becomes a parallel performance bottleneck. A good property of our proposed approach is that it enables easy and effective load balancing. Since the resulting subgraphs G_u^+ and G_u^- after partitioning G_v are independent, the computation on the vertex v can be easily parallelized. In practice, recursive functions usually take only a few iterations (no more than 5–6 iterations in our experiments in Sect. 5.2) to transform a big G_v into many sufficiently small subgraphs. With sufficiently small tasks, effective load balancing can be achieved by sending some tasks on the computing nodes with heavy workload to others with lighter one.

Generally, workload can be balanced across computing nodes by repeatedly invoking the *compute-shuffle* cycle. In the *compute* phase, every computing node performs partitioning operations on the graphs it has received; in the *shuffle* phase, all the non-clique graphs on the nodes are reshuffled so that every node receives roughly the same number of unfinished graphs. The workload limit of each *compute* phase can be quantified by the number of partitioning operations executed or CPU time consumed.

In summary, the parallel algorithm for maximal clique enumeration with load balancing consists of the following two phases:

1. **Subgraph Retrieval:** For every vertex v in the graph G, retrieve the induced graph G_v whose vertices are relevant to v's maximal clique computations in parallel;

2. Iterative Computation:

- *compute* **phase:** For each computing node, compute maximal cliques of the graphs assigned to it by recursive graph partitioning;
- *shuffle* **phase:** Evenly reshuffle all the unfinished graphs across the nodes;

4.2   MapReduce Implementation

Based on the observation that non-trivial cliques consist of triangles, we propose to use the technique of triangle enumeration proposed in [12], which is more efficient than 2-hop retrieval, to implement the process of subgraph retrieval. In the rest of this subsection, we briefly describe the implementation details of iterative clique computation performed by reducers.

Suppose that the initial induced G_{v_i} subgraphs (for each v_i in G) are maintained by a queue Q_c. The process of maximal clique computation in the *Reduce* phase is sketched in Algorithm 2. It dequeues a G_u subgraph from Q_c and iteratively partitions it in a depth-first manner. If the resulting G_w^+ has a small size, which means that its maximal clique computation can be finished in short time, it is recursively partitioned to the end (Lines 6–8). Otherwise, it is temporarily enqueued into Q_c if it is not a clique (Line 13). It then continues to partition G_w^- in the same manner as G_u (Line 17). Each subgraph in the queue is represented by its `anchor`, `cand` and `not` sets. A *Reducer* iteratively dequeues a subgraph from Q_c and processes it until Q_c becomes empty or it reaches the predefined workload limit. Finally, it writes all the left subgraphs in Q_c (if any) to disk. A new MapReduce cycle is then iteratively triggered to process the unfinished subgraphs. For more details on the MapReduce implementation, please refer to our technical report [28].

5   Experimental Evaluation

We empirically evaluate the performance of our new approach, denoted by GP, by a comparative study. We compare our approach with a variant of the BK algorithm proposed in [11], which employs the same pruning methods as BK but has been reported to be faster than BK. The typical parallel BK approach confines the computation on an individual vertex to a computing node. We also compare the GP approach with the parallel BK approach enhanced with the dynamic load balancing technique (denoted by BK-L), which was proposed in [23]. It was implemented by MPI in [23]. We have instead implemented a MapReduce version. Each reducer is set to have a predefined workload limit. After every reducer reaches its workload limit, the unfinished subgraphs are evenly redistributed across computing nodes.

Our experiments are conducted on both real and synthetic graph datasets. The details of the real and synthetic graph datasets are summarized in Table 1. The evaluation on real datasets can show the efficiency of the proposed algorithms in real applications while the evaluation on synthetic datasets can clearly demonstrate their sensitivity to varying graph characteristics. The real graph

---

Algorithm 2. Maximal Clique Computation in *Reducer*

Input: A queue of unfinished subgraphs Q_c;

```
1:  while (Q_c is not empty) and (workload limit is not reached) do
2:      Dequeue a subgraph G_u from Q_c;
3:      while (G_u is not a clique) and (workload limit is not reached) do
4:          Choose the vertex w with the minimal degree in G_u as the anchor;
5:          Partition G_u into G_w^+ and G_w^-;
6:          if |cand(G_w^+)| ≤ k then
7:              Recursively partition G_w^+ using Algorithm 1 to the end;
8:          else
9:              if G_w^+ can not be pruned then
10:                 if G_w^+ is a clique then
11:                     Output G_w^+;
12:                 else
13:                     Enqueue G_w^+ into Q_c;
14:                 end if
15:             end if
16:         end if
17:         G_u=G_w^-;
18:     end while
19:     if G_u can not be pruned then
20:         if G_u is a clique then
21:             Output G_u;
22:         else
23:             Enqueue G_u into Q_c;
24:         end if
25:     end if
26: end while
```

---

data are selected from [2], which are in various domains including email communication networks, social networks, web graphs and Wiki communication networks. The synthetic datasets are generated by the SSCA#2 generator and the power-law generator R-MAT from the popular graph generator suite GTgraph [7]. A SSCA#2 graph is directed, and made up of random-sized cliques, with a hierarchical inter-clique distribution of edges based on a distance metric. We vary the values of the *TotVertices* and *MaxCliqueSize* parameters, which specify the number of vertices and the size of the maximum clique respectively. The R-MAT generator applies the Recursive Matrix (R-MAT) graph model to produce the graphs with power-law degree distributions and small-world characteristics, which are common in many real life graphs. We vary two parameter values, the number of vertices and the number of edges.

The experiments are executed on a ten-machine cluster. Each machine runs the Ubuntu Linux (version 10.04), has a memory size of 16 G, disk storage of 160 G and four Intel Xeon E5502 CPUs with the frequency of 1.87 GHz. The evaluation of sequential algorithms is conducted on a JVM (Java Virtual Machine) running on a machine. In case that processing the entire graphs listed in Table 1

Table 1. Details of the Real and Synthetic Graph Datasets

Dataset	Data Description	Number of Vertexes	Number of Edges
D_1	Email Network from a EU Research Institution	265,214	364,481
D_2	Web graph from Google	875,713	4,322,051
D_3	Web graph of Berkeley and Stanford	685,230	7,600,595
D_4	Wikipedia communication network	1,928,669	3,494,674
D_5	Pokec online social network	1,632,803	30,622,564
D_6	Social circles from Twitter	11,316,811	85,331,846
R-MAT	Synthetic graphs with power-law degree distributions and small-world characteristics	Two parameters used: the number of vertices and the ratio of edges to vertices	
SSCA#2	Synthetic graphs with a hierarchical inter-clique distribution of edges based on a distance metric	Two parameters used: the number of vertices and the size of maximum clique	

is beyond the capability of a single machine or even the ten-machine cluster (e.g., maximal clique enumeration over the Twitter dataset), we randomly select some vertices [18] and compute their maximal cliques over the entire graphs.

5.1 Evaluation of Sequential Algorithms

In this subsection, we evaluate the performance of the sequential algorithm on both real and synthetic graphs. The performance is evaluated on two metrics, the size of search tree and the runtime. For fair comparison, in the BK algorithm, search tree size corresponds to the total number of extracted subgraphs, whose set of vertices should be computed; it corresponds to the total number of extracted G_u^+ subgraphs and G_u^- cliques in the GP algorithm. While the runtime of an algorithm depends on its implementation details, search tree size accurately measures search space and is independent of algorithmic implementations.

The evaluation results on the real graphs (D_1-D_5) are presented in Fig. 2(a) and (b). Note that running the dataset D_6 is beyond the capability of a single computing node. Therefore, they will be used later for parallel evaluation. It can be observed that on search tree size, the GP algorithm consistently outperforms the BK algorithm by big margins. Compared with BK, GP uses a more aggressive strategy to filter out unnecessary trees. However, the filtering operation by the not set consumes extra time. Additionally, GP has to update $G(\text{cand})$ and transform it into $G(\text{cand}^-)$. As a result, GP usually takes more time per

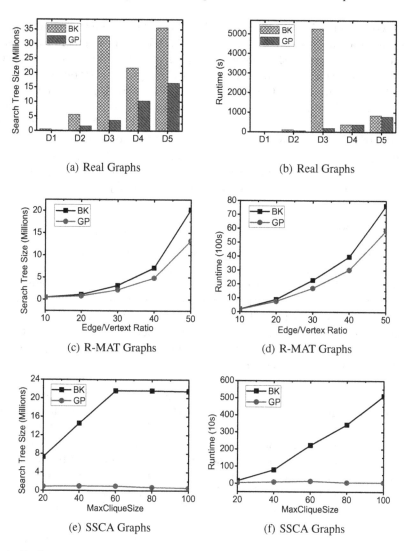

(a) Real Graphs

(b) Real Graphs

(c) R-MAT Graphs

(d) R-MAT Graphs

(e) SSCA Graphs

(f) SSCA Graphs

Fig. 2. Evaluation of sequential algorithms for maximal clique enumeration: GP performs considerably better than BK.

traversal than BK. Correspondingly, GP still performs better than BK but the margins tend to become smaller.

We also evaluate their performance on synthetic graphs to investigate how their performance vary with different graph characteristics and densities. For R-MAT graphs, the number of vertices is set to be 1 million and the edge-to-vertex ratio varies from 10 to 50. The results are shown in Fig. 2(c) and (d). Similar to what were observed on real graphs, GP outperforms BK on both search space size and runtime. It is interesting to observe that the performance gap between

BK and GP steadily increases with graph density. The evaluation results on the SSCA graphs are also shown in Fig. 2(e) and (f). We set the number of vertices to be 2^{20} and vary the size of the maximum clique from 20 to 100. It can be observed that compared with the results on real and R-MAT graphs, GP outperforms BK by the largest margins on the SSCA graphs. The SSCA graphs have larger-sized maximal cliques. As a result, GP is able to achieve bigger save on search space size. Similar to what were observed on R-MAT graphs, the performance advantage of GP steadily increases with the sizes of maximal cliques.

Our experiments show that the sequential GP algorithm performs considerably better than the BK algorithm and its performance advantage tends to increase with graph density and the sizes of maximal cliques.

5.2 Evaluation of Parallel Solution

In this subsection, we compare the GP approach against the BK and BK-L approaches. Since all the parallel approaches use the same method of subgraph retrieval, we exclude its cost from performance evaluation in our comparative study. We use triangle enumeration to implement subgraph retrieval. We specify the parameter k in Algorithm 2 by the number of vertices contained by a graph. It is set to be 50. The maximal execution time per *Reduce* phase is set to be 300 s. The parameter k and the workload limit of execution time per *Reduce* phase are similarly set for the BK-L approach.

Since the real graphs (D_1-D_5) in Table 1 can be efficiently processed on a single worker, we do not evaluate the performance of parallel approaches on them. On SSCA graphs, even the BK approach manages to evenly distribute the workload across workers. As a result, the parallel evaluation results on SSCA graphs are similar to what were observed in sequential evaluation. We do not present their results here either. Also note that processing the entire graph of the Twitter dataset (D_6) takes too long on our ten-machine cluster. We therefore generate 5 random test tasks, denoted by $D_6^1, \ldots, D_6^5$, by randomly choosing 1 % of the vertices in the graph and enumerating their maximal cliques over the entire graph. The tested sets include both the vertices with large degrees and the vertices with small degrees. The observations coming from the experiments can be applied to the entire graphs.

The comparative results on the Twitter and R-MAT graphs are presented in Fig. 3(a)-(d). On the Twitter graph, Fig. 3(a) shows that the typical BK approach performs very poorly in some cases (e.g., D_6^2 and D_6^3). It can not finish computation within the 3-hour runtime limit. Closer scrutiny reveals that there exist some vertices heavily connected to other vertices in these graphs. The maximal clique computation on these vertices are extremely expensive and thus become parallel performance bottleneck if without dynamic load balancing. The experiments show that both the BK-L and GP approaches can effectively break the performance bottleneck by redistributing the computation on an individual vertex across multiple computing nodes. For instance on D_6^2, both of them reduce the runtime to less than 1 hour. However, it can be observed that the

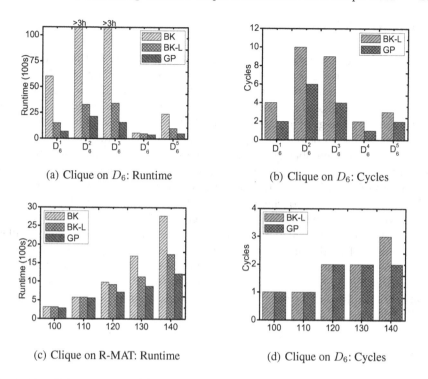

(a) Clique on D_6: Runtime

(b) Clique on D_6: Cycles

(c) Clique on R-MAT: Runtime

(d) Clique on D_6: Cycles

Fig. 3. Parallel evaluation: (1) the BK approach w/o dynamic load balancing may perform very poorly; (2) the GP approach can perform considerably better than the BK-L approach.

GP approach achieves overall better parallel performance than the BK-L approach. Compared with BK-L, GP usually generates much smaller traversal trees and is able to partition big graphs into sufficiently small subgraphs with less iterations. With more effective load balancing mechanism, GP achieves better performance than BK-L in terms of number of required MapReduce cycles, as shown in Fig. 3(b). The evaluation results on the R-MAT graphs, as shown in Fig. 3(c) and (d), are similar except that the performance difference among the three approaches appears less significant. Compared with the Twitter graph, a R-MAT graph has more balanced edge distribution among its vertices. As a result, the effect of dynamic load balancing becomes less dramatic. On the denser graphs (e.g., when the edge/vertex ratio is equal to 140), GP still outperforms BK-L by considerable margins (more than 30 %) in terms of runtime. It is also worthy to point out that similar to what was observed in sequential evaluation, the performance advantage of GP over BK-L increases with the density of R-MAT graphs.

6   Related Work

Maximal clique enumeration have been studied extensively in the literature [6, 11,16,27,32]. Focusing on centralized search algorithms, they usually rely on global state and cannot be easily implemented in the parallel setting. Besides the classical BK algorithms, another category of algorithms [20,25,31] use a *reverse search* strategy. One key feature of these reverse search algorithms is that it is possible to define an upper bound on their runtime as a polynomial with respect to the number of maximal cliques in a graph. There are also some work [3,29] studying the closely related problem of detecting maximum clique. They used variants of the existing algorithms for maximal clique enumeration.

Existing parallel approaches [15,22,23,26,33] for maximal clique enumeration were based on either MPI or MapReduce. They distribute the vertices across computing nodes and compute every vertex's dense subgraphs in parallel. They used the BK algorithm or its variants to compute maximal cliques containing an individual vertex in a graph. The dynamic load balancing technique of redistributing subtree searches was first proposed in [23]. [17] also proposed a scalable and fault-tolerant parallel solution for maximum clique detection using MapReduce.

7   Conclusion

In this paper, we have proposed a novel approach based on recursive graph partitioning to address the problem of maximal clique enumeration over graph data. Compared with previous approaches, it achieves smaller search space and is also inherently more parallelizable. Its better parallelizability enables more effective load balancing and ultimately results in more efficient parallel performance. Our extensive experiments have validated its efficacy.

References

1. Mapreduce. http://en.wikipedia.org/wiki/MapReduce
2. Real graph datasets. http://snap.stanford.edu/data/
3. McClosky, B., Hicks, I.V.: Combinatorial algorithms for the maximum k-plex problem. J. Comb. Optim. **23**, 29–49 (2012)
4. On, B.W., Elmacioglu, E., et al.: Improving grouped-entity resolution using quasi-cliques. In: ICDM (2006)
5. Bron, C., Kerbosch, J.: Algorithm 457: finding all cliques of an undirected graph. Commun. ACM **16**(9), 575–577 (1973)
6. Cheng, J., Ke, Y., et al.: Finding maximal cliques in massive networksby H*-graph. In: SIGMOD (2010)
7. Bader, D.A., Madduri, K.: GTgraph: a synthetic graph generator suite (2006). http://www.cse.psu.edu/madduri/software/GTgraph/
8. Eppstein, D., Strash, D.: Listing all maximal cliques in large sparse real-world graphs. In: Pardalos, P.M., Rebennack, S. (eds.) SEA 2011. LNCS, vol. 6630, pp. 364–375. Springer, Heidelberg (2011)

9. Eppstein, D., Löffler, M., Strash, D.: Listing all maximal cliques in sparse graphs in near-optimal time. In: Cheong, O., Chwa, K.-Y., Park, K. (eds.) ISAAC 2010, Part I. LNCS, vol. 6506, pp. 403–414. Springer, Heidelberg (2010)
10. Akkoyunlu, E.A.: The enumeration of maximal cliques of large graphs. SIAM J. Comput. **2**(1), 1–6 (1973)
11. Tomita, E., Tanaka, A., Takahashi, H.: The worst-case time complexity for generating all maximal cliques and computational experiments. Theor. Comput. Sci. **363**(1), 28–42 (2006)
12. Cazals, F., Karande, C.: A note on the problem of reporting maximal cliques. Theor. Comput. Sci. **407**(1–3), 564–568 (2008)
13. Pavlopoulos, G.A., Secrier, M., et al.: Using graph theory to analyze biological networks. BioData Min. **4**(10), 1–10 (2011)
14. Malewicz, G., Austern, M.H., et al.: Pregel: a system for large-scale graphprocessing. In: SIGMOD (2010)
15. Cheng, J., Zhu, L.H., et al.: Fast algorithms for maximal clique enumeration with limited memory. In: KDD (2012)
16. Cheng, J., Ke, Y.P., et al.: Finding maximal cliques in massive networks. TODS **36**(4), Article No. 21, 1–34 (2011)
17. Xiang, J.G., Guo, C., Aboulnaga, A.: Scalable maximum clique computation using mapreduce. In: ICDE (2013)
18. Leskovec, J., Faloutsos, C.: Sampling from large graphs. In: SIGKDD (2006)
19. Wang, J.Y., Zeng, Z.P., Zhou, L.Z.: CLAN: an algorithm for mining closed cliques from large dense graph databases. In: ICDE (2006)
20. Makino, K., Uno, T.: New algorithms for enumerating all maximal cliques. In: Hagerup, T., Katajainen, J. (eds.) SWAT 2004. LNCS, vol. 3111, pp. 260–272. Springer, Heidelberg (2004)
21. Leskovec, J., Lang, K.J., et al.: Statistical properties of community structure in large social and information networks. In: WWW, pp. 695–704 (2008)
22. Lu, L., Gu, Y., et al.: dMaximalCliques: a distributed algorithm for enumerating all maximal cliques and maximal clique distribution. In: IEEE International Conference on Data Mining Workshops, pp. 1320–1327 (2010)
23. Schmidt, M.C., Samatova, N.F., et al.: A scalable, parallel algorithm for maximal clique enumeration. J. Parallel Distrib. Comput. **69**, 417–428 (2009)
24. Haraguchi, M., Okubo, Y.: A method for pinpoint clustering of web pages with pseudo-clique search. In: Jantke, K.P., Lunzer, A., Spyratos, N., Tanaka, Y. (eds.) Federation over the Web. LNCS (LNAI), vol. 3847, pp. 59–78. Springer, Heidelberg (2006)
25. Chiba, N., Nishizeki, T.: Arboricity and subgraph listing algorithms. SIAM J. Comput. **14**(1), 210–223 (1985)
26. Du, N., Wu, B., et al.: A parallel algorithm for enumerating all maximal cliques in complex network. In: ICDM Workshops (2006)
27. Modani, N., Dey, K.: Large maximal cliques enumeration in sparse graphs. In: CIKM, pp. 1377–1378 (2008)
28. Chen, Q., Fang, C., et al.: Parallelizing clique and quasi-clique detection over graph data. Technical report, Northwestern Polytechnical University, (2014). http://wowbigdata.cn/paper/clique.pdf
29. Rossi, R.A., Gleich, D.F., et al.: Fast maximum clique algorithms for large graphs. In: WWW (2014)
30. Hanneman, R.: Introduction to social network methods, Chap. 11:cliques (2005). http://faculty.ucr.edu/~hanneman/nettext/

31. Tsukiyama, S., Ide, M., Shirakawa, I.: A new algorithm for generating all the maximal independent sets. SIAM J. Comput. **6**(3), 505–517 (1977)
32. Stix, V.: Finding all maximal cliques in dynamic graphs. Comput. Optim. Appl. **27**, 173–186 (2004)
33. Wu, B., Yang, S., et al.: A distributed algorithm to enumerate all maximal cliques in mapreduce. In: International Conference on Frontier of Computer Science and Technology, pp. 45–51 (2009)
34. Yang, S., Wang, B., et al.: Efficient dense structure mining using mapreduce. In: IEEE International Conference on Data Mining Workshops, pp. 332–337 (2009)
35. Zhang, Y., Abu-Khzam, F.N., et al.: Genome-scale computational approaches to memory-intensive applications in systems biology. In: ACM/IEEE Supercomputing (2005)

Miscellaneous

Hyrise-NV: Instant Recovery for In-Memory Databases Using Non-Volatile Memory

David Schwalb[1(✉)], Girish Kumar B.K.[2], Markus Dreseler[1], Anusha S.[2],
Martin Faust[1], Adolf Hohl[2], Tim Berning[1], Gaurav Makkar[2], Hasso Plattner[1],
and Parag Deshmukh[2]

[1] Hasso Plattner Institute, Potsdam, Germany
david.schwalb@hpi.de
[2] NetApp, Sunnyvale, USA
girishkumar.bk@netapp.com

Abstract. Emerging non-volatile memory technologies (NVM) offer
fast and byte-addressable access, allowing to rethink the durability mechanisms of in-memory databases. In this paper, we present Hyrise-NV, a
database storage engine that maintains table and index structures on
NVM. Our architecture updates the database state and index structures
transactionally consistent on NVM using multi-version data structures,
allowing to instantly recover databases independent of their size. For
index structures, we present nvBTree using multi-versioning to provide
failure-atomic tree updates on NVM. We evaluate Hyrise-NV both on
DRAM and with hardware-based emulation of NVM using the TPC-
C benchmark. Hyrise-NV recovers databases independent of their size,
allowing the recovery of a table with 10 million rows in less than 100 ms.

1 Introduction

In-memory database systems [10,12,13,21,22] use main memory as their primary
location of data, but require the use of write-ahead logging and checkpointing
on storage for durability and recovery [8,15,16,23,28]. Startup and recovery
times are proportional to the database size as the dataset needs to be loaded
from storage into memory. Additionally, recovery requires log files to be replayed
and index structures to be recreated, resulting in typical load times of multiple
hours for large enterprise systems. Lazy data loading shifts the load cost to
query processing, but makes response times unpredictable. Long recoveries are
problematic even in replicated setups as the critical time until reestablishing
redundancy after system updates and restarts should be minimized.

Recent announcements by hardware vendors indicate that byte-addressable
and non-volatile memory technologies (NVM) will soon be available on the memory bus. These systems will offer fast and fine-grained access to durable memory
and blur the boundary between storage and memory.

Contributions. (1) We present the NVM-based storage engine Hyrise-NV that
persists all table data and index structures directly on NVM and enables failure-
atomic and durable updates on NVM using multi-version data structures and the

© Springer International Publishing Switzerland 2016
S.B. Navathe et al. (Eds.): DASFAA 2016, Part II, LNCS 9643, pp. 267–282, 2016.
DOI: 10.1007/978-3-319-32049-6_17

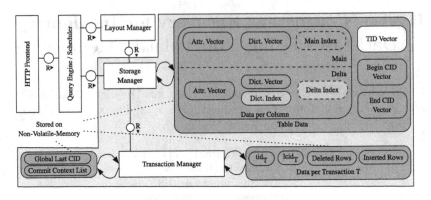

Fig. 1. Overview of the Hyrise-NV architecture. The highlighted parts in green are stored on NVM, the dashed index structures are optional. Dictionary and delta index are a nvBTree (shown in light green) (Color figure online).

persistent index structure called nvBTree, see Sect. 3. (2) We evaluate our system and the tree using two methods of hardware-based emulation, see Sect. 4. Hyrise-NV allows instant recovery independent of the database size in less than 100 ms. Maximum transactional throughput is currently 20 % lower than with a log-based approach. However, we show that future systems with the *clwb* instruction will likely reduce the cost to ≈10 %.

2   HYRISE Architecture

Hyrise[1] is an in-memory storage engine specifically targeted at mixed workload scenarios [10]. It features a balanced execution of both analytical and transactional workloads using task-based scheduling, optimized for the set-processing nature of business applications. We present our extension Hyrise-NV, see Sect. 3, that persists all table data and index structures directly on NVM. We distinguish it from Hyrise-Log, which uses a write-ahead log, and Hyrise-None without any durability guarantees.

Figure 1 gives a high level overview of the Hyrise system architecture. Each column consists of two partitions: main and delta partition. The main partition is dictionary-compressed using an ordered dictionary, replacing tuple values with encoded values from the dictionary. In order to minimize the overhead of maintaining the dictionary's sort order in the case of new values, updates are accumulated in the write-optimized delta partition [14,26]. In contrast to the main partition, data in the delta partition is stored using an unsorted dictionary. To ensure a small size of the delta partition, Hyrise executes a periodic merge process to combine all data from the main partition as well as the delta partition into a new main partition which then serves as the primary data store [14].

---

[1] Source available at https://github.com/hyrise/hyrise.

Data modifications follow the insert-only approach [10] using Multi-Version Concurrency Control (MVCC). Hyrise maintains separate index structures for the main and delta partition. While the main index uses group-key indices [9], the delta index is implemented as a tree-based multi-map of values and positions. This efficiently supports range queries and the insertion of new values as commonly required in enterprise workloads [23].

Hyrise-Log implements logging, checkpointing and recovery mechanisms. We discuss it as a baseline to compare Hyrise-NV against. The logging mechanism [33] leverages the applied dictionary compression and only writes redo information to the log. Undo information is not required due to the lack of in-place updates. New log entries are buffered in a ring-buffer and written to disk using a group commit mechanism [16,23]. Checkpoints create a consistent snapshot of the database as a binary dump on disk in order to speed up recovery, allowing to directly load the checkpoint from disk without the need of expensive delta log replays. In contrast to disk-based database systems where a buffer manager can flush all dirty pages in order to create a snapshot, Hyrise-Log supports checkpoints by persisting the complete delta partitions of all tables plus begin and end timestamps for main and delta to disk.

3 Hyrise-NV: Adding NVM-support to Hyrise

This section presents Hyrise-NV and focuses on changes required to the architecture of Hyrise in order to leverage NVM as the primary persistence domain. Hyrise-NV stores the complete database on NVM, including all table data and index data structures, to allow for instant restarts. We show how using NVM as the primary persistence domain does not interfere with the ACID criteria and how to ensure that changes reach NVM in an atomic and durable way.

We envision systems to support hybrid combinations of volatile DRAM and persistent NVM at the same level in the memory hierarchy with byte-level load and store access. Using NVM comes with the following challenges: (i) When executing stores to NVM, applications typically require one set of writes to be durable before another set of writes [5]. To guarantee durability and ordering, applications currently need to explicitly trigger the flush of modified cache lines and enforce their completion using memory fences, e.g. using *clflush* and *sfence* instructions. (ii) Dynamic memory management on NVM must provide means to safely allocate durable memory. In contrast to allocators for volatile memory, NVM allocators must be failure-atomic to avoid inconsistent states in allocator metadata. Hyrise-NV uses a custom memory allocator for NVM that allows to differentiate between volatile and non-volatile memory regions and to implement application specific recovery strategies to find and restore objects from NVM [25].

3.1 Modifications to Transaction Handling

In order to keep the database state directly on NVM and to recover based on the NVM data, Hyrise-NV stores the following data structures on NVM as outlined in Fig. 1: (i) the table data including main and delta partitions with attribute vectors, dictionaries and MVCC vectors (ii) the state of the transaction manager including the global last visible commit id, and (iii) table metadata. For instant recovery, without expensive recreations of index structures, the system also persists (iv) all index structures and (v) metadata for each running transaction.

To guarantee the consistency of these data structures during updates, a careful system design is necessary, with explicit barriers in the system that guarantee the write order on NVM, as well as mechanisms to provide atomicity and durability for transaction management on NVM. Columns, dictionaries and indices are designed to be append-only and do not execute in-place updates of values. Indices are stored using the nvBTree. Managing consistent updates of the nvBTree structures is outlined in Sect. 3.2. Alternatively, NVM-aware hashmaps [24] can be used. The insert-only approach avoids complications coming from in-place updates of rows and ensures atomicity of transactions using MVCC visibility mechanisms. Building on this, a transaction can work on its private data space on NVM until it has finished and all changes are made visible atomically by incrementing the last visible commit id on NVM.

Figure 2 shows the steps during transaction processing in our system and the needed explicit barriers that guarantee consistency on NVM. A barrier consists of *clflush* instructions that writes all modified cache lines to NVM and a *sfence* instruction that waits for their completion.

As an example, let us assume a transaction T with id tid_T and the last visible commit id $lcid_T$ that updates a single row, resulting in one invalidation and one insert. During active processing, T invalidates one row and inserts the updated version. The row to be updated is marked as invalid in the context of T by setting the MVCC end timestamp *vend* of the row (1) and is added to the local list of T's deleted rows (2). After invalidating the old version, the updated version is

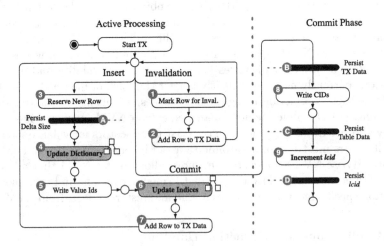

Fig. 2. Additional barriers are required during transaction processing on NVM to explicitly flush data from caches and to guarantee write ordering.

inserted as a new row, see steps (3) to (7). For this, an isolated write space owned by T is reserved in the delta by atomically incrementing the delta size (3). In case the allocated memory of the delta is exhausted, a more complex mechanism guarantees the atomic resize. For every column, the value is encoded as a value id using the delta dictionary (4). If no entry exists in the delta dictionary for the new value, it is inserted into a persistent, unordered dictionary as described in Sect. 3.2. Next, all new value ids are written into the attribute vectors (5) and existing index structures are updated to reflect the newly inserted row (6). Finally, the newly inserted row is added to the local list of inserted rows from T (7). T commits by finalizing all modified rows and setting the commit id (8) and incrementing the global last visible commit id $lcid$ (9).

To ensure the consistency on NVM, Hyrise-NV requires four barriers A, B, C and D during the processing of transactions: *Barrier A* ensures that the size of the delta partition is flushed to NVM before the transaction proceeds to populate the reserved rows. The barrier is required to avoid index structures referencing rows that do not yet exist. When T enters the commit phase, *Barrier B* ensures that the transaction context is persisted on NVM, containing a list of all inserted and updated rows. This is required in order to repair in-flight transactions in case of failures. *Barrier C* persists all table data and flushes all dirty changes on attribute vectors and $vbeg$ and $vend$. Therefore, it is guaranteed that all transaction changes are persisted before incrementing the $lcid$. Incrementing $lcid$ makes all changes of T visible for subsequent transactions. The final *Barrier D* ensures that the $lcid$ is written to NVM before returning T as committed to the client. The combination of append-only updates, explicit barriers and consistent tree structures guarantees a consistent state on NVM at any time.

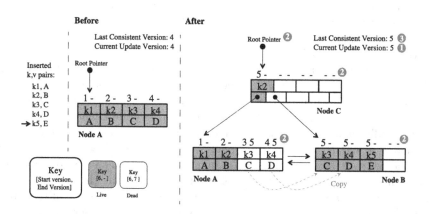

Fig. 3. Example of a multi-versioned nvBTree on NVM, outlining a split operation of node A. Steps 1–3 are ordered by explicit barriers.

3.2   nvBTree: An Index Structure for NVM

Hyrise-NV stores and persists all index structures directly on NVM to allow for instant database restarts and recovery. nvBTree is designed to be stored on NVM and implements a multi-map of values and positions used in the delta partition. It extends the existing tree implementation[2] by adding multi-versioning to make consistent and atomic updates on NVM, requiring no additional logs [29]. The tree allows for parallel read operations, but requires exclusive write locking. It is designed to support failure-atomic inserts directly on NVM by enforcing a consistent write order and using multi-versioning.

The multi-versioning mechanism uses a monotonically increasing version number to mark the most recent consistent version of a nvBTree, called the last consistent version LCV. This version number is used by any thread that reads from the data structure to evaluate the visibility of entries. Additionally, each tree maintains a current update version CUV that is written before updates are started. Each node, both inner tree nodes and data leaf nodes, has n slots with an assigned start and end version that determines the validity of the respective key-value pair. Writes to the data structures never perform in-place updates and every update results in the creation of a new version. Therefore, updates are effectively executed in a sandbox and are made visible by atomically incrementing the consistent version number. After all the modifications for an update have been made persistent, the LCV is atomically incremented, making the changes visible to subsequent read operations. Nodes are internally sorted by keys and multiple values for the same key are sorted by version. This allows optimized insertions to detect duplicates and efficiently validate primary key constraints. The involved right shifts generate additional *clflush* operations leading to write amplification on the memory bus [3]. However, multiple values for a single key are sorted by version to avoid costly right-shift operations.

For brevity, we define the keys K as the set K=$\{k_1,....k_m\}$ where $k_i > k_{(i-1)}$ and the value set V as V=$\{A,.....,Z\}$. Assume a nvBTree with $n = 4$ and the keys k_1 to k_5 to be inserted in a sorted order, as outlined in Fig. 3. The insertion of k_5 leads to a split operation of node A. First, the CUV is incremented and persisted using an explicit barrier. Then, step 1 allocates and populates two new nodes B and C. All keys greater than k_2 are copied to node B and k_5 is inserted into B. Additionally, node C is prepared to be the new root node, linking the split nodes A and B with k_2 as an inner value. All newly inserted and copied key-value pairs are assigned the version number 5. Additionally, Step 2 updates node A, invalidating the old versions of k_3 and k_4 by setting the end version 5. Then, node C is installed as the new root node of the tree. Finally, step 3 atomically increments the LCV. After each step, an explicit barrier, consisting of a *clflush* instruction followed by an *sfence* instructions, ensures that all changes are flushed to NVM and the ordering of the steps. Now consider the case where k_3, k_4 and k_5 are inserted in order before a second value for k_3 is inserted. Figure 4 shows the required right shift operation, first shifting k_5 and then k_4

---

[2] STX B$^+$-Tree: http://idlebox.net/2007/stx-btree/.

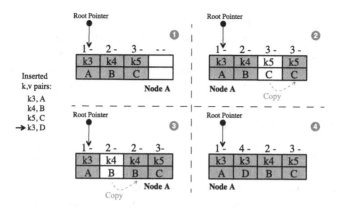

Fig. 4. Example outlining an insert into a sorted nvBTree node and the resulting right shift operation and its associated write amplification. Steps 1–4 are ordered by explicit barriers.

to the right in order to insert k_3 into the second slot. Individual shifts for each key-value pair are required in order to avoid overwriting values in between. Each value shift is internally a copy operation followed by an explicit barrier and the version numbers of copied values are maintained.

During recovery, tree structures potentially have to be verified and repaired before transaction processing can be continued. Verification is required in case a partial update has left the tree in an inconsistent version and $LCV < CUV$. For verification, the tree is scanned in order to identify and garbage collect entries with invalid versioning information. Partial right shift operations are identified as to subsequent identical key-value pairs within a node. Although verifying the tree on recovery requires a full scan, typically only a very small subset of database indices are required to be verified and delta indices are limited in size.

3.3 Recovery from NVM

Having the complete database state persisted on NVM, Hyrise-NV is able to recover from this information in case of failures without requiring write-ahead logs. The recovery process works by (i) re-initializing the system with persisted data structures, (ii) rolling back in-flight transactions and (iii) recovering index structures. The state of the transaction manager, metadata and table data are reinitialized in the first step. The system might crash while a transaction T has already written its commit id to $vbeg$ or $vend$, but before it incremented $lcid$. In that case, T needs to be reverted during recovery in order to allow further increments of $lcid$ without making the partial changes form T visible. To avoid scanning the complete vectors $vbeg$ and $vend$ of all tables during recovery to find changes of in-flight transactions that have to be reverted, the transaction context of committing transactions is persisted when entering the commit phase. Based on this, the recovery can easily traverse all in-flight transactions and revert potential changes by iterating through the lists of inserted and deleted rows.

Technically, our recovery mechanism depends on the number of tables and their columns, the number of index structures with partial updates and their size, and the number of in-flight transactions in the system. Practically, the recovery mechanism minimizes the necessary work during recovery and system restarts, allowing to recover databases virtually independent of their number of rows.

4   Experimental Evaluation

This section presents the experimental evaluation of Hyrise-NV. We evaluate the (i) recovery time, (ii) influence of NVM latency and processor operations on runtime performance and (iii) micro-benchmarks of the presented nvBTree on NVM. We compare Hyrise-NV with the traditional log-based version Hyrise-Log, and with Hyrise-None without any durability mechanisms.

All benchmarks, unless stated otherwise, were executed on a server with four Intel Xeon E7-8870 processors with 10 cores running at 2.4 GHz and 1.5 TB of DDR3 1067 MHz RAM. Benchmarks used a single NUMA node to avoid any NUMA effects. Hyrise-Log uses a PCIe flash drive, with a theoretical read bandwidth of 1.5 GB/s and a read latency of 68 μs.

4.1   Recovery Time

The main design goal of Hyrise-NV is an instant system restart. This is why all data structures are directly stored and manipulated on NVM. Hyrise-NV achieves instant recovery times of ~100 ms, independent of the table's main and delta sizes. As data is directly persisted on NVM, a system restart only requires re-mapping the respective data structures and potentially rolling back in-flight transactions as outlined in Sect. 3.3. Figure 5 shows the recovery times in seconds for different delta sizes and varying sizes of the TPC-C stock table.

We differentiate between the two log-based recovery cases with a delta checkpoint and without. The delta checkpoint is created directly before initiating a

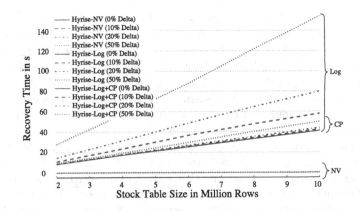

Fig. 5. Recovery times from NVM vs. logs with and without checkpoints.

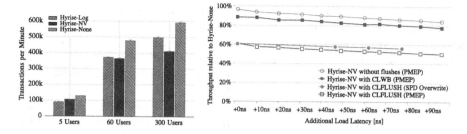

Fig. 6. TPC-C Evaluation: (a) throughput for varying number of users, using *clflush* and (b) throughput with hardware emulated NVM and both *clflush* and the upcoming *clwb*.

database failure leading to the recovery. In this optimal case, it contains the latest snapshot of the delta and no log-replay is required.

If no delta checkpoint exists, recovery takes up to 150 s with 5 million rows in the main and 5 million rows in the delta, as the required data structures for the main partition need to be loaded and the complete delta log needs to be replayed. Recovering from an existing delta checkpoint does not require to replay a delta log, since the delta checkpoint reflects the complete state of the delta as a binary dump on SSD. This reduces the recovery time to ~50 s for loading the checkpoint, still depending linearly on the total table size. Although the delta checkpoint reduces the number of transactions that are replayed from the delta log, the delta size still influences recovery times by adding up to 23 % for a 50 % delta, as index structures for the delta need to be recreated. With an empty delta, the checkpoint has no effect and the "Log" and "Log+CP" lines overlap. In summary, Hyrise-NV allows to instantly recover data-bases independent of their number of rows in contrast to log-based approached that depend linearly on the database size and the available bandwidth. The shown performance improvements are significant both after crashes and in the case of system reboots. While physically rebooting the machine might take several minutes, the dominating factor of reloading the data is eliminated.

4.2   Runtime Performance

In order to evaluate the runtime performance of Hyrise-NV, we compare it to Hyrise-Log (which uses a traditional log-based approach) and differentiate the identified overhead for system throughput into architectural and hardware overhead. *Architectural overhead* describes the additional complexity of ensuring the required write atomicity and ordering to support the described NVM-only architecture. We quantify the architectural overhead by comparing Hyrise-NV to Hyrise-Log using normal DRAM without any increased latencies. In contrast, *hardware overhead* describes the respective overhead that is introduced by increased memory latencies of future NVM hardware. Additionally, we compare the results to Hyrise-None without any persistence mechanisms.

We use TPC-C with 20 warehouses, executed in burst mode without any think-times. Throughput is reported as the total number of completed transactions per minute. Each user reflects one database connection and executes a stream of transactions by triggering a transaction and waiting for its commit. Clients and server are running on the same machine and queries are transmitted via HTTP using a modified Apache *ab* tool.

Figure 6a reports the total number of successfully completed TPC-C transactions per minute for 5, 60 and 300 parallel users. The used group-commit window was tuned to perform best with the given number of parallel users, resulting in a 10 ms window for 300 users, whereas the window size was reduced to 4 ms for 60 users and 1 ms for one and five users. With 300 users, Hyrise-None achieves a maximum of 600 K transactions per minute, whereas Hyrise-Log and Hyrise-NV reach 500 K and 400K transactions per minute, respectively. For 300 parallel users, Hyrise-NV achieves a 20 % lower throughput than Hyrise-Log.

In contrast, if the number of parallel users in the system is low, the overhead of the log-based approach increases as the efficiency of batching transactions is limited by the small number of parallel users. For 5 users, this results in a relative throughput that is 70 % of the throughput with Hyrise-None, and decreased further to 60 % for a single user. Comparing Hyrise-NV and Hyrise-Log, Hyrise-NV has a 14 % and 18 % higher throughput for 5 and 1 user(s) respectively.

4.3   Hardware Emulation

The results of the previous benchmarks were obtained using regular DRAM to simulate NVM and thus do not reflect differences in hardware. Further evaluation showed that most of the experienced overhead is caused by the *clflush*es that unnecessarily invalidate cache lines, causing avoidable cache misses. Future processors will use optimized instructions like *clwb*[3] that write the data to memory without invalidating the cache line, thus allowing future accesses to be served from cache. Using Intel's Persistent Memory Evaluation Platform (PMEP), we can estimate the performance gains foreseeable with *clwb*. Furthermore, it allows us to simulate the higher latencies of NVM and measure the impact of this on our performance. For comparison, we also evaluate Hyrise-NV on a second evaluation platform (as used in [11]), which can emulate slower NVM, but not *clwb*. In this second, evaluation system, the latency is changed by modifying the Serial Presence Detect (SPD) of the memory DIMMs using a proprietary method. This results in an increased real memory latency t_M as shown in Fig. 7.

As two different systems were used in the benchmarks, Fig. 6b shows results relative to Hyrise-None. The benchmark without flushes uses the modifications needed for NVM, such as a custom allocator, mmapped data regions on PMFS (Persistent Memory File System, [7]), and additional methods in the transaction manager, but does not explicitly flush to NVM. This helps distinguishing between the cost of the adaptions and the cost of *clflush*/*clwb*. Two learnings can

---

[3] https://software.intel.com/sites/default/files/managed/07/b7/319433-023.pdf.

Configuration	A	B	C	D
Total Latency t_T [ns]	188.5	234.7	250.2	256.6
Memory Latency t_M [ns]	36.0	82.2	97.7	113.1
System Overhead t_S [ns]	152.5	152.5	152.5	143.5
Total Latency Factor	1.0X	1.2X	1.3X	1.4X
Memory Latency Factor	1.0X	2.3X	2.7X	3.1X

Fig. 7. Modified latencies and their different configurations. This emulator only supports the four mentioned configurations. The real hardware latency t_M was modified whereas the system overhead t_S for requesting and transferring the data was unchanged, resulting in an observable total latency t_T.

be taken from this. First, as expected, the experienced performance drop can be largely explained with by the delicacies of *clflush*. By replacing it with *clwb*, the costs of flushing are reduced from to 8 %. Second, for the given workload, the impact of higher-latency NVM is fairly low, costing only 13 % even with NVM that has a 100 ns higher hardware latency than current DRAM.

4.4   Index Micro Benchmarks

This section presents the evaluation of nvBTree as described in Sect. 3.2. We present micro-benchmark results of the index performance to evaluate the impact of explicit flushes and memory fences that are required to achieve consistent updates on NVM. All nvBTree micro-benchmarks were executed on a machine with six Intel(R) Xeon(R) E5-2440 processors running at 2.40 GHz with 128 GB of memory, using PMFS to emulate 80 GB of NVM.

Figure 8a shows the insert latency for varied value size from 32 bytes to 4 K bytes. The values are inserted into a generated tree with 10 million key-value pairs and a constant key size of 4 bytes. In a second dimension, the number

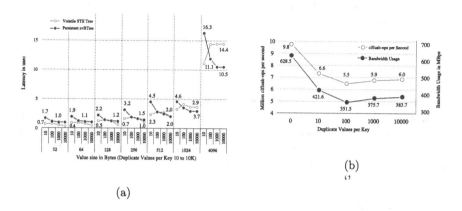

(a)

(b)

Fig. 8. (a) Latency of inserts into a nvBTree with 10 million entries for varying value sizes and varying number of duplicate values per key. (b) Memory bandwidth usage and *clflush* operations/s for inserting 64B values into an nvBTree.

of duplicate values per key in the tree is varied from 10 to 10 K. We compare our implementation nvBTree with the STX B$^+$-Tree without multi-versioning and without explicit flushes and memory fences. Looking at the value size of 32 bytes, inserts into the nvBTree start at 1.7 μs and then gradually taper to 1 μs as the number of duplicate values per key increase. This pattern holds for all the value sizes albeit at slightly higher latencies as the value size increase. The reason behind the peaks with ten duplicate values per key is the write amplification created during insertion due to node-internal sorting. If a key's insertion position happens to be between two existing keys in one leaf node, then the new insert shifts all existing keys to the right, see Fig. 4. These shifts during insertion create additional *clflush* operations, result in additional bandwidth usage and also increase the average latency of inserts. However, as the number of duplicate values per key increases, the write amplification is reduced because nodes are internally sorted by keys but not by value. Therefore, inserts for existing keys can be appended at the end of the existing values for the same key and require fewer shift operations.

Figure 8b shows the memory bandwidth usage of nvBTree with 64 byte values and a varied number of duplicates per key from 0 to 10 K. The experiment was executed starting with an empty tree and inserting 10 million key-value pairs using a single thread. We visualize the number of *clflush* operations per second and the total memory bandwidth usage. The number of *clflush* operations per second ranges from 9.8 million for unique keys in the tree to 6 million for 10 K duplicate values per key. We varied the number of modified cache lines embedded between two memory fences using the *sfence* instruction. As the number of cache lines is increased, the latency per cache line decreases and flattens after 128 cache lines with around 0.2μs. We conclude that our benchmark setup achieves the optimal flush-latency per cache line with 128 cache lines, or 8 K of modified data, embedded between two memory fences. We analyzed the distribution of flushes triggered between two memory fences for inserting 10 million 4 KB-values into a nvBTree. Almost 45 % of the time, a single cache line was flushed to update the LCV of the tree. Flushing the actual data accounted for 40 % of the flushes and flushed 73 cache lines in average, 64 data cache lines plus additional metadata. The rest of the flushes have a varying number of cache lines, ranging from 1 to 225 and are triggered due to tree balancing. We draw two conclusions: (i) The data structure layout of a key-sorted nvBTree directly determines the number of cache lines changed per insert and hence impacts the overall latency. Optimizing the layout for NVM should therefore be the prime design consideration in order to minimize explicit flush operations. (ii) When explicitly flushing multiple cache lines, applications should group as many cache lines between memory fences as possible for better write latencies and throughput.

5   Related Work

Recent work evaluates emerging NVM technologies and how they can be leveraged in the context of database systems. DeBrabant et al. [6] present an evaluation of using NVM for OLTP database systems and evaluate hybrid DRAM-NVM architectures as well as NVM-only architectures using a hardware emulator, concluding that neither of the two approaches is ideally suited for an NVM-based storage hierarchy. In contrast to their work, we focus on a hybrid architecture and leverage multi-versioning inside the database to avoid in-place updates on NVM and to achieve an instant restart of the system.

Oukid et al. [19] present work similar to our approach, describing a prototypical storage engine that leverages full capabilities of NVM by removing the traditional log and updating the persisted data in-place. We describe how barriers are placed in our system and explain the purpose and necessity of each. In addition to being able to reproduce their results, we (1) provide further evaluations comparing Hyrise-NV against log- and checkpoint-based approaches, (2) show the database's performance with NVM using two different emulators, and (3) provide an estimation of the benefits of *clwb*. Höppner et al. [11] present a technique for hybrid-memory scaling of columnar in- memory databases, focusing on NVM as an additional storage tier and evaluating the performance implications when storing the read-optimized main partitions of tables on NVM. Pelley et al. [20] reconsider storage management with NVM to optimize recovery performance and forward-processing throughput. In contrast to our approach, they focus on disk-based databases and use NVM to store the log file instead of using a NVM-only architecture. Additionally, their evaluation is using software-based emulation, whereas we present a hardware-based evaluation. Fang et al. [8] describe an optimized logging approach exploiting storage class memory achieving high performance by a simplified system design and better concurrency support. Wang and Johnson [28] present passive group commits as a scalable and distributed logging technique using NVM and identify logging as a major source of overhead. In contrast, we evaluate a log-free approach and store the primary persistence of data directly on NVM. Narayanan and Hodson [17] outline a hardware supported approach using a flush-on-fail mechanism that guarantees to flush volatile data from processor registers and caches to NVM on failure, eliminating the need of expensive flushes during runtime. Chatzistergiou et al. [1] propose a usermode library approach to manage transactional updates to NVM.

Early work discusses persistent in-memory data-bases, e.g. Wu et al. [32] propose eNVy as a non-volatile, main memory storage system using flash memory directly attached to the memory bus using a dedicated controller. Thatte [27] presents a storage architecture for object-oriented database systems based on a uniform memory abstraction using persistent memory. Venkataraman et al. [29] present data structures designed for byte addressable NVM. They eliminate the need for a write-ahead log by providing consistent updates of durable data structures. nvBTree builds on those mechanisms, integrating and evaluating them in an in- memory database system. Chen et al. [3] present improved database algorithms to reduce the execution time on PCM while increasing write endurance. Further work [2] presents a Phase Change Memory-friendly B^+-Tree structure with unsorted leaf nodes and Yang et al. [34] propose a consistent and cache-optimized

B^+-Tree structure with reduced CPU cacheline flushes. These may be beneficial in a database context and evaluation shall be subject to future work.

Additionally, multiple systems and techniques have been proposed to integrate NVM into existing systems and simplify its usage. Mnemosyne [30] and NV-Heaps [4] provide transactional semantics using software transactional memory. NVMalloc [31] addresses the challenges of allocating persistent memory and deals with wear-leveling challenges. Specialized file systems are designed for the characteristics of NVM like PMFS [7] or BPFS [5]. We see the discussed techniques as orthogonal to the work presented in this paper, focusing on NVM as the primary persistence in database systems.

6   Conclusion and Future Work

In this paper, we presented Hyrise-NV, a columnar in-memory database engine using NVM as the primary persistence for tables and index structures. Our architecture enforces atomicity and ordering guarantees and performs database changes directly on NVM using multi-version data structures without requiring a write-ahead log. We evaluate our approach using the TPC-C benchmark and hardware emulated NVM by overwriting the SPD values of the DIMMs. We report instant database recovery independent of the number of rows, coming by the price of an architectural throughput overhead of up to 20 %. This overhead can be reduced to 8 % with the upcoming *clwb* instruction. Future work includes research on optimized data structures for NVM, reducing the need to flush caches and to enforce barriers, and further evaluations using more sophisticated emulations and real NVM hardware. Besides optimizing the performance of NVM-backed databases, ensuring their correctness and preventing durable errors is a challenge in itself. Possible approaches are theoretical models to prove correctness or experimental methods with e.g. simulated fault injection testing to test how well the database recovers after simulated system crashes. The topics of high availability and disaster recovery are also of great interest, raising the question if log structures are still required or if other mechanisms are more appropriate, as log files are no more required for durability and atomicity purposes. Many in-memory databases already use replication on the peer node's DRAM for the dual purpose of availability and parallelism. With the predicted trends of decreasing latencies for high speed interconnects [18], this will be increasingly adopted. For example, recent work by Zhang et al. [35] proposes a system that provides reliability and availability based on the use of NVM. With the emerging hardware and networking technologies for future data centers, this will be an attractive option for continuous availability for enterprise applications.

Acknowledgments. We thank Konrad Büker, Jürgen Schrage and Ahmad Waizy from Fujitsu and Rami Akkad, Bernhard Höppner and Jürgen Müller from the SAP Innovation Center Potsdam for their support and hardware access. We also thank Subramanya Dulloor from Intel Labs for access to the PMEP emulator.

References

1. Chatzistergiou, A., et al.: REWIND: recovery write-ahead system for in-memory non-volatile data-structures. In: VLDB (2015)
2. Chen, S., Jin, Q.: Persistent B+-trees in non-volatile main memory. In: VLDB (2015)
3. Chen, S., et al.: Rethinking database algorithms for phase change memory. In: CIDR (2011)
4. Coburn, J., et al.: NV-Heaps: making persistent objects fast and safe with next-generation, non-volatile memories. SIGPLAN **46**, 105–118 (2011)
5. Condit, J., et al.: Better I/O through byte-addressable, persistent memory. In: SOSP (2009)
6. DeBrabant, J., et al.: A prolegomenon on OLTP database systems for non-volatile memory. In: VLDB (2014)
7. Dulloor, S.R., et al.: System software for persistent memory. In: EuroSys (2014)
8. Fang, R., et al.: High performance database logging using storage class memory. In: ICDE (2011)
9. Faust, M., et al.: Fast lookups for in-memory column stores: group-key indices, lookup and maintenance. In: ADMS (2012)
10. Grund, M., et al.: HYRISE—a main memory hybrid storage engine. In: VLDB (2010)
11. Höppner, B., et al.: An approach for hybrid-memory scaling columnar in-memory databases. In: ADMS (2014)
12. Kallman, R., et al.: H-store: a high-performance, distributed main memory transaction processing system. In: VLDB (2008)
13. Kemper, A., Neumann, T.: HyPer: a hybrid OLTP&OLAP main memory database system based on virtual memory snapshots. In: ICDE (2011)
14. Krüger, J., et al.: Fast updates on read-optimized databases using multi-core CPUs. VLDB **5**, 61–72 (2011)
15. Malviya, N., et al.: Rethinking main memory OLTP recovery. In: ICDE (2014)
16. Mohan, C., et al.: ARIES: a transaction recovery method supporting fine- granularity locking and partial rollbacks using write-ahead logging. In: TODS (1998)
17. Narayanan, D., Hodson, O.: Whole-system persistence. In: ASPLOS (2012)
18. Ongaro, D., et al.: Fast crash recovery in RAMCloud. In: SOSP (2011)
19. Oukid, I., et al.: Instant recovery for main-memory databases. In: CIDR (2015)
20. Pelley, S., et al.: Storage management in the NVRAM era. In: VLDB (2013)
21. Plattner, H., et al.: The impact of columnar in-memory databases on enterprise systems. In: VLDB (2014)
22. Raman, V., et al.: DB2 with BLU acceleration: so much more than just a column store. In: VLDB (2013)
23. Schwalb, D., et al.: Efficient transaction processing for hyrise in mixed workload environments. In: IMDM (2014)
24. Schwalb, D., et al.: NVC-hashmap: a persistent and concurrent hashmap for non-volatile memories. In: IMDM (2015)
25. Schwalb, D., et al.: nvm malloc: memory allocation for NVRAM. In: ADMS (2015)
26. Stonebraker, M., et al.: C-Store: a column-oriented DBMS. In: VLDB (2005)
27. Thatte, S.M.: Persistent memory: a storage system for object-oriented databases. In: OODS (1991)
28. Tianzheng, W., Ryan, J.: Scalable logging through emerging non-volatile memory. In: VLDB (2014)

29. Venkataraman, S., et al.: Consistent and durable data structures for non-volatile byte-addressable memory. In: FAST (2011)
30. Volos, H., et al.: Mnemosyne: lightweight persistent memory. In: ASPLOS (2011)
31. Wang, C., et al.: NVMalloc: exposing an aggregate SSD store as a memory partition in extreme-scale machines. In: IPDPS (2012)
32. Wu, M., Zwaenepoel, W.: eNVy: a non-volatile, main memory storage system. In: ASPLOS (1994)
33. Wust, J., et al.: Efficient logging for enterprise workloads on column-oriented in-memory databases. In: CIKM (2012)
34. Yang, J., et al.: NV-Tree: reducing consistency cost for NVM-based single level systems. In: FAST (2015)
35. Zhang, Y., et al.: Mojim: a reliable and highly-available non-volatile memory system. In: ASPLOS (2015)

Triangle-Based Representative Possible Worlds of Uncertain Graphs

Shaoying Song[1], Zhaonian Zou[1]([⊠]), and Kang Liu[2]

[1] Harbin Institute of Technology, Harbin, China
znzou@hit.edu.cn
[2] Harbin Engineering University, Harbin, China

Abstract. Uncertain graph data has been collected, processed and analyzed in a wide range of applications. Under the possible world model, an uncertain graph represents a probability distribution over all its possible worlds. Each possible world is a deterministic graph in which the uncertain graph may be present in practice. To deal with the hardness of computations on uncertain graphs, Parchas et al. first proposed the concept of a degree-based representative possible world. This approach is distinguished from the sampling approach in that it computes on one representative possible world instead of on a large number of possible worlds sampled at random. However, the degree-based representative possible world only tries to preserve vertex degrees. In this paper, we are motivated by the fact that motif structures such as triangles can affect the structural properties of graphs and propose the concept of a triangle-based representative possible world. We also develop an algorithm for finding the triangle-based representative possible world of an uncertain graph. We conducted extensive experimental evaluations and show that the triangle-based representative possible world outperforms the degree-based representative possible world in preserving the structural characteristics of an uncertain graph in expectation.

1 Introduction

Graph is a general data structure for representing complicated relationships between entities. Recently, vast amount of *graph-structured data* (*graph data* for short) has been collected, processed and analyzed in a wide range of applications such as social networks, biological networks, sensor networks and road networks. In many of these applications, graph data may contain errors or may be incomplete [23] and therefore is inherently *uncertain*. *Uncertainty* of graph data is usually caused by noisy measurements, data integration, prediction models, or privacy preserving perturbation processes [9].

A graph with uncertainty is called an *uncertain graph*. In recent years, considerable studies have been carried out on the representation, storage, query processing, analysis and mining of uncertain graph data [4,19,20,22]. To express

This work was partially supported by the NSF of China (No. 61532015 and No. 61173023) and the 973 Program of China (No. 2011CB036202).

S.B. Navathe et al. (Eds.): DASFAA 2016, Part II, LNCS 9643, pp. 283–298, 2016.
DOI: 10.1007/978-3-319-32049-6_18

the semantics of an uncertain graph, the *possible world model* [21] has been proposed and widely used. Under this model, an uncertain graph essentially represents a probability distribution over all its *possible worlds*. A possible world is a deterministic graph in which the uncertain graph may be present in practice. An uncertain graph usually implies an exponentially large number of possible worlds. This makes many graph querying and mining problems computationally prohibitive on uncertain graphs, even though their counterparts on exact graphs are solvable in polynomial time.

Sampling has been shown to be an effective approach to dealing with the hardness of managing and mining uncertain graphs [7]. In general, for a problem instance on an uncertain graph $\mathcal{G}$, we first obtain N possible worlds of $\mathcal{G}$ independently at random, where N is a large number. Then, we compute the problem on each of these sampled possible worlds. Finally, we compute an approximation of the exact solution by synthesising the results computed on all these sampled possible worlds. To reduce the variance of the estimator, Jin et al. [5] take advantages of the constraints of the problem they studied. Li et al. [8] use stratified sampling to reduce the variance of the estimator. However, the sampling approach generally has two major drawbacks: (1) Sampling a possible world takes $O(m)$ time, where m is the number of edges in the uncertain graph; (2) To guarantee that the approximation error exceeds ϵ with probability at most δ, we have to sample $O(\frac{1}{\epsilon^2} \ln \frac{1}{\delta})$ possible worlds, where ϵ and δ are small numbers in $[0, 1]$.

To deal with the above mentioned drawbacks of the sampling approach, Parchas et al. [13] proposed the concept of a *representative possible world*. The fundamental idea is to compute over one representative possible world instead of on a large number of randomly sampled possible worlds, which can certainly improve efficiency. Naturally, the representative possible world should best preserve the structural characteristics of the uncertain graph. Parchas et al. [13] found that by preserving the degree of each vertex, the representative possible world well captures the essence of the underlying uncertain graph.

The notion of a representative possible world proposed by Parchas et al. [13] only considers vertex degrees, so we designate their concept by the *degree-based representative possible world*. In fact, besides vertex degrees, many *motif structures* such as *triangles* can affect the structural properties of graphs [1,2]. Motifs are subgraphs that significantly repeat themselves in a specific graph. They are of notable significance because they can reflect functional properties. Most networks display the tendency that two neighbors of the same vertex also tend to be neighbors of one another, thus constituting a triangle [2,12,17]. Triangle is a special network motif that occurs in almost all graphs, which is increasingly prevalent for graph analytics for theoretical as well as practical reasons [3].

In this paper, we first introduce triangles in the formulation of a representative possible world and propose the concept of a *triangle-based representative possible world*. In particular, the triangle-based representative possible world of an uncertain graph is the possible world that minimizes the error of both vertex degrees and *triangle degrees*, where the triangle degree of a vertex is the number of triangles that contains this vertex.

It is hard to find the optimal triangle-based representative possible world of an uncertain graph. In this paper, we propose an algorithm, called TRPW, to approximate the triangle-based representative possible world of an uncertain graph. The algorithm consists of two phrases. In the first phase, we generate a seed possible world by selecting the expected number of edges at random. In the second phase, we iteratively refine the possible world to decrease error.

We performed extensive experiments to verify the effectiveness of triangle-based representative possible world. Our experimental results show that the triangle-based representative possible world and the degree-based one have comparable performance in approximating the expected degrees and the triangle degrees of vertices in an uncertain graph. However, the triangle-based representative method outperforms the degree-based one in approximating other structural characteristics, including the expected number of triangles, the expected clustering coefficients of vertices and the expected shortest-path distances between pairs of vertices. It thus verifies that, by taking triangles into account, we can find better representative possible world for an uncertain graph.

The contributions of this paper are summarized as follows.

1. We propose the concept of a triangle-based representative possible world.
2. We propose the TRPW algorithm for finding the triangle-based representative possible world of an uncertain graph.
3. We carried out extensive experiments and verified that the triangle-based representative possible world outperforms the degree-based representative possible world since it takes more meaningful motif structures into consideration.

2   Preliminaries

In this section we introduce the model of uncertain graphs and review the concept of degree-based representative possible worlds.

Uncertain Graphs. An *uncertain graph* is a triple (V, E, P), where V is a set of vertices, E is a set of edges, and $P : E \rightarrow (0, 1]$ is a function that assigns an existence probability value in $(0, 1]$ to every edge in E. The existence probability, $P(e)$, of an edge $e = (u, v)$ is the probability that e exists between vertices u and v in practice. As a convention, we denote an uncertain graph by a written letter, e.g., $\mathcal{G}$, and denote an exact graph by a printed letter, e.g., G. Moreover, we use $V(\mathcal{G})$, $E(\mathcal{G})$ and $P_{\mathcal{G}}$ to denote the vertex set, the edge set, and the existence probability function of an uncertain graph $\mathcal{G}$, respectively. Analogously, we use $V(G)$ and $E(G)$ to denote the vertex set and the edge set of an exact graph G, respectively.

An exact graph G is a *possible world* of an uncertain graph $\mathcal{G}$ if $V(G) = V(\mathcal{G})$ and $E(G) \subseteq E(\mathcal{G})$. A possible world represents a deterministic structure in which the uncertain graph may exist in practice. Clearly, there are totally $2^{|E(\mathcal{G})|}$ possible worlds for an uncertain graph $\mathcal{G}$. We use $\mathcal{G} \Rightarrow G$ to denote the event that an uncertain graph $\mathcal{G}$ exists in the form of one of its possible worlds G in practice.

Following the literature [23], we assume that the existence probabilities of edges of an uncertain graph are independent. Therefore, the probability of $\mathcal{G} \Rightarrow G$ is

$$\Pr(\mathcal{G} \Rightarrow G) = \prod_{e \in E(\mathcal{G})} P_{\mathcal{G}}(e) \prod_{e \in E(\mathcal{G}) \backslash E(G)} (1 - P_{\mathcal{G}}(e)). \tag{1}$$

Expected Semantics. Evaluating the structural characteristics of a graph, e.g., degree distribution, diameter, or vertex centrality, is a fundamental issue in graph mining, which encompasses many applications in the Internet, the WWW, social networks, and so on. Note that there are significant differences between an exact graph and an uncertain graph. The structural characteristics defined for an exact graph cannot be applied to an uncertain graph. To bridge the gap, the *expected semantics* for uncertain graphs has been proposed and widely adopted.

Without loss of generality, we illustrate the expected semantics by considering the vertex-centric clustering coefficient metric. Let $c_G(v)$ denote the clustering coefficient of a vertex v in an exact graph G. Normally, v has different clustering coefficients in different possible worlds of an uncertain graph $\mathcal{G}$. It is usually inconvenient to manipulate the distribution of clustering coefficients of v across all possible worlds of $\mathcal{G}$. Under the expected semantics, we evaluate the expected value of the clustering coefficients of v across all possible worlds of $\mathcal{G}$, that is, $\sum_{G:\mathcal{G} \Rightarrow G} d_G(v) \Pr(\mathcal{G} \Rightarrow G)$.

Degree-Based Representative Possible Worlds. Since it is infeasible to enumerate all possible worlds of a large uncertain graph $\mathcal{G}$, people often estimate the expected value of a structural characteristic metric on N possible worlds chosen at random, where $N \ll 2^{|E(\mathcal{G})|}$. At the extreme when $N = 1$, Parchars et al. [13] proposed the concept of a representative possible world, which is based on the fact that the vertex degree distribution of a graph has a great effect on its topology properties [11]. Hence, a possible world is said to be representative if it can well preserve the vertex degree distribution of the uncertain graph in expectation. To be more precise, let $d_G(v)$ denote the degree of a vertex v in an exact graph G and $\bar{d}_{\mathcal{G}}(v)$ the expected value of the degrees of vertex v across all possible worlds of $\mathcal{G}$. We call $\bar{d}_{\mathcal{G}}(v)$ the *expected degree* of v. More precisely,

$$\bar{d}_{\mathcal{G}}(v) = \sum_{G:\mathcal{G} \Rightarrow G} d_G(v) \Pr(\mathcal{G} \Rightarrow G). \tag{2}$$

The degree-based representative possible world of an uncertain graph $\mathcal{G}$ is therefore the possible world whose vertex degrees best approximate the expected degrees of the vertices of $\mathcal{G}$. It is formally defined as follows.

Definition 1. *The* degree-based representative possible world *of an uncertain graph $\mathcal{G}$ is the possible world G^* of $\mathcal{G}$ that minimizes $\sum_{v \in V(\mathcal{G})} |d_{G^*}(v) - \bar{d}_{\mathcal{G}}(v)|$.*

3   Triangle-Based Representative Possible Worlds

The degree-based representative possible world of an uncertain graph only preserves the expected degrees of vertices. In fact, many other structural properties except vertex degrees can affect the underlying characteristics of a graph. Among them, triangle plays an important role. For example, triangles are lying at the heart of clustering coefficient and of transitivity ratio. Specifically, a triangle is a cycle consisting of three vertices, which is an important structural motif that has been frequently observed in a wide variety of real graphs. For example, in a social network, two friends of a given person are highly likely to be friends themselves, so the given person and its two friends constitute a triangle. Preserving the distribution of triangles can be a promising approach to well capture the structural characteristics of the graph. Therefore, we propose the novel concept of a *triangle-based representative possible world*. Throughout this paper, we denote a triangle with vertices u, v and w by $\Delta(u, v, w)$.

Definition 2. *The* triangle degree *of a vertex v in an exact graph G, denoted by $D_G(v)$, is the number of triangles in G that contains v. More formally,*

$$t_G(v) = |\{\Delta(v, u, w)|u \in V(G), w \in V(G)\}|.$$

Definition 3. *The* expected triangle degree *of a vertex v in an uncertain graph $\mathcal{G}$ is the expected value of triangle degrees of v across all possible worlds of $\mathcal{G}$. More formally,*

$$\bar{t}_\mathcal{G}(v) = \sum_{G:\mathcal{G}\Rightarrow G} t_G(v) \Pr(\mathcal{G} \Rightarrow G). \tag{3}$$

The triangle-based representative possible world of an uncertain graph is the possible world that best preserves the expected vertex degrees as well as the expected triangle degrees of all vertices.

Definition 4. *The* triangle-based representative possible world *of an uncertain graph $\mathcal{G}$ is the possible world G^* of $\mathcal{G}$ that minimizes*

$$\sum_{v \in V(\mathcal{G})} |d_{G^*}(v) - \bar{d}_\mathcal{G}(v)| + \sum_{v \in V(\mathcal{G})} |t_{G^*}(v) - \bar{t}_\mathcal{G}(v)|. \tag{4}$$

Figure 1(a) shows an uncertain graph $\mathcal{G}$. The number beside every edge is the existence probability of the edge. The pair of numbers beside each vertex are the expected degree and the expected triangle degree of the vertex. Figure 1(b) illustrates a possible world of $\mathcal{G}$. The error of the possible world computed by Eq. (4) is 4.48. The optimal triangle-based representative possible world is shown in Fig. 1(c). The error of the optimal solution is 2.52. Intuitively, the optimal solution better captures the structure of $\mathcal{G}$ in expectation.

4   Finding Triangle-Based Representative Possible Worlds

In this section we propose the algorithm for finding the triangle-based representative possible world G^* of an uncertain graph $\mathcal{G}$. Section 4.1 presents the methods for computing the metrics such as expected triangle degree that will be used in the algorithm. Section 4.2 describes the algorithm.

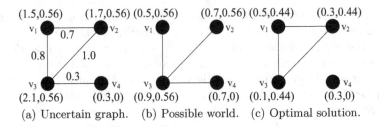

Fig. 1. Illustration of triangle-based representative possible worlds.

4.1 Basics

Reference [13] gives a method for computing the expected degree of a vertex in an uncertain graph, which can be rephrased in the following lemma.

Lemma 1. *The expected degree, $\overline{d}_{\mathcal{G}}(v)$, of a vertex v in an uncertain graph $\mathcal{G}$ can be computed by*

$$\overline{d}_{\mathcal{G}}(v) = \sum_{w:(v,w)\in E(\mathcal{G})} P_{\mathcal{G}}((v,w)). \tag{5}$$

The expected triangle degree is a structural property that the triangle-based representative possible world tries to preserve for every vertex of $\mathcal{G}$. The following lemma gives us a method for computing the expected triangle degree of a vertex.

Lemma 2. *The expected triangle degree, $\overline{t}_{\mathcal{G}}(v)$, of a vertex v in an uncertain graph $\mathcal{G}$ can be computed by*

$$\overline{t}_{\mathcal{G}}(v) = \sum_{\Delta(v,u,w)\subseteq\mathcal{G}} P_{\mathcal{G}}((u,v))P_{\mathcal{G}}((v,w))P_{\mathcal{G}}((u,w)). \tag{6}$$

Proof. Let $X_{u,v,w}$ be a random variable such that $X_{u,v,w} = 1$ if the triangle $\Delta(u,v,w)$ exists in practice and 0 otherwise. Then, we have $E[X_{u,v,w}] = P_{\mathcal{G}}((u,v))P_{\mathcal{G}}((v,w))P_{\mathcal{G}}((u,w))$. Let X_v be the number of triangles containing v that exists in practice. We have $X_v = \sum_{\Delta(v,u,w)\subseteq\mathcal{G}} X_{u,v,w}$. By the linearity of expectation,

$$E[X_v] = \sum_{\Delta(v,u,w)\subseteq\mathcal{G}} E[X_{u,v,w}] = \sum_{\Delta(v,u,w)\subseteq\mathcal{G}} P_{\mathcal{G}}((u,v))P_{\mathcal{G}}((v,w))P_{\mathcal{G}}((u,w)).$$

Moreover, $\overline{t}_{\mathcal{G}}(v) = E[X_v]$. Thus, the lemma holds. □

Our algorithm for finding the triangle-based representative possible world also relies on the expected number of edges and the expected number of triangles across all possible worlds of uncertain graph $\mathcal{G}$. Let $\overline{m}(\mathcal{G})$ be the expected number of edges of $\mathcal{G}$ and $\overline{t}(\mathcal{G})$ be the expected number of triangles of $\mathcal{G}$. Specifically,

$$\overline{m}(\mathcal{G}) = \sum_{G:\mathcal{G}\Rightarrow G} |E(G)|\Pr(\mathcal{G}\Rightarrow G) \text{ and } \overline{t}(\mathcal{G}) = \sum_{G:\mathcal{G}\Rightarrow G} t(G)\Pr(\mathcal{G}\Rightarrow G),$$

where $t(G)$ is the number of triangles in possible world G.

Reference [13] gives us a method for computing the expected number of edges of $\mathcal{G}$, which is rephrased in the following lemma.

Lemma 3. *The expected number of edges in a random possible world of an uncertain graph $\mathcal{G}$ is computed by*

$$\overline{m}(\mathcal{G}) = \sum_{e \in E(\mathcal{G})} P_{\mathcal{G}}(e). \tag{7}$$

We also have the following lemma on the expected number of triangles.

Lemma 4. *The expected number of triangles in a random possible world of an uncertain graph $\mathcal{G}$ is*

$$\overline{t}(\mathcal{G}) = \sum_{\Delta(v,u,w) \subseteq \mathcal{G}} P_{\mathcal{G}}((u,v)) P_{\mathcal{G}}((v,w)) P_{\mathcal{G}}((u,w)). \tag{8}$$

Proof. Let $X_{u,v,w}$ be the random variable as defined in the proof of Lemma 2. We have $E[X_{u,v,w}] = P_{\mathcal{G}}((u,v)) P_{\mathcal{G}}((v,w)) P_{\mathcal{G}}((u,w))$. Let X be the number of triangles in a randomly chosen possible world of $\mathcal{G}$. We have $X = \sum_{\Delta(v,u,w) \subseteq \mathcal{G}} X_{u,v,w}$. By the linearity of expectation,

$$E[X] = \sum_{\Delta(v,u,w) \subseteq \mathcal{G}} E[X_{u,v,w}] = \sum_{\Delta(v,u,w) \subseteq \mathcal{G}} P_{\mathcal{G}}((u,v)) P_{\mathcal{G}}((v,w)) P_{\mathcal{G}}((u,w)).$$

Thus, the lemma holds. □

4.2 Algorithm

We now propose the algorithm for finding the triangle-based representative possible world G^* of an uncertain graph $\mathcal{G}$. The algorithm consists of two phases.

1. Let $i \leftarrow 0$. Generate a seed graph G_0 for the triangle-based representative possible world G^*. Specifically, G_0 consists of all the vertices of $\mathcal{G}$ and $\lfloor \overline{m}(\mathcal{G}) \rfloor$ edges selected at random from $E(\mathcal{G})$, where $\overline{m}(\mathcal{G})$ is the expected number of edges of $\mathcal{G}$ that can be computed by Eq. (5).
2. We adjust the structure of G_i to produce a new graph G_{i+1} so that the value of the loss function, i.e., Eq. (4), for G_{i+1} is less than that for G_i. Then, we increase i by 1 and continue to optimize the structure of the new G_i. This phase terminates as long as no changes to G_i can decrease the loss any more or i reaches a user-specified threshold.

The pseudocode of the algorithm is illustrated in Algorithm TRPW. We describe the detailed steps of TRPW as follows.

Phase 1. The first phase corresponds to lines 1–8 of TRPW. For ease of notation, let E_i be the set of edges of the possible world G_i constructed in the ith iteration

Algorithm 1. TRPW

Input: an uncertain graph $\mathcal{G}$
Output: an approximate triangle-based representative possible world of $\mathcal{G}$
1: $i \leftarrow 0$; $E_0 \leftarrow \emptyset$
2: $\overline{m}(\mathcal{G}) \leftarrow \sum_{e \in E(\mathcal{G})} P_{\mathcal{G}}(e)$
3: sort the edges in $E(\mathcal{G})$ in non-increasing order of their existence probabilities
4: **while** $|E_0| < \lfloor \overline{m}(\mathcal{G}) \rfloor$ **do**
5:     $e \leftarrow$ an edge in $E(\mathcal{G}) \setminus E_0$ with the largest existence probability
6:     $r \leftarrow$ a random number in $[0, 1]$
7:     **if** $r \leq P_{\mathcal{G}}(e)$ **then**
8:         $E_0 \leftarrow E_0 \cup \{e\}$
9: **for** $i \leftarrow 1$ **to** N **do**
10:     pick an edge $e_1 = (u_1, v_1)$ from E_i uniformly at random
11:     pick an edge $e_2 = (u_2, v_2)$ from $E(\mathcal{G}) \setminus E_i$ uniformly at random
12:     $d_1 \leftarrow$ the value of Eq. (9)
13:     $d_2 \leftarrow$ the value of Eq. (10)
14:     **if** $d_1 + d_2 < 0$ **then**
15:         $E_{i+1} \leftarrow (E_i \setminus \{e_1\}) \cup \{e_2\}$
16:     pick an edge $e_3 = (u_3, v_3)$ from $E(\mathcal{G}) \setminus E_i$ uniformly at random
17:     $d_3 \leftarrow$ the value of Eq. (11)
18:     **if** $d_3 < 0$ **then**
19:         $E_{i+1} \leftarrow E_{i+1} \cup \{e_3\}$
20: **return** $(V(\mathcal{G}), E_i)$

of TRPW. In phase 1, we first initialize E_0 to be empty (line 1). Then, we add $\lfloor \overline{m}(\mathcal{G}) \rfloor$ edges to E_0 (lines 3–8). Particularly, we sort all edges of $\mathcal{G}$ in non-increasing order of their existence probabilities (line 3). Then, we iteratively select an edge e in $E(\mathcal{G}) \setminus E_0$ with the largest existence probability and add e to E_0 with probability $P_{\mathcal{G}}(e)$ (lines 4–8). Phase 1 is completed when $|E_0| = \lfloor \overline{m}(\mathcal{G}) \rfloor$.

Phase 2. The second phase corresponds to lines 9–19 of TRPW, which are executed in iterations. In the ith iteration, we revise the structure of G_i to get a new possible world G_{i+1} so that the error computed by Eq. (4) is decreased. Particularly, we perform two kinds of operations to update the structure of G_i, namely, *edge substitution* and *edge insertion*.

Operation 1 (Edge Substitution). In this operation, we try to substitute an edge in E_i with an edge not in E_i to decrease total error. If the substitution can decrease error, we confirm the substitution. The edge substitution operation is carried out in lines 10–15 of TRPW. Particularly, we first select an edge $e_1 = (u_1, v_1)$ from E_i and an edge $e_2 = (u_2, v_2)$ not in E_i both uniformly at random. To determine whether e_1 can be replaced by e_2 or not, we need to compute the change in the value of Eq. (4) supposing that e_1 is substituted by e_2 (lines 12–13). If the change is less than 0, we substitute e_1 by e_2 (line 15).

We now describe the computations in lines 12–13. For ease of notation, let $\delta_{G_i}(v) = d_{G_i}(v) - \overline{d}(\mathcal{G})$ and $\tau_{G_i}(v) = t_{G_i}(v) - \overline{t}(\mathcal{G})$ for all $v \in V(\mathcal{G})$. Suppose we obtain G_{i+1} after substituting e_1 by e_2. The removal of e_1 only decreases the degrees of u_1 and v_1, in turn only affecting the degree errors of u_1 and v_1. More precisely, $\delta_{G_{i+1}}(u_1) - \delta_{G_i}(u_1) = |\delta_{G_i}(u_1) - 1| - |\delta_{G_i}(u_1)|$ and $\delta_{G_{i+1}}(v_1) - \delta_{G_i}(v_1) = |\delta_{G_i}(v_1) - 1| - |\delta_{G_i}(v_1)|$.

The insertion of e_2 only increases the degrees of u_2 and v_2, in turn only affecting the degree errors of u_2 and v_2. More precisely, $\delta_{G_{i+1}}(u_2) - \delta_{G_i}(u_2) =$

$|\delta_{G_i}(u_2)+1|-|\delta_{G_i}(u_2)|$ and $\delta_{G_{i+1}}(v_2)-\delta_{G_i}(v_2)=|\delta_{G_i}(v_2)+1|-|\delta_{G_i}(v_2)|$. For all other vertices $v \notin \{u_1, v_1, u_2, v_2\}$, we have $\delta_{G_{i+1}}(v)=\delta_{G_i}(v)$. Thus, the change of total degree error in Eq. (4) after substituting e_1 with e_2 is

$$
\begin{aligned}
&|\delta_{G_i}(u_1)-1|-|\delta_{G_i}(u_1)|+|\delta_{G_i}(v_1)-1|-|\delta_{G_i}(v_1)| \\
&+|\delta_{G_i}(u_2)+1|-|\delta_{G_i}(u_2)|+|\delta_{G_i}(v_2)+1|-|\delta_{G_i}(v_2)|.
\end{aligned}
\tag{9}
$$

Similarly, the removal of e_1 decreases the triangle degrees of u_1, v_1 and all vertices that are adjacent to both u_1 and v_1. Let c_1 be the number of vertices that are adjacent to both u_1 and v_1. We have $\tau_{G_{i+1}}(u_1)-\tau_{G_i}(u_1)=|\tau_{G_i}(u_1)-c_1|-|\tau_{G_i}(u_1)|$ and $\tau_{G_{i+1}}(v_1)-\tau_{G_i}(v_1)=|\tau_{G_i}(v_1)-c_1|-|\tau_{G_i}(v_1)|$. For every vertex w that is adjacent to both u_1 and v_1, the removal of e_1 decreases the triangle degree of w by 1. Hence, $\tau_{G_{i+1}}(w)-\tau_{G_i}(w)=|\tau_{G_i}(w)-1|-|\tau_{G_i}(w)|$.

The insertion of e_2 increases the triangle degrees of u_2, v_2 and all vertices that are adjacent to both u_2 and v_2. Let c_2 be the number of vertices that are adjacent to both u_2 and v_2. We have $\tau_{G_{i+1}}(u_2)-\tau_{G_i}(u_2)=|\tau_{G_i}(u_2)+c_2|-|\tau_{G_i}(u_2)|$ and $\tau_{G_{i+1}}(v_2)-\tau_{G_i}(v_2)=|\tau_{G_i}(v_2)+c_2|-|\tau_{G_i}(v_2)|$. For every vertex w that is adjacent to both u_2 and v_2, the insertion of e_2 increases the triangle degree of w by 1. Hence, $\tau_{G_{i+1}}(w)-\tau_{G_i}(w)=|\tau_{G_i}(w)+1|-|\tau_{G_i}(w)|$. For all other vertices $v \notin \{u_1, v_1, u_2, v_2\}$ or v is not adjacent to both u_1 and v_1 or v is not adjacent to both u_2 and v_2, we have $\tau_{G_{i+1}}(v)=\tau_{G_i}(v)$. Thus, the change of total triangle degree error in Eq. (4) after replacing e_1 by e_2 is

$$
\begin{aligned}
&|\tau_{G_i}(u_1)-c_1|-|\tau_{G_i}(u_1)|+|\tau_{G_i}(v_1)-c_1|-|\tau_{G_i}(v_1)| \\
&+|\tau_{G_i}(u_2)+c_2|-|\tau_{G_i}(u_2)|+|\tau_{G_i}(v_2)+c_2|-|\tau_{G_i}(v_2)| \\
&+ \sum_{w \in N_{G_i}(u_1) \cap N_{G_i}(v_1)} |\tau_{G_i}(w)-1|-|\tau_{G_i}(w)| \\
&+ \sum_{w \in N_{G_i}(u_2) \cap N_{G_i}(v_2)} |\tau_{G_i}(w)+1|-|\tau_{G_i}(w)|.
\end{aligned}
\tag{10}
$$

Consequently, the sum of Eqs. (9) and (10) is the change of Eq. (4) after substituting e_1 by e_2. If the sum is less than 0, we confirm the substitution of e_1 by e_2.

Operation 2 (Edge Insertion). In our experiments, we observe that the number of triangles in the degree-based representative possible world is far fewer than the expected number of triangles in the uncertain graph. This is due to the lack of edges, so this operation inserts new edges without increasing total error. The edge insertion operation is carried out at lines 16–19. In this operation, we select an edge e_3 not in E_i uniformly at random (line 16) and try to add e_3 to G_i. If the insertion of e_3 can decrease the value of Eq. (4), then we add e_3 to G_i (line 19). Let $e_3=(u_3, v_3)$ and c_3 be the number of vertices that are adjacent to both u_3 and v_3. The insertion of e_3 changes the value of Eq. (4) by

$$
\begin{aligned}
&|\tau_{G_i}(u_3)+c_3|-|\tau_{G_i}(u_3)|+|\tau_{G_i}(v_3)+c_3|-|\tau_{G_i}(v_3)| \\
&+ \sum_{w \in N_{G_i}(u_3) \cap N_{G_i}(v_3)} |\tau_{G_i}(w)+1|-|\tau_{G_i}(w)|.
\end{aligned}
\tag{11}
$$

If the value of Eq. (11) is less than 0, the insertion of e_3 can decrease the error, so we insert e_3 to G_i (line 19).

When Phase 2 completes N iterations, we return the obtained graph G_N as an approximation of the optimal triangle-based representative possible world of the input uncertain graph $\mathcal{G}$.

Complexity Analysis. We now analyze the time complexity of the TRPW algorithm. Let $n = |V(\mathcal{G})|$ and $m = |E(\mathcal{G})|$. In Phase 1, computing $\overline{m}(\mathcal{G})$ requires $O(m)$ time. Sorting all edges of $\mathcal{G}$ in non-increasing order of their existence probabilities takes $O(m \log m)$ time. Then, it takes $O(m)$ time to compute the expected degrees of all vertices. Moreover, it takes $O(m^{1.5})$ time to compute the expected triangle degrees of all vertices. Indeed, we can adapt the triangle counting algorithm in [16] to compute the expected triangle degrees of all vertices. Particularly, every time when we enumerate a triangle $\Delta(u, v, w)$, we accumulate $P_\mathcal{G}((u, v))P_\mathcal{G}((v, w))P_\mathcal{G}((v, w))$ to all of the expected triangle degrees of u, v and w. Note that the time complexity of the triangle counting algorithm is $O(m^{1.5})$. Therefore, we can compute the expected triangle degrees of all vertices in $O(m^{1.5})$ time. In every iteration of Phase 2, it takes $O(1)$ time to compute Eq. (9). Let d be the average degree of vertices of $\mathcal{G}$. It takes $O(d)$ time to compute both Eqs. (10) and (11). Since Phase 2 is carried out in N iterations, where N is a large number. Thus, the time complexity of Phase 2 is $O(dN)$. Consequently, the running time of TRPW is $O(m^{1.5} + dN)$.

5 Experimental Evaluation

We conducted extensive experiments to compare triangle-based representative possible worlds with degree-based representative possible worlds and evaluated the practical performance of the proposed algorithm TRPW.

Experimental Setting. We implemented the TRPW algorithm proposed in this paper and the algorithms ADR and ABM proposed in [13]. We ran all experiments on a machine with 4-core 3.10 GHz Intel Core i5 CPU and 8 GB of RAM, running Windows 8. All experiments were conducted on three publicly available uncertain graphs, namely *Flicker*, *DBLP* and *BioMine* [13].

- **Flicker.** This uncertain graph represents a social network, where users can join common-interest groups and become friends. The probability of an edge represents homophily, meaning similar interests manifest social bonds [14]. The number of vertices and edges are 24125 and 300836, respectively.
- **DBLP.** This uncertain graph represents a co-authorship network, where vertices represent authors. If two authors co-authored a paper, there is an edge between them. The larger the probability of the edge, the more papers the corresponding two authors have co-authored in the past, and the more likely they are to cooperate in the future [14]. The number of vertices and edges of *DBLP* are 684911 and 2284991, respectively.

– **BioMine.** This uncertain graph represents a snapshot of biological interactions of the BioMine project. Probabilities are labeled on edges representing the reliability of interactions between vertices [15]. The number of vertices and edges of *BioMine* are 1008201 and 6722503, respectively.

Experimental Results on Degree Distributions. In this experiment, we compare the distribution of expected vertex degrees of an uncertain graph with the distributions of vertex degrees of the degree-based representative possible world and the triangle-based representative possible world, respectively. Figures 2(a)–(c) plot the distributions obtained on the datasets *Flicker, DBLP* and *BioMine*. Specifically, Fig. 2(d)–(f) magnify the degree distributions within interval [0, 10]. Figure 2 shows that the degree distribution computed by TRPW, ADR and ABM are all very close to the distribution of expected degrees. Particularly, the degree distribution computed by TRPW is slightly closer to the expected degree distribution. This verifies that the distribution of expected degrees of an uncertain graph can be well preserved by the distribution of vertex degrees of the triangle-based representative possible world.

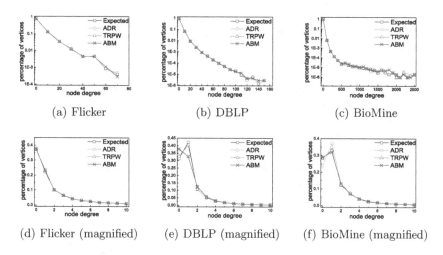

(a) Flicker   (b) DBLP   (c) BioMine

(d) Flicker (magnified)   (e) DBLP (magnified)   (f) BioMine (magnified)

Fig. 2. Comparison of vertex degree distributions.

Experimental Results on Triangle Degree Distributions. Similar to the previous experiment, we compare the distribution of expected triangle degrees of an uncertain graph with the distributions of triangle degrees of the degree-based representative possible world and the triangle-based representative possible world, respectively. Figures 3(a)–(c) show the distributions obtained on the datasets *Flicker, DBLP* and *BioMine*. Figures 3(d)–(f) magnify the triangle degree distributions within interval [0, 10]. From Fig. 3, we can see that the triangle degree distribution computed by TRPW is closer to the expected triangle degree distribution than those computed by ADR and ABM. Hence, the distribution of expected triangle degrees can be better preserved in the triangle-based representative possible world.

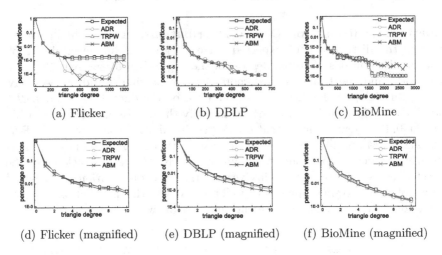

(a) Flicker   (b) DBLP   (c) BioMine

(d) Flicker (magnified)   (e) DBLP (magnified)   (f) BioMine (magnified)

Fig. 3. Comparison of triangle degree distributions.

Experimental Results on Triangle Counts. In this experiment, we compare the expected number of triangles in an uncertain graph with the number of triangles in the degree-based representative possible world and that in the triangle-based representative possible world. Figure 4 illustrates that, on all the datasets *Flicker*, *DBLP* and *BioMine*, the number of triangles in the triangle-based representative possible world computed by TRPW is significantly closer to the expected number of triangles than the numbers of triangles in the degree-based representative possible worlds computed by ADR and ABM.

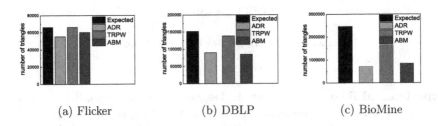

(a) Flicker   (b) DBLP   (c) BioMine

Fig. 4. Comparison of triangle counts.

Experimental Results on Clustering Coefficients. In an exact graph G, the *clustering coefficient* of a vertex v is defined by the ratio of the number

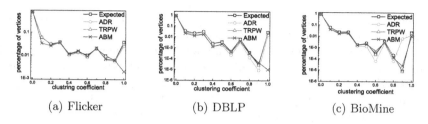

(a) Flicker   (b) DBLP   (c) BioMine

Fig. 5. Comparison of distributions of clustering coefficients.

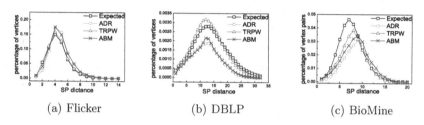

(a) Flicker   (b) DBLP   (c) BioMine

Fig. 6. Comparison of distributions of shortest path distances.

of triangles that contain v to $\binom{d_G(v)}{2}$, where $d_G(v)$ is the degree of v in G. In an uncertain graph $\mathcal{G}$, the expected clustering coefficient of a vertex v is the expected value of the clustering coefficient of v across all possible worlds of $\mathcal{G}$.

Figure 5 indicates that the distribution of clustering coefficients computed by TRPW is significantly closer to the distribution of expected clustering coefficients. Thus, the expected clustering coefficients of all vertices in an uncertain graph can be well preserved by the triangle-based representative possible world.

Experimental Results on Shortest-Path Distances. Since shortest-path distances involves all-pair shortest-path computations which leads to expensive computational cost, we conducted experiments on subgraphs which were produced by forest fire [7]. Figure 6 shows that the distribution of shortest distances computed by TRPW is significantly closer to the distribution of expected shortest path distances than those computed by ADR and ADM. Thus, the distribution of expected shortest distances between all pairs of vertices in an uncertain graph can be well captured by the distribution of shortest distances between all pairs of vertices in the triangle-based representative possible world.

6   Related Work

Uncertain Graph Management and Mining. In recent years many efforts have been made on managing and mining uncertain graphs, e.g., subgraph search [18], frequent pattern mining [23], reliability search [6], shortest path queries [5], and graph clustering [10,24]. The possible world model of uncertain graphs has been widely used to represent uncertainty in graphs. Under this model, an uncertain graph encodes a probability distribution over all its possible worlds. The exponential number of possible worlds often makes many querying processing and mining problems computationally prohibitive on uncertain graphs.

To tame the hardness of managing and mining uncertain graphs, an effective approach is sampling [7]. Among all the sampling methods, the Monte Carlo method is the most widely used one. In this method, a possible world G is sampled with probability $\Pr(\mathcal{G} \Rightarrow G)$. To reduce the variance of the estimator, Jin et al. [5] take advantage of the special constraints of the problem they studied. Li et al. [8] use stratified sampling to reduce the variance of estimator. However, the sampling approach generally has two major drawbacks: (1) Sampling a possible world takes $O(m)$ time, where m is the number of edges in the uncertain graph; (2) To guarantee that the approximation error exceeds ϵ with probability at most δ, we have to sample $O(\frac{1}{\epsilon^2} \ln \frac{1}{\delta})$ possible worlds, where $0 < \epsilon, \delta < 1$.

Representative Possible Worlds. To deal with the above mentioned drawbacks of the sampling approach, Parchas et al. [13] proposed the concept of a representative possible world. The basic idea is to process a query on one representative possible world instead of on a large number of randomly sampled possible worlds. To make query evaluation accurate, they argued that the representative possible world should preserve the underlying characteristics of the uncertain graph. Vertex degree distribution is one of the most fundamental properties of graphs. Hence, they proposed to capture the intrinsic properties of the uncertain graph by preserving the degree of each vertex in expectation. However, they only consider vertex degrees. In this paper, we discover that we can capture the structural properties even better by considering more complex motif structures such as triangles.

Motif Structures. Network motifs are subgraphs that repeat themselves in a specific graph or even among various graphs. They are of notable significance because they may reflect functional properties [1,2]. Triangle is a special network motif that occurs in almost all graphs. For example, in a social network, two friends of a given person are highly likely to be friends themselves, so the given person and its two friends constitute a triangle [17]. The estimation of the clustering coefficient and the transitivity ratio can be translated to counting the number of triangles. Newman [12] proposed a model which specified both the number of edges and the number of triangles. Motivated by these discoveries, we study how to extract better representative possible worlds by preserving both vertex degrees and triangle counts.

7   Conclusions

In this paper we first introduce triangle motifs in the formulation of a representative possible world and propose the concept of a triangle-based representative possible world. We propose the TRPW algorithm for finding the triangle-based representative possible world of an uncertain graph. Our experimental results verify that the triangle-based representative possible world outperforms the degree-based representative possible world in preserving the structural characteristics of an uncertain graph in expectation.

References

1. Becchetti, L., Boldi, P., Castillo, C., Gionis, A.: Efficient semi-streaming algorithms for local triangle counting in massive graphs. In: KDD, pp. 16–24 (2008)
2. Dobrin, R., Beg, Q.K., Barabási, A.L., Oltvai, Z.N.: Aggregation of topological motifs in the escherichia coli transcriptional regulatory network. BMC Bioinf. **5**(1), 10 (2004)
3. Elenberg, E.R., Shanmugam, K., Borokhovich, M., Dimakis, A.G.: Beyond triangles: a distributed framework for estimating 3-profiles of large graphs. In: KDD, pp. 229–238 (2015)
4. Jin, R., Liu, L., Aggarwal, C.C.: Discovering highly reliable subgraphs in uncertain graphs. In: KDD, pp. 992–1000 (2011)
5. Jin, R., Liu, L., Ding, B., Wang, H.: Distance-constraint reachability computation in uncertain graphs. PVLDB **4**(9), 551–562 (2011)
6. Khan, A., Bonchi, F., Gionis, A., Gullo, F.: Fast reliability search in uncertain graphs. In: EDBT, pp. 535–546 (2014)
7. Leskovec, J., Faloutsos, C.: Sampling from large graphs. In: KDD, pp. 631–636 (2006)
8. Li, R., Yu, J.X., Mao, R., Jin, T.: Efficient and accurate query evaluation on uncertain graphs via recursive stratified sampling. In: ICDE, pp. 892–903 (2014)
9. Liben-Nowell, D., Kleinberg, J.M.: The link-prediction problem for social networks. JASIST **58**(7), 1019–1031 (2007)
10. Liu, L., Jin, R., Aggarwal, C.C., Shen, Y.: Reliable clustering on uncertain graphs. In: ICDM, pp. 459–468 (2012)
11. Mahadevan, P., Krioukov, D.V., Fomenkov, M., Dimitropoulos, X.A., Claffy, K.C., Vahdat, A.: The internet as-level topology: three data sources and one definitive metric. Comput. Commun. Rev. **36**(1), 17–26 (2006)
12. Newman, M.E.J.: Random graphs with clustering. Phys. Rev. Lett. **103**(5), 058701 (2009)
13. Parchas, P., Gullo, F., Papadias, D., Bonchi, F.: The pursuit of a good possible world: extracting representative instances of uncertain graphs. In: SIGMOD, pp. 967–978 (2014)
14. Potamias, M., Bonchi, F., Gionis, A., Kollios, G.: k-nearest neighbors in uncertain graphs. PVLDB **3**(1), 997–1008 (2010)
15. Sevon, P., Eronen, L., Hintsanen, P., Kulovesi, K., Toivonen, H.: Link discovery in graphs derived from biological databases. In: Leser, U., Naumann, F., Eckman, B. (eds.) DILS 2006. LNCS (LNBI), vol. 4075, pp. 35–49. Springer, Heidelberg (2006)
16. Chu, S., Cheng, J.: Triangle listing in massive networks. ACM Trans. Knowl. Disc. Data **6**(4), 3123 (2012)
17. Tsourakakis, C.E.: Fast counting of triangles in large real networks without counting: algorithms and laws. In: ICDM, pp. 608–617 (2008)
18. Yuan, Y., Wang, G., Chen, L., Wang, H.: Efficient subgraph similarity search on large probabilistic graph databases. PVLDB **5**(9), 800–811 (2012)
19. Zou, Z., Gao, H., Li, J.: Discovering frequent subgraphs over uncertain graph databases under probabilistic semantics. In: KDD, pp. 633–642 (2010)
20. Zou, Z., Li, J.: Structural-context similarities for uncertain graphs. In: ICDM, pp. 1325–1330 (2013)
21. Zou, Z., Li, J., Gao, H., Zhang, S.: Frequent subgraph pattern mining on uncertain graph data. In: CIKM, pp. 583–592 (2009)

22. Zou, Z., Li, J., Gao, H., Zhang, S.: Finding top-k maximal cliques in an uncertain graph. In: ICDE, pp. 649–652 (2010)
23. Zou, Z., Li, J., Gao, H., Zhang, S.: Mining frequent subgraph patterns from uncertain graph data. IEEE Trans. Knowl. Data Eng. **22**(9), 1203–1218 (2010)
24. Züfle, A., Emrich, T., Schmid, K.A., Mamoulis, N., Zimek, A., Renz, M.: Representative clustering of uncertain data. In: KDD, pp. 243–252 (2014)

Efficient Query Processing with Mutual Privacy Protection for Location-Based Services

Shushu Liu[1], An Liu[1(✉)], Lei Zhao[1], Guanfeng Liu[1], Zhixu Li[1],
Pengpeng Zhao[1], Kai Zheng[1], and Lu Qin[2]

[1] School of Computer Science and Technology, Soochow University, Suzhou, China
anliu@suda.edu.cn
[2] University of Technology, Sydney, Australia

Abstract. Data privacy in location-based services involves two aspects. The location of a user is a kind of private data as many sensitive information can be inferred from it given some background knowledge. On the other hand, the POI database is a great asset to the LBS provider as its construction requires many resources and efforts. In this paper, we propose a method of protecting mutual privacy (i.e., the location of the user issuing a query and the POI database of the LBS provider) for location-based query processing. Our approach consists of two steps: *data preparation* and *query processing*. Data preparation is conducted by LBS itself and is totally an offline computation, while query processing involves some online computation and multiple rounds of communication between LBS and the user. We implement the query processing by two rounds of oblivious transfer extension (OT-Extension) on two small key sets, resulting an immediate response even on some big POI databases. We also theoretically prove the security and analyze the complexity of our approach. Compared with two state-of-the-art methods, our approach has several orders of magnitude improvement in response time, at the expense of little and acceptable communication cost.

Keywords: Privacy · Query processing · Location-based services

1 Introduction

The prevalence of GPS equipped mobile devices and the fast development of wireless communication technologies stimulate the emergence of location-based services (LBS). A LBS receives a users' location and provides that user with information or services tailored to that location. For example, with the help of LBS, a user can get answers to various location-based queries, such as the nearest ATM, restaurant, or retail store to his/her current geographical position. LBS has a wide range of promising applications [3,11,13,20,21,25,26], however, it also poses significant privacy risks, as the location information collected from users can reveal far more than a user's latitude and longitude. Knowing where a user is plus some background knowledge can infer many sensitive information about individuals, such as home address, health condition, lifestyle habits, and

© Springer International Publishing Switzerland 2016
S.B. Navathe et al. (Eds.): DASFAA 2016, Part II, LNCS 9643, pp. 299–313, 2016.
DOI: 10.1007/978-3-319-32049-6_19

political attitude. To boost further development of LBS, user privacy must be protected. On the other hand, a LBS provider typically holds a Points of Interest (POI) database, based on which the location-based queries can be processed. The POI database is a great asset to the LBS provider, as building such a database generally requires many resources and is by no means a trivial task. It is therefore expected that the privacy of a LBS provider, that is, its POI database, should also be kept secret. That is, a LBS should only send necessary POI data to authorized users as the answers to their location-based queries.

A variety of approaches can be used to provide a certain degree of privacy preserving for location-based queries. Methods such as access control [1,22], mix zone [2,6], k-anonymity [4,5,14] and dummy location [9,10,23] can prevent LBS provider from learning the exact position of a user. Specifically, the first three methods introduce a trusted third party who maintains all users' location. When a user wants to do a query, a cloaked area, instead of the exact location of the user, is generated by the third party and sent to the LBS provider for subsequent processing. Clearly, this kind of methods is vulnerable to misbehavior of the third party. To overcome this shortcoming, the method of dummy location abandons the trusted third party and fills a user's query with both real and fake POIs. To achieve a high level of security, however, a lot of extra POIs need to be send to the LBS provider, resulting in heavy unnecessary communication cost. Besides, all these approaches ignore the privacy requirement of the LBS provider.

Private Information Retrieval (PIR) is another popular privacy-preserving technique for location-based queries. It allows a user to retrieve the data he/she wants from a database, without disclosing the index of the data to the database server. When simply applying PIR to location-based queries, the LBS provider cannot know the location of the user, but the user will get more POIs than the answer to his/her query. To deal with this, Yi et al. [7] present a solution based on two encryption schemes, Paillier and Rabin, which only allows the user to get the exact POIs to his/her query. Paulet et al. [18] also achieve this goal through a two-stage operation of Oblivious Transfer (OT) and PIR. However, it is remarkable that, these approaches have a linear computation cost with the number of POIs, resulting in a prohibitive long query processing time when the POI database is large.

Following the model presented in [18], we propose an efficient approach for location-based queries with mutual privacy protection. Our approach has a significant improvement in online performance by executing two rounds of OT extension on two small key sets. To reduce online query processing cost, the LBS provider constructs a private grid and a public grid index for its POI database. Every cell in both grids is encrypted by a different symmetric key. During the online phase of query processing, the user obtains the key of the cell of the public grid where he/she is located. With this target cell, the user can get the key of the cell of the private grid where he/she is located, and thus retrieves the required POIs eventually. Notice that in this procedure, the user can only obtain one specific key from the LBS provider, so he can only decrypt the necessary POIs. Hence, our approach guarantees data privacy for both the user and the LBS provider.

The rest of the paper is organized as follows. Section 2 discusses some recent achievements in privacy preserving LBS. Section 3 presents our approach for location-based queries processing with mutual privacy protection. Section 4 analyses the computation cost and the communication cost of our approach. We report experimental results in Sect. 5 and conclude our work in Sect. 6.

2   Related Work

The public concern over privacy stimulated lots of research efforts in privacy preserving LBS. And the privacy-preserving computation of trajectory [12] is an extension of LBS. An early step in protecting user's location privacy is notifying and requesting user for the usage of their location. Information access control [1,22] is a technique used to protect location information gathered by location tracking systems. It requires the location of user are gathered and relies on LBS provider to restrict access to stored location information through rule-based polices. But it's vulnerable when the third party who maintains all user locations misbehaves.

Mix zone [2,6], k-anonymity [4,5,14] and dummy location [9,10,23] solve this problem by hiding user's location into some bigger zone or more records so that LBS provider can not locate the exact position of user. Both mix zone and k-anonymity use a trusted third party which is in charge of user's location and assign user with cloaked query by which user can query LBS provider without revealing the exact location of them. In this situation, k-anonymity is effected heavily on the distribution and density of the users, which, are out of control and the balance of privacy and precision is another difficult problem need to be solved. Dummy location is completed by sending many random locations along with user's query, though dummy location does not rely on any third party, the LBS provider still can restrict the user in a small space of the total domain, which leads to a weak privacy.

Private information retrieval (PIR) [7,17,18] based LBS is a new scheme which can provide a stronger cryptographic guarantees than former techniques. Both access control, mix zone and k-anonymity are vulnerable for the employ of third party who maintains all user locations. Also, k-anonymity is effected heavily on the distribution and density of the users, which, are out of control. LBS queries based on dummy locations in [23] incurs both computation and communication overhead for client. However, most of issues overhead can be solved by the introduction of PIR-based LBS queries. PIR is a technique which allows a user to retrieve data from a database, without disclosing the index to the database server. A PIR-based LBS queries usually require two stages. In the first stage, the mobile user retrieves the index of his location from the LBS provider. In the second stage, the mobile user retrieves the POIs according to the index from the LBS provider.

In [7], Ghinita et al. used Computational PIR [8] which is based on the Quadratic Residuosity Assumption which states that it is computationally hard to find the quadratic residues in modulo arithmetic of large composite number

n where the factorization of n is unknown. The proposed method for nearest neighbors consists of two stages just as discussed before: determining the index of a public grid and retrieving the value of target cell with PIR. It is a secure and efficient approach for privacy preserving LBS queries if we only care about the privacy of user, but it is not appropriate when we not only care about LBS's not learning anything about user's query, but also want user not to learn anything about other data in LBS's database other than the one she queried. Since with PIR scheme, client can also infer the data which in the same column with the target item.

One way to deal with the problem of data leakage is to encrypt all the data needed to be transferred. [24] proposed by Yi et al. accomplishes this objective by the combination of Paillier Homomorphic encryption scheme [16] and the Rabin encryption scheme [19]. Initially, LBS divides the location based database into cells with the same size, and collects k nearest POIs in each cell. After that, user just need to retrieval the data of target grid without compromising the privacy of both user and LBS. At first, user generates an encrypted query with Paillier and Rabin and sends it to LBS. LBS generates response by linearly computation on all data and sends the encrypted response back. Then, user can and only can get the target item decrypted with his private key. However, this method has a big computation cost for the usage of two encryption scheme and is not practical.

Another way realized this goal is introduced by Paulet et al. in [18], same to [7], this protocol is also organized according to two stages. In the first stage, the user determines his/her location within a public grid, using oblivious transfer (OT)[15]. This data contains both the ID and associated symmetric key for the block of data in the private grid. In the second stage, the user executes a communicational efficient PIR, to retrieve the appropriate block in the private grid which can be decrypted with the symmetric key obtained in the first stage. The property of oblivious transfer and PIR preserved the privacy of user. The data privacy of LBS can also be preserved, for all data of LBS are encrypted and user only has the key for target block.

3   Privacy-Preserving Location-Based Query Processing

3.1   System Model

In this paper, we simply adopt the system model proposed in [18] as the foundation of our work. As shown in Fig. 1, there are three types of entities in the model: a set of users, a mobile service provider (SP), and a location server (LS). LS holds a POI database and provides some location-based services that are consumed by the users. The POI database consists of ρ POI records, each of which gives the coordinate (i.e., longitude and latitude) of a position and a description about what is at the position. SP is responsible for establishing and maintaining the communication between LS and the users. We also take the reasonable assumption made in [18] where SP is considered to be a passive entity that is not allowed to collude with LS. This is because, if a user wants to consume some LBS, he/she has to endow SP with the authority to collect the whereabouts of

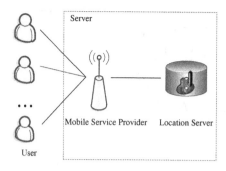

Fig. 1. System model

his/her mobile device. In this case, any method for user privacy protection will be definitely overthrew provided that SP is allowed to collude with LS.

By excluding the possibility of collusion between SP and LS, we consider only two possible adversaries. When LS is an adversary, he/she tries to get the position of the users based on their location-based queries. When a user is an adversary, he/she tries to obtain more POI data than the necessary answer to his/her location-based queries. Hence, the objective of our work presented in this paper is to design a method that can protect mutual privacy (i.e., the user's location and LS's POI data) when processing location-based queries.

3.2 Approach Overview

Figure 2 gives an overview of our approach to the problem of location-based query processing with mutual privacy protection. Our approach consists of two steps: *data preparation* and *query processing*. Data preparation is conducted by LS itself and is totally an offline computation, while query processing involves some online computation and multiple rounds of communication between LS and the user. Table 1 summarizes the notations in our paper.

Data preparation. LS first constructs a quadtree as the spatial index of his/her POI database, based on which the quadtree subdivision of the whole region, denoted as grid Q, can be made directly. Every leaf of the quadtree has at most d POIs, which means every cell in Q has also at most d POIs. For these cells having fewer than d POIs, LS adds dummy POIs as placeholders. Besides the private grid Q, LS also generates a public grid P by dividing the whole region into $m \times m$ cells where $m = 2^{h-1}$ and h is the height of the quadtree. Clearly, every cell P_i in grid P must be located in exactly one cell Q_j in grid Q. Based on this observation, every cell P_i in P stores the index j of its corresponding cell Q_j in Q. To keep his/her POI data secret, LS encrypts both the private grid Q and the public grid P. In particular, every cell Q_j (i.e., the d POIs in that cell) is encrypted by a key QK_j and every cell P_i (i.e., the index of its corresponding cell in grid Q) is encrypted by a key PK_i. More details about data encryption will be

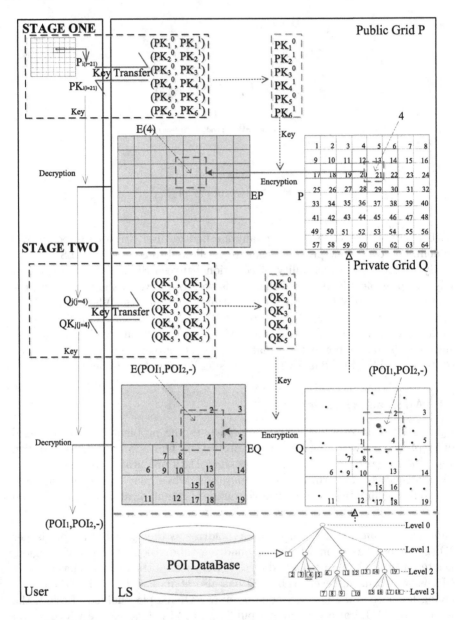

Fig. 2. Overview of privacy preserving location-based query processing

given in the next subsection. The two encrypted grids $E(P)$ and $E(Q)$, together with m, are sent to all users who want to consume location-based services.

Query Processing. According to the value of m and his/her coordinates (note that the coordinates are just the input of location-based queries), it is quite easy for a user u to determine P_i, the cell of the public grid P where he/she is located

Table 1. Summary of notations

Notation	Definition	Notation	Definition
P	public grid	Q	private grid
M	number of cells in gird P	N	number of cells in grid Q
$E(P)$	encrypted public grid	$E(Q)$	encrypted private grid
α	bit length of data in a cell of P	β	bit length of data in a cell of Q
PK	keys to encrypt data in P	QK	keys to encrypt data in Q
d	leaf capacity of quadtree	ϕ	$\log_2 M + \log_2 N$

in. Holding the index of cell P_i, the user is able to obtain the encryption key PK_i for P_i by running an oblivious key transfer protocol with LS. Note that, this key is also the decryption key for P_i (more details will be discussed later), so the user can immediately perform decryption locally and get the index of Q_j, the cell of the private grid Q where he/she is located in. Once the index of cell Q_j is decided, the user performs the oblivious key transfer protocol again to get the encryption key QK_j for cell Q_j. By decrypting $E(Q_j)$ with the key QK_j, the user can obtain the POIs contained in cell Q_j, which is just the answer to his/her location-based query. Note that, the oblivious key transfer protocol ensures LS cannot learn the index of the cell where the user is located in, while the user cannot obtain more keys than the one for the cell where he/she is located in, thus protecting mutual privacy.

3.3   Protocol

Algorithm 1 shows our protocol of location-based query processing with mutual privacy protection. Recall that grid P has $M = m * m$ cells, each of which contains an α-bits integer P_I, indicating the index of the corresponding cell in the private grid Q. The number of cells in grid Q depends on the POI database and is assumed to be N. Each cell in grid Q has d POIs (including dummy POIs), and their information such as the coordinates is represented as a string of bits, denoted as an β-bits integer Q_J. To encrypt P_I for $1 \leq I \leq M$, LS generates l random pairs of keys $(PK_1^0, PK_1^1), (PK_2^0, PK_2^1), \cdots, (PK_l^0, PK_l^1)$ where $l = \log_2 M$. For a given I, its binary representation $(i_1, i_2, \cdots, i_l)$ is used as the selection bits to select l keys from the l pairs of keys. The exclusive-OR of the l selected keys, $\bigotimes_{k=1}^{l} PK_k^{i_k}$, is given to a pseudorandom generator H (which is typically implemented as secure hash function such as SHA-1) to produce an α-bits pseudorandom string. The encryption of P_I is completed by computing the exclusive-OR of P_I and the pseudorandom string. Q_J can be likewise encrypted, as seen from lines 6-7. The encrypted grids $E(P_I)$ and $E(Q_J)$ are sent to u when u has a stable and fast network such as Wi-Fi, which completes the offline stage of our protocol.

In the online stage, u can determine the cell in the public grid P where he/she is located in based on his/her current coordinates. Based on the index of the cell, u can obtain the l keys that are used to encrypt the cell by running an obvious key transfer protocol (see Algorithm 2), which is built on the oblivious transfer extension protocol presented in [15]. The pseudorandom string used for encryption can be reproduced by the same H function provided that the same seed is given to H, so the decryption can be easily done after u gets these l keys. Once the index of the cell where u is located in the private grid Q is known by u, he/she can run the key transfer protocol again to get the keys for that cell, based on which Q_J can be decrypted to retrieve the POIs in that grid, which is just the answer to u's query.

Algorithm 1. Query Processing with Mutual Privacy Protection

Input: LS holds a POI database, user u holds his/her coordinates
Output: u gets a set of POIs which are close to his/her position

1: **Offline:**
2: LS constructs a quadtree based on the POI database
3: LS builds the private grid Q and public grid P based on the quadtree
4: LS generates l random pairs of keys $(PK_1^0, PK_1^1), (PK_2^0, PK_2^1), \cdots,$
   (PK_l^0, PK_l^1) where $l = \log_2 M$
5: LS computes $E(P_I) \leftarrow H(\bigotimes_{k=1}^{l} PK_k^{i_k}, \alpha) \otimes P_I$ for $1 \leq I \leq M$, where
   $(i_1, i_2, \cdots, i_l)$ is the binary representation of I
6: LS generates l' random pairs of keys $(QK_1^0, QK_1^1), (QK_2^0, QK_2^1), \cdots,$
   $(QK_{l'}^0, QK_{l'}^1)$ where $l' = \log_2 N$
7: LS computes $E(Q_J) \leftarrow H(\bigotimes_{k=1}^{l'} QK_k^{j_k}, \beta) \otimes Q_{I'}$ for $1 \leq I' \leq N$, where
   $(j_1, j_2, \cdots, j_{l'})$ is the binary representation of J
8: LS sends $E(P_I)$ and $E(Q_J)$ to u

9: **Online:**
10: LS and u run the key transfer protocol (see Algorithm 2) to let u get l keys
    $PK_1^{i_1}, PK_2^{i_2}, \cdots, PK_l^{i_l}$
11: u gets j by decrypting $E(P_i)$, that is, computing $H(\bigotimes_{k=1}^{l} PK_k^{i_k}, \alpha) \otimes E(P_i)$
12: LS and u run the key transfer protocol again to let u get l' keys
    $QK_1^{j_1}, QK_2^{j_2}, \cdots, QK_{l'}^{j_{l'}}$
13: u gets Q_j (i.e., a set of POIs) by decrypting $E(Q_j)$, that is, computing
    $H(\bigotimes_{k=1}^{l} QK_k^{i_k}, \beta) \otimes E(Q_j)$

3.4 Correctness and Security Analysis

We first prove the utility of our proposed protocol which means the POIs user obtained are exactly the nearest POIs to user. Then, we prove that our protocol is secure for both user and LBS. LBS can not infer any information about user, the same, user can get no more information about LBS than the POIs he/she has requested.

Theorem 1 (Correctness). Assume that user and LBS follow Alogrithm 2 correctly. LBS holds l pairs of keys $(K_1^0, K_1^1), (K_2^0, K_2^1), \cdots, (K_l^0, K_l^1)$ and user

---

Algorithm 2. Key Transfer

Input: LS holds l pairs of keys $(K_1^0, K_1^1), (K_2^0, K_2^1), \cdots, (K_l^0, K_l^1)$;
   User u holds an index I whose binary representation is $(i_1, i_2, \cdots, i_l)$
Output: $(K_1^{i_1}, K_2^{i_2}, \cdots, K_l^{i_l})$ hold by u

Preliminaries: This algorithm operates over a group Z_q of prime order. More
specifically, G_q can be a subgroup of order q of Z_p^*, where p is prime and $q|(p-1)$. Let
g be a generator group, for which the computational Diffie-Hellman assumption holds.

1: LS chooses a random element $e \in Z_q$ and sends it to u
2: u selects l random elements $\{\gamma_1, \gamma_2, \cdots, \gamma_l\}$, where $1 \le \gamma_j \le q$ for $1 \le j \le l$
3: u sets two keys $\mathbf{K}_j^{i_j} = g^{\gamma_j}$ and $\mathbf{K}_j^{1-i_j} = e/\mathbf{K}_j^{i_j}$ for $1 \le j \le l$
4: u sends all $\mathbf{K}_j^0$ to LS, for $1 \le j \le l$
5: LS chooses a random μ, and computes e^μ and g^μ
6: LS computes $(\mathbf{K}_j^0)^\mu$ and $(\mathbf{K}_j^1)^\mu = e^\mu/(\mathbf{K}_j^0)^\mu$, for $1 \le j \le l$
7: LS computes $E_j^0 = H((\mathbf{K}_j^0)^\mu, \beta) \otimes K_j^0$ and $E_j^1 = H((\mathbf{K}_j^1)^\mu, \beta) \otimes K_j^1$, for $1 \le j \le l$
8: LS sends g^μ and the encrypted message $E_j = (E_j^0, E_j^1)$ to u, for $1 \le j \le l$
9: u decrypts $K_j^{i_j} = H((g^\mu)^{\gamma_j}, \beta) \otimes E_j^{i_j}$, for $1 \le j \le l$

---

u holds an index I whose binary representation is $(i_1, i_2, \cdots, i_l)$, we will prove
that the result $(K_1^{i_1}, K_2^{i_2}, \cdots, K_l^{i_l})$ which is obtained by user u according to
protocol is equal to $(QK_1^{i_1}, QK_2^{i_2}, \cdots, QK_l^{i_l})$ which is exactly the value user u
wants.
Proof. Firstly, according to line 9 of key transfer,

$$K_k^{i_k} = H((g^\mu)^{\gamma_k}, \beta) \otimes E_k^{i_k}$$

for the value of $i_k (1 \le k \le l)$, the binary representation of I, is either 0 or 1,
and we will discuss the process with two cases:

1:when $i_k = 0$, the value user u wants is $QK_l^{i_k} = K_k^0$. Next, our job is to
prove that the value $K_k^{i_k}$ which user obtained according to the protocol is K_k^0
too. Firstly, user u sets the value of $\mathbf{K}_k^0$ and $\mathbf{K}_k^1$ as (line 3):

$$\mathbf{K}_k^0 = g^{\gamma_k}, \mathbf{K}_k^1 = e/\mathbf{K}_k^0$$

then u sends $\mathbf{K}_k^0 = g^{\gamma_k}$ to LS (line 4), so we have $H((g^\mu)^{\gamma_k}, \beta) = H((\mathbf{K}_k^0)^\mu), \beta)$, furthermore,

$$H((g^\mu)^{\gamma_k}, \beta) \otimes E_k^{i_k} = H((g^\mu)^{\gamma_k}, \beta) \otimes H((\mathbf{K}_k^0)^\mu), \beta) \otimes K_k^0$$

according to the property of exclusive,

$$H((g^\mu)^{\gamma_k}, \beta) \otimes H((\mathbf{K}_k^0)^\mu), \beta) \otimes K_k^0 = K_k^0$$

eventually, we have $K_k^{i_k} = K_k^0 = QK_k^{i_k}$.
2:when $i_k = 1$, the proof is just a copy to ahead, so we give the analytical
procedure directly:

$$K_k^{i_k} = H((g^\mu)^{\gamma_k}, \beta) \otimes E_k^{i_k} = H((g^\mu)^{\gamma_k}, \beta) \otimes H((\mathbf{K}_k^1)^\mu), \beta) \otimes K_k^1 = K_k^1 = QK_k^{i_k}$$

It is proved that for the k-th query of I, $K_k^{i_k}$ is the right response for either $i_k = 0$ or $i_k = 1$. So, for all $1 \leq k \leq l$, $K_k^{i_k} = QK_k^{i_k}$.

Based on the correctness of key transfer, it's easy to verify the correctness of our proposed privacy preserving query processing. According to Algorithm 1, LBS sends both encrypted grids P and Q to u. If u gets the right key from LS to decrypt the target grid within P, then he gets the index of Q. The same, right key for Q_j, guarantees that u can decrypt the right POI records, and get the right POIs eventually.

Security Model. We assume that LS and client are both *semi-honest*, also known as "honest but curious". They run the protocol exactly as specified (no deviations, malicious or otherwise), but may try to learn as much as possible about the input of the other party from their views of the protocol.

It should be noted that though secure protocols against malicious adversaries exist, they are far too inefficient to implement and be used in practice. Secure protocols against semi-honest adversaries, however, are not only useful in practice but also the foundation of designing secure protocols against malicious adversaries. As we mentioned ahead, by excluding the possibility of collusion between SP and LS, we consider only two possible adversaries.

Security of u. When LS is an adversary, he/she tries to get the position of the users based on their location-based queries. However, the privacy of u is preserved, since the value that he/she sends to LS is out of distinguishing whether it was chosen directly at random or as $e/\mathbf{K}_j^0$ for random $\mathbf{K}_j^1$. LS gets no more information than random value $\mathbf{K}_j^0$.

Security of LS. In the preliminaries, we have the requirement that u cannot know the discrete logarithms of both $\mathbf{K}_j^0$ and $\mathbf{K}_j^1$, since this would reveal to u the discrete logarithm of e. Therefore, the Diffie-Hellman assumption implies that u cannot predict any value of $(\mathbf{K}_j^0)^\mu$ and $(\mathbf{K}_j^1)^\mu$ too. Based on the random oracle assumption, it ensures that u can not distinguish $H((\mathbf{K}_j^0)^\mu)$ either $H((\mathbf{K}_j^1)^\mu)$ from random, and so is the encrypted message E_j^0 and E_j^1. Thus u can get no more POI data except the one he/she has required. The privacy of LS can be preserved.

4   Performance Analysis

In this chapter, we analyze the computation and communication cost of our approach. As the most expensive operations in our approach are modular exponentiation (EXP) and hash functions (HASH), we focus on the number of times they are required. Table 2 summarizes the performance of our approach where ϕ equals to $\log_2 M + \log_2 N$.

4.1   Computation Cost

According to the protocol, our privacy preserving location-based query processing consists of two steps, data preparation and query processing. In the step of

Table 2. Performance analysis of our protocol

Computation cost			Communication cost
	LS	user	$M\alpha + N\beta + 2v + (2w + v)\phi$
Data Preparation	N+M HASH, 4 EXP	0	
Query Processing	2ϕ HASH, ϕ EXP	ϕ HASH, 2ϕ EXP	

data preparation, the main computation cost comes from the encryption of two grids, P and Q. Recall that P and Q have M and N cells respectively, and each cell is encrypted via a secure hash function. Therefore, the computation cost of data preparation is $M + N$ hashes.

The step of query processing requires two rounds of key transfer. To simplify discussion, we first assume the query of user u can be represented by 1-bit integer i. As seen from line 3 of Alogrithm 2, either $\mathbf{K}^0$ or $\mathbf{K}^1$ is set to g^γ by one exponentiation. After that, u sends the $\mathbf{K}^0$ to LS, and LS reconstructs the key $(\mathbf{K}^0)^\mu$ by one exponentiation (see line 6 of Alogrithm 2). Next, 2 hashes are used to encrypt original messages $(\mathbf{K}^0, \mathbf{K}^1)$ by LS (see line 7 of Alogrithm 2). To extract message $\mathbf{K}^i$ from E^i (see line 9 of Alogrithm 2), one exponentiation is needed for the computation of $(g^\mu)^\gamma$ and one hash for key.

Recall that each grid of P and Q is represented by an α-bits and β-bits integer, respectively. Consequently, the number of key pairs, l, of P is $\log_2 M$ and $\log_2 N$ for Q. Hence, to transfer $\log_2 M$ keys for P and $\log_2 N$ keys for Q, LS needs 4 exponentiations offline and ϕ exponentiations adds 2ϕ hashes online, while u needs 2ϕ exponentiations and ϕ hashes, where $\phi = \log_2 M + \log_2 N$.

4.2   Communication Cost

At the end of data preparation, LS sends the encrypted grids $E(P)$ and $E(Q)$ to u (see line 8 in Algorithm 1). Since P has M grids with α bits for each grid, and Q has N grids with β bits of each grid, the size of encrypted grids P and Q are $M\alpha$ and $N\beta$ respectively. By taking advantage of hash function, the length of encrypted grids $E(P)$ and $E(Q)$ remains $M\alpha$ and $N\beta$. Therefore, the communication cost of data preparation is $M\alpha + N\beta$.

Before discussing the communication cost of query processing, it is important to distinguish between the length of the input element, and the length of group element v (which is typically 1024 bits long). With l pairs of keys as the input, the data from user to LS is composed of l group elements $\mathbf{K}_j^0 (1 \leq j \leq l)$ (see line 4 in Alogrithm 2), and the data from LS to user is composed of a group element g^μ and $2l$ encrypted messages $E = (E_j^0, E_j^1)$ in the size of input (see line 8 in Alogrithm 2).

In the two rounds of key transfer, the length of key which is used to encrypt grid P and Q is set to w (which is typically 80 bits), which means the length of input element is w bits. As the number of key pairs, l, during key transfer is $\log_2 M$ and $\log_2 N$, the communication cost is $v\phi$ bits from user to LS and $2v + 2w\phi$ bits from LS to user.

5   Experimental Evaluation

In this section, we present experimental evaluations of our approach on a synthetic dataset. We implement the methods proposed in [18,24], and compare them with our approach. All experiments are performed on a PC with 3.2 GHz CPU, 8 GB RAM, JDK 7 and Win 7. The synthetic dataset is a 1000*1000 region with uniformly distributed POIs, each of which is represented by a 64-bits integer. The key size to pseudo-random function H is set to 80. We have two variables in the experiment, the number of POIs, N, and the leaf capacity of quadtree d. d is set to 15 while N varies from $20k$ to $100k$ with a step of $20k$. N is set to $50k$ when d varies from 5 to 25 with a step of 5.

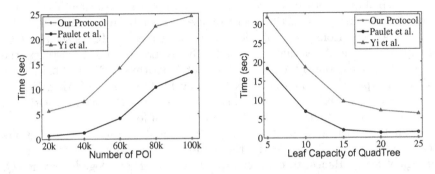

Fig. 3. Online response time of three approaches

It is clear from Fig. 3 that our approach has the best performance in terms of online response time. Using our approach, queries on all datasets can be responded in less than 0.1 s, even for the biggest dataset with 100k POIs. In contrast, though the method of Paulet et al. can return answers within 1 s on the dataset with 20k POIs, it takes more time when N becomes bigger or d becomes smaller. For example, when $N = 50k$ and $d = 5$, it takes about 18 s which is unacceptable in practice. The method proposed by Yi et al. needs 5 s even on the smallest dataset with 20k POIs.

Clearly, with the increasing number of POIs or decreasing number of leaf capacity, both grids Q and P have more cells, which results in the increase of both computation and communication cost. However, our approach in the phase of online computation works only on $(log_2 M + log_2 N)$ pairs of keys, so the increase of POIs can be largely overlooked which means our approach has a good scalability. Besides, LS needs to send users the encrypted $E(P)$ and $E(Q)$ whose size is linear with $M + N$. Hence, the increase of POIs or the decrease of leaf capacity will both lead to a linear increase of communication cost.

Figure 4 shows the communication cost of three approaches. Clearly, the communication cost of our approach is not the best, but it is the cost for a significant improvement in online response time. Further, the increase of communication cost is totally acceptable in practice. For example, for the biggest dataset with

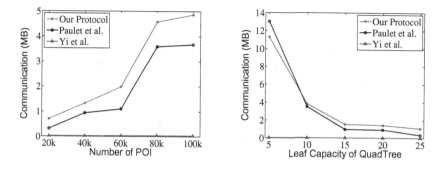

Fig. 4. Communication cost of three approaches

100k POIs, the communication cost of our approach is only about 12 MB, which is affordable in a stable and fast network such as Wi-Fi.

We also notice that our approach and the method of Paulet et al. both need to encrypt some data beforehand to protect the data privacy of LS. Figure 5 shows the time of data preparation. Clearly, our approach needs less time than the method of Paulet et al. In particular, our approach can be finished in 10 s for a dataset with 100k POIs. That is, even with the time of data preparation, a new user can get response in seconds using our approach.

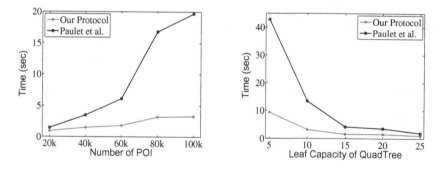

Fig. 5. Data preparation cost

In conclusion, the empirical study shows that our approach based on two rounds of OT transfer has a great improvement in online response time. Specifically, our approach can provide mutual privacy protection for location-based queries in less than 0.1 s with a communication cost less than 12 MB even for a dataset of 100k POIs. Therefore, our protocol is both computational efficient and communicational efficient.

6 Conclusion

In this paper, we propose an efficient approach to protecting mutual privacy in location-based queries. We achieve the goal by performing two rounds of OT extension on two small key sets. We theoretically prove the security and analyze the complexity of our approach. Compared with state-of-the-art work, our approach has several orders of magnitude improvement in online response time, only at the expense of little and acceptable communication cost. Empirical study shows that our approach is both computational efficient and communicational efficient.

Acknowledgment. This work was partially supported by Natural Science Foundation of China (Grant Nos. 61572336, 61572335, 61532018, 61402313, 61402312, 61303019), and Natural Science Foundation of Jiangsu Province (Grant No. BK20151223).

References

1. Beresford, A.R., Stajano, F.: Location privacy in pervasive computing. IEEE Pervasive Comput. **2**(1), 46–55 (2003)
2. Bettini, C., Wang, X.S., Jajodia, S.: Protecting privacy against location-based personal identification. In: Jonker, W., Petković, M. (eds.) SDM 2005. LNCS, vol. 3674, pp. 185–199. Springer, Heidelberg (2005)
3. Bettini, S.M.C.: A comparison of spatial generalization algorithms for LBS privacy preservation. In: MDM, pp. 258–262 (2007)
4. Bamba, B., Liu, L., Pesti, P., Wang, T.: Supporting anonymous location queries in mobile environments with privacygrid. In: WWW, pp. 237–246 (2008)
5. Chow, C.Y., Mokbel, M.F., Liu, X.: A peer-to-peer spatial cloaking algorithm for anonymous location-based service. In: SIGSPATIAL, pp. 171–178 (2006)
6. Gedik, B., Liu, L.: Location privacy in mobile systems: a personalized anonymization model. In: ICDCS, pp. 620–629 (2005)
7. Ghinita, G., Kalnis, P., Khoshgozaran, A., Shahabi, C., Tan, K.L.: Private queries in location based services: anonymizers are not necessary. In: SIGMOD, pp. 121–132 (2008)
8. Kushilevitz, E., Ostrovsky, R.: Replication is not needed: Single database, computationally-private information retrieval. In: FOCS, p. 364 (1997)
9. Kido, H., Yanagisawa, Y., Satoh, T.: An anonymous communication technique using dummies for location-based services. In: ICPS, pp. 88–97 (2005)
10. Krumm, J.: A survey of computational location privacy. Pers. Ubiquit. Comput. **13**(6), 391–399 (2009)
11. Liu, G., Wang, Y., Orgun, M.A.: Optimal social trust path selection in complex social networks. AAAI **10**, 1397–1398 (2010)
12. Liu, A., Zhengy, K., Liz, L., Liu, G., Zhao, L., Zhou, X.: Efficient secure similarity computation on encrypted trajectory data. In: ICDE, pp. 66–77 (2015)
13. Myles, G., Friday, A., Davies, N.: Preserving privacy in environments with location-based applications. JPCC **2**(1), 56–64 (2003)
14. Mokbel, M.F., Chow, C.Y., Aref, W.G.: The new Casper: query processing for location services without compromising privacy. In: VLDB, pp. 763–774 (2006)

15. Naor, M., Pinkas, B.: Oblivious transfer with adaptive queries. In: Wiener, M. (ed.) CRYPTO 1999. LNCS, vol. 1666, pp. 573–590. Springer, Heidelberg (1999)
16. Paillier, P.: Public-key cryptosystems based on composite degree residuosity classes. In: Stern, J. (ed.) EUROCRYPT 1999. LNCS, vol. 1592, pp. 223–238. Springer, Heidelberg (1999)
17. Papadopoulos, S., Bakiras, S., Papadias, D.: Nearest neighbor search with strong location privacy. PVLDB 3(1–2), 619–629 (2010)
18. Paulet, R., Kaosar, M.G., Yi, X., Bertino, E.: Privacy-preserving and content-protecting location based queries. TKDE 26(5), 1200–1210 (2014)
19. Rabin, M.O.: Digitalized signatures and public-key functions as intractable as factorization, MIT Lab for Computer Science, Technical report (1979)
20. Shang, S., Yuan, B., Deng, K., Xie, K., Zheng, K., Zhou, X.: PNN query processing on compressed trajectories. GEOINFORMATICA 16(3), 467–496 (2012)
21. Shang, S., Ding, R., Zheng, K., Jensen, C.S., Kalnis, P., Zhou, X.: Personalized trajectory matching in spatial networks. VLDB J 23(3), 449–468 (2014)
22. Youssef, M., Atluri, V., Adam, N.R.: Preserving mobile customer privacy: an access control system for moving objects and customer profiles. In: MDM, pp. 67–76 (2005)
23. Yiu, M.L., Jensen, C.S., Huang, X., Lu, H.: SpaceTwist: managing the trade-offs among location privacy, query performance, and query accuracy in mobile services. In: ICDE, pp. 366–375 (2008)
24. Yi, X., Paulet, R., Bertino, E., Varadharajan, V.: Practical k nearest neighbor queries with location privacy. In: ICDE, pp. 640–651 (2014)
25. Zheng, K., Zheng, Y., Yuan, N.J., Shang, S., Zhou, X.: Online discovery of gathering patterns over trajectories. TKDE 26(8), 1974–1988 (2014)
26. Zheng, K., Zhou, X., Fung, P.C., Xie, K.: Spatial query processing for fuzzy objects. VLDB J 21(5), 729–751 (2012)

Semantic-Aware Location Privacy Preservation on Road Networks

Yanhui Li[1]([✉]), Ye Yuan[1], Guoren Wang[1], Lei Chen[2], and Jiajia Li[3]

[1] Northeastern University, Shenyang, China
yliboneu@gmail.com
[2] Hong Kong University of Science and Technology, Hong Kong, China
[3] Shenyang Aerospace University, Shenyang, China

Abstract. In this paper, we address the topic of location privacy preservation of mobile users on road networks. Most existing techniques of privacy preservation rely on structure-based *spatial cloaking*, but pay little attention to location semantic information. Yet, location semantic information may disclose sensitive information about mobile users. Thus, we propose *CloSed*, a semantic-awareness privacy preservation model to protect users' privacy from violation. We design *cloaked sets* that should cover different semantic regions of road networks as well as satisfy quality of service (QoS). As the problem of calculating the optimal cloaked set is NP-hard, we design a greedy algorithm that balances QoS and privacy requirements. Extensive experiments evaluations demonstrate the efficiency and effectiveness of our proposed algorithm in providing privacy guarantees on large real-world datasets.

1 Introduction

Advances in positioning technologies along with the tremendous popularity of mobile devices have resulted in the widespread adoption of location-based services (LBS) on road networks. Examples of these applications include navigation services, identification of points of interest (POIs), and receiving traffic alerts or notifications. While enjoying the convenience of LBS, however, users also face significant risks of privacy leakage [23]. Adversaries can exploit user location information for such nefarious purposes as stalking, spamming, and inferring political/religious affiliations or alternative lifestyles.

The state-of-the-art for protecting the positions of LBS users over road networks is based on the model of *segment l-diversity* [3,26]. In this model, the actual user position is replaced by a set of *segments*, i.e., edges in a road network, and the number of segments indicates the degree of diversity. Though this solution can satisfy most privacy-preservation requirements, it cannot resist the types of *semantic homogeneity attacks* illustrated by the following example.

Example 1. Consider a scenario in Fig. 1. A patient, named Bob, asks for services through his GPS-enabled mobile phone from road e_1. To prevent Bob's location from leakage, the approach based on segment l-diversity cloaks Bob's walking

© Springer International Publishing Switzerland 2016
S.B. Navathe et al. (Eds.): DASFAA 2016, Part II, LNCS 9643, pp. 314–331, 2016.
DOI: 10.1007/978-3-319-32049-6_20

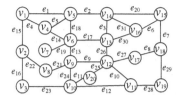

Edge id	S_label	Edge id	S_label	Edge id	S_label
e_1	hospital	e_5	hospital	e_9	shop
e_2	bank	e_6	church	e_{10}	market
e_3	school	e_7	police	e_{11}	church
e_4	hospital	e_8	club	...	...

(a) Simplified Road Network   (b) Mapping Table

Fig. 1. Semantic road network

road with other nearby roads [3,26]. In our example, we assume that $l = 3$, and then the *cloaked set* may be $\{e_1, e_4, e_5\}$. Unfortunately, it is easy for an adversary to infer that Bob is in the hospital, since all roads e_1, e_4, and e_5 have the homogeneous semantics, namely hospital. Hence, even though Bob's location is seemingly obfuscated, it can be inferred by a semantic homogeneity attack.

Although some techniques have been proposed to resist semantic homogeneity attacks [6,16,28] over road networks, they have different limitations. The solutions proposed in [16] have a deterministic property for its cloaked areas, so it is subject to *reverse engineering attacks*, e.g., a *replay privacy attack* [26]. The offline approaches of [6,28] cannot support the privacy requirement update due to their cloaked sets being generated a priori for a particular privacy requirement. Changes in mobile users' privacy requirements are frequent, thus seriously threatening the applicability of these approaches. The work [28] also presents an online cloaking algorithm to protect sensitive semantic information. However, the cloaking cost is expensive due to considering velocity-based linkage attacks which is out of our scope.

To solve the problem highlighted in Example 1 and overcome the drawbacks of methods above, we propose *CloSed*, a semantic-aware privacy-preserving model, and design a new solution to protect the location privacy of mobile users on road networks against semantic homogeneity attacks. We illustrate the model using Example 1 as a running example throughout the paper. In our approach, instead of cloaking Bob's current road e_1 with other nearby roads $\{e_4, e_5\}$ of the hospital, CloSed generates a cloaked road set consisting of Bob's current *semantic road* and other nearby semantic roads, e.g., the cloaked set is $\{e_1, e_2, e_3\}$. In this example, the semantic of Bob's current road may be hospital, bank or school. Therefore, the adversary can no longer infer the exact semantic of Bob's location.

Our primary goal is to protect location privacy while guaranteeing the quality of location-based services for snap-shot queries. Our strategy focuses on semantic diversity, which guarantees that it would be difficult to associate a specific user with a specific semantic with a high possibility. It also regards QoS as a critical measure for designing privacy preservation solutions, and supports personalized privacy requirements.

To achieve this goal, in *CloSed*, each mobile user can designate his location privacy requirement as *l-semantic diversity*. That is, rather than *l-segment* diversity, we focus on the cloaked road set possessing at least l different semantic types. Thus, mobile users can use location-based services without the need to reveal their private location and location semantic information. To implement l-semantic diversity over road networks, our solution, named *EIRank*, consists of two steps: pre-processing and online cloaking. The pre-processing phase reduces the cloaked space by roughly grouping roads into different clusters, called *buckets*, while the online cloaking phase generates the desirable cloaked set in each bucket. In the pre-processing phase, to guarantee the efficient generation of buckets, structure and semantic information should be integrated. The major challenge lies in how to combine them together seamlessly. We propose the concepts of *edge interaction (EI) network* and *virtual nodes* to embed structure and semantic information together.

In designing our solution to the problem of privacy preservation for mobile users on road networks, we thus make a number of contributions, as follows:

- CloSed's semantic-aware model extends existing solutions by offering protection against semantic homogeneity attacks.
- CloSed's approximation algorithm EIRank naturally balances privacy requirements with QoS.
- EIRank integrates structure and semantic information seamlessly by transforming road networks into EI networks and leveraging the idea of virtual nodes.
- EIRank's evaluation over large real-world datasets demonstrate its efficiency at cloaking the optimal road set and guaranteeing that exact location and semantic information cannot be leaked.

The remainder of the paper is organized as follows. We introduce our road network and privacy preservation model in Sect. 2. In Sects. 3 and 4, we describe the technique and algorithm for location anonymization. In Sect. 5, we report extensive experimental results. We briefly review the related work in Sect. 6. Finally, Sect. 7 concludes the paper.

2   Problem Definition

We begin this section by presenting the road network model. We then formally define our privacy preservation model, and the goals of the associated techniques. Finally, we present our algorithm framework.

2.1   Road Network

Definition 1 *(Semantic Road Network). A road network is modeled as an undirected graph $G = (V, E, \xi)$ with a node set V and an edge set E, such that (i) a node $v \in V$ represents a road intersection or a location (e.g., hospital); (ii) an edge $e = (u, v) \in E$, also called a **segment**, connects two nodes u and v; and (iii) L represents a semantic function, i.e., for each edge $e \in E$, $\xi(e)$ is the sensitive semantic label of segment e.*

Example 2 (Semantic Road Network). Fig. 1(a) shows an example of a semantic road network, in which each edge is associated with a semantic ID. Figure 1(b) gives semantic labels corresponding to the IDs. Nodes v_1, v_4 and v_5 in Fig. 1(a) are different buildings within the same hospital. Edges e_1, e_4 and e_5 connecting these three nodes would then have the same sensitive semantic label "hospital". Thus, the area represented by the triangle (v_1, v_4 and v_5) would indicate the hospital.

2.2   Privacy Preservation Model

To resist against semantic homogeneity attacks as given in the introduction, we propose the following privacy preservation model.

Definition 2 *(<u>Cloaked</u> <u>Set</u> l-<u>se</u>mantic <u>d</u>iversity (**CloSed**)). A user's published cloaked segment set $S_c = \{e_1, e_2, ..., e_i, ...\}$ is said to have l-semantic diversity, if (i) S_c contains at least l different types of semantic labels, i.e., $|\xi(S_c)| = |\bigcup_{\forall e \in S_c} \xi(e)| \geq l$, and (ii) the possibility of distinguishing a user's semantic label among other semantic labels in S_c does not exceed $\frac{1}{l}$.*

Returning to Example 1, to achieve *CloSed*, Bob's published cloaked segment set can be $S_c = \{e_1, e_2, e_3\}$ or $S_c = \{e_1, e_2, e_6\}$. The cloaked segment set $S_c = \{e_1, e_2, e_3\}$ indicates that Bob may be in a hospital, a bank or a school. The cloaked segment set $S_c = \{e_1, e_2, e_6\}$ indicates that Bob may be in a hospital, a bank or a church.

Architecture. Similar to existing works [2,4,10,13,20,26], we adopt the classical centralized privacy-preserving architecture. In this architecture, the location anonymizer is a trusted entity that lies between mobile users and service providers (SP), and performs location anonymization and result filtering operations. More specifically, the location anonymizer first removes identity labels (e.g., id) and transforms the original query with an accurate location to another query with a cloaked set, according to users' privacy requirements. Next, SP computes and forwards the produced candidate results to the location anonymizer. At last, the location anonymizer extracts the exact answers from the candidate results by adequately filtering false hit information.

Based on the processing framework, in the CloSed model, a mobile user should specify his/her privacy profile (l, σ_t), where l indicates l-semantic diversity and σ_t is the maximum temporal tolerance to guarantee QoS. To preserve user privacy, we identify an important property that is sufficient for a cloaking technique.

Definition 3 *(Segment Oblivious). For a query user u in segment e, given u's profile (l, σ_t), his/her published cloaked segment set S_c satisfies the segment oblivious property iff (i) S_c contains at least l semantic labels; (ii) $e \in S_c$; and (iii) a query initiating in any segment of the cloaked set S_c will return the same cloaked set S_c as the cloaked set for the given l.*

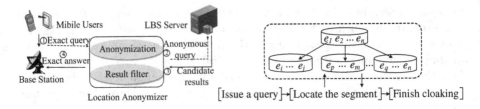

318   Y. Li et al.

Fig. 2. Privacy-preserving architecture   **Fig. 3.** Algorithmic framework

From Definition 3, it can be shown that any solution to Definition 2 satisfies the following theorem.

Theorem 1. *A cloaking technique for a road network G_r can achieve l-semantic diversity, if every cloaked set S_c published in G_r satisfies the segment oblivious property.*

Proof. According to Definitions 2 and 3, it is obviously to reach Theorem 1.

In addition to the preservation of cloaked set l-semantic diversity, the other objectives of our cloaking technique are that: (1) The cloaked set should not reveal the exact segment of any user; and (2) The cloaking technique should not compromise the QoS.

2.3 Algorithmic Framework of Anonymization

To achieve privacy preservation, in the location anonymizer (Fig. 2), the technique employed needs to blur the exact active segment of each mobile user to a cloaked set that satisfies user's privacy profile. A segment is marked as active segment if it is associated with at least one query.

To meet the requirement of privacy (i.e., Theorem 1) and achieve high QoS, our anonymization algorithm consists of two stages: an offline pre-processing phase and an online cloaking phase, as shown in Fig. 3. In the offline pre-processing phase, we allocate all segments of a road network to different buckets, so that we can perform anonymization in one bucket rather than search the entire road network in the cloaking process. In the online cloaking phase, we locate the buckets of active segments and anonymize segments based on user privacy profiles.

3 Segment Allocation

This section presents the offline pre-processing phase as introduced in the algorithmic framework. Specifically, the segments of a road network are allocated to different buckets according to users' privacy requirements. To achieve most user privacy requirements, we make the following observation.

Observation 1: The location semantic privacy requirements L of user privacy profiles follow a Gaussian distribution $L \sim N(\mu, \sigma^2)$, i.e., most user privacy requirements fall in the middle range, and fewer have higher privacy requirements. The parameter μ is the mean of the distribution, and the parameter σ is its standard deviation.

It follows that we can leverage the 3σ rule, also known as the 68-95-99.7 empirical rule, which states that about 99.7 % of values drawn from a Gaussian distribution are within three standard deviations from the mean. We accordingly set the semantic number of a bucket to $\mu + 3\sigma$ to satisfy all user location anonymization in one bucket. Definition 4 states the goal of segment allocation.

Definition 4 *(Segment Allocation). The segments of a road network $G = (V, E, \xi)$ are allocated to p buckets, $G_1, G_2, ...G_p$, where $p > 1$ and $G_i = (V_i, E_i, \xi_i)$, such that $V = \bigcup_{1 \leq i \leq p} V_i$, $E = \bigcup_{1 \leq i \leq p} E_i$, $\xi(E) = \bigcup_{1 \leq i \leq p} \xi_i(E_i)$, and the following conditions are satisfied.*
(i) The segments of all buckets are disjoint, i.e., $\forall 1 \leq i, j \leq p$, $E_i \cap E_j = \phi$.
(ii) The semantic number of a bucket must exceed the threshold $\mu + 3\sigma$, i.e., $|\xi(E_i)| = |\bigcup_{\forall e \in E_i} \xi(e)| \geq \mu + 3\sigma$.

In addition to protecting the location privacy of mobile users, the cloaking algorithm should not compromise QoS, which mainly depends on communication cost. We use the number of candidate results to measure communication cost, which is formulated in Definition 5. Without loss of generality, we focus our attention on k-nearest neighbors (kNN) queries.

Definition 5 *(LBS Server Processing). [26] For a query q with associated anonymous segment set S_c, the candidate results of q consists of two parts: (1) the POIs on the segments of S_c, and (2) the results as q is issued on the boundary nodes of the boundary set S_{bn}, where the boundary set is a set of nodes whose some connected edges are not included in S_c. Formally,* $CR(q, S_c) = (\bigcup_{s \in S_c} O(q, s)) \bigcup (\bigcup_{v \in S_{bn}} O(q, v))$

Based on this query processing model, it can be seen that the communication cost $CR(q, S_c)$ is significantly influenced by parameters $|S_c|$ and $|S_{bn}|$. However, reducing $|S_c|$ and $|S_{bn}|$ imposes conflicting demands on $CR(q, S_c)$. This is explained by the fact that segments that are near each other tend to possess similar semantic labels.

For a user privacy profile, our objective is to find the optimal cloaked set which is minimized in terms of communication cost, while satisfying l-sematic diversity. In summary, our problem is equivalent to the following optimization problem:

Minimize $CR(q, S_c)$, subject to $|\xi(S_c)| = |\bigcup_{\forall e \in S_c} \xi(e)| \geq l$.

According to the paper [27], the problem of computing an optimal cloaked set is NP-hard.

Solution. Based on above analysis, we propose a greedy solution called *EIRank*. Intuitively, cloaking adjacent segments with different semantic labels provides a compact structure and semantic preference simultaneously. In other words, we prefer cloaking the segments exhibiting *structure similarity* and *semantic label dissimilarity*. To measure the similarity of linkage structures and the dissimilarity of semantic labels, we introduce two scoring functions: $S(n_1, n_2)$ and $Diff(e_p.\varphi, e_q.\varphi)$, respectively.

In many applications, objects are considered similar if they are related to similar objects. Based on this intuition, we adopt a general similarity metric called SimRank to measure the similarity of linkage structures. SimRank is calculated by Eq. 1.

$$S(n_1, n_2) = \begin{cases} 1 & n_1 = n_2 \\ \frac{C}{|I_{n_1}||I_{n_2}|} \sum_{j \in I_{n_2}} \sum_{i \in I_{n_1}} S(i,j) & n_1 \neq n_2 \end{cases} \tag{1}$$

where C refers to as a decay factor, is a constant between 0 and 1, and I_n represents the set of neighbors of n. Note that Eq. 1 is defined to be 0 when $I_{n_1} = \emptyset$ or $I_{n_2} = \emptyset$.

To evaluate the dissimilarity of semantic labels of segments, we use the normalized edit distance. In this case, the dissimilarity of semantic labels $Diff(e_p.\varphi, e_q.\varphi)$ is measured by the edit distance between the semantic labels with regard to the length of the semantic label. The edit distance, $Edit(e_p.\varphi, e_q.\varphi)$, between two semantic labels, $e_p.\varphi$ and $e_q.\varphi$, is defined as the minimum number of basic operations required to transform one semantic label into the other. In this paper, the basic operations are defined as insertion, deletion and substitution of symbols, which is formalized as follows.

Let $T_s(b|a)$ represents the substitution of symbol a by symbol b ($a \neq b$), $T_i(a)$ represents the insertion of symbol a, and $T_d(a)$ represents the deletion of symbol a. Then,

$$Diff(e_p.\varphi, e_q.\varphi) = \frac{Edit(e_p.\varphi, e_q.\varphi)}{Max(|e_p.\varphi|, |e_q.\varphi|)} \tag{2}$$

where $e_p.\varphi$ denotes the label function of e_p, and $Max(|e_p.\varphi|, |e_q.\varphi|)$ represents a function that computes the larger length of the two labels $e_p.\varphi$ and $e_q.\varphi$.

To combine linkage structure and segment semantic information for segment allocation, we propose a solution, called *EIRank*, for simultaneously representing link-based similarity and semantic-based dissimilarity. Our solution consists of four steps: *EI network construction, label clustering, Augmented EI Network Construction* and *segment allocation*. Next, we will discuss each step in details.

EI Network Construction. For simplicity, we assume that the semantic label of each edge is unique. To integrate linkage structure and segment semantic, the semantic road network is transformed into an edge interaction (EI) network. An EI network node, called *e-node*, represents an edge in the original semantic road network, and two e-nodes are adjacent if their corresponding edges share a common node in the original semantic road network. The labels of e-nodes in the EI network are given by the semantic labels of the corresponding edges in the road network. For example, edges e_1 and e_2 share a common node v_2 in the semantic road network (Fig. 4(a)), and thus e-nodes e_1 and e_2 are linked together in the EI network (Fig. 4(b)). Since the segment id itself represents the semantic label of the segment, we do not mark the labels of the e-nodes anymore in the EI network.

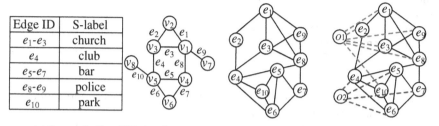

Edge ID	S-label
e_1-e_3	church
e_4	club
e_5-e_7	bar
e_8-e_9	police
e_{10}	park

(a)Semantic Road Network   (b)EI Network   (c)Augmented EI Network

Fig. 4. Example of EIRank strategy

Label Clustering. The problem of computing the dissimilarity of two segment labels is equivalently converted into the one of computing the dissimilarity of two e-node labels in the EI network. We use the method mentioned above to achieve this goal. Take the labels of the two e-nodes e_1 and e_4 in Fig. 4(b) as an example. By performing four basic operation $T_s(l|h)$, $T_s(b|r)$, $T_d(c)$ and $T_d(h)$, the label of e-node e_1 is transformed to the label of e-node e_4. Therefore, the dissimilarity of the two e-node labels is $Diff(e_1.\varphi, e_4.\varphi) = \frac{2}{3}$.

Based on the dissimilarity of the labels of e-nodes in the EI network, we perform a generalized k-medians clustering [19] for the labels of the e-nodes in the EI network. The result of label clustering for Fig. 4(b) is $\{church, police, park\}$ and $\{bar, club\}$.

Augmented EI Network Construction. In this step, we create a virtual node for each cluster and connect the e-nodes whose labels are in the same cluster to the virtual node. This new generated network is called *augmented EI network*. The original e-nodes in a label cluster have higher structure similarities by adding the virtual nodes. For example, Fig. 4(c) shows the updated EI network corresponding to Fig. 4(b). Two virtual e-nodes o_1 and o_2 are added to represent the clusters $\{church, police, park\}$ and $\{bar, club\}$, respectively. Then, the e-nodes in the set $\{e_1, e_2, e_3, e_8, e_9, e_{10}\}$ are connected to the virtual node o_1. In the same manner, virtual node o_2 is connected to the e-nodes in the set $\{e_4, e_5, e_6, e_7\}$.

Segment Allocation. As stated above, the segments of a cloaked set needs to have the structure similarity and semantic label dissimilarity. Based on the above steps, the dissimilarity of the original e-node labels has been transformed to the similarity of the linkage structure. This is consistent with the similarity of the

---

Algorithm 1. Baseline Algorithm

Input: Semantic road network $G = (V, E, \xi)$, Bucket scale N_l
Output: Buckets $G_1, G_2, ..., G_p$
1 Transform the G into EI network;
2 Execute the label clustering for e-nodes;
3 Compute $S(e_p, e_q)$ for all e-node pairs;
4 Allocate(e_p, e_q, G_S);
5 Merge buckets G_i where $|\xi(G_i)| < N_l$;
6 return non-empty buckets $G_1, G_2, ..., G_p$;

---

linkage structure. Next, we use the function $S(e_p, e_q)$ to measure the similarity for every pair of non-virtual e-nodes.

To compute SimRank efficiently, we adopt the method in [7]. In this case, the similarity of e-nodes is measured by Eq. 3, which states that the similarity of two e-nodes is the expectation of the total time which is the time taken by two random walkers starting from two different nodes to finally meet.

$$S(e_p, e_q) = E(C^{\tau(e_p, e_q)}) \tag{3}$$

Once the similarity has been computed for all e-node pairs, we use the single-linkage hierarchical clustering [24] to perform the segment allocation. The function Allocate (e_p, e_q, G_S) is used to describe this process.

The complete description of our EIRank strategy is given in Algorithm 1.

4   Online Cloaking Phase

In the previous section, we have described the pre-processing phase of our approach. Once partitioned buckets have been obtained, the remaining work is to generate a cloaked set according to a user's online request. Before detailing our cloaking algorithm, we present several index structures, namely *Ordered Locating Index* (*OLI*), semantic-aware order preserving list (*SOPlist*), and cloaked l-diverse segment set (*Cloaked l-maplist*), used in the online cloaking.

4.1   Index Structure

Ordered Locating Index. In order to quickly locate the position of a segment in a segment allocation, we design a novel index structure called *OLI* based on the hash table for organizing the segments in order. We keep a record of each entry in the form of (*Seg,Sid,Cid,Pointer*) where Seg is the segment identifier, Sid is the bucket identifier of the segment Seg, Cid is the position identifier of segment Seg in bucket Sid, and Pointer is a pointer to the next entry. We use Eq. 4 to compute the sequence of segment $Seq(e_{i,j})$ in the ordered linked list to obtain the Sid and Cid of $e_{i,j}$. The first three entries of this equation are used to compute the number of segments before segment $e_{i,j}$. Note that $e_{i,j}$ connects the nodes i and j.

$$Seq(e_{i,j}) = \sum_{k=1}^{i-1} degree(k) - |S_{overlap}|_{S_{overlap}=\{e_{lt}\},t<i,l<i}$$

$$+ |S_{prior}|_{S_{prior}=\{e_{ip}\},i<p<j} + 1 \tag{4}$$

Set the segment $e_{6,9}$ in Fig. 1 as an example. Using Eq. 4, we compute its segment sequence $Seq(e_{6,9}) = degree(v_1) + degree(v_2) + degree(v_3) + degree(v_4) + degree(v_5)$ - $|\{e_{1,2}, e_{1,4}, e_{1,5}, e_{2,3}, e_{4,5}\}|) + |\{e_{6,7}\}| + 1 = 3+3+2+3+4-5+1+1 = 12$. Then, searching for the 12th record in OLI which is shown in Fig. 5, we get Sid = 1 and Cid = 2. We conclude that the segment $e_{6,9}$ is in bucket 1, at position 2.

SOPlist and Cloaked l-maplist. To facilitate the execution of the cloaking algorithms, we also propose two other data structures. *SOPlist* is a 2-semantic diversity index whose objective is to speed up the computation of the cloaked set. Each record of SOPlist is represented as ((seman1, n_1), (seman2, n_2), Pointer), where (seman1, n_1) ((seman2,n_2)) denotes n_1 (n_2) adjacent segments of semantic label seman1 (seman2), while Pointer is a pointer to the next record.

The role of *Cloaked l-maplist* is to record the cloaked sets that have been formed for distinct semantic requirements so far. This is achieved by re-using the mapping between segments and cloaked sets. A basic cell of Cloaked l-maplist is represented as(l_i, npointer, spointer) and l_i_set, where l_i indicates l_i-semantic diversity, *npointer* and *spointer* are pointers to the next basic cell and l_i_set, respectively, and l_i_set records the last position of each cloaked set with regard to semantic requirement l_i. l_i_set is dynamically maintained to keep track of the current maximum position of cloaked sets of semantic requirement l_i in a bucket.

Example 3 (SOPlist and Cloaked l-maplist). Suppose the content of a bucket is $\{e_{23}, e_{13}, e_{22}, e_{21}, e_{17}, e_4, e_1, e_5, e_{18}, e_{14}, e_{19}\}$. Then, Fig. 5 shows the SOPlist and Cloaked l-maplist corresponding to the bucket.

4.2   The Cloaking Algorithm

We introduce our online cloaking algorithm, which is summarized in Algorithm 2. It mainly uses of the segment oblivious property which is stated in Definition 3.

The algorithm first initiates an empty cloaked set and computes the sequence of specified segments to locate the position of the segment in the segment allocation (lines 1–2). The algorithm then finds the maximum value l_{max} in Cloaked l-maplist and compares the segment location Cid in the bucket with l_{max} (line 3). If the value l_{max} is larger than Cid, the algorithm simply searches for the Cloaked l-maplist to find the range of the cloaked set (lines 4–5). In this case, it means that the cloaked set has been computed. Otherwise, it is necessary to execute the operations of lines 6–12. Finally, the algorithm searches for the segments in the bucket range from $[x_1, x_2]$, and returns the corresponding cloaked set.

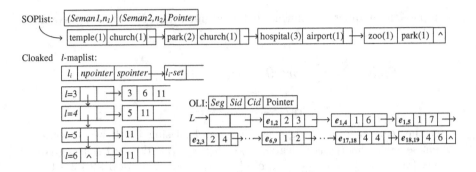

Fig. 5. Index structure

Example 4 (Online Cloaking). Continuing with Example 3, we assume that $l_3\_set=\{3\}$, $l_{max} = 3$, and two users u_1 and u_2 with the same privacy profile (3, 1) located in segment e_{13} and e_{17}, respectively. Since e_{13}.Cid=2< $l_{max} = 3$, we can compute $x_1 = 0, x_2 = 3$ according to Cloaked l-maplist and return $S_c = \{e_{23}, e_{13}, e_{22}\}$. Since e_{17}.Cid=5> $l_{max} = 3$, we cannot compute the interval $[x_1, x_2]$ directly. So we continue to cloak from $l_{max} + 1 = 4$ in the SOPlist, and obtain $l_{max} = 6$. When we checks the item zoo(1), we stops traversing in the SOPlist. Then,we can conclude that the residual semantic number exceeds 3. So we can safely set $l_{max} = 6$. As the condition $e_{17}.Cid < l_{max}$ is satisfied, we set $x_1 = 3$, $x_2 = 6$, and obtain $S_c=\{e_{21}, e_{17}, e_4\}$.

5   Experimental Evaluation

In this section, we evaluate the performance of our proposed location anonymization algorithms through extensive experiments. Our methods are implemented on a machine with CPU Inter(R) Core(TM)i7-2600, 8.00 GB memory, 3.40 GHz frequency, 500 GB hard disk. All programs are coded in C++.

5.1   Experimental Setup

(1) Datasets. We use two real road network datasets[1]: California and Oldenburg road networks. These datasets contain POIs of various categories, e.g., church, hospital, airport, which we used as query objects in our experiment. Table 1 gives the parameters of the two real road networks.

(2) Query Generator. For each real dataset, we randomly pick 2000 query points from the positions of trajectories. To simulate different traffic condition, these trajectories are derived from real trajectories and synthetic trajectories which are generated by a traffic simulator[2]. The parameters of queries are listed in Table 2. In each experiment, we run 2000 queries and report the average result.

---

[1] http://www.cs.utah.edu/~lifeifei/SpatialDataset.htm.
[2] http://www.fh-oow.de/institute/iapg/personen/brinkhoff.

Algorithm 2. Online Cloaking

Input: Location$(x,y) \epsilon e_i$, Privacy profile(l, σ_t), OLI OSI, Soplist SL, Cloaked
l-maplist CL
Output: Cloaked set S_c
1 Initialize $S_c = \Phi$;
2 Compute Seq(e_i) to acquire Sid_0,Cid_0 in OSI ;
3 Compute maximum value l_{max} of l_i_set where $l_i = l$;
4 **if** $l_{max} \geq Cid_0$ **then**
5 $\quad\lfloor$ compute interval(x_1, x_2) in Cloaked l-maplist CL ;

6 **else**
7 $\quad\mid$ **while** $l_{max} < Cid_0$ **do**
8 $\quad\mid\quad\mid$ $l_{old-max} = l_{max}$;
9 $\quad\mid\quad\mid$ update l_{max}=cloak(l_{max}+1,l, SL);
10 $\quad\mid\quad\mid$ **if** $residualsemantic(l_{max}, SL) < l$ **then**
11 $\quad\mid\quad\mid\quad\lfloor$ update l_{max}=end position of Soplist SL ;
12 $\quad\mid\quad\lfloor$ insert x_2 into CL ;
13 $\quad\lfloor$ $x_1 = l_{old-max}$, $x_2 = l_{max}$;
14 $S_c = \bigcup \{e_k\}_{x_1 < e_k.Cid \leq x_2, Seq(e_k).sid = Sid_0}$;
15 **return** S_c ;

Table 1. Real dataset parameters

Name of dataset	Vertex count	Edge count	Semantic types count	POIs count
OLdenburg (OL) road network	6,105	7,035	10	600
California (CA) road network	21048	21693	62	104,771

(3) Algorithms. We evaluate the following algorithms. (a) EIRank: The algorithm is our proposed solution for protecting location privacy on road networks. (b) SA: This is an algorithm proposed in [17]. To compare with our approach, we modify this solution. That is, we don't consider identity protection (k-anonymity), and are only interested in protecting location and location semantic information. More specially, the algorithm first achieves a *Voronoi-partition graph* from the road network. Then, it determines the initial vertex's Voronic-partition according to the query user's location. Next, it gradually merges neighboring vertex's Voronic-partitions until the semantic requirement is satisfied.

Table 2. Parameters setting

Parameters	Default values	Range
l: semantic diversity	5	[2,10]
k: kNN query	5	[2,10]
t: semantic type count	62	[62,100]

(4) Metrics. In our experiments, we evaluate the following metrics. (a) Cloaking Size: this metric measures the size of a cloaked set. It is defined as the count of the segment that contains in a cloaked set. (b) Relative Semantic level: this metric measures the achieved semantic diversity l' for the cloaking algorithm normalized by the user specified sematic diversity level l, i.e., $\frac{l'}{l}$. (c) Cloaking Time: the cloaking time is used to measure the runtime of the cloaking algorithm.

Besides, we also use the following two metrics to measure QoS. (c) Query Time (PT): this metric is measured by the execution time of processing a query at the server side. (d) Communication Cost (CC): we use the size of the candidate results set to measure the communication cost.

5.2   Experimental Results

In the first three experiments, we examine the efficiency of our cloaking algorithms. In the last two experiments, we examine PT and CC.

Cloaking Size. Figure 6(a) shows the effect of varying semantic diversity on the cloaking size. From the results, we can observe that with the increase of semantic diversity, the cloaking sizes all increase. In addition, the cloaking size of SA is always larger than that of EIRank. The main reason is that the cloaking strategies of the two algorithms are different. EIRank performs segment-based perturbation, which stops just after obtaining user specified semantic requirement. In contrast, SA performs vertex Voronoi-based perturbation. Based on this difference, a cloaked set of SA contains more segments than that of EIRank.

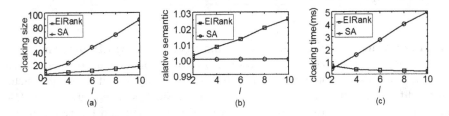

Fig. 6. The efficiency of the cloaking algorithms on the California road network

Relative Semantic Level. Figure 6(b) shows the relative semantic level with regard to semantic diversity. It can be seen that as the semantic diversity increases, the relative semantic level of SA remains unchanged and that of *EIRank* increases. This is because the semantic number of a cloaked set exactly equals to the user-defined semantic diversity for SA algorithm. To resist reverse engineering attacks, the lastest cloaked set of each bucket contains more than l semantics for EIRank.

Cloaking Time. Figure 6(c) shows the impact of varying semantic diversity on the cloaking time for the two algorithms. From the figure, we observe that with the increase of semantic diversity, the time cost of EIRank drops significantly and the time cost of SA increases dramatically. It also can be seen that the cloaking time cost of EIRank is always less than that of SA.

A large semantic diversity l results in a relative large cloaked set. For the other segments other than query segment, the cloaked sets are generated by searching the *Cloaked l-maplist* directly. As we do not need to reconstruct the cloaked sets, which greatly decreases the cloaking time. In contrast, each cloaked set of SA is generated completely dependently. With the increase of semantic diversity, SA needs to search more vertex's Voronoi-partition to achieve the cloaked set, which increases the cloaking time. As the cloaked set of SA is larger than that of EIRank, EIRank runs faster than SA.

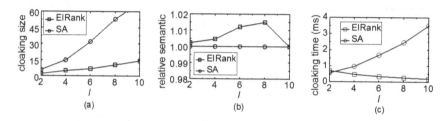

Fig. 7. The efficiency of the cloaking algorithms on the Oldenburg road network

Figure 7 shows the performance of the cloaking algorithms on the Oldenburg road network. We observe that the trendy is the same as that of California road network. Based on this fact, in the following experiments, we just show the performance of the algorithms on the California road network.

Query Processing Cost. Figure 8(a) illustrates the query time of the two algorithms with different values of semantic diversity. From the results, it is clear that the query time all increases as the semantic diversity increases. Furthermore, the query processing of SA run quite slowly in comparison to EIRank as expected. The results above are reasonable, the query time mainly depends on *cloaking size*. For the same semantic diversity, cloaking size of EIRank is smaller than that of SA. The cloaked set becomes large for a big semantic diversity, and hence the query time increases.

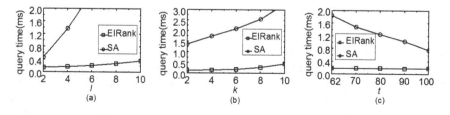

Fig. 8. Query time vs parameters l, k and t.

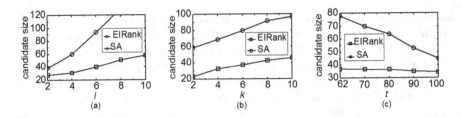

Fig. 9. Communication cost vs parameters l, k and t

Figure 8(b) shows the effect of varying k on the query time. It can be seen that with the increase of k, the query costs of two algorithms increase. We also observe that the algorithm EIRank outperforms SA in most cases. The reasons are as follows. On the one hand, based on our query processing model, a larger k needs to search more segments to acquire the k-nearest neighbors for boundary nodes. On the other hand, the cloaked set size of EIRank is far smaller than that of SA.

Figure 8(c) shows the effect of semantic type count t on query time. As observed, with the semantic type count increases, the query time of two algorithms degrades. We also notice that the parameter t has stronger influence on SA algorithms than on EIRank algorithms.

These phenomena are explained by the following facts: (1) With the increase of semantic types, a part of semantic types are replaced by smaller granularity semantic types. As cloaking of SA algorithm is based on vertex's Voronoi-partition, the size of cloaked sets is smaller than before. and (2) The parameter t has little impact on the cloaked set size of EIRank, and hence it almost have no effect on query time of EIRank.

Communication Cost. Figure 9 shows the impact of different parameters on communication cost. As mentioned above, we measure communication cost indirectly in terms of the size of the candidate results sets, since each result set must be transmitted from server to location annoymizer. From these graphs, we can see that the trend is the same as for query processing cost, and can be explained in a similar way.

6   Related Work

Our work relates to two main streams of research, concerning location privacy and location semantics, respectively.

Location privacy. Location anonymization has attracted much interest as a solution to protect user location privacy in LBS. It mainly makes use of location obfuscation techniques to hide an user's exact location. Examples include space transformation [2,9,25], fake location [14,29], mix-zones [21], and spatial cloaking [1,5,8,11–13,20]. Among various anonymization techniques, spatial cloaking is the prominent. It enlarges an user's exact location to a cloaked region until some privacy conditions are satisfied, such as k-anonymity [8]. Unfortunately, most existing cloaking techniques are no longer applicable in road networks, because the area granularity of measurement tends to fail.

Recently, there exists several research on location privacy over road networks [3,15,16,18,26]. The most famous technique is based on the model of segment l-diversity [3,26]. As mentioned above, this solution cannot prevent the location semantic information leakage.

Location Semantics. In generally, the sensitive information is disclosed using two kinds of published information: query semantics [22,27] and location semantics. In the first case, it means that the query contents issued from a cloaked set are at least l different types. Our paper concentrates on protecting the sensitive information using location semantics over road networks.

Location l-diversity is first introduced in [1], However, it doesnt distinguish the place type. Lee et al. [16] proposes mining the place semantics using Earth Mover's Distance to avoid location semantic leakages, but it does not consider the road networks environment. Yigitoglu et al. [28] extends the _semantic location cloaking model_ [6] to protect semantic location in urban settings. Due to the cloaked sets being generated a priori for a particular privacy requirement, this approach cannot support the privacy requirement updating. As the limitations mentioned above, we don't make comparison with them. Li et al. [17] solves the location semantic leakages in road networks based on the vertex Voronoi-partition. Unfortunately, this solution is subject to _reverse engineering attacks_. Our solution overcomes these drawbacks.

7 Conclusion

In this paper, we propose a semantic-aware privacy preservation model named _CloSed_ to preserve user privacy on road networks. In our model, the cloaked set provides semantic protection without compromising QoS. To achieve this goal, we design an advanced algorithm to balance the privacy requirement and QoS. Extensive experiment evaluations show the efficiency and effectiveness of our proposed algorithms on large-scale real datasets.

Acknowledgments. This research is supported by the National Natural Science Foundation of China under Grant Nos. 61572119, 61173029, 61332006, 61332014, 61328202, 61502317 and U1401256.

References

1. Bamba, B., Liu, L., Pesti, P., Wang, T.: Supporting anonymous location queries in mobile environments with privacygrid. In: WWW, pp. 237–246. ACM (2008)
2. Chor, B., Kushilevitz, E., Goldreich, O., Sudan, M.: Private information retrieval. IEEE Symp. Found. Comput. Sci. **45**(6), 41–50 (1998)
3. Chow, C., Mokbel, M.F., Bao, J., Liu, X.: Query-aware location anonymization for road networks. GeoInformatica **15**(3), 571–607 (2011)

4. Chow, C., Mokbel, M., Aref, W.: Casper*: query processing for location services without compromising privacy. Trans. Database Syst. (TODS) **34**(4), 24 (2009)
5. Chow, C., Mokbel, M., Liu, X.: A peer-to-peer spatial cloaking algorithm for anonymous location-based services. In: GIS, pp. 171–178 (2006)
6. Damiani, M., Silvestri, C., Bertino, E.: Fine-grained cloaking of sensitive positions in location-sharing applications. Pervasive Comput. **10**(4), 64–72 (2011)
7. Fogaras, D., Rácz, B.: A scalable randomized method to compute link-based similarity rank on the web graph. In: Lindner, W., Fischer, F., Türker, C., Tzitzikas, Y., Vakali, A.I. (eds.) EDBT 2004. LNCS, vol. 3268, pp. 557–567. Springer, Heidelberg (2004)
8. Gedik, B., Liu, L.: Location privacy in mobile systems: a personalized anonymization model. In: Distributed Computing Systems, pp. 620–629. IEEE (2005)
9. Ghinita, G., Kalnis, P., Khoshgozaran, A., Shahabi, C., Tan, K.: Private queries in location based services: anonymizers are not necessary. In: SIGMOD, pp. 121–132. ACM (2008)
10. Ghinita, G., Zhao, K., Papadias, D., Kalnis, P.: A reciprocal framework for spatial k-anonymity. Inf. Syst. **35**(3), 299–314 (2010)
11. Gruteser, M., Grunwald, D.: Anonymous usage of location-based services through spatial and temporal cloaking. In: Proceedings of the 1st International Conference on Mobile Systems, Applications and Services, pp. 31–42. ACM (2003)
12. Hu, H., Xu, J.: Non-exposure location anonymity. In: ICDE, pp. 1120–1131. IEEE (2009)
13. Kalnis, P., Ghinita, G., Mouratidis, K., Papadias, D.: Preventing location-based identity inference in anonymous spatial queries. TKDE **19**(12), 1719–1733 (2007)
14. Kido, H., Yanagisawa, Y., Satoh, T.: An anonymous communication technique using dummies for location-based services. In: ICPS, pp. 88–97 (2005)
15. Ku, W., Zimmermann, R., Peng, W., Shroff, S.: Privacy protected query processing on spatial networks. In: Data Engineering Workshop, pp. 215–220 (2007)
16. Lee, B., Oh, J., Yu, H., Kim, J.: Protecting location privacy using location semantics. In: SIGKDD (2011)
17. Li, M., Qin, Z., Wang, C.: Sensitive semantics-aware personality cloaking on road-network environment. Int. J. Secur. **8**(1), 133–146 (2014)
18. Li, P., Peng, W., Wang, T., Ku, W., Xu, J., Hamilton, J., et al.: A cloaking algorithm based on spatial networks for location privacy. In: Sensor Networks, Ubiquitous and Trustworthy Computing, pp. 90–97 (2008)
19. Martłnez-Hinarejos, C.D., Juan, A., Casacuberta, F.: Generalized k-medians clustering for strings. In: Perales, F.J., Campilho, A.J.C., de la Blanca, N.P., Sanfeliu, A. (eds.) IbPRIA 2003. LNCS, vol. 2652, pp. 502–509. Springer, Heidelberg (2003)
20. Mokbel, M., Chow, C., Aref, W.: The new casper: query processing for location services without compromising privacy. In: VLDB, pp. 763–774. VLDB Endowment (2006)
21. Palanisamy, B., Liu, L.: Mobimix: protecting location privacy with mix-zones over road networks. In: ICDE, pp. 494–505. IEEE (2011)
22. Pan, X., Wu, L., Piao, C., Xu, X.: P³RN:personalized privacy protection using query semantics over road networks. In: Li, F., Li, G., Hwang, S., Yao, B., Zhang, Z. (eds.) WAIM 2014. LNCS, vol. 8485, pp. 323–335. Springer, Heidelberg (2014)
23. Pedreschi, D., Bonchi, F., Turini, F., Verykios, V.S., Atzori, M., Malin, B., Moelans, B., Saygin, Y.: Privacy protection: regulations and technologies, opportunities and threats. In: Giannotti, F., Pedreschi, D. (eds.) Mobility, Data Mining and Privacy, pp. 101–119. Springer, Heidelberg (2008)

24. Sibson, R.: Slink: an optimally efficient algorithm for the single-link cluster method. Comput. J. **16**(1), 30–34 (1973)
25. Stavros, P., Spiridon, B., Dimitri, P.: Nearest neighbor search with strong location privacy. PVLDB **3**, 619–629 (2010)
26. Wang, T., Liu, L.: Privacy-aware mobile services over road networks. PVLDB **2**(1), 1042–1053 (2009)
27. Xiao, Z., Xu, J., Meng, X.: p-Sensitivity: a semantic privacy-protection model for location-based services. In: MDMW 2008, pp. 47–54. IEEE (2008)
28. Yigitoglu, E., Damiani, M., Abul, O., Silvestri, C.: Privacy-preserving sharing of sensitive semantic locations under road-network constraints. In: MDM, pp. 186–195. IEEE (2012)
29. Yiu, M., Jensen, C., Huang, X., Lu, H.: Spacetwist: managing the trade-offs among location privacy, query performance, and query accuracy in mobile services. In: ICDE, pp. 366–375. IEEE (2008)

Advanced Applications(2)

An Efficient Location-Aware Top-k Subscription Matching for Publish/Subscribe with Boolean Expressions

Hanhan Jiang[1], Pengpeng Zhao[1,2]([✉]), Victor S. Sheng[3], Jiajie Xu[1,2],
An Liu[1,2], Jian Wu[1,2], and Zhiming Cui[1,2]

[1] School of Computer Science and Technology, Soochow University, Suzhou, China
ppzhao@suda.edu.cn
[2] Collaborative Innovation Center of Novel Software Technology
and Industrialization, Suzhou 215006, China
[3] Computer Science Department, University of Central Arkansas, Conway, USA
ssheng@uca.edu

Abstract. Location-aware publish/subscribe (pub/sub) has attracted a lot of attentions with the booming of mobile Internet technologies and the rising popularity of smart-phones. Subscribers subscribe their interests with their locations as subscriptions, and publishers publish geo-information as events. Many state-of-art applications with a massive amount of geo-information, such as location-aware targeted advertising systems, face this situation. Existing related work mainly focuses on unstructured geo-textual information. However, many online-to-offline applications have enormous geo-information with different structured descriptions. To handle such structured information, a new type of location-aware pub/sub approach is needed. In this paper, we handle these subscriptions using boolean expressions. Since the number of publishers and subscribers can be enormous, it is extremely important to improve the matching effectiveness and efficiency of top-k query processing. In this paper, we develop a novel solution named RR^t-trees. RR^t-trees integrates R^t-tree and a predicate index structure together to return top-k best matched subscriptions from a great number of events. Our experimental results on synthetic and real-world datasets show that RR^t-trees achieve better performance than baseline methods.

Keywords: Location-aware pub/sub · Top-k · Boolean expressions

1 Introduction

With the rapid progress of mobile Internet technologies and the growing popularity of smart phones, a great amount of geo-information is generated at an unprecedented scale. In social network applications (e.g., Facebook or Twitter), there are a great number of users. Their personal information can be described

S.B. Navathe et al. (Eds.): DASFAA 2016, Part II, LNCS 9643, pp. 335–350, 2016.
DOI: 10.1007/978-3-319-32049-6_21

by a set of attribute-value pairs and their geo-locations revealed by GPS. In a online-to-offline system, there are millions of users browsing products on the system, such products can be described by a set of attribute-value pairs and associated with geo-locations. In this paper, we refer to such data with both attribute-value pairs and geo-location information as *geo-tagged attribute-value pair objects.*

In a location-aware publish/subscribe system, subscribers subscribe their interests and publishers publish events with geo-information. This kind of systems has many real-world applications. In a location-aware targeted advertising application system, advertisers are subscribers, who specify the properties of their interested users. For example, they can have a subscription like *(e.g., "16≤ age≤28,hobby∈{ Tennis, basketball}","51.16145,0.14123").* The system will display corresponding advertisements on users screens. This is a well-known pushing advertising model. Users of social network systems (such as Facebook, Twitter, etc.) act as publishers. Their personal information, such as age, hobby and locations, becomes an event *(e.g., "age=20, sex=female, hobby=tennis,school=Harvard", "51.16515,0.14123").* Advertisements can be displayed on these users screen if there is a high relevance between an event and a subscription. This kind of information pushing model is useful for online-to-offline commerce platforms, such as Groupon[1], Sellers or service providers in Groupon system are subscribers, who may want to accurately push their advertisements to potential customers by specifying both users properties and a series of their product information *(e.g., "hobby=smart-phone, item∈{Iphone6s, Iphone6}, price≤499$", "51.25643, 0.14845")* as a subscription. Users of this system are publishers. When a user click a product link, the information of this product and the users properties can become an event *(e.g., "hobby=smart-phone, item=Iphone6s, price=469", "51.2612, 0.12545").* In such applications, only a few advertisements can be displayed due to the limited screen size.

Existing unstructured location-aware publish/subscribe systems [1,3,5,8,19, 20] can support subscriptions with geo-textual descriptions very well. For example, users of Twitter can register their interests with a geo-textual description *(e.g., "Cheapest iphone6s", "31.4522,51.4451").* The system has to ensure a timely delivery of relevant geo-textual objects to the user. However, this kind of location-aware pub/sub systems cannot support *geo-tagged attribute-value pair objects*, which need a structured description to capture attributes and values. Existing structured location-aware pub/sub systems [5,13] use a boolean expression presenting a subscription. They can efficiently retrieve all matched information. Therefore, a user may be overwhelmed. To address these issues, we propose a new type of top-k subscriptions matching with boolean expressions, referred as *Location-aware Top-k Subscription Matching with Boolean Expressions* (in short TSMB-loc). The latest solution proposed for top-k subscription matching with boolean expression [7] focused on fuzzy matches. We will develop a solution ensuring strict boolean semantics of expressions.

---

[1] http://www.Groupon.com.

There are two challenges on developing solutions for location-aware top-k subscription matching with boolean expressions. First, how can we filter out the candidates of top-k best subscriptions from millions of subscriptions with a large amount of attributes, values and geo-locations. Second, it needs to retrieve the top-k best subscriptions over tons of candidates. Thus, an efficient and effective solution to cope with the TMSB-loc problem is necessary.

To efficiently and effectively process the TSMB-loc problem, we propose a novel R^t-tree based index structure, ranked R^t-tree (called RR^t-trees since then), by integrating the R^t-tree [3] index structure with a predicate index structure together and using a subscription partitioning scheme. When an event with location information arrives, our method can quickly retrieve its top-k best matched subscriptions. To summarize, we make the following contribution:

- We propose a new problem, location-aware top-k subscription matching with boolean expressions(TSMB-loc).
- We propose a new index structure, called RR^t-trees as the solution of TSMB-loc, which can efficiently retrieve top-k best subscriptions from millions of subscriptions.
- We conduct experiments on a synthetic dataset and a real-world dataset to evaluate the performance of our proposed RR^t-trees.

The remaining of this paper is organized as follows. In Sect. 2, we overview related work. Then, we formalize the problem in Sect. 3. In Sect. 4, we propose a baseline solution by extending a boolean expression index structure and a state-of-the-art spatial index R-tree. In Sect. 5, we propose an advanced solution, the RR^t-trees index structure. In Sect. 6 we give the similarity upper bound of RR^t-trees. In Sect. 7, we present the matching algorithm of RR^t-trees. Extensive experimental results are reported in Sect. 8, and we conclude this paper in Sect. 9.

2   Related Work

This research topic is closely related to two main research branches: structured pub/sub systems and location-aware unstructured pub/sub systems. We will briefly review these two branches as follows.

Structured Pub/Sub. There are some researches on structured pub/sub with boolean expressions [2,4,9,11,12,18,21]. Guo et al. [5] proposed a new location-aware pub/sub system named *Elaps*. *Elaps* can continuously detect moving subscriptions of users over event streams. Hu et al. [7] proposed a R^I-tree for top-k subscription matching over structured information. They are all different to our problem since *Elaps* cannot maintain a top-k best matched subscriptions and R^I-tree is a partial matching in a boolean expression. To adapt to different workloads, Sadoghi and Jacobsen [10] presented BE*-tree index structure. BE*-tree allows the values of attributes to be continuous. It combines a bi-directional tree expansion mechanism with an overlap-free splitting strategy. D et al. [21] proposed Op-index, which builds an inverted index over the pivot attributes of subscriptions and developed a two-level partitioning scheme to handle subscriptions with high dimensional attributes. However, BE*-tree does not make the location dimension into consideration and Op-index can not return top-k subscriptions.

Location-Aware Unstructured Pub/Sub. There are many related researches on unstructured location-aware pub/sub over geo-textual data. [1,5, 6,8,14–17,19,20,22–24]. To study the location-aware pub/sub problem for parameterized spatio-textual subscriptions, Hu et al. [6] presented a filter-verification framework by integrating prefix filtering and spatial pruning techniques together. To efficiently filter geo-textual data, Li et al. [8] proposed R^t-tree,which loads the selected token from subscriptions into different R-tree nodes. To support ranking semantics, Yu et al. [20] make an extension of R^t-tree. These three works [6,8,20] only focus on geo-textual information, which cannot retrieve top-k subscription matching. Note that pub/sub focusing on geo-textual cannot support geo-tagged attribute-value pair objects. We propose RR^t-trees, which has two distinguishing features. First, RR^t-trees allows user to specify their interests in form of boolean expressions. A boolean expression is much more expressive than that of a textual content. Second, RR^t-trees focus on retrieving top-k best matched subscriptions.

3   Problem Definition

In this section we formally define the problem of location-aware top-k subscription matching with boolean expressions.

Subscription: Subscribers register their interests as subscriptions. A subscription s is consisted by three elements: $s.B$, $s.loc$, α, where $s.B$ is a boolean expression to describe the interests of subscribers, $s.loc$ is the spatial location of a subscriber, and α is a parameter to balance the relative importance between spatial similarity and boolean expression similarity (we call it BE similarity since then). The boolean expression is a combination of predicates in conjunctive normal form. A predicate is a constraint specified by users to represent the relationship between an attribute and its value. A predicate contains three elements: an attribute A, an operator f_{op}, and a value v. That is, p(A,f_{op},v) denotes a predicate p. The operator can be a relational operator $(<,>,\leqslant,\geqslant,=,\neq)$, or a set $(\in,\notin)$. Each predicate has a weight ω_s, where $\sum_{i=1}^{n} \omega_{si} = 1$. Thus, the subscription can be modeled as follows:

$$s : \{[(p_1, \omega_{s1}) \wedge (p_2, \omega_{s2}) \wedge (p_i, \omega_{si}) \wedge ... \wedge (p_n, \omega_{sn})],\ loc, \alpha\}$$

Event: An event e contains a collection of attribute-value pairs denoted as $e.V$ and a geo-position denoted as $e.loc$. The attribute-value pairs $e.V$ are represented in the form of conjunction of predicates with equality operator. That is, $\nu(A, v)$ denotes an attribute-value pair ν. Each attribute-value pair has a weight ω_e, where $\sum_{i=1}^{n} \omega_{ei} = 1$. Thus, an event can be denoted as follows:

$$e : \{[(\nu_1, \omega_{e1}) \wedge (\nu_2, \omega_{e2}) \wedge (\nu_3, \omega_{e3}) \wedge ... \wedge (\nu_n, \omega_{en})], loc\}$$

The weight ω_s is given by subscribers and is used to represent users' preference among predicates in a subscription. The weight ω_e signifies the relevance between value-pair and its predicate. It is generated according to the appearing frequency of the attribute-value pair in the whole dataset.

Definition 1. *(Predicate Match) Given an attribute-value pair ν, for a predicate p appears in a subscription, we said that there is a predicate match if $p.A = \nu.A$ and $p_i(\nu_i.v) = true$.*

Definition 2. *(Boolean Expression Match) A boolean expression $s.B$ is said to match a collection of attribute-value pairs $e.V$ if each of the predicates in $s.B$ has a match in $e.V$.*

Definition 3. *(Similarity Function ϕ) Given an event $e=e:\{[(\nu_1, \omega_{e1}) \wedge (\nu_2, \omega_{e2}) \wedge (\nu_3, \omega_{e3}) \wedge ... \wedge (\nu_n, \omega_{en})], loc\}$ and a subscription $s=s: \{[(p_1, \omega_{s1}) \wedge (p_2, \omega_{s2}) \wedge (p_i, \omega_{si}) \wedge ... \wedge (p_n, \omega_{sn})], loc, \alpha\}$, we define the similarity function $\phi(e, s)$ as follows:*

$$\phi(e, s) = s.\alpha \cdot \varphi_{BE}(e, s) + (1 - s.\alpha) \cdot \varphi_s(e, s) \tag{1}$$

where φ_{BE} is a BE similarity function and the φ_S is a spatial similarity function.

$$\varphi_{BE}(e, s) = \sum_{p_j \in s, \nu_i \in e, p_j.A = \nu_i.A, p_i(\nu_i.v) = true}^{n = s.\varsigma} \omega_{sj} \cdot \omega_{ei} \tag{2}$$

where $s.\varsigma$ is the size of subscription s(number of predicates in a subscription).
The spatial similarity is given by:

$$\varphi_s(e, s) = 1 - \frac{distance(e.loc, s.loc)}{MaxDistance} \tag{3}$$

where distance(e.loc,s.loc) is the Euclidian distance between s and e, and the MaxDistance is the maximum distance among subscriptions.

As shown in Fig. 1 for the event $e=\{A=3(0.1) \wedge B=3(0.5) \wedge C=4(0.2) \wedge F=2(0.2), e.loc\}$, the boolean expression subscription S_1 matches the attribute-value pairs of e according to Definition 2. However, the subscription boolean expression S_4 does not match attribute-value pairs of e as there is no attribute-value pair in e that matches the predicate G$\geq$4. Based on Definition 3, the spatial similarity $\varphi_S(e,s_1)$ is 0.35, and the BE similarity $\varphi_{BE}(e,s_1)$ is 0.25. Therefore, the balanced similarity $\phi(e,s_1)$ is 0.30. Similarly, since the spatial similarity $\varphi_S(e,s_9)$ is 0.15, and the BE similarity is $\varphi_{BE}(e,s_9)$ is 0.18, $\phi(e,s_9)$ is 0.18. Thus, if we want to retrieve top-1 best subscription for event e, the answer is S_1.

Location-Aware Top-k Subscription Matching: Given a set of subscriptions S, TSMB-loc aims to finds the top-k most relevant strict matched subscriptions $S^k \in S$. For any subscription $s \in S^k$ and $s^* \in (S-S^k)$, $\phi(e,s^*) < \phi(e,s)$.

S_1	$B=3(0.3) \wedge A \in (3,2,5)(0.4) \wedge C \geq 2(0.3)$	0.5
S_2	$D \leq 2(0.6) \wedge C=7(0.4)$	0.6
S_3	$E=2(0.3) \wedge A=4 (0.6) \wedge B \leq 2(0.1)$	0.8
S_4	$A=3(0.3) \wedge F \in (5,2,1) (0.4) \wedge G \geq 4(0.3)$	0.3
S_5	$F \leq 5(0.5) \wedge D=2(0.2) \wedge F=5(0.1) \wedge G=4(0.2)$	0.5
S_6	$E=3 (0.3) \wedge A=5(0.1) \wedge C \geq 2(0.6)$	0.7
S_7	$E \geq 3 (0.5) \wedge B \in (3,2,5)(0.3) \wedge C=4(0.2)$	0.5
S_8	$A \leq 12(0.4) \wedge B=3(0.2) \wedge C \geq 2(0.3) \wedge E \geq 3(0.1)$	0.4
S_9	$A \leq 3(0.5) \wedge B \leq 3 (0.1) \wedge C=5(0.1) \wedge F \leq 2(0.3)$	0.2
S_{10}	$E=3(0.7) \wedge B \in (3,2,5)(0.2) \wedge C=4(0.1)$	0.9
e	$A=3 (0.1) \wedge B=3(0.5) \wedge C=4 (0.2) \wedge F=2(0.2)$	

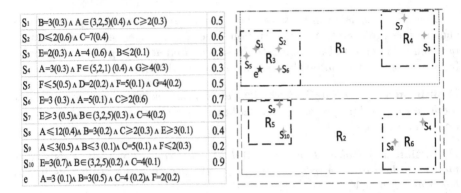

Fig. 1. An example of subscriptions and events

4   A Baseline Solution

In this section, we extend two state-of-art index structures (Op-index and R-tree) to cope with the TMSB-loc problem. These extensions will be used as baseline solution to evaluate our advanced solutions proposed in Sect. 5.

Op-index is a well-known pub/sub index structure for boolean expressions, which builds an inverted index structure over a pivot attribute[2] and designs a two level partitioning scheme to handle the pub/sub problem with high-dimensions attribute. Based on op-index and a well-known spatial information index structure R-tree, we can integrate op-index with R-tree together (called OPR-tree in short) to cope with the TSMB-loc problem. We first build an R-tree for all the locations of subscriptions. When the subscriptions fall into leaf nodes of the R-tree, we organize subscriptions inside each leaf node using the Op-index structure. The construction and the query processing of OPR-tree will be explained in details.

OPR-Tree Construction: For each subscription s, we retrieve its leaf node n in the R-tree using its spatial point information $s.loc$ and then select its pivot attribute δ_A. We partition subscriptions inside each leaf node into groups according to their pivot attributes. We can present a list of subscriptions with the same pivot attribute δ_A as a list denoted as $L(n, \delta_A)$. Each list (e.g., $L(n, \delta_A)$) can be further partitioned based on the operators $(<, >, \leq, \geq, =, \neq)$ of predicates. The predicates with the same operator are organized into a sub-list. A sub-list of $L(n, \delta_A)$ can be presented as $L(n, \delta_A, op)$, where op is a specific operator. For each group of predicates $L(n, \delta_A, op)$, we use a signature segment to map the predicates by a hash function. We compute the hash value of each predicate using a hash function $h(p.A)$ to select a bit from the signature segment and set this bit to 1. Besides, there is a collection of counter arrays, corresponding with each subscription list $L(n, \delta_A, op)$. The value of the counter array is initialed as the

---

[2] The attribute with the least appearing frequency in the whole dataset becomes the pivot attribute.

size of the subscriptions. For each predicate in $L(n, \delta_A, op)$, there is a pointer to point to its corresponding counter array value.

OPR-Tree Query Processing: For an event e, we search the R-tree using the spatial point information $e.loc$ to find a corresponding leaf node n. Then we extract each candidate pivot attribute $\nu_i.A$ ($\nu_i \in e.V$) from the set of distinct attribute-value pairs $e.V$. If $\nu_i.A$ is indeed a pivot attribute, we extract each attribute-value pair ν_i to search the predicate lists $L(n, \delta_A, op)$ in $L(n, \delta_A)$. For each attribute-value pair ν_i, we first calculate the hash value of a candidate attribute, i.e., $h(\nu_i.A)$. If the corresponding bit of the signature segment is 1, we search the corresponding predicate list $L(n, \delta_A, op)$. Then, the BE similarity $\varphi_{BE}(e,s)$ is calculated if $p_j(\nu_i.v=true)$, where $p_j \in L(n, \delta_A, op)$. If the corresponding value of the counter array goes to 0, we calculate the spatial similarity $\varphi_S(e,s)$. The final balanced similarity $\phi(e,s)$ is calculated and added into the temporary result set as a candidate top-k result. The upper bound of a given leaf node n to a given event e is a balanced similarity between the minimum Euclidian distance from e to n and the maximum weight of attribute-value in e.

OPR-tree partitions the subscriptions by the region of leaf nodes and organizes these subscriptions using Op-index. However, the region of a leaf node may be very small, which results in a poor pruning ability in high spatial dimensions. Therefore, OPR-tree is not very efficient. In order to avoid this problem, we proposed RR^t-trees method in the next section.

5 RR^t-trees Solution

In this section, we present a framework that integrates R^t-tree and a predicates index structure into a new index, named RR^t-tree. Based on RR^t-tree, we develops a partitioning scheme to organize subscriptions into disjointed RR^t-trees. Finally, we introduce the upper bound of efficiently and effectively filtering out top-k best subscriptions over this framework.

5.1 RR^t-tree Index Structure

As we discussed in Sect. 4, OPR-tree is not very efficient to meet TSM-loc since the poor pruning ability in spatial dimension.

R^t-tree [3] is an unstructured location-aware pub/sub index structure, which integrates so-called high-quality representative tokens selected from subscriptions into the nodes of R-tree. Based on R^t-tree, we propose a method, ranked R^t-tree (RR^t-tree) to cope with the TSMB-loc problem. The basic idea of RR^t-tree is to convert the tokens of R^t-tree into predicates of a subscription, which will be loaded into the ancestor nodes of a leaf node in which a subscription locates. Then, predicates in each node are indexed using a predicate index structure.

RR^t-Tree Construction: We build an R-tree for all the locations of subscriptions. For a given subscription s, we first extract the distinct predicate p_i, containing its weight, where $p_i \in s$. Then, we load p_i into different nodes at different levels, which is determined by the spatial location $s.loc$ of the subscription s.

Considering a built R-tree, its height is H, the size of a given subscription s is s.ς. If s.ς>H, we directly insert the last $s.\varsigma - H+1$ predicates. If $s.\varsigma<H$, only the ancestors in the first $s.\varsigma$ level contain the corresponding predicates of s. Letting $s.p_i$ denote the i-th predicate in s, then $s.p_i$ is loaded in an ancestor node at i-th level. For each node n on i-th level, there is a set of predicates denoted as P. We build inverted lists over the attributes of the predicates in P to organize predicates with the same attribute. To track the number of matched predicates of a subscription during an event query processing, we assign a hash map M for each subscription and initial each hash value $M[s]$ to be 0. When a predicate p_i matches an attribute-value pair, we increase its corresponding hash value by 1. Based on the number of matched predicates $M[s]$, we can efficiently filter the subscription. To explain this, we have the following lemmas.

Lemma 1. Consider an event e and a node n at the i-th level. If M[s]<i, s cannot be a candidate top-k result.

Proof. As s appears in the i-th level, it contains at least i predicates. For each node on path from the root to node n, s.B must have a predicate which cannot match all attribute-value pairs in event e. According Definition 2, s cannot be the candidate top-k result of e.

Lemma 2. Consider an event e and a node n at the i-th level, if M[s]=s.ς, s must be a candidate top-k result

Proof. If M[s]=s.ς, all the predicates of s.B are matched by e. According to Definition 2, s must be a candidate top-k subscription of e.

Lemma 3. Consider an event e and a node n at the i-th level, if M[s]=i<s.ς, and n is a leaf node, s cannot be a candidate top-k result of e.

Proof. Since n is a leaf node, we have M[s]=i=H<s.ς. That is, the rest s.ς- H+1 predicates are loaded in the node n, which cannot match all attribute-value pairs in event e. According to Definition 2, s cannot be the candidate top-k result of e.

Lemma 4. Consider an event e and a node n at the i-th level. For any $p_i.A \in p_i \in P \in n$, if $p_i.A$ does not appear in e, we can directly pruning the node n.

Proof. If $p_i.A$ does not appear in e, according to Definition 1, no predicate in node n matches any attribute-value pair. According to Definition 2, subscriptions on node n cannot be candidate top-k results of e.

Predicate Index Structure: In each node on RR^t-tree, there are a set of predicates P, a weight of each predicate, the maximum alpha value α_{max} and minimum alpha value α_{min}. To efficiently retrieve matched predicates in P, we design an index structure for P. We index the predicates of P in two partitioning steps. In the first step, predicates are partitioned into disjointed predicate lists based on attributes as follows:

$$P = L(A_1) \cup L(A_2) \cup L(A_i) \cup ... \cup L(A_n) \qquad (4)$$

For each predicate in list $L(A_i)$, there is a pointer to point to the number of matched predicates M[s] of its corresponding subscription. In second step, the

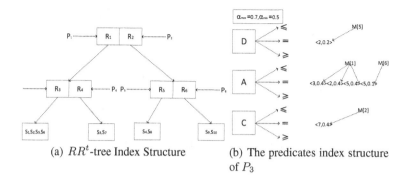

(a) RR^t-tree Index Structure   (b) The predicates index structure of P_3

Fig. 2. An RR^t-tree index for the subscriptions shown in Fig. 1

predicates list $L(A_i)$ is further partitioned by their operators (we only use standard operators, such as $\leq, \geq, =$) into its corresponding value list $L(A_i, op)$ as follows:

$$L(A_i) = L(A_i, \leq) \cup L(A_i, =) \cup L(A_i, \geq) \cup ... \cup L(A_n, op) \qquad (5)$$

Figure 2(a) shows the RR^t-tree index structure for the subscriptions shown in Fig. 1. Figure 2(b) shows the predicate index structure for P_3.

5.2 RR^t-trees Index Structure

Since the number of subscriptions can be very large, it is necessary to improve the efficiency of RR^t-tree. To address this problem, we partition subscriptions according to their pivot attributes (discussed in Sect. 4) into N^3 subscription lists and organize the subscription lists using disjointed RR^t-trees. We simply name this method RR^t-trees since then. Given a set of subscriptions S, we partition these subscriptions according to their pivot attributes δ_A and organize them using RR^t-trees as follows

$$S = RR^t - trees = RR^t - tree(\delta_{A1}) \cup RR^t - tree(\delta_{A2}) \cup ... \cup RR^t - tree(\delta_{An}) \quad (6)$$

From Definitions 1 and 3, we can conclude that if an event e matches a subscription s, then all the attributes in s must appear in e. Obviously, if there is an attribute in s but not in e, e wont be matched by s. Thus, given an event e, we only consider the subscriptions whose pivot attributes appear in e. Attributes with a low frequency in a whole dataset has low probabilities to appear in subscriptions. Thus, we choose the lowest frequency attribute in a subscription as the pivot attribute.

The index structure of RR-trees for the subscriptions shown in Fig. 1 is shown in Fig. 3. According to the rule of selecting pivot attributes mentioned above, A, D, E and G are selected as a pivot attribute respectively. Given an event e in Fig. 1, subscriptions in both $L(E)$ and $L(G)$ don't match e definitely.

---

[3] The number of distinct attributes in a whole dataset.

S₁	A
S₂	D
S₃	E
S₄	G
S₅	G
S₆	E
S₇	E
S₈	A
S₉	E
S₁₀	A

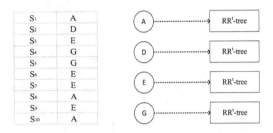

Fig. 3. The index structure of RR^t-trees

6   Similarity Upper Bound of RR^t-tree and RR^t-trees Solution

After having described the RR^t-tree and RR^t-trees index structure, it is the time to introduce the upper bound of similarity.

Definition 4. *($UB_{BE}(e,n)$) For a given event e and a node n in RR^t-tree, the upper bound of the BE similarity $UB_{BE}(e,n)$ is defined as follows. The BE similarity bound of a given event e to a node n is:*

$$UB_{BE}(e,n) = Max\left\{ s \in n.parent\left[\sum_1^{i-1}\omega_{si} \cdot \omega_{ej} + \omega^*_{emax} \cdot \left(1 - \sum_1^{i-1}\omega_{si}\right)\right]\right\} \quad (7)$$

where $\sum_1^{i-1}\omega_{si} \cdot \omega_{ej}$ is the total score of matched predicates of s appearing in level 1 to level i-1 where $i > 1$, and ω^*_{emax} is the maximum weight of unmatched attribute-value pairs in e for subscription s. And $1 - \sum_1^{i-1}\omega_{si}$ is the remaining total weights of unmatched predicates in subscription s.

Definition 5. *($UB_S(e,n)$) For a given event e and a node n, the upper bound of the spatial similarity $UB_S(e,n)$ is defined as follows:*

$$UB_S(e,n) = (1 - \frac{MinDistane(e.loc, n.MBR)}{MaxDistance}) \quad (8)$$

where MaxDistance is the maximum distance between subscriptions, n.MBR it the minimum bounding rectangle of node n and $MinDistance(e.loc, n.MBR)$ is the minimum Euclidian distance between $e.loc$ and any point on $n.MBR$.

Definition 6. *($UB(e,n)$) According to Eqs. 7 and 8, for a given event e and a node n, the total upper bound $UB(e,n)$ is defined as follows:*

$$UB(e,n) = Max\{\forall \alpha \in (\alpha_{min},\alpha_{max})\min[1 - \alpha, UB_{BE}(e,n)] + \alpha \cdot UB_S(e,n)\} \quad (9)$$

where $\alpha_{min}, \alpha_{max}$ are the minimum and maximum α of subscriptions in node n.

According to Definition 6, we have the following lemma:

Lemma 5. Given an event e and a node n whose MBR encloses a set of subscriptions S^n. For any subscription s where $s \in S^n$, there is:

$$\phi(e, s) < UB(e, n) \qquad (10)$$

We omit the proof due to space constraints.

7   Matching Algorithm

How can we use RR^t-tree to retrieve top-k best matched subscriptions with boolean expressions? We show the processing of retrieving top-k best matched subscriptions in Algorithm 1. We use a bound-queue to store the nodes that have not been visited in the algorithm. The nodes in the bound-queue is ordered by their $UB(e, n)$ in a descending order, which is calculated according to Eq. 9 from its parent node. For the root node, the bound is 1. Given an event e, we traverse all the RR^t-tree(ν_i.A) in RR^t-trees from the root node, where $\nu_i \in e$. The algorithm will return candidate subscriptions for the top-k best matched subscriptions. It will stop under two situations as follows.

- When k subscriptions are found and its minimum similarity is larger than the maximum $UB(e, n)$ in the bound-queue.
- When the bound-queue is empty.

8   Experiments

In this section, we will evaluate three indexes (OPR-tree, RR^t-tree, RR^t-trees) on synthetic and real-world datasets. All the methods are implemented in java (JDK7) and experiments are running on a machine with 3.2 GHz Intel(R) (TM) Core i5-3470 CPU and 16 GB of RAM.

8.1   Experimental Setup

Two datasets are used in our experiments, one synthetic dataset and one real-world dataset (eBay dataset), shown in Table 1. To generate the synthetic dataset, we implement a data generator, which can generate attributes, operators and values. For the set operator $\in, \notin$, we rewritten them into standard operator $=, \geq, \leq$. For the weights of attribute-value pairs in an event are generated base on this equation $\omega_{ei} = \frac{\nu.f}{\sum_{i=0}^{n} \nu_i.f}$, where $\nu_i.f$ is the frequency of an attribute-value pair in the whole dataset. And n is the number of attribute-value pairs in an event. We totally generate 5M synthetic subscriptions corresponding with 10k events for event matching tests in the synthetic dataset. For the real-world dataset (eBay), we generate 10M subscriptions and 10k events based on 10k product messages, and we extract the spatial information from Twitter to generate final subscriptions and events.

Algorithm 1. Matching(e,k)

---

1 **Input** *An event e and the value k*
2 **Output** *Top-k best matched subscriptions S*
3 **Initialize** :a *bound-queue*,a hash map M, a candidate top-k subscription list R
  and a temporary similarity storage of subscriptions *Temp*
4 Extract each attribute-value pair ν_j from e
5 **for** *each RR^t-tree(δ_{Ai})* **do**
6   **if** ($\nu_j.A == \delta_{Ai}$) **then**
7   Search RR^t-tree(δ_{Ai});
8   **for** *each node n in RR^t-tree(δ_{Ai})* **do**
9   **if** *(n is a root node)* **then**
10   *bound-queue*.Push(n, 1);
11   Visit *bound-queue*.Pop();
12   **for** *each predicate $p_i(\nu_j.v)==true$* **do**
13            $Temp[s] = Temp[s] + \omega_{si} \cdot \omega_{ej}$;
14   **if** $(++M[s] == s.\varsigma)$ **then**
15              $\phi(e,s) = (1 - s.\alpha) \cdot Temp[s] + s.\alpha \cdot \varphi_S(e,s)$;
16   **if** *(R.size < k)* **then**
17   R.add(s);
18   **else**
19   **if** $\phi(e,s) > $ R.*min* **then**
20   R.add(s);
21   **if** *(++M[s] < L)* **then**
22   /*L is the i-th level*/
23   *Temp*.remove(s);
24   **if** *(n is visited and node n is not a leaf node)* **then**
25   *bound-queue*.Push(n.child,UB(e,n.child));
26   **if** *(n is a leaf node)* **then**
27   Visit *bound-queue*.Pop();
28   **if** *(*bound-queue.*Empty or (*R.*min>* bound-queue.*Pop().UB and*
  R.*size=k))* **then**
29   Return R;

8.2   Experimental Results

In this section, we will evaluate three indexes on synthetic datasets and real-world datasets. For synthetic datasets we evaluate the performance of three indexes from different perspectives, a varied number of subscriptions and distinct attributes, varied average size of subscriptions, varied average size of events, varied parameter k and α.

Matching Time under Various k: The value of k is an important parameter for the TSMB-loc problem. Figure 4 shows the performance of the three methods

Table 1. Parameters and settings

Parameters	Synthetic dataset	eBay
Number of subscription	1M, 2M, 3M, 4M, 5M	2M, 4M, 6M, 8M, 10M
Average subscription size	4~20	2~10
Average event size	5 25	8
Max value of α	0.2, 0.4, 0.6, 0.8, 1.0	0.2, 0.4, 0.6, 0.8, 1.0
Number of distinct attribute	5k, 10k, 15k, 20k, 25k	10k
Top-k parameter	1, 5, 10, 20	10, 20

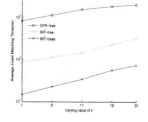 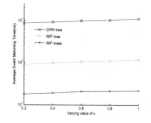 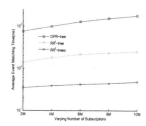

Fig. 4. Varying value of k on ebay dataset

Fig. 5. Varying value of α on eBay dataset

Fig. 6. Varying number of subscriptions on eBay dataset

with the increment of k on ebay dataset. We can clearly see that the average matching time of all the three methods increases with the increment of k. However, RR^t-trees achieves the best performance because of its powerful pruning ability. It is nearly 3 times faster than the next best method RR^t-tree. OPR-tree performs the worst, using the largest average matching time.

Matching Time under Various α: We also conduct several experiments to investigate the impact of varying the maximum balance value α. Our experimental results are shown in Fig. 5. Figure 5 shows that the average matching time of all the three methods grows slowly as the value of α increases. RR^t-trees still achieve the best performance over the real-world dataset.

Matching Time on Varying Number of Subscription: From Fig. 6, we can see that all the three methods are sensitive to the number of subscriptions. And RR^t-trees achieves the lowest event matching time, followed by RR^t-tree. OPR-tree performs the worst, having the highest matching time. RR^t-trees is 3.5 times faster than RR^t-tree. This is because RR^t-trees uses pivot attributes to partition subscriptions. This causes that indexing the subscriptions using RR^t-trees much more efficient than that using a single RR^t-tree. Furthermore, the upper bounds over a small number of subscriptions are easier to calculate than those over a larger scale of subscriptions, which causes the matching time of RR^t-trees grows much more smoothly than that of the other two methods.

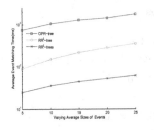
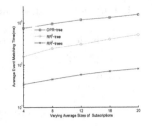
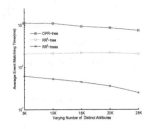

Fig. 7. Average sizes of events on synthetic dataset

Fig. 8. Average sizes of subscriptions on ebay dataset

Fig. 9. Varying number of distinct attributes on synthetic dataset

Matching Time under Different Sizes of Events: Again, since the size of each event in the real-world dataset eBay is fixed, we couldnt conduct these experiments on the dataset eBay. We conduct experiments over the synthetic dataset. The experimental results are shown in Fig. 7. We can see that the average matching time of the three methods increases with the increment of the size of events. This is because the number of candidate subscriptions increases when the size of events increases. RR^t-trees scales much better than the other two methods (RR^t-tree and OPR-tree). It is about 3 times faster than the next best method RR^t-tree.

Matching Time under Different Sizes of Subscriptions: The size of subscriptions can affect the matching performance. The average event matching time on different sizes of subscriptions of the three methods on ebay dataset is reported in Fig. 8. From this figure, we can see that all the three methods are sensitive to the size of subscriptions. But, both RR^t-trees and RR^t-tree scale better than OPR-trees. It is because than both in RR^t-trees and RR^t-tree, subscriptions are pruned by each predicate on the nodes of R-tree, as we described in Sect. 5. RR^t-trees scales better than RR^t-tree because of its pruning ability of pivot attributes. However, OPR-tree only prunes subscriptions by the spatial bounds.

Matching Time under Different Numbers of Distinct Attributes: Since the number of distinct attributes is fixed in the real-world dataset eBay, we conduct experiments with various numbers of distinct attributes on our synthetic dataset. As Fig. 9 shows, when the number of attributes increases, the average matching time of both RR^t-trees and OPR-tree decreases. However, RR^t-tree performs oppositely. With the increment of number of attributes, its average matching time increases a little. $RR^t - trees$ deceases gradually, because it generates many more narrowed partitions when the number of distinct attributes increases during partitioning subscriptions according to pivot attributes.

9   Conclusion

In this paper, we tackled the problem of location-aware top-k subscription matching, which is significant for location-aware publish/subscribe systems with a stream of geo-tagged attribute-value pair objects. Facing the challenge of efficiently and effectively delivering events to the top-k subscribers, we proposed a novel index structure called RR^t-trees, which integrates R^t-tree and a predicate index structure. In addition, we developed an efficient filtering strategy to reduce the searching space. Extensive experiments conducted in both synthetic and real-world datasets demonstrate the effectiveness and efficiency of our algorithms.

Acknowledgment. This work was partially supported by Chinese NSFC project (61472263, 61402312, 61402311), and the US National Science Foundation (IIS-1115417).

References

1. Chen, L., Cong, G., Cao, X., Tan, K.L.: Temporal spatial-keyword top-k publish/subscribe. In: 2015 IEEE 31st International Conference on Data Engineering (ICDE), pp. 255–266 (2015)
2. Cugola, G., Margara, A.: High-performance location-aware publish-subscribe on GPUs. In: Narasimhan, P., Triantafillou, P. (eds.) Middleware 2012. LNCS, vol. 7662, pp. 312–331. Springer, Heidelberg (2012)
3. Eugster, G.: Location-based publish/subscribe. In: 2013 IEEE 12th International Symposium on Network Computing and Applications, pp. 279–282 (2005)
4. Fontoura, M., Sadanandan, S., Shanmugasundaram, J., Vassilvitski, S., Vee, E., Venkatesan, S., Zien, J.: Efficiently evaluating complex Boolean expressions. In: Proceedings of the 2010 ACM SIGMOD International Conference on Management of Data, pp. 3–14. ACM (2010)
5. Guo, L., Zhang, D., Li, G., Tan, K.L., Bao, Z.: Location-aware pub/sub system: when continuous moving queries meet dynamic event streams. In: Proceedings of the 2015 ACM SIGMOD International Conference on Management of Data, pp. 843–857. ACM (2015)
6. Hu, H., Liu, Y., Li, G., Feng, J., Tan, K.L.: A location-aware publish/subscribe framework for parameterized spatio-textual subscriptions. ICDE **2015**, 711–722 (2015)
7. Hu, J., Cheng, R., Wu, D., Jin, B.: Efficient top-k subscription matching for location-aware publish/subscribe. In: Claramunt, C., Schneider, M., Wong, R.C.-W., Xiong, L., Loh, W.-K., Shahabi, C., Li, K.-J. (eds.) SSTD 2015. LNCS, vol. 9239, pp. 333–351. Springer, Heidelberg (2015)
8. Li, G., Wang, Y., Wang, T., Feng, J.: Location-aware publish/subscribe. In: Proceedings of the 19th ACM SIGKDD International Conference on Knowledge Discovery and Data Mining, pp. 802–810. ACM (2013)
9. Machanavajjhala, A., Vee, E., Garofalakis, M., Shanmugasundaram, J.: Scalable ranked publish/subscribe. Proc. VLDB Endow. **1**(1), 451–462 (2008)
10. Sadoghi, M., Jacobsen, H.-A.: Relevance matters: Capitalizing on less (top-k matching in publish/subscribe). In: 2012 IEEE 28th International Conference on Data Engineering, pp. 786–797 (2012)
11. Sadoghi, M., Burcea, I., H.a.J: Gpx-matcher: A generic Boolean predicate-based xpath expression matcher. In: EDBT 2011, pp. 45–56 (2011)

12. Sadoghi, M., Jacobsen, H.-A.: Be-tree: an index structure to efficiently match Boolean expressions over high-dimensional discrete space. In: ACM Conference on Management of Data, pp. 637–648 (2011)
13. Sadoghi, M., Jacobsen, H.A.: Location-based matching in publish/subscribe revisited. In: Proceedings of the Posters and Demo Track, p. 9. ACM (2012)
14. Shang, S., Deng, K., Xie, K.: Best point detour query in road networks. pp. 71–80. In: ACM (2010)
15. Shang, S., Ding, R., Yuan, B., Xie, K., Zheng, K., Kalnis, P.: User oriented trajectory search for trip recommendation. In: 15th International Conference on Extending Database Technology, EDBT 2012, pp. 156–167 (2012)
16. Shang, S., Ding, R., Zheng, K., Jensen, C.S., Kalnis, P., Zhou, X.: Personalized trajectory matching in spatial networks. VLDB J. **23**(3), 449–468 (2014)
17. Shang, S., Yuan, B., Deng, K., Xie, K., Zheng, K., Zhou, X.: PNN query processing on compressed trajectories. Geoinformatica **16**(3), 467–496 (2012)
18. Whang, S.E., Garcia-Molina, H., Brower, C., Shanmugasundaram, J., Vassilvitskii, S., Vee, E., Yerneni, R.: Indexing Boolean expressions. Proc. VLDB Endow. **2**(1), 37–48 (2009)
19. Xiang Wang, Y.Z., Xuemin Line, W.W.: Ap-tree: Efficiently support continuous spatial-keyword queries over stream. In: 2015 IEEE 31st International Conference on Data Engineering (ICDE), pp. 1107–1118 (2015)
20. Yu, M., Li, G., Wang, T., Feng, J., Gong, Z.: Efficient filtering algorithms for location-aware publish/subscribe. IEEE Trans. Knowl. Data Eng. **27**(4), 950–963 (2015)
21. Zhang, D., Chan, C.Y., Tan, K.L.: An efficient publish/subscribe index for e-commerce databases. Proc. VLDB Endow. **7**(8), 613–624 (2014)
22. Zheng, B., Yuan, N.J., Zheng, K., Xie, X., Sadiq, S., Zhou, X.: Approximate keyword search in semantic trajectory database. In: 2015 IEEE 31st International Conference on Data Engineering (ICDE), pp. 975–986. IEEE (2015)
23. Zheng, K., Huang, Z., Zhou, X., et al.: Discovering the most influential sites over uncertain data: a rank based approach. IEEE Trans. Knowl. Data Eng. **99**, 1 (2011)
24. Zheng, K., Zhou, X., Fung, P.C., Xie, K.: Spatial query processing for fuzzy objects. VLDB J. **21**, 729–751 (2012)

Predicting the Popularity of *DanMu*-enabled Videos: A Multi-factor View

Ming He[1,2], Yong Ge[2], Le Wu[3], Enhong Chen[1(✉)], and Chang Tan[4]

[1] University of Science and Technology of China, Hefei, China
mheustc@gmail.com, cheneh@ustc.edu.cn
[2] University of North Carolina at Charlotte, Charlotte, USA
yong.ge@uncc.edu
[3] Hefei University of Technology, Hefei, China
lewu.ustc@gmail.com
[4] Anhui Radio and Television Information Network CO., LTD, Beijing, China
tanchang1986@gmail.com

Abstract. Recent years have witnessed the prosperity of a new type of real-time user-generated comment, or so-called *DanMu*, in many recent online video platforms. These *DanMu*-enabled video platforms present scrolling marquee comments overlaid directly on top of the videos by synchronizing these comments to specific playback times. In this paper, we study the prediction of video popularity in these platforms, which may benefit a lot of applications ranging from online advertising for website holders to popular video recommendation for audiences. Different from traditional online video platforms where only traditional reviews are available, these *DanMu*s make viewers easily see other viewers' opinions and communicate with each other in a much more direct way. Consequently, viewers are easily influenced by others' behaviors over time, which is considered as the herding effect in social science. However, how to address the unique characteristics (i.e., the herding effect) of *DanMu*-enabled online videos for more accurate popularity prediction is still under-explored. To that end, in this paper, we first explore and measure the herding effect of *DanMu*-enabled video popularity from multiple aspects, including the popular videos, the popular *DanMu*s and the newly updated videos. Also, we recognize that the uploaders' influence and video quality affect the video popularity as well. Along this line, we propose a model that incorporates the herding effect, uploaders' influence and video quality for predicting the video popularity. An effective estimation method is also proposed. Finally, experimental results on real-world data show that our proposed prediction model improves the prediction accuracy by 47.19 % compared to the baselines.

1 Introduction

Recent years have witnessed the rapid development of online video platforms and the prosperity of real-time user generated comment, or so-called *DanMu*, in many

S.B. Navathe et al. (Eds.): DASFAA 2016, Part II, LNCS 9643, pp. 351–366, 2016.
DOI: 10.1007/978-3-319-32049-6_22

online video platforms such as Acfun[1] and Bilibili[2]. Different from traditional online reviews that are displayed in a separate space outside the video (e.g., Youtube.com[3]), *DanMu* as a new type of comment is overlaid directly on the top of videos by synchronizing the comment to specific playback times. As a matter of fact, the *DanMu*-enabled video has activated user's behaviors, such as comments, views and so on. For instance, a recent report by a leading Chinese *DanMu*-enabled platform reveals that *DanMu* has improved the online user activity by 100 times[4]. Understanding *DanMu*-enabled video's popularity growth is of great importance for a broad range of services, such as online ads. in video platforms, video recommendation for users, and other commercial opportunities.

In the literature, many efforts have been devoted to predict the popularity of online videos. For example, some works have shown that the popularity of online video is positively correlated to the historical views and the number of comments generated by users [6,7]. Intuitively, a viewer could directly see others' interactions (e.g., views and comments) with videos, which makes whether the viewer watches a particular video is easily affected by other users' previous interactions with this video. This phenomenon is known as the herding effect in social science and it is widely studied in financial markets [1,10,12]. Generally, the herding effect is defined as *everyone doing what everyone else doing, even when their private information suggests doing something quite different* [3]. For example, in the stock market, if few people begin to sell a certain type of stock, it may lead to the overall crowd panic and selling spree. Similarly, in the online video platforms, people's decision on whether to watch a particular video is also influenced by others' behaviors, thus we argue that the herding effect should play an important role for video popularity prediction.

In fact, there exists considerable works on predicting the popularity of videos based on the herding effect, which empirically show that a video will attract new views at a rate proportional to the number of views already acquired [8,16]. However, compared to traditional online videos, the *DanMu* makes the herding effect stronger and more dynamic, as the simultaneously displayed *DanMu* comments convey interesting information about the content of videos and make viewers communicate with each other in a much more direct way. Thus users are more easily affected by other users to view videos or not on *DanMu*-enabled video sites. However, few of existing approaches could be directly applied to this prediction task due to the following two challenges caused by the unique characteristics of *DanMu*-enabled videos. First, compared to traditional videos, people interact with *DanMu*-enabled videos more frequently from various aspects, e.g., the views of videos and the *DanMus* associated with videos. In the meantime, due to the temporal variation of these videos, the herding effect is dynamic and changes from time to time. How to capture the dynamic herding effect from multiple aspects? Second, how to combine the dynamic herding effect with other

---

[1] http://www.acfun.tv/.

[2] http://www.bilibili.com/.

[3] http://www.youtube.com/.

[4] http://digi.163.com/14/0915/17/A66VE805001618JV.html.

features (e.g., the video quality and the uploaders' characteristics) that may influence the popularity of *DanMu*-enabled videos for the prediction?

To this end, as a pilot study, we aim at predicting the popularity growth of *DanMu*-enabled videos by leveraging the unique characteristics of *DanMu*-enabled videos. Specifically, we first propose a measurement to quantify the herding effect from multiple aspects of *DanMu*-enabled videos, including the popular videos, the popular *DanMu*s and the newly updated videos. Then we propose a prediction model to combine the herding effect with other factors that may influence the video popularity. After that an efficient estimation method is proposed to automatically learn the herding effect and other parameters. Finally, we conduct experiments with a real-world data set collected from a *DanMu*-enabled online video platform. The experimental results show that our proposed model improves the prediction accuracy by 47.19 % compared to the baselines.

2 Related Work

To the best of our knowledge, no prior work has considered the dynamic herding effect and *DanMu* information to predict the popularity of videos. However, there have been considerable prior works on predicting the popularity of videos.

Many prior works have analyzed different aspects of video's metrics such as total views, total comments, total collects and so on [6,7]. Cha et al. [7] analyzed the intrinsic statistical properties of video popularity distributions and studied the popularity lifetime of videos and the relationship between requests and video age. Mitra et al. [15] found the presence of "invariant" among video's characteristics, such as heavy-tailed total views distributions and the positive correlation between total views and total ratings.

Some researchers made preliminary studies on predicting the popularity growth of videos. For instance, Borghol et al. [5] developed a methodology that was able to assess accurately, both qualitatively and quantitatively, the impacts of various content-agnostic factors on video popularity. What's more, there exist some prediction models [8,16] based on the herding effect [4]. Such as Szabo et al. [16] leveraged the observation that the total views received soon after a video was uploaded provided a strong indication of its total future views to develop a prediction model for video views; this method is applied by some works to build predictive popularity based on applying regression to different feature spaces [2,11,14,17]. Recently, Le et al. [13] presented an adoption model that considered multiple aspects for product adoption. Our model advances their work by considering the unique characteristics of Danmu enabled videos.

Although there exist some works considering the herding effect in the popularity growth model, none of them considered the dynamic herding effect from multiple aspects over time. Especially, these works just suggested that a video would attract new views at a rate proportional to the number of views already acquired. They have not considered what the impact of the popular videos based on different aspects' herding effect to other videos over time.

To the emerging type of user-generated comment *DanMu*,some recent works focused on this new DanMu phenomenon. Among them, there are only two works that have used *DanMu* data, but neither of them use the *DanMu* information to predict the popularity growth of videos. Specifically, Wu *et al.* [19] leveraged the textual content of *DanMu*s to extract time-sync video tags automatically and Wu *et al.* [20] investigated the co-relation between the volume of one particular *DanMu* and popularity measures such as the number of replays and bookmarks of videos. Our work differs from studies as we put emphasis on predicting the popularity growth of *DanMu*-enabled videos.

In summary, in our work, we combine the unique nature of *DanMu* and the dynamic herding effect from multiple aspects to predict the popularity of videos. Besides using the *DanMu* information and the dynamic herding effect, we also make use of video quality and uploaders' influence to the proposed prediction model, which makes the prediction model more accurate and effective.

3   Data and Statistics

In this section, we first illustrate the nature of *DanMu* comments in *DanMu*-enabled websites and introduce the collected data set, and then present some unique characteristics of this data set.

3.1   *DanMu* Illustration and Data Collection

In Fig. 1, we show two snapshots of a sample video[5] that include several Chinese *DanMu*s on top of the video. We pick up four *DanMu*s by different viewers, translate them to English and show them at the bottom. The green axis at the bottom indicates the video time, where the selected four *DanMu*s are aligned based on the associated video time. The *DanMu*s "God, Norton" and "Edward Norton" written by user A and user B are about the actor of the officer in the snapshot, which are very close at the video time. Also we can observe that previous *DanMu*s have direct influence on future *DanMu*s. For instance, after users A and B mention the name of the actor in their *DanMu*s, user C mentions the movie "Fight Club" acted by the same actor, and then user D mentions another movie (i.e., "Red Dragon") acted by the same actor. From this example, we can see that *DanMu* enables much more intensive communication among users than traditional reviews.

We crawled a dataset from acfun.tv, a leading *DanMu* website in China of each day, which makes that we can capture the dynamics of the dataset. The collected dataset contains videos that are uploaded during early November to the end of May, which lasts for about half a year. For each video, we collected users' behaviors of a video that include: *Views* (the number of views), *DanMu*s (the number of *DanMu*s), *TReviews* (the number of traditional reviews), *Collects* (the number of collects), and *Coins* (the number of coins). Specifically, when a viewer thinks the video is valuable and interesting, she can give some coins to the uploader by buying it from the website. Table 1 shows a summary of

---

[5] Available at http://www.acfun.tv/v/ac1731008.

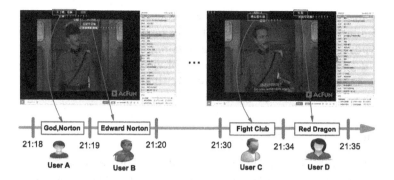

Fig. 1. Two snapshots of a *DanMu*-enabled video (Color figure online).

<table>
<tr><td colspan="3">Table 1. Video features.</td></tr>
</table>

Feature	Total count	Avg count
Videos	3,623	Null
Views	73,059,811	20187.85
DanMus	883,637	243.89
TReviews	60,956	16.84
Collects	308,448	85.23
Coins	220,578	60.95

Table 2. Uploader information.

# Uploaders	745
# Videos	3,623
Avg # of Videos per uploader	4.86
# Followers	278,520
Avg # of followers per uploader	373.85

video features. As can be seen from this table, there are much more user activities on Views than other user behaviors. It is reasonable as the view behavior does not need extra action from viewers compared to other features. And the *DanMu* activities are more active than TReviews, Coins and Collects. All these comparisons reflect that the *DanMu* function has attracted much more users' contribution and that online viewers prefer to write *DanMus* rather than do other behaviors for *DanMu*-enabled videos.

Also, for each video, we collected the uploader information as shown in Table 2. We can find that some uploaders upload more than 1 video as the average number of videos per uploader is 4.86. And the average number of followers per uploader is closed to 400, which confirms that there exists strong social atmosphere in acfun.tv. It enhances that we need consider the social influence in predicting the popularity growth in our model.

As a preliminary, we first provide the formal definition of *the popularity for videos*. Intuitively, as the videos can be viewed by viewers anytime after they are uploaded, views of videos at different time compose the popularity of videos. Formally, we set v_{mt} as the views of video m until day t and v_t as the total views of all videos until day t. For simplicity, we adopt the proportion of views as the popularity of video m until day t denoted as $p_{mt} = \frac{v_{mt}}{v_t}$. Due to the herding effect towards to popular videos, then we provide the statement of *the popular videos*. Similar to the traditional definition of popular videos, we define that the popular videos are those videos which are more probably chosen by viewers. For example, if a video has the largest number of views or *DanMus*, viewers choose this video to view with a greater probability than other videos. Also, if a video is newly uploaded, viewers prefer to choose this video to watch as well.

3.2   Volume Distributions

Video Features. Figure 2 shows the distribution of the number of views per video. As can be seen from this figure, a few videos attract much more views, which reflects that there is a strong herding effect on user's viewing behavior as most viewers toward to view few popular videos. We also draw histograms of other four video statistics (i.e., TReviews, Collects, Coins and *DanMus*) in Fig. 3. As depicted by this figure, about 71 % videos have less than 100 *DanMus* and 85 % videos have less than 1000 *DanMus*. Compared to *DanMu* volume distribution, nearly 99 % videos have less than 1000 TReviews, Collects and Coins. Thus, we empirically conclude that *DanMu* behavior is more active than other three behaviors.

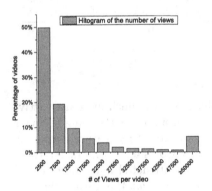

Fig. 2. Histograms of views.

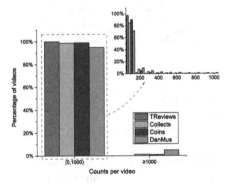

Fig. 3.   Histograms of other four features.

To explore whether there are correlation between video popularity and other video features, we represent each video feature statistics in a vector $VF_f = (VF_{f1}, \cdots, VF_{fM})$ and the popularity in a vector $VP = (VP_1, \cdots, VP_M)$, where VF_{fm} is the value of feature f for video m, and VP_m is the value of popularity for video m. Then we compute the correlation of VF_f and VP by using Pearson correlation measure [9]. This measure is widely used to measure the liner dependency between two vectors. Detailed Pearson correlation values of video features and the video popularity is $(TReviews, Collects, Coins, DanMus) = (0.72, 0.75, 0.59, 0.70)$. We find that all Pearson values are non-negative and larger than 0.55. Based on this observation, we can conjecture that the feature statistics are positively correlated with the popularity of a video. Thus all of these four features should be leveraged when modeling the popularity of videos.

Uploader Features. Also we demonstrate a simple correlation analysis between the number of uploaded videos and followers as the correlation value will more directly reflect the uploader's activity and social influence. The Pearson correlation between the number of uploaded videos and followers is 0.47, which enhances our conjecture that the uploader who has uploaded more videos has stronger social influence.

4 The Proposed Model

As illustrated above, due to the unique characteristics of *DanMu*-enabled videos, the popularity of videos has a large dynamic herding effect from multiple aspects. In this section, we propose a model that utilizes the dynamic herding effect to predict the popularity of videos. Also, to fully leverage all the factors that influence a video's popularity, we also consider *DanMu* information, video quality and uploaders' influence. Table 3 lists the notations used in this paper.

4.1 Dynamic Herding Effect from Multiple Aspects

As we know, compared with traditional online videos, the *DanMu* makes the herding effect stronger and more dynamic, which means that users are more

Table 3. Mathematical notations.

Symbol	Description
k	The aspect of popular videos
HE_{mt}	A combination of herding effect from multi aspects of video m on day t
$o_{k,t}$	The center of aspect k's popular videos on day t
$dis_{m,o_{k,t}}$	The distance between video m and $o_{k,t}$
$\theta_{k,t}$	k-th aspect's parameter of herding effect on day t
VQ_m	A vector representing video m's quality
UI_m	A vector representing video m's uploader's influence
p_{mt}	The popularity of video m on day t
a_i	The coefficient of popularity between p_{mt} and $p_{m(t-i)}$
β	The coefficient of herding effect
γ	The coefficient vector of video quality
δ	The coefficient vector of uploaders' influence
M_{t-1}	The number of videos on day $t-1$
v_{mt}	The number of views of video m until day t
v_t	The number of views of all videos until day t

inclined to view popular videos and more easily affected by other users to view videos or not on *DanMu*-enabled video sites. Specially, when viewers have no specific videos to view, they usually rank the videos by the number of views and choose top ranking videos to view. Therefore, we conjecture that the popularity of videos is affected by the dynamic herding effect of views. Besides the dynamic herding effect of views, we also find that more users tend to view newly uploaded videos and popular videos ranked by the number of *DanMus*. Based on these observations, we propose three types of dynamic herding effect: the dynamic herding effect of views, the dynamic herding effect of *DanMus* and the dynamic herding effect of uploaded date. Specifically, we set the aspect k as the type of popular videos: $k = 1$ stands for the popular videos measured by the number of views, $k = 2$ represents the popular videos measured by the number of *DanMus* and $k = 3$ represents the popular videos measured by the uploaded date of videos. And we set HE_{mt} as a combination of dynamic herding effect from these three aspects of video m on day t as $HE_{m,t} = \prod_{k=1}^{K} [1 + dis_{m,o_{k,t}}]^{-\theta_{k,t}}$, where $o_{k,t}$ is the popular videos' center of aspect k on day t and $\theta_{k,t}$ stands for k-th aspect's parameter of herding effect on day t. Then we give details of $dis_{m,o_{k,t}}$, which stands for the distance between video m and $o_{k,t}$ as follows:

$$dis_{m,o_{1,t}} = \frac{|\overline{v}_{o_{1,t}} - v_{m(t)}|}{\overline{v}_{o_{1,t}}}; \quad dis_{m,o_{2,t}} = \frac{|\overline{d}_{o_{2,t}} - d_{i(t)}|}{\overline{d}_{o_{2,t}}}; \quad dis_{m,o_{3,t}} = \frac{|\overline{u}_{o_{3,t}} - u_{m(t)}|}{\overline{u}_{o_{3,t}}}.$$

where $\overline{v}_{o_{1,t}}$ is the average number of views of popular videos, $\overline{d}_{o_{2,t}}$ is the average number of *DanMus* of popular videos and $\overline{u}_{o_{3,t}}$ is the average uploaded date of popular videos on day t.

4.2   Predicting Model on Videos

Except for the dynamic herding effect from multiple aspects, the generation of a new video's view is also driven by video quality and uploaders' influence.

Video Quality. Intuitively, if the quality of a video is high, the popularity of this video may be greater in the next time. So, we import a vector $\boldsymbol{VQ}_m$ representing video m's quality. Based on our application, we choose five features for video quality: VQ_{m1}, total views; VQ_{m2}, total TReviews; VQ_{m3}, total collects; VQ_{m4}, total coins; VQ_{m5}, total *DanMus*.

Uploaders' Influence. We have found that many users choose videos uploaded by influential uploaders to watch. In view of this observation, we conjecture that the growth of popularity is also affected by the uploaders' influence. Intuitively, if an uploader has more followers, it means that the uploader has greater influence. Similarly, the number of uploaded videos and the average views of uploaded videos also have a positive correlation to the influence of the uploader. To this end, we set the vector $\boldsymbol{UI}_m$ as video m's uploader's influence and choose these three features representing uploaders' influence: UI_{m1}, total uploaded videos; UI_{m2}, total followers; UI_{m3}, the average views of uploaded videos.

While features' values of video quality and uploaders' influence are not in same scale, so we use the transformation $\frac{Max(f)-f}{Max(f)-Min(f)}$ to process the original values of each feature to eliminate the extreme features' influence.

Inspired by the model introduced in [18], which investigated the influence of the prevailing consensus on current analysts' recommendations' choices (strong buy, buy, hold, sell, and strong sell) in a stock, and predicted analysts' recommendations' choices based on the herding effect of the prevailing consensus in the stock by a new proposed statistical model (the prevailing consensus was defined as popular recommendations' choices), we can draw a close analogy to the context in [18]: a user chooses a video to view, which is also affected by the prevailing consensus (popular videos), and then we propose a general framework to predict the popularity p_{mt} for video m on day t as follows:

$$\tilde{p}_{mt} = \sum_{i=1}^{I} a_i \frac{p_{m(t-i)}(\beta * HE_{m,t-i} + \boldsymbol{\gamma}^T * \boldsymbol{VQ}_{m,t-i} + \boldsymbol{\delta}^T * \boldsymbol{UI}_{m,t-i})}{D_{t-i+1}}, \quad (1)$$

where D_t ensures that all videos' popularity sum to 1, a_i is the coefficient of popularity between p_{mt} and $p_{m(t-i)}$, β stands for the coefficient of herding effect, $\boldsymbol{\gamma}$ stands for the coefficient vector of video quality, $\boldsymbol{\delta}$ stands for the coefficient vector of video uploader's influence and M_{t-1} represents the number of videos on day $t-1$. The popularity $p_{m(t-1)}$ of video m on day t is equal to $v_{m(t-1)}/v_{t-1}$ and p_{mt} is v_{mt}/v_t. While our application adds new videos over time, we adjust the $p_{mt} = \frac{v_{mt}}{v_t - v_{new,t}}$ to eliminate the effect of new adding videos, where $v_{new,t}$ is total views of new adding videos on day t.

Comparing to the original model [18]: $p_{i,j}(\sigma, \tau) = p_{i,j}(0) \left\{ \frac{[1+(j-\tau)^2]^{-\sigma}}{D_i} \right\}$, where τ is the popular recommendation choice, σ is the parameter of herding effect, we have improved the original model to our model Eq. 1 on four aspects: First, the parameter of herding effect is dynamic and changes from time to time in our model, while the parameter of herding effect is constant over time in [18]. Second, we import the dynamic herding effect of multiple aspects (views, *DanMu*s and uploaded date), while the original model only adopted one aspect (the popular recommendation choice). Third, the popularity for a video correlates with several previous days in our work, while the proportion only correlates with the only day before in [18]. Fourth, we combine five features (total views, total TReviews, total collects, total coins and total *DanMu*s) of video quality and three features (total uploaded videos, total followers and the average views of uploaded videos) of uploaders' influence to predict the popularity of videos. These improvements make our prediction model more effective and more generalized.

4.3 Parameters Learning Algorithm

Optimizing Functions. The value of $\theta_{k,t}$ stands for the t-th day's herding effect of aspect k on other videos, if $\theta_{k,t} = 0$, it means that the popular videos regarding aspect k have no effect on other videos. If $\theta_{k,t} > 0$, it means that the popular videos have a positive effect on other videos. And if $\theta_{k,t} < 0$, it means that the popular videos have a negative effect on other videos. As we know p_{mt}, VQ_m, UI_m and $dis_{m,o_{k,t-1}}$, we use the Eq. 2 to learn parameters as follows:

$$J = \min_{\{\theta_{k,t-i}, a_i, \beta, \gamma_j, \delta_u\}} \frac{1}{2} \left[\sum_{t=I+1}^{T} \sum_{m=1}^{M_{t-1}} (p_{mt} - \tilde{p}_{mt})^2 + \sum_{i=1}^{I} a_i^2 + \beta^2 + \sum_{j=1}^{J} \gamma_j^2 + \sum_{u=1}^{U} \delta_u^2 \right] \tag{2}$$

where T is longevity of the dataset by day.

We use the gradient descent algorithm to learn all parameters, which is very effective for learning parameters. For simplicity, we define symbols $C_{m,t-i}$, $d_{mk(t-i)}$ as Eq. 3.

$$C_{m,t-i} = p_{m(t-i)}(\beta * HE_{m,t-i} + \gamma^T * VQ_{m,t-i} + \delta^T * UI_{m,t-i})$$
$$d_{mk(t-i)} = 1 + dis(m, o_{k,t-i}) \tag{3}$$

At first, we calculate the partial derivatives of all parameters. For space limitation, we omit the inferencing process of the partial derivatives and directly give the partial derivatives' equations of $a_i, \beta, \gamma_j, \delta_u$ and $\theta_{k,t-i}$ as follows:

$$J_{a_i}^{(1)} = \sum_{t=I+1}^{T} \sum_{m=1}^{M_{t-1}} (\tilde{p}_{mt} - p_{mt}) * \frac{C_{m,t-i}}{D_{t-i+1}} + a_i \tag{4}$$

$$J_{\beta}^{(2)} = \sum_{t=I+1}^{T} \sum_{m=1}^{M_{t-1}} (\tilde{p}_{mt} - p_{mt}) \sum_{i=1}^{I} \{ \frac{a_i}{D_{t-i+1}^2} [D_{t-i+1} p_{m,t-i} HE_{m,t-i}$$
$$- C_{m,t-i} \sum_{m_1=1}^{M_{t-i}} (p_{m_1,t-i} HE_{m_1,t-i})] \} + \beta \tag{5}$$

$$J_{\gamma_j}^{(3)} = \sum_{t=I+1}^{T} \sum_{m=1}^{M_{t-1}} (\tilde{p}_{mt} - p_{mt}) \sum_{i=1}^{I} \{ \frac{a_i}{D_{t-i+1}^2} [D_{t-i+1} p_{m,t-i} VQ_{mj,t-i}$$
$$- C_{m,t-i} \sum_{m_1=1}^{M_{t-i}} (p_{m_1,t-i} VQ_{m_1 j,t-i})] \} + \gamma_j \tag{6}$$

$$J_{\delta_u}^{(4)} = \sum_{t=I+1}^{T} \sum_{m=1}^{M_{t-1}} (\tilde{p}_{mt} - p_{mt}) \sum_{i=1}^{I} \{ \frac{a_i}{D_{t-i+1}^2} * [D_{t-i+1} p_{m,t-i} UI_{mu,t-i}$$
$$- C_{m,t-i} \sum_{m_1=1}^{M_{t-i}} (p_{m_1,t-i} UI_{m_1 u,t-i})] \} + \delta_u \tag{7}$$

$$J^{(5)}_{\theta_{k,t-i}} = \sum_{t=I+1}^{T} \sum_{m=1}^{M_{t-1}} (\tilde{p}_{mt} - p_{mt}) \frac{a_i}{D^2_{t-i+1}} \left[C_{m,t-i} \sum_{m_1=1}^{M_{t-i}} (\beta p_{m_1,t-i} \ln d_{m_1 k(t-i)} \right.$$

$$\left. d_{m_1 k(t-i)}^{-\theta_{k,t-i}}) - D_{t-i+1} \beta p_{m,t-i} d_{mk(t-i)}^{-\theta_{k,t-i}} \ln d_{mk(t-i)} \right] \qquad (8)$$

Updating Parameters. We have generated the partial derivatives of all parameters, then we obtain their update rules of a_i, β, γ_j, δ_u and $\theta_{k,t-i}$ as follows:

$$(a_i, \beta, \gamma_j, \delta_u, \theta_{k,t-i})^{(n+1)} = (a_i, \beta, \gamma_j, \delta_u, \theta_{k,t-i})^{(n)} - \eta(J^{(1)}_{a_i}, J^{(2)}_{\beta}, J^{(3)}_{\gamma_j}, J^{(4)}_{\delta_u}, J^{(5)}_{\theta_{k,t-i}})^{(n)}$$
$$(9)$$

where η is the learning rate.

Initialization. To set a proper starting point for learning, we set all features equal value at start. Under this assumption, we have the following settings:

$$\begin{cases} \beta_m = 1 \\ \theta_{k,t-i} = 1/K, & \text{where } K \text{ is multible aspects;} \\ a_i = 1/I, & \text{where } I \text{ is linear regression days;} \\ \gamma_j = 1/J, & \text{where } J \text{ is video's featurs;} \\ \delta_u = 1/U, & \text{where } U \text{ is uploader's features;} \end{cases} \qquad (10)$$

After initialization, we iterate the learning algorithms updating parameters until the objective function converges.

5 Evaluation

To evaluate the effectiveness of the proposed prediction model, we present an empirical evaluation. We use the real video data introduced in Sect. 3.1 to validate our proposed models.

5.1 Comparison Models

We compare the performance of our prediction model to five additional popularity growth models:

(a) Constant Scaling model (CSM), which has been introduced in [16]. The CSM leveraged the observation that the total views received soon after a video was uploaded provided a strong indication of its total future views to develop a prediction model for video popularity.

(b) Linear regression model about p_{mt} (pmtREG). The pmtREG is linearly correlated to previous I days' popularity as $\tilde{p}_{mt} = \sum_{i=1}^{I} a_i * p_{m(t-i)}$.

(c) Linear regression model about video quality (qualityREG). The qualityREG is linearly correlated to video quality as $\tilde{p}_{mt} = \sum_{n=1}^{N} \lambda_n * VQ_{mn,t-1}$, where λ_1 stands for the coefficient of Views, λ_2 stands for TReviews' coefficient, λ_3 stands for Collects' coefficient, λ_4 stands for Coins' coefficient and λ_5 stands for *DanMus*' coefficient.

(d) Fixed Herding Effect model (FM). The FM only adopts fixed herding effect, which means that each day has the same herding effect about parameter θ_k, without combining video quality and uploaders' influence.

(e) Dynamic Herding Effect model (DM). The DM only uses dynamic herding effect, which means that each day has different herding effect of parameter θ_k, without combining video quality and uploaders' influence, too.

We implemented all the methods in C# and conducted experiments on a Windows 8 system with a 3.4 GHz Intel i7 CPU and 32 GB memory. For comparing the learning ability of growth models, we dynamically adjust the length of training days including 80 days, 95 days, 110 days, 125 days and 140 days and use the following days for testing.

5.2   Validating Popularity Growth Models

In this first step of experiment, we validate the effectiveness of different popularity growth models. As we predict the popularity of the video m, the most direct and efficient evaluation index is to compute the average absolute difference ratio between observation values and prediction values as $aadr_t = \frac{\sum_{m=1}^{M_t} \frac{Abs(\tilde{p}_{mt} - p_{mt})}{p_{mt}}}{M_t}$.

The experimental results of $aadr$ are exhibited in Fig. 4. We find that popularity growth models (e.g. DM, HVUM) adopting dynamic herding effect have smaller $aadr$ than other popularity growth models (e.g. CSM, FM) across all training days, which reflects that the dynamic herding effect contributes significantly to reducing the prediction error of popularity growth. And as training days increasing, the $aadr$ of each growth models becomes decrease due to more training information conduces to learn parameters of growth models accurately.

To exhibit the improvement and difference among different growth models clearly, we add the relative improvement of $aadr$ (denoted as RelativeImp), which calculates the rate of each model's improvement compared to CSM been extensively adopted in many previous popularity growth models. And Fig. 5 demonstrates the average values of $aadr$ and RelativeImp. We observe that the $aadr$ 13.71 % of FM is the worst one among all growth models, which shows that the fixed herding effect can not accurately capture the dynamics of popularity growth. And regarding to RelativeImp, both linear regression models (pmtREG: -30.58 %, qualityREG: -108.18 %) have no advantages on predicting popularity dynamic growth compared to CSM. It is noticed that the HVUM has the best prediction ability on popularity dynamic growth with the minimum 3.27 % of $aadr$ and the maximum 47.19 % of RelativeImp. Particularly, our proposed model HVUM has improved 5 % accuracy compared to DM on RelativeImp, which demonstrates that the video quality and the uploaders' influence contribute to improving the prediction accuracy on popularity dynamic growth.

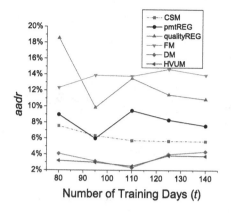

Fig. 4. *aadr* with training days.

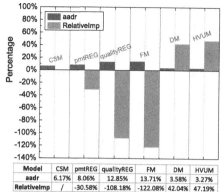

Model	CSM	pmtREG	qualityREG	FM	DM	HVUM
aadr	6.17%	8.06%	12.85%	13.71%	3.58%	3.27%
RelativeImp	/	-30.58%	-108.18%	-122.08%	42.04%	47.19%

Fig. 5. Average *aadr* and RelativeImp.

We can thus conclude that the HVUM faithfully captures the growth dynamics of video popularity with rich and insufficient training information.

5.3 Impact of Different Video Features

Next, we take a deep analysis about the training values of video quality's coefficients $\gamma_1 = 0.0015, \gamma_2 = -0.0055, \gamma_3 = -0.0036, \gamma_4 = 0.0029, \gamma_5 = -0.0017$ to explore the impact of different features of video quality. We find that the next day's popularity is most relevant to previous day's Coins as $\gamma_4 = 0.0029$, which reflects that the only paid feedback Coins is the most relevant feature to the dynamic growth of popularity among all features. And the feature Views also plays an important role in predicting the next day's popularity as $\gamma_1 = 0.0015$. While the *DanMus*' coefficient γ_4 equals to -0.0017, it shows that the feature of *DanMus* plays a significantly negative effect on predicting the dynamic growth of popularity. Nevertheless, in previous works, none work adopts the *DanMu* information to improve the accuracy of popularity models.

5.4 Describing Herding Effect

In Sect. 5.2, the dynamic herding effect contributes significantly on improving the accuracy of popularity growth models. In this section, we take an in-depth analysis about fixed and dynamic herding effect respectively.

At first, we give training values of $(\theta_1, \theta_2, \theta_3) = (-0.0036, -0.0076, 0.0077)$ of FM to analyze different aspects' fixed herding effect. We find that both θ_1 and θ_2 are less than 0, which shows that the popular videos measured by the aspects of views and *DanMus* have a negative effect on other videos. However, θ_3 is larger than 0, which shows that each day's new videos excite users to view other videos. Based on these two observations, we know that the popular videos according to the number of views and *DanMus* inhibit users to view other videos

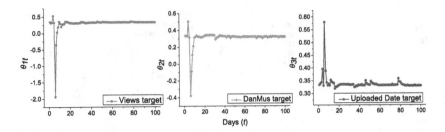

Fig. 6. Herding effect parameters' dynamic change.

and if the website holders add new videos to their owning websites everyday, the traffic of videos' services will be improved.

At the end, we demonstrate the training values of HVUM to explore the effect of the dynamic herding effect of different aspects on predicting the popularity growth in Fig. 6. Based on Fig. 6, we can draw several implications: First, the popular videos according to each aspect have different effect on other videos on different days. For example, to the dynamic herding effect of the views aspect, the popular videos have a positive effect on other videos on the day 1 as $\theta_{11} = 0.34$, while the popular videos have a negative effect on other videos on the day 6 as $\theta_{16} = -1.94$. Second, at the initial few days, the fluctuation of parameters is stronger than subsequent days. The main reason of this observation is that the videos are not enough to learn the accurate value of θ at first few days. Third, over time, we find that the values of herding effect parameters are stable as the training videos are sufficient.

6   Conclusions

In this paper, we introduced a model for predicting the popularity growth of *DanMu*-enabled videos, which combines the dynamic herding effect, *DanMu* information, video quality and uploaders' influence. We collected a large set of data from a *DanMu*-enabled online video system (i.e., acfun.tv) that includes 3,623 videos, 73,059,811 views, 883,637 *DanMus* and 745 uploaders of each day. We first analyzed the distributions of video features and uploader features over time. Then we proposed to measure the herding effect of *DanMu*-enabled video popularity from multiple aspects, including the popular videos, the popular DanMus and the newly updated videos. We also recognized that the uploaders' influence and video quality affect the *DanMu*-enabled video popularity as well. Therefore, we combined the dynamic herding effect, uploaders' influence and video quality in a unified framework to predict the popularity of *DanMu*-enabled videos. After that, we designed an efficient estimation method to automatically learn the herding effect and other parameters. Finally, experimental results demonstrated the effectiveness of our prediction model. We believe that the successful prediction of video popularity provides valuable commercial and technical implications to improve various online video-based services.

Acknowledgements. This research was partially supported by grants from the National Science Foundation for Distinguished Young Scholars of China (Grant No. 61325010), the National High Technology Research and Development Program of China (Grant No. 2014AA015203) and the Fundamental Research Funds for the Central Universities of China (Grant No. WK2350000001). This research was supported in part by NIH (1R21AA023975-01), NSFC (71571093, 71372188, 61572032), and National Center for International Joint Research on E-Business Information Processing (2013B01035). Truly appreciate Jinmei Lin's help and suggestions in user experience on *DanMu*-enabled videos.

References

1. Andersson, M., Lee, C., Hedesström, T.M., Gärling, T.: Effects of reward system on herding in a simulated financial market. Interaction on the Edge, pp. 12 (2006)
2. Bandari, R., Asur, S., Huberman, B.A.: The pulse of news in social media: forecasting popularity. In: ICWSM, pp. 26–33 (2012)
3. Banerjee, A.V.: A simple model of herd behavior. The Quarterly J. Econ. **107**, 797–817 (1992)
4. Barabási, A.-L., Albert, R.: Emergence of scaling in random networks. Science **286**(5439), 509–512 (1999)
5. Borghol, Y., Ardon, S., Carlsson, N., Eager, D., Mahanti, A.: The untold story of the clones: content-agnostic factors that impact YouTube video popularity. In: Proceedings of the 18th ACM SIGKDD, pp. 1186–1194. ACM (2012)
6. Cha, M., Kwak, H., Rodriguez, P., Ahn, Y.-Y., Moon, S.: I tube, you tube, everybody tubes: analyzing the world's largest user generated content video system. In: Proceedings of the 7th ACM SIGCOMM Conference on Internet Measurement, pp. 1–14. ACM (2007)
7. Cha, M., Kwak, H., Rodriguez, P., Ahn, Y.-Y., Moon, S.: Analyzing the video popularity characteristics of large-scale user generated content systems. IEEE/ACM Trans. Networking (TON) **17**(5), 1357–1370 (2009)
8. Cha, M., Mislove, A., Gummadi, K.P.: A measurement-driven analysis of information propagation in the flickr social network. In: Proceedings of the 18th international conference on World wide web, pp. 721–730. ACM (2009)
9. Cohen, J., Cohen, P., West, S.G., Aiken, L.S.: Applied multiple regression/correlation analysis for the behavioral sciences. Routledge (2013)
10. Hey, J.D., Morone, A.: Do markets drive out lemmingsor vice versa? Economica **71**(284), 637–659 (2004)
11. Hogg, T., Lerman, K.: Social dynamics of digg. EPJ Data Sci. **1**(1), 1–26 (2012)
12. Hsieh, S., Tai, Y.Y., Vu, T.B.: Do herding behavior and positive feedback effects influence capital inflows? evidence from asia and latin america. Int. J. Bus. Finance Res. **2**(2), 19–34 (2008)
13. Le, W., Qi, L., Chen, E., Xie, X., Chang, T.: Product adoption rate prediction: A multi-factor view
14. Lerman, K., Hogg, T.: Using a model of social dynamics to predict popularity of news. In: Proceedings of the 19th International Conference on World Wide Web, pp. 621–630. ACM (2010)
15. Mitra, S., Agrawal, M., Yadav, A., Carlsson, N., Eager, D., Mahanti, A.: Characterizing web-based video sharing workloads. ACM Trans. Web (TWEB) **5**(2), 8 (2011)

16. Szabo, G., Huberman, B.A.: Predicting the popularity of online content. ACM Commun. **53**(8), 80–88 (2010)
17. Tsagkias, M., Weerkamp, W., de Rijke, M.: News comments:exploring, modeling, and online prediction. In: Gurrin, C., He, Y., Kazai, G., Kruschwitz, U., Little, S., Roelleke, T., Rüger, S., van Rijsbergen, K. (eds.) ECIR 2010. LNCS, vol. 5993, pp. 191–203. Springer, Heidelberg (2010)
18. Welch, I.: Herding among security analysts. ACM Commun. **58**(3), 369–396 (2000)
19. Wu, B., Zhong, E., Tan, B., Horner, A., Yang, Q.: Crowdsourced time-sync video tagging using temporal and personalized topic modeling. In: Proceedings of the 20th ACM SIGKDD, pp. 721–730. ACM (2014)
20. Wu, Z., Ito, E.: Correlation analysis between user's emotional comments and popularity measures. In: IIAI 3rd International Conference on Advanced Applied Informatics (IIAIAAI), pp. 280–283. IEEE (2014)

Integrating Human Mobility and Social Media for Adolescent Psychological Stress Detection

Li Jin[1,2]($\boxtimes$), Yuanyuan Xue[1,2], Qi Li[1,2], and Ling Feng[1,2]

[1] Tsinghua National Laboratory for Information Science and Technology (TNList),
Department of Computer Science and Technology,
Tsinghua University, Beijing 100084, China
l-jin12@mails.thu.edu.cn
[2] Centre for Computational Mental Healthcare Research,
Institute of Data Science, Tsinghua University, Beijing 100084, China
{xue-yy12,liqi13}@mails.thu.edu.cn, fengling@tsinghua.edu.cn

Abstract. Leveraging social media to detect psychological stress is an emerging research topic, as it addresses one of the most common mental health issues. One of the notable challenges in this area, however, is data sparsity: users with high stress level tend to reduce their activities on social networks. While teenagers' mobility behavior always appears some outliers for stress release in physical world, a question arises: can we identify the stress-related outlier features from daily trajectories to facilitate stress detection? In this paper, we propose a co-training-based semi-supervised learning approach that consists of two separated classifiers. One classifier is conditional random field (CRF), which takes outlier features from GPS trajectories as input to model the daily moving behavior correlation of stress. The other classifier is deep neural network (DNN) involving tweet features to model the social media behavior correlation of stress. We evaluate our approach with an over 6-month user study on 57 teenagers from Beijing, and demonstrate effectiveness of the proposed model compared to state-of-the-art methods .

Keywords: Stress detection · Co-training · Human mobility · Social media

1 Introduction

Due to the growing pace of modern life in the competitive society, psychological stress has become one of the major factors causing health problems, especially for teenagers who are not mature enough to deal with it properly and effectively. An online survey[1] of 1018 U.S. teens (aged 13–17) made by the American Psychological Association in 2013 found that teens were suffering from stress in all areas of their lives, from school to friends, work and family, where more than a quarter (27 %) said they experienced "extreme stress" during the school year. Adolescence is a critical period for teens' growth and development. Bearing

---

[1] http://www.apa.org/news/press/releases/stress/2013/.

© Springer International Publishing Switzerland 2016
S.B. Navathe et al. (Eds.): DASFAA 2016, Part II, LNCS 9643, pp. 367–382, 2016.
DOI: 10.1007/978-3-319-32049-6_23

too much stress without being released timely hurts teens both physically and mentally, leading to clinical depressions, insomnia, and even suicide. Currently, about 20 % of teens have psychological illness around the world[2]. According to China's Center for Disease Control and Prevention, suicide has become the top cause of death among Chinese youth[3], and excessive stress is considered to be a major factor of suicide. Therefore, it is of great significance to detect and deal with stress before it causes severe consequence.

Leveraging social media to develop stress detection tools has drawn many research interests in recent years. Previous work [9,10,19] detects psychological stress through the tweets and posting patterns from microblog. However, limitations exist in tweet-based stress detection. Firstly, tweets are limited to maximally 140 characters on social microblog platforms, and teens do not always express their stressful states directly in tweets. Secondly, users with high stress level may exhibit low activeness on social networks, as reported by a recent study from the Pew Research Center[4]. These phenomena incur the inherent data sparsity and ambiguity problem, which may hurt stress detection performance. Besides social media behavior, teens' mobility behavior always appears some outliers for stress release in physical world. The increasing availability of GPS-enabled devices motivates us to expand the tweet-wise investigation scope by incorporating useful cues from recorded GPS trajectories.

In this paper, we propose a co-training-based stress detection model based on teens' daily GPS trajectories and microblog data. Rather than treating stress-related features equally from cross-domain data sources, we explore how to incorporate these heterogeneous features into a data analytics model effectively in view of limited labeled data. The contribution can be summarized as follows:

- We propose a co-training-based semi-supervised learning approach, which leverages unlabeled data to improve the detection accuracy. Additionally, the approach consists of two classifiers respectively modeling the teens' trajectory outlier features and tweet features that correlate to the stress.
- We identify positional and temporal outlier features from teens' daily trajectories to facilitate stress detection. To the best of our knowledge, this is the first attempt in the literature to investigate and model the inner relationship between psychological stress and trajectory outliers.
- We build a system called T-Sensor, and conduct an over 6-month user study on 57 teens from Beijing to demonstrate its effectiveness.

The rest of our paper is organized as follows. In Sect. 2, we discuss the related work. Then we formulate the stress detection problem and present the framework of T-Sensor in Sects. 3 and 4. We describe the stress-related feature extraction and detail the learning model in Sects. 5 and 6. Finally, we show our experimental results in Sect. 7, and draw a conclusion in Sect. 8.

---

[2] http://learning.sohu.com/s2012/shoot/.

[3] http://theweek.com/articles/457373/rise-youth-suicide-china.

[4] http://www.pewinternet.org/files/2015/01/PI_Social-media-and-stress_0115151.pdf.

2   Related Work

Psychological stress detection techniques can be divided into three main kinds based on different analysis methods [14]. The first kind uses subjective questionnaires (e.g., Perceived Stress Scale [1]) or individual/group meetings with psychologists to analyze users' stress situations. This kind of methods requires high cooperation from users and sometimes relies on people's ability to recall their experiences. Observing human's physiological and physical signals change with the variation of psychological stress status, many researchers use various sensors to objectively monitor the changes of physiological and physical signals. For example, [4] found skin conductivity and heart rate metrics have close correlations with driver's stress level, and integrated multiple physiological signals (i.e., skin conductance, electromyogram, electrocardiogram, and respiration) to detect stress. [8] measured mental stress by observing physical signals changes caused by head and mouth movements as well as eye movements including eye gaze, eye blink, and pupil dilation. [11] treated smartphone as a kind of sensor to detect people's mental stress by analyzing their voice variation in diverse conversational situations. Compared with these body contact and invasive stress measurements, the social media (e.g., BBS, Facebook and Microblog) arises as another low-cost sensing channel to obtain people's self-expressed contents and behaviors, where some emotional signals could be captured and analyzed. For example, [15] constructed a two-stage supervised learning framework to identify potential depression candidates, based on the content and temporal features extracted from their write-ups on BBS. [12] adopted negative binomial regression analysis to evaluate college students' Facebook disclosures which met DSM[5] criteria for a depression symptom or a major depressive episode. [19] investigated a number of teens' typical tweeting behaviors that might reveal adolescent stress, and applied five classifiers to detect teens' stress. [9,10] trained a deep sparse neural network to detect users' stress from cross-media microblog.

3   Problem Formulation

In this section, we firstly give some preliminary concepts, and then present a formal definition of the problem.

Definition 1. Trajectory: A spatial trajectory tr is a sequence of time-ordered spatial points, $tr : p_1 \rightarrow p_2 \rightarrow \ldots \rightarrow p_n$, where each point consists of a latitude, longitude and a time stamp, e.g., $p = \{lat, long, t\}$.

Definition 2. POI: A point of interest POI is a venue (like a school and shopping mall) in the physical world, having a name, address, coordinates, category, and other attributes.

---

[5] https://en.wikipedia.org/wiki/Diagnostic_and_Statistical_Manual_of_Mental_Disorders.

Definition 3. Road Network: A road network RN is a directed graph $G(V, E)$, where E is a set of edges defining the road segments, and V is a set of vertices defining the intersections and terminal points for E.

Definition 4. Psychological Stress: The psychological stress of a teen at time t is represented as a triple (s, c, t) (or briefly s_c^t), where $s \in \{0, 1\}$ is a binary value indicating stressed or non-stressed, $c \in \mathcal{C}$ is the stress category.

Problem Definition. Given a sequence of a teen's tweets Tw and daily GPS trajectories Tr within a time period $I = \{t_1, t_2, \ldots, t_n\}$, the objective is to learn a stress mapping function $f : (\{Tw, Tr, I\}) \rightarrow \{s_c^{t_1}, s_c^{t_2}, \ldots, s_c^{t_n}\}$ to determine a teen's stress existing period.

4   Framework Overview

Fig. 1 presents the framework of our T-Sensor system, which consists of three major layers: (1) feature extraction, (2) stress detection model; (3) service providing. We will detail the first two layers in the following sections respectively.

5   Feature Extraction

In this section, we define and extract several stress-related features from daily GPS trajectories and tweets based on previous psychological principles [6].

5.1   Trajectory Outlier Feature Space

Our approach to extract stress-related outlier features from daily GPS trajectories consists of three main steps. We firstly calibrate the raw trajectory data using road network and POIs. Then we partition teens' historical trajectories to discover their regular lifestyle. Finally, we model the stress-related anomalous behavior and conduct feature extraction on teens' trajectory outliers.

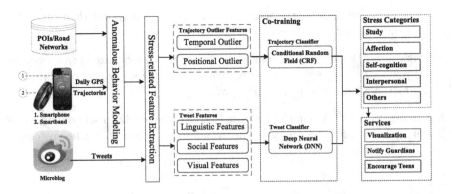

Fig. 1. Framework of T-Sensor system

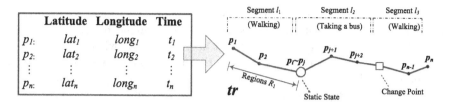

	Latitude	Longitude	Time
p_1:	lat_1	$long_1$	t_1
p_2:	lat_2	$long_2$	t_2
$\vdots$	$\vdots$	$\vdots$	$\vdots$
p_n:	lat_n	$long_n$	t_n

Fig. 2. GPS log, segment and change point

Considering efficient map-matching of teens' daily trajectories, we adopt the anchor-based calibration system [16] after partitioning the road network using grid. Inspired by the change point-based segmentation method, inference model and graph-based post-processing algorithm proposed by [20], we then conduct a two-level trajectory partitioning strategy to identify teens' mobility patterns as shown in Fig. 2. In the first level, we cluster the stay points from the same region in coarse granularity. If the Euclidean distance d among a set of time-ordered spatial points $p_i \rightarrow p_{i+1} \rightarrow \ldots \rightarrow p_j$ is less than a threshold $\tau_d = 200\,m$, we cluster them and partition trajectory tr into static and moving state. In the second level, we partition the moving sub-trajectories based on different motion modes (e.g., walking, running, taking a bus and riding a bicycle) in fine granularity. Each partitioned segment can be represented by a feature vector $\mathbf{f} = (t_s, t_e, \sigma, m, R, P)$, where t_s and t_e are the start and end time stamp, σ identifies the static or moving state, m corresponds to the motion mode, R and P are the passed regions and POIs.

To discover teens' regular lifestyle, a frequent segment mining algorithm [3] is conducted on partitioned trajectory dataset $\mathcal{L}$, and the support of the ith segment l_i can be calculated by $\mathcal{S}(l_i) = \frac{|\{l|s(l,l_i) > \tau_s, l.\sigma = l_i.\sigma, l.m = l_i.m, l \in \mathcal{L}\}|}{k}$, where k is the number of days; $s(l, l_i)$ is the similarity between two segments; and τ_s is the threshold. Considering feature vector $\mathbf{f}$, we adopt a linear weighted formula to define the similarity function $s(l, l_i)$ as follows:

$$s(l, l_i) = w_t \cdot \gamma_t(l, l_i) + w_R \cdot \gamma_R(l, l_i) + w_P \cdot \gamma_P(l, l_i) \tag{1}$$

where $\gamma_t(l, l_i) = \frac{[l.t_s, l.t_e] \bigcap [l_i.t_s, l_i.t_e]}{[l.t_s, l.t_e] \bigcup [l_i.t_s, l_i.t_e]}$ is the time overlap ratio; $\gamma_R(l, l_i) = \frac{|l.R \bigcap l_i.R|}{|l.R \bigcup l_i.R|}$ is the region overlap ratio; $\gamma_P(l, l_i) = \frac{|l.P \bigcap l_i.P|}{|l.P \bigcup l_i.P|}$ is the POI overlap ratio; w_t, w_R and w_P are the weight parameters. Given a threshold $\tau_{\mathcal{L}}$, we can further choose the frequent segments at different time of day with the support $\geq \tau_{\mathcal{L}}$ to constitute the regular lifestyle Γ. We initially set the weights $w_t = w_R = w_P = \frac{1}{3}$, the thresholds $\tau_s = 0.8$, and $\tau_{\mathcal{L}} = 0.7$. The adjustment for the parameters to determine Γ is detailed in Sect. 6.2. For the ith frequent segment l_i in Γ, we describe its feature vector $\mathbf{f}$ by Gaussian distribution $f(l_i.x) \sim N(\mu(l_i.x), \sigma(l_i.x)^2)$, where the mean $\mu(l_i.x) = \frac{\sum_{j=1}^{N_i} l_i^j.x}{N_i}$ and the standard deviation $\sigma(l_i.x) = \sqrt{\frac{1}{N_i} \sum_{j=1}^{N_i} (l_i^j.x - \mu(l_i.x))}$ are calculated based on N_i similar fragments from $\mathcal{L}$.

To model the stress-related anomalous behaviors, we design a psychological questionnaire and call for 620 teens to fill in it online by SOJUMP website[6]. Through asking questions like "what do you often do when suffering from stress?", 46.3 % of teens choose to change the surrounding environment and go to some unusual places (e.g., park, bookstore, cinema, shopping mall); 37.9 % of teens choose to stay at familiar places and do something to relax (e.g., sleep, cry, exercise, play computer games). In addition, 88.2 % of teens express that they get confused about how to deal with serious stress properly, which always hinder them in the current work. Based on the empirical investigation, we study to extract stress-related outliers from teens' mobility behaviors, which refer to a partitioned segment that is grossly different from or inconsistent with the corresponding segment from teens' regular lifestyle Γ. In this paper, we mainly consider two stress-related outliers, which can be defined as follows:

- **Positional outlier feature:** F_p. The passed regions of segment l do not belong to the τ_p confidence interval of $f(l.R) \sim N(\mu(l.R), \sigma(l.R)^2)$ in Γ, which is represented by $F_p = (l.R, \tau_p, \Gamma)$.
- **Temporal outlier feature:** F_t. The occurring time interval of segment l does not belong to the τ_t confidence interval of $f(l.t) \sim N(\mu(l.t), \sigma(l.t)^2)$ in Γ, which is represented by $F_t = (l.t_s, l.t_e, \tau_t, \Gamma)$.

5.2   Tweet Feature Space

Compared with other user groups, teens exhibit different microblog behaviors in tweeting/retweeting content when suffering from stress. Besides using short linguistic sentences, young teens many times like to add some pictorial emoticons (like laughing, shy, angry, and crazy symbols) to vividly express their minds, or use multiple exclamation and question marks for emphasis purpose. In addition, a majority of teens are fond of posting and forwarding multi-media information like music, pictures, etc.

To incorporate different features from the cross-media microblog data to enhance stress detection, we extract a set of features from the tweets including words from texts, colors from images, and the frequency of social actions. The definitions are as follows:

- **Linguistic features:** F_l. Based on the psycholinguistic dictionary LIWC [2], we extract 5 features (11 dimensions) defined in [19] consisting of linguistic association between stress category and negative emotion words (1 dimension), number of positive and negative emotion words/emoticons (4 dimensions), punctuation marks with associated emotion words (4 dimensions), emotional degree (1 dimension) and shared music genres (1 dimension) to describe the stress-related tweeting text.
- **Visual features:** F_v. Images convey emotions like excitement and sadness, and psychological experiments have shown that color themes are crucial for

---

[6] http://www.sojump.com/.

the recognition of emotions. Based on previous work on affective image classification [17,18] and color psychology theories[7], we extract 5 features (21 dimensions) consisting of five-color theme (15 dimensions), saturation (2 dimensions), brightness (2 dimensions), warm or cool color (1 dimension), clear or dull color (1 dimension) as visual representation for the tweeting images when teens are suffering from stress.

- **Social features:** F_s. Besides the text and image of a tweet, some additional features like comments, retweets and likes indicate the tweet's social attention from teens' friends. Considering an apparently stressful tweet may attract more attention from teens' friends in microblog, we extract the number of comments, retweets and likes (3 dimensions) of a tweet to measure its social attention degree into social features.

6   Model and Learning

In this section, we firstly introduce a co-training-based stress detection model to leverage the extracted trajectory outlier features and tweet features, then respectively detail trajectory-based CRF classifier and tweet-based DNN classifier.

6.1   Co-Training

Co-training is a semi-supervised learning technique that requires two views of the data. It applies to datasets that have a natural separation of their features into two disjoint sets, which provide different and complementary information for an instance. Ideally, the two feature sets of each instance are conditionally independent given the category, and each set of features is sufficient for classification. Co-training can generate a better inference result as one of the classifiers correctly labels data that the other classifier previously misclassified [13].

According to the co-training framework, we propose a trajectory-based CRF classifier (CRF-Tr) to model the daily moving behavior correlation and a tweet-based DNN classifier (DNN-Tw) to model the social media behavior correlation of psychological stress. The two models are integrated into a co-training-based learning framework presented in Algorithm 1. As shown in line 4 and 5, we firstly train the two classifiers with trajectory outlier features (i.e., F_p and F_t) and tweet features (i.e., F_l, F_v and F_s). The trained CRF-Tr and DNN-Tw classifier are then used to infer unlabeled stress values iteratively, adding the most confidently classified instances to the labeled dataset S_l for the next round of training, until S_u becomes empty or the number of rounds has reached to the threshold θ. When the iterative process ends, CRF-Tr and DNN-Tw classifier are returned. At the inference time, we apply two classifiers to the corresponding features separately, determining the daily stress state by the product of the two probability scores.

---

[7] https://en.wikipedia.org/wiki/Color_psychology.

Algorithm 1. T-Sensor Co-Training

---

input : A set of features $(F_p, F_t, F_l, F_v, F_s)$, some labeled stress values S_l, and
a set of unlabeled stress values S_u, a threshold θ controlling the rounds

output: CRF-Tr and DNN-Tw

1 **begin**
2 $\quad$ $i \leftarrow 0$;
3 $\quad$ **while** $i < \theta$ && $S_u \neq \emptyset$ **do**
4 $\quad\quad$ CRF-$Tr \leftarrow$ CRF.Learning(F_p, F_t, S_l);
5 $\quad\quad$ DNN-$Tw \leftarrow$ DNN.Learning(F_l, F_v, F_s, S_l);
6 $\quad\quad$ Apply trained CRF-Tr and DNN-Tw on S_u;
7 $\quad\quad$ **foreach** $\varepsilon \in S_u$ **do**
8 $\quad\quad\quad$ $\varepsilon.c = arg_{c_i \in \mathcal{C}} Max(Pro_{CRF}^{c_i} \times Pro_{DNN}^{c_i})$;
9 $\quad\quad$ Add n_i self-labeled stress values with highest probability to S_l;
10 $\quad\quad$ $i{+}{+}$;
11 $\quad$ **return** CRF-Tr and DNN-Tw;

6.2 Trajectory-Based CRF Classifier

The trajectory-based CRF classifier is a discriminative undirected probabilistic graphical model to infer teens' daily stress state. The primary advantage of CRFs over hidden Markov models is the conditional nature, resulting in the relaxation of the independence assumptions to ensure tractable inference. Additionally, CRFs avoid the label bias problem, a weakness exhibited by maximum entropy Markov models [5]. We adopt the graphical structure $\mathcal{G}$ from a linear-chain CRF, which consists of two kinds of nodes $\mathcal{G} = (\mathcal{X}, \mathcal{Y})$ as shown in Fig. 3. The gray nodes $\mathcal{Y} = \{Y_1, Y_2, \ldots, Y_n\}$ represent hidden state variables to be inferred given the sequence of observations denoted by white nodes $\mathcal{X} = \{X_1, X_2, \ldots, X_n\}$, where $X_i = \{F_p, F_t, t\}$ corresponds to trajectory outlier features. Considering the weights (i.e., w_t, w_R and w_P) and the thresholds (i.e., τ_s and τ_L) determine the observing outliers F_p and F_t, we iteratively assign different initial values to calculate the optimal parameters based on detection performance of CRF-Tr classifier. Meanwhile, the outlier features with the same time stamp t (i.e., day) are aggregated together. The $Y_i \in \mathcal{Y}$ is structured to form a chain with an edge connecting each Y_{i-1} and Y_i, as well as having the labeled stress values belonging to $\mathcal{C}$. When conditioned on $\mathcal{X}$, the random variable Y_i obey the Markov property regarding the graph $\mathcal{G}$:

$$p(Y_i|\mathcal{X}, Y_j, i \neq j) = p(Y_i|\mathcal{X}, Y_j, i \sim j) \tag{2}$$

where $i \sim j$ means that Y_i and Y_j are neighbors in $\mathcal{G}$.

The probability of a particular label sequence $\mathcal{Y}$ given observation sequence $\mathcal{X}$ is defined as a normalized product of potential functions as follows:

$$\exp(\sum_j \lambda_j t_j(Y_{i-1}, Y_i, \mathcal{X}, i) + \sum_k \mu_k s_k(Y_i, \mathcal{X}, i)) \tag{3}$$

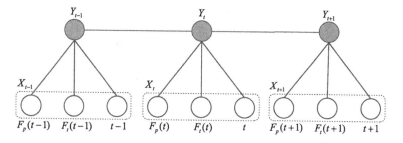

Fig. 3. The graphic presentation of the trajectory-based CRF classifier

where $t_j(Y_{i-1}, Y_i, \mathcal{X}, i)$ is a transition feature function of the entire observation sequence and the labels at positions i and $i-1$; $s_k(Y_i, \mathcal{X}, i)$ is a state feature function of the label at position i and the observation sequence; λ_j and μ_k are parameters to be estimated from training data.

By writing $s(Y_i, \mathcal{X}, i) = s(Y_{i-1}, Y_i, \mathcal{X}, i)$, we can simply transform Eq. (3) to:

$$p(\mathcal{Y}|\mathcal{X}, \lambda) = \frac{1}{Z(\mathcal{X})} exp(\sum_j \sum_{i=1}^n \lambda_j f_j(Y_{i-1}, Y_i, \mathcal{X}, i)) \qquad (4)$$

where $Z(\mathcal{X})$ is a normalization factor; $f_j(Y_{i-1}, Y_i, \mathcal{X}, i)$ is either a state function $s(Y_{i-1}, Y_i, \mathcal{X}, i)$ or a transition function $t(Y_{i-1}, Y_i, \mathcal{X}, i)$.

Given k sequences of training data $\{\mathcal{X}^{(k)}, \mathcal{Y}^{(k)}\}$, the parameters λ can be determined by maximum likelihood learning $p(\{\mathcal{Y}^{(k)}\}|\{\mathcal{X}^{(k)}\}, \lambda)$ as follows:

$$L(\lambda) = \sum_k [log \frac{1}{Z(\mathcal{X}^{(k)})} + \sum_j \sum_{i=1}^n \lambda_j f_j(Y_{i-1}^{(k)}, Y_i^{(k)}, \mathcal{X}^{(k)}, i)] \qquad (5)$$

where λ can be identified using gradient-based methods. Finally, we can combine observed features $\mathcal{X}$ and λ to determine the probability of a certain value for Y_i.

6.3   Tweet-Based DNN Classifier

The tweet-based DNN classifier is designed to learn the stress values by incorporating the cross-media features from tweets. We firstly define and extract a set of low-level features based on psychological and art theories: linguistic features F_l from tweeting text, visual features F_v from tweeting image, and social features F_s from comments, retweets and favorites as input. To solve the problem of missing modalities in cross-media tweet data, we use a cross auto-encoder (CAE) [9] to learn the modality-invariant representation of each single tweet with different modalities, which can be formulated as follows:

$$\begin{cases} u = f(w_l F_l + w_v F_v + w_s F_s + b) \\ (\widetilde{F}_l, \widetilde{F}_v, \widetilde{F}_s) = f(\widetilde{w}_l u + \widetilde{w}_v u + \widetilde{w}_s u + \widetilde{b}) \end{cases} \qquad (6)$$

where u is the modality-invariant representation; w_l, w_v, w_s and b are parameters in the encoder, whereas $\widetilde{w}_l$, $\widetilde{w}_v$, $\widetilde{w}_s$ and $\widetilde{b}$ are parameters in the decoder; $f(\cdot)$ is

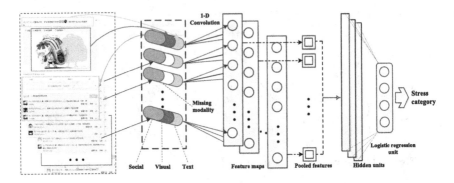

Fig. 4. Framework of the tweet-based DNN classifier

the activation function. We use a sigmoid activation function $f(\varphi) = \frac{1}{1+\exp(-\varphi)}$ in our model. $\widetilde{F_l}$, $\widetilde{F_v}$ and $\widetilde{F_s}$ are the reconstructed input modalities.

The basic idea of CAE is to force the model to reconstruct missing modalities in the training stage and to learn cross-modality correlation from the data. While training the cross auto-encoder, we choose training data that contains all the three modalities. We then adopt the stochastic gradient descent to train the CAE with a cropped set of data $\{\overline{F_l}, \overline{F_v}, \overline{F_s}\}$ that inputs from one or two modalities are absent, while requiring it to reconstruct all the three. Through denoting all the parameters in the CAE as θ, the energy function is defined as follows:

$$\mathcal{J}(\overline{F_l}, \overline{F_v}, \overline{F_s}; \theta) = \frac{1}{2}\left(\sum_{\sigma \in \{l,v,s\}} ||\widetilde{F_\sigma} - F_\sigma||^2\right) + \frac{\lambda}{2}\left(\sum_{\sigma \in \{l,v,s\}} ||\widetilde{F_\sigma}||^2 + ||F_\sigma||^2\right) \quad (7)$$

where the first term measures the reconstruction accuracy; the second term is the weight decay regularization term that prevents parameters in the model from diverging arbitrarily; λ is the regularization weight. Using data with different modalities as input, the CAE can be trained and learn a modality-invariant representation u in Eq. (6).

The features of tweets, which come from a teenager's daily tweets in timeline, form a time series. To model teens' tweeting behavior to describe daily stress state, we apply convolutional neural networks (CNNs) [7], which focus on learning the stationary local features from series like images (pixel series), audio, and other time series. CNNs have a large learning capacity, but much fewer connections and parameters to learn than similar-size standard network layers. We use a 1-dimension CNN in our model, where the CAE units are used as filters and convolute over the sequence of tweets to form modality-invariant feature maps. Based on mean-over-time (MOT) pooling operations, we summarize the feature maps into fewer feature instances by summing up the activations since they are sampled in the same length of time (i.e., day). Finally, the teens' daily stress is classified by a logistic regression unit. A 5-layer architecture is used in this paper. Figure 4 demonstrates the overall framework of tweet-based DNN classifier.

7   Experiments

In this section, we first describe the datasets collected by T-Sensor system, then evaluate the performance of co-training-based stress detection model.

7.1   Experiment Setting

We calls for 57 teenagers from middle/high school of Beijing to conduct an over 6-month user study (Mar-Sep, 2015), where each teenager wear a smartband provided by us to connect with T-Sensor system installed on their smartphones. By default T-Sensor runs in the background of smartphone to collect GPS trajectories. Meanwhile, T-Sensor is permitted to crawl teens' posting tweets from microblog (i.e., Sina Weibo[8]) as described in DB1. To avoid the cold-start problem, we pre-train DNN by crawling 1014 teens from Sina Weibo with a tag named "The Generation After 96s" to establish a weibo-stress dataset, where we manually label the stress value of each tweet as described in DB2. For efficient map-matching of teens' daily trajectories, we use the road networks and POIs of Beijing. The statistics of dataset is shown in Table 1.

Table 1. Statistics of dataset

	Data duration	Mar-Sep, 2015
Teenagers	# of high school students	37
	# of middle school students	20
Trajectories	# of effective days	169
	Avg. time of recording	13.6 hours/day
DB1: Sina Weibo	# of tweets	26247
DB2: Sina Weibo	# of teenagers	1014
	# of tweets	448,324
Road Networks	# of road segments	162,246
	# of road nodes	121,771
POIs	# of POIs	369,668
	# of categories	602

Comparison Methods. We compare the following classification methods for user-level psychological stress detection, with our co-training model respectively.

- **Naive Bayes (NB)**: it is a simple probabilistic classifier based on the Bayes theorem with strong (naive) independent assumptions.

---

[8] http://open.weibo.com/.

- **Support Vector Machine (SVM)**: it tries to find a hyperplane that divides training samples into their classes with maximum margin. In our problem we use SVM with RBF kernel which can handle most nonlinear binary classifications better.
- **Random Forest (RF)**: it is an ensemble learning method by building a set of decision trees with random subsets of features and bagging them for final classification results.
- **Gradient Boosted Decision Tree (GBDT)**: it trains a gradient boosted decision tree model with features associated with each user. GBDT is an ensemble method, which constructs an additive regression model, utilizing decision trees as the weak learner.
- We employ scikit-learn[9] to implement the above comparison methods, where the trajectory outlier thresholds are initially set to $\tau_p = \tau_t = 0.95$.

Evaluation Measures. We evaluate effectiveness of the methods by comparing the performance metrics Accuracy, Recall, Precision and F1-Measure. In the following experiments, we adopt 5-fold cross validation over 10 randomized experimental runs, where DNN is trained by NVIDIA TESLA K20 GPU.

7.2 Experimental Results

Comparison of Detection Performance. To evaluate effectiveness of our model, we first conduct a test using different models based on the collected trajectory and microblog dataset by T-Sensor system. In this experiment, we use all the features that are described in previous section: trajectory outlier features (i.e., F_p and F_t) and tweet features (i.e., F_l, F_v and F_s). For comparison methods, we respectively use conditional random field (CRF) and deep neural network (DNN) to run all the features, denoted as CRF-All and DNN-All. Table 2 shows the experimental results, and the results are significant due to the large amount of instances for evaluation. We see that co-training performs the best with an accuracy of 88.92 % and F1-Measure of 91.09 % against other comparison methods. DNN-ALL gains better performance than CRF-ALL with an improvement of 4.33 % in accuracy and 3.87 % in F1-Measure because it is more suitable to process the cross-media features from tweets. NB and SVM do not work well with an lower accuracy < 76 % and F1-Measure < 81 % since it is inappropriate for them to learn local features from our time-series data. The results demonstrate that our proposed model can effectively leverage the human mobility and social media behaviors for stress detection. We further perform *t-test* and all the *p-values* are < 0.01, which indicates that the improvement of co-training over the comparison methods are statistically significant.

Co-Training Process Analysis. Figure 5 further reveals the co-training progress of our approach, where we add an instance into the training data when it fulfills one of following conditions: (1) CRF-Tr and DNN-Tw predict it as a class with a probability score > 0.7; (2) the probability score of the instance is

---

[9] http://scikit-learn.org.

Table 2. Comparison results of effectiveness using different models (%)

Method	Accuracy	Precision	Recall	F1-Measure
NB	74.60	76.23	80.61	78.36
SVM	75.16	78.08	83.14	80.53
RF	78.44	81.63	85.09	83.32
GBDT	80.68	84.17	81.40	82.76
CRF-ALL	80.22	82.35	84.63	83.47
DNN-ALL	84.55	86.02	88.71	87.34
Co-Training	**88.92**	**90.04**	**92.16**	**91.09**

ranked in the top 10 results. The unlabeled data gradually improves the detection performance, justifying the ability of the co-training-based learning framework in dealing with data sparsity. When the number of iterations reaches to 90, co-training has the indication of convergence in precision and recall.

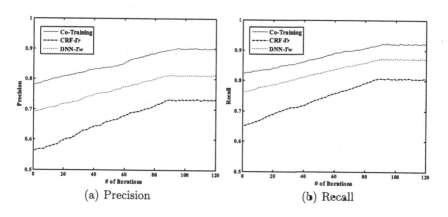

(a) Precision   (b) Recall

Fig. 5. Learning progress of co-training

Feature Contribution Analysis. To evaluate the contribution of different features in stress detection, we test the proposed model by respectively removing one of the pre-defined features: positional outlier feature F_p, temporal outlier feature F_t, linguistic features F_l, visual features F_v and social features F_s, denoted as Co-POF, Co-TOF, Co-LF, Co-VF and Co-SF. The results of this experiment are shown in Fig. 6. Considering trajectory outlier feature space, we see that Co-POF drops to 79.76 % in comparison with 81.27 % of Co-TOF in F1-Measure, which is consistent with the online questionnaire results by SOJUMP website in Sect. 5.1. More teens prefer to change the surrounding environment and go to some unusual places for stress release compared to stay at familiar environment. Considering tweet feature space, we see that Co-LF drops to 69.54 % in

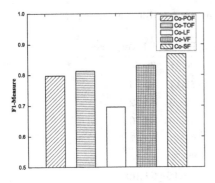
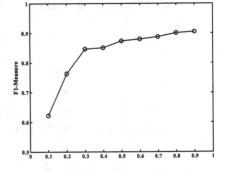

Fig. 6. Feature contribution   **Fig. 7.** Data scale analysis

F1-Measure because linguistic features of tweets is the main sensing channel to obtain teens' self-expressed contents concerning the stress. In case of removing visual features, Co-VF drops to 83.11 % in F1-Measure although not every posting tweet contains images. Meanwhile, Co-SF drops to 86.85 % in F1-Measure, which verifies the effect of social attention for stress detection.

Co-Training Data Scale Analysis. To evaluate the data scalability of the proposed model, we train the model with different scales of training data, and compare the final detection performance in F1-Measure. In this test, we use all the features as input to the model. Figure 7 shows the trend of detection performance with different proportions (i.e., 10 %-90 %) of training data. It is clear that when adopting approximately 30 % of all training data, our model can obtain a competitive performance of 84.68 % in F1-Measure. Moreover, the performance keeps increasing given more training data. These results verify that our model can leverage unlabeled data to improve the detection performance when there is limited training data.

8   Conclusion

In this paper, employing real-world GPS trajectory data as the basis, we mainly study the correlation between teens' psychological stress states and their mobility behaviors. Based on the discovered patterns, we propose a co-training-based stress detection model which combines CRF-Tr classifier with DNN-Tw classifier: (1) Firstly, we model the stress-related anomalous behavior based on psychological theories and our empirical investigations; (2) We then define and extract a set of trajectory outlier features and tweet features as input to each classifier; (3) Finally, we construct a co-training-based learning framework to leverage the extracted features for stress detection. With T-Sensor system installed on the smartphones, we conduct an over 6-month user study of 57 teenagers from Beijing by providing them the smartbands to wear. The results demonstrate effectiveness of our proposed model based on teens' daily GPS trajectories and microblog data.

Acknowledgement. The work is supported by National Natural Science Foundation of China (61373022, 61532015, 71473146) and Chinese Major State Basic Research Development 973 Program (2015CB352301).

References

1. Cohen, S., Kamarck, T., Mermelstein, R.: A global measure of perceived stress. J. Health Soc. Behav. **24**, 385–396 (1983)
2. Gao, R., Hao, B., Li, H., Gao, Y., Zhu, T.: Developing simplified chinese psychological linguistic analysis dictionary for microblog. In: Imamura, K., Usui, S., Shirao, T., Kasamatsu, T., Schwabe, L., Zhong, N. (eds.) BHI 2013. LNCS, vol. 8211, pp. 359–368. Springer, Heidelberg (2013)
3. Han, J., Kamber, M., Pei, J., Mining, D.: Concepts and Techniques. Morgan Kaufmann Publisher, San Francisco (2011)
4. Healey, J., Picard, R.: Detecting stress during real-world driving tasks using physiological sensors. IEEE Trans. Intell. Transp. Syst. **6**(2), 156–166 (2005)
5. Lafferty, J., McCallum, A., Pereira, F.: Conditional random fields: probabilistic models for segmenting and labeling sequence data. In: Proceedings of ICML (2001)
6. Lazarus, R.S.: Stress and Emotion: A New Synthesis. Springer Publisher, New York (2006)
7. LeCun, Y., Bengio, Y.: Convolutional networks for images, speech, and time series. In: Arbib, M.A. (ed.) The Handbook of Brain Theory and Neural Networks, pp. 255–258. MIT Press, Cambridge (1995)
8. Liao, W., Zhang, W., Zhu, Z., Ji, Q.: A real-time human stress monitoring system using dynamic bayesian network. In: Proceedings of CVPR Workshop (2005)
9. Lin, H., Jia, J., Guo, Q., Xue, Y., Huang, J., Cai, L., Feng, L.: Psychological stress detection from cross-media microblog data using deep sparse neural network. In: Proceedings of ICME (2014)
10. Lin, H., Jia, J., Guo, Q., Xue, Y., Li, Q., Huang, J., Cai, L., Feng, L.: User-level psychological stress detection from social media using deep neural network. In: Proceedings of MM (2014)
11. Lu, H., Rabbi, M., Chittaranjan, G., Frauendorfer, D., Mast, M., Campbell, A., Gatica-Perez, D., Choudhury, T.:Stresssense: detecting stress in unconstrained acoustic environments using smartphones. In: Proceedings of UbiCompp (2012)
12. Moreno, M., Jelenchick, L., Egan, K., Cox, E., Young, H., Gannon, K., Becker, T.: Feeling bad on facebook depression disclosures by college students on a social networking site. Depression Anxiety **28**(6), 447–455 (2011)
13. Nigam, K., Ghani, R.: Analyzing the effectiveness and applicability of co-training. In: Proceedings of CIKM (2000)
14. Sharma, N., Gedeon, T.: Objective measures, sensors and computational techniques for stress recognition and classification: a survey. Comput. Methods Programs Biomed. **108**(3), 1287–1301 (2012)
15. Shen, Y.-C., Kuo, T.-T., Yeh, I.-N., Chen, T.-T., Lin, S.-D.: Exploiting temporal information in a two-stage classification framework for content-based depression detection. In: Pei, J., Tseng, V.S., Cao, L., Motoda, H., Xu, G. (eds.) PAKDD 2013, Part I. LNCS, vol. 7818, pp. 276–288. Springer, Heidelberg (2013)
16. Su, H., Zheng, K., Wang, H., Huang, J., Zhou, X.: Calibrating trajectory data for similarity-based analysis. In: Proceedings of SIGMOD (2013)
17. Wang, X., Jia, J., Liao, H., Cai, L.: Affective image colorization. J. Comput. Sci. Technol. **27**(6), 1119–1128 (2012)

18. Wang, X., Jia, J., Yin, J., Cai, L.: Interpretable aesthetic features for affective image classification. In: Proceedings of ICIP (2013)
19. Xue, Y., Li, Q., Jin, L., Feng, L., Clifton, D.A., Clifford, G.D.: Detecting adolescent psychological pressures from micro-blog. In: Zhang, Y., Yao, G., He, J., Wang, L., Smalheiser, N.R., Yin, X. (eds.) HIS 2014. LNCS, vol. 8423, pp. 83–94. Springer, Heidelberg (2014)
20. Zheng, Y., Li, Q., Chen, Y., Xie, X., Ma, W.: Understanding mobility based on GPS data. In: Proceedings of UbiCompp (2008)

Collaborative Learning Team Formation: A Cognitive Modeling Perspective

Yuping Liu[1], Qi Liu[1], Runze Wu[1], Enhong Chen[1(✉)], Yu Su[2],
Zhigang Chen[1], and Guoping Hu[3]

[1] University of Science and Technology of China, Hefei 230026, China
{liuyup,wrz179}@mail.ustc.edu.cn,
{qiliuql,cheneh}@ustc.edu.cn, zgchen@iflytek.com
[2] Anhui University, Hefei, China
yusu@iflytek.com
[3] Anhui USTC IFLYTEK Co., Ltd., Hefei, China
gphu@iflytek.com

Abstract. With a number of students, the purpose of collaborative learning is to assign these students to the right teams so that the promotion of skills of each team member can be facilitated. Although some team formation solutions have been proposed, the problem of extracting more effective features to describe the skill proficiency of students for better collaborative learning is still open. To that end, we provide a focused study on exploiting cognitive diagnosis to model students' skill proficiency for team formation. Specifically, we design a two-stage framework. First, we propose a cognitive diagnosis model SDINA, which can automatically quantify students' skill proficiency in continuous values. Then, given two different objectives, we propose corresponding algorithms to form collaborative learning teams based on the cognitive modeling results of SDINA. Finally, extensive experiments demonstrate that SDINA could model the students' skill proficiency more precisely and the proposed algorithms can help generate collaborative learning teams more effectively.

1 Introduction

Collaborative learning is the instructional use of small heterogeneous group of students who team together (e.g., 5 to 8 students [10]) to work on a structured activity. Over the past decades, many researches have confirmed the effectiveness of this type of learning [23]. By maximizing the promotion of skills of students, collaborative learning can not only help stdd,ents exhibit higher academic achievement, but also can reduce the workload of instructors [22].

Indeed, the success of collaborative learning can be only guaranteed by assigning each student to the right team. Generally speaking, two types of solutions, based on manual decisions or automatical algorithms, have been studied for this team formation problem. In manual approaches, students may select their own teammates [11], or teams are assigned by instructors [20]. Unfortunately, by

S.B. Navathe et al. (Eds.): DASFAA 2016, Part II, LNCS 9643, pp. 383–400, 2016.
DOI: 10.1007/978-3-319-32049-6_24

self-selecting best students tend to cluster and leave weak ones to shift for themselves. By instructors-assigning, it is almost impossible for instructors to assign all students effectively. Thus, it is necessary to form learning teams automatically. Based on students' specific characteristics, some researchers focus on generating heterogenous teams which mix students of different levels [14], while others try to form teams which can quantize and maximize the gain of students [1]. Usually, characteristics of students in these studies are directly extracted from the biographical data, simple performance attributes or personality traits [27]. Meanwhile, without the real-world data, simulated values on these characteristics are used to design the team formation algorithms [2,12,13,18,27]. In spite of the importance of the existing research, features in these studies are too simple to capture students' skill proficiency very well, and thus, the performance of the corresponding team formation solutions could be further improved. Actually, one of the best ways to get the skill proficiency of students is to model the cognitive information of them, e.g., from their performance in the exams [4,8]. However, there are still two challenges to be addressed for exploiting cognitive diagnosis. Firstly, how to precisely quantify the skill proficiency of students? Secondly, how to design the appropriate algorithms to automatically get collaborative learning teams based on this type of feature?

To conquer these two challenges, we propose a two-stage framework to apply cognitive diagnosis for collaborative learning team formation. Specifically, in the first stage we propose a novel cognitive diagnosis model Soft-DINA (SDINA). Compared to the existing diagnosis model DINA [4,21], which quantifies the students' skill proficiency in binary values (either 0 or 1), SDINA is able to model students with continuous values. Then, the output of SDINA is further exploited to generate collaborative learning teams in the second stage. Following the views that students in the same team should be diverse and the improvement of each student should be maximized, we consider two optimization objectives – dissimilarity based objective and gain based objective, and we propose effective algorithms to generate collaborative learning teams for each of these objectives. Finally, the results of extensive experiments demonstrate that (1) SDINA could model the students' skill proficiency by predicting their performance (i.e., exam scores) more precisely, and (2) the proposed algorithms can help generate collaborative learning teams effectively under several evaluation metrics. The main contributions of this paper could be summarized as:

- To the best of our knowledge, this is the first comprehensive attempt for the problem of collaborative learning team formation by introducing cognitive diagnosis to extract features of students' skill proficiency.
- We propose a novel cognitive diagnosis model SDINA, which improves existing model DINA. SDINA automatically quantifies students' skill proficiency in continuous values for more accurate analysis of students.
- Given students' skill proficiency, we propose two objectives following the existing research achievements, and then we design effective algorithms to generate collaborative learning teams for each of these objectives.

2   Related Work

In this section, we will introduce the related studies in two categories.

2.1   Student Modeling

Here we focus on team formation-oriented student modeling methods. In traditional collaborative team formation problems, the students were modeled by features extracted directly from biographical data like age, gender, or some simple performance attributes like grades, self-evaluation, peer-assessment, or personality traits like learning styles [27]. For instance, Hwang et al. [15] considered the number of already known concepts and scores of a pre-test to model students.

Though few of existing team formation studies explored examination records, this kind of data has been widely used for other student modeling tasks, e.g., performance prediction. Many data mining efforts have been conducted, for instance, matrix factorization (MF) technique [19] has been adopted by considering student as user, problem as item, and student's score on a problem as rating. E.g., Toscher et al. [25] utilized singular value decomposition (SVD) to model and predict students. But, latent factors of students inferred by MF are unexplainable which limits the applicability of MF in scenarios where the explanation of skills need to be specified.

To model students with examination data in an interpretative way, psychometricians in education psychology have developed a series of cognitive diagnosis models (CDMs) [8]. By capturing the students' cognitive characteristics, CDMs can predict students' performance and obtain targeted remedy plan for each student. Generally speaking, there are two main categories of CDMs: continuous ones and discrete ones. The fundamental continuous CDMs are *item response theory (IRT)* models [9], which characterize a student by a continuous variable, i.e., latent trait. However, IRT is unable to get the latent cognitive character like students' skill proficiency. For discrete CDMs, the basic method is *deterministic inputs, noisy "and" gate model (DINA)* [4,21]. DINA describes a student by a latent binary vector variable which denotes whether she masters the skills required by the exam or not. And the specific relationship between problems and skills is a prior knowledge given by education experts (e.g. the exam designer). E.g., based on fuzzy system, Runze Wu et al. [26] proposed a solution for cognitive examinee modeling from both objective and subjective problems.

2.2   Team Formation

Existing studies on collaborative learning team formation can be broadly split into two basic types. The first type focuses on forming heterogenous or homogeneous teams considering multiple student characteristics. E.g., authors in [18] designed an algorithm named SPOS to form heterogeneous teams. Similarly, the fuzzy c-means and random selection algorithm were used in [2] for formulating homogenous and heterogeneous teams. Unfortunately, all these works implemented their algorithms based on simulated student characteristics. The second type focuses on forming teams which can maximize the gain of students. E.g., Rakesh *et al.* [1] proposed a framework for grouping students in order to maximize the overall gain of students. Given specific objectives, there are also researches which formed teams through optimization methods. E.g.,Virginia *et al.* [27] proposed a deterministic crowing evolutionary algorithm to form teams. However, few of the existing team formations exploit cognitive diagnosis results as features to better model students. Therefore, the problem of how to apply this type of feature to design more suitable team formation solutions is still open.

Besides collaborative learning team formation, the traditional team formation problem usually focused on forming a team from a large set of candidates of experts such that the resulting team is best suited to perform the assignment. This kind of team formation entails set-cover and is modeled by experts' skills and their collaboration network [16]. In contrast, the collaborative team formation problem is a partition problem rather than a set-cover one.

3   Collaborative Learning Team Formation

We will introduce details of our two-stage framework in this section. The flowchart is shown in Fig. 1. Given students' examination results, we propose a cognitive diagnosis model SDINA to automatically quantify the skill proficiency of students. Then, we propose two objectives with heuristic algorithms to form collaborative learning teams which can facilitate the promotion of all students.

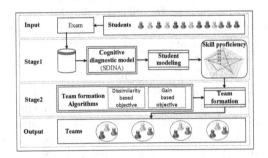

Fig. 1. Our two-stage framework.

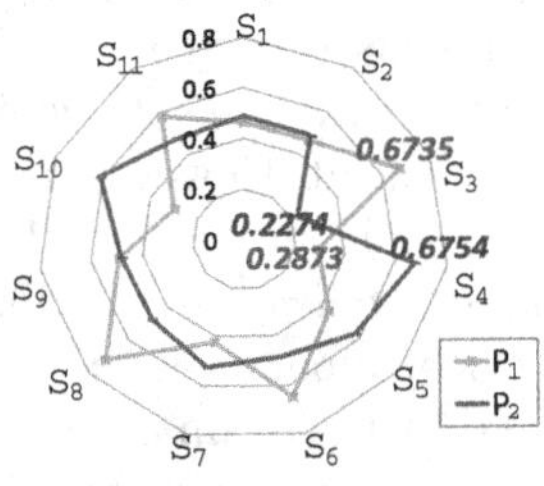

Fig. 2. Example of students' proficiency in skills S_1 to S_{11}.

3.1   Cognitive Modeling of Students' Skill Proficiency

In the following, we assume there are U students $\mathcal{P} = \{P_1, P_2, \ldots, P_U\}$, participating in the same course, e.g., math. We also assume $\mathcal{S} = \{S_1, S_2, \ldots, S_V\}$ to be a universe of V skills in this course, e.g., skills in a math course may include math concepts like set, formulas like computing sphere volume, process skills like calculation or induction. After cognitive modeling, each P_i will be associated with a vector of skill proficiency $\alpha_i = \{\alpha_{i1}, \alpha_{i2}, \ldots, \alpha_{iV}\}$, where α_{ij} ranges from 0 to 1 and represents that P_i' proficiency in S_j is α_{ij}. We summarize the proficiency of all students in all skills as $\alpha = \{\alpha_1, \alpha_2, \ldots, \alpha_U\}$, which need to be evaluated according to an examination. Actually, Fig. 2 shows the proficiency of 2 real-world students in skills S_1 to S_{11}, which is the output by our SDINA. Although the two students have similar average abilities (about 0.5), there are still distinctively differences in their proficiency in certain skills. If these two students are put into a same team, they can learn from others' strong points to offset their own weakness. We will first briefly review the traditional cognitive diagnosis model DINA, and then show the way to get α by our SDINA.

DINA Review. *DINA* [4] *(the deterministic inputs, noisy and gate model)* assumes that each problem in an exam is involved with multiple predefined skills (tagged by education experts in advance), and then it characterizes a student by a binary vector variable, which denotes whether she has mastered the skills required in the exam. Specifically, for an exam designed to assess V skills of students, given a problem l, a student i, we observe a dichotomous response which is a binary variable R_{il} with a value in $\{0, 1\}$. The response indicates the correctness of the answer provided by the student i to the problem l, i.e., 1 represents true and 0 represents false. Then, DINA is defined as:

$$P(R_{il} = 1|\alpha_i, s_l, g_l) = (1 - s_l)^{\eta_{il}} g_l^{1-\eta_{il}}. \tag{1}$$

Here, student i is characterized by a latent binary vector variable $\alpha_i = (\alpha_{i1}, \alpha_{i2}, \ldots, \alpha_{iV})$, i.e., $\alpha_{ij} = 1$ represents that student i has mastered skill j and vice versa, and problem l is characterized by two parameters: s_l represents carelessness or slipping; g_l is a guessing parameter. η_{il} is a latent variable that indicates whether student i is able to solve the problem l, and is defined as $\eta_{il} = \prod_{j=1}^{V} \alpha_{ij}^{q_{lj}}$ where q_{lj} indicates whether problem l requires skill j. It means student i is disable to solve problem l unless all of the skills required for problem l have already been mastered by her. As for parameter estimation, we could maximize the marginalized likelihood of Eq. (1), which can be implemented using *EM* algorithm [3]. Then, with the estimated parameters $\hat{s}_1, \hat{s}_2, \cdots, \hat{s}_Z$ and $\hat{g}_1, \hat{g}_2, \cdots, \hat{g}_Z$ (Z is the number of problems), α_i can be determined via maximizing the posterior probability given student i's response vector:

$$\begin{aligned} \hat{\alpha}_i &= argmax_\alpha P(\alpha|R_i) = argmax_\alpha L(R_i|\alpha, \hat{s}_l, \hat{g}_l)P(\alpha) \\ &= argmax_\alpha L(R_i|\alpha, \hat{s}_l, \hat{g}_l) = argmax_\alpha \prod_{l=1}^{Z} P(R_i|\alpha, \hat{s}_l, \hat{g}_l), \end{aligned} \tag{2}$$

Thus, for an exam with V skills, α has 2^V possible patterns and each pattern is assumed to be with an equal prior distribution without loss of generality.

Soft-DINA. In DINA, binary skill vector α_i of each student i can be found by maximizing the posterior probability. Though it is intuitive, this binary representation of students' skill proficiency is too coarse to characterize the mastery degree (cognitive level) of students. For instance, a student with a mastery degree of 0.9 in a specific skill and a student with 0.6 may have the same binary value of 1 based on DINA, while the significant difference is missing. For the sake of keeping as much information of vectors of proficiency as possible, we propose another cognitive model *Soft-DINA* or *SDINA* for short, which is an improvement of DINA, to get vectors of proficiency with continuous values.

Specifically, considering the 2^V kinds of α from all zeros $(0, 0, \dots, 0)$ to all ones $(1, 1, \dots, 1)$ and given the responses of a student i, R_i, each of these α is involved with one posterior probability $P(\alpha|R_i)$, DINA only chooses the specific skill vector α which can maximize the posterior probability. To precisely measure the probability that the student i masters a specific skill j, we propose to consider the posterior probability from all the possible α. Formally, in SDINA, we redefine the estimated skill vector $\tilde{\alpha}_i$ and calculate the posterior probability that student i masters skill j as follows:

$$\tilde{\alpha}_{ij} = P(\alpha_{ij} = 1|R_i) = \frac{\sum_{\alpha_{xj}=1} P(\alpha_x|R_i)}{\sum_{x=1}^{2^V} P(\alpha_x|R_i)}$$
$$= \frac{\sum_{\alpha_{xj}=1} L(R_i|\alpha_x, \hat{s}_l, \hat{g}_l) P(\alpha_x)}{\sum_{x=1}^{2^V} P(\alpha_x|R_i)} = \frac{\sum_{\alpha_{xj}=1} \prod_{l=1}^{Z} L(R_{il}|\alpha_x, \hat{s}_l, \hat{g}_l) P(\alpha_x)}{\sum_{x=1}^{2^V} P(\alpha_x|R_i)}, \tag{3}$$

where $x = 1, 2, \dots, 2^V$, represents the 2^V kinds of possible α_i, and the numerator part computes the probability of $\alpha_{ij} = 1$ in these 2^V kinds of possible α_x. To simplify the formulation, we also assume each α_x has an equal prior probability, then the equation above can be rewritten as follows:

$$\tilde{\alpha}_{ij} = \frac{\sum_{\alpha_{xj}=1} \prod_{l=1}^{Z} L(R_{il}|\alpha_x, \hat{s}_l, \hat{g}_l)}{\sum_{x=1}^{2^V} \prod_{l=1}^{Z} L(R_{il}|\alpha_x, \hat{s}_l, \hat{g}_l)}. \tag{4}$$

In this way, we get $\tilde{\alpha}_i = (\tilde{\alpha}_{i1}, \tilde{\alpha}_{i2}, \dots, \tilde{\alpha}_{iV})$, a vector of continuous values between 0 and 1, where $\tilde{\alpha}_{ij}$ represents student i's proficiency(i.e. probability) in skill j. That is to say, $\tilde{\alpha}_{ij} = 1$ means i has fully mastered skill j, and $\tilde{\alpha}_{ij} = 0$ means i has not mastered skill j at all.

Note that, although the skills accessed by each problem are manually labeled, it is feasible and commonly used in pedagogy [8]. In fact, to construct an examination, designers must clearly delineate the assement purpose, specically describe what skills are measured and develop proper assessment tasks[1].

3.2   Collaborative Learning Team Formation

In this subsection, we show the way to form teams based on α. Assume $\mathcal{G} = \{G_1, G_2, \dots, G_M\}$ to be a set of M teams. We will put students into teams, with two basic constraints. Firstly, only one team is assigned to a student, making sure that a student only belongs to one team. Secondly, the team size is better to be equal, with a difference of no more than one student, to ensure fairness and

---

[1] There are also studies about the automatic labeling of skills [7], which is beyond the scope of our research.

team balance. Formally, given the students' skill proficiency, the collaborative learning team formation problem can be formulated as follows.

Problem 1 *(Collaborative Learning Team Formation): Given a set of students* $\mathcal{P}$*, each student* P_i*'s proficiency* $\alpha_i = (\alpha_{i1}, \alpha_{i2}, \ldots, \alpha_{iV})$*, and a set of teams* $\mathcal{G} = \{G_1, G_2, \ldots, G_M\}$*. Under the following constraints,*

$$
\begin{aligned}
G_{k_1} \cap G_{k_2} &= \emptyset, \\
-1 \leq |G_{k_1}| - |G_{k_2}| &\leq 1, \\
k_1, k_2 = 1, 2, \ldots, M, &\ k_1 \neq k_2
\end{aligned}
\tag{5}
$$

assign every student P_i *to a team* G_k*,* $P_i \subseteq G_k$*,* $k = 1, 2, \ldots, M$*. In order that, the promotion of all students in these* V *skills can be facilitated.*

With this definition, we should first clarify the measurements/objectives of a good team, and then, the specific algorithms to generate effective collaborative learning teams in terms of these objectives could be designed. Generally, there are two types of different objectives in existing studies, i.e., the skill proficiency of students in the same team should be heterogeneous [14] and the improvement of each student should be maximized [1]. In terms of these objectives and our extracted feature, we propose a *dissimilarity based objective* to form teams by maximizing the average dissimilarity of students within a team and a *gain based objective* to form teams by maximizing the average gain of students, respectively.

Dissimilarity Based Objective. According to [13], in a reasonably heterogeneous group student-scores reveal a combination of low, average and high student-scores. However, this measurement is limited to 3 discrete classes of only one attribute value (student-score). Indeed, a better mechanism is to use continuous values of several attributes, e.g., the continuous value of students' proficiency in several skills. Inspired by the heterogeneity definition in [12], we also use the average dissimilarity between team members as the metric of heterogeneous degree. Without loss of generality, the difference between P_{i_1} and P_{i_2} in the proficiency of S_j, $D_j(P_{i_1}, P_{i_2})$, is defined as $|\alpha_{i_1 j} - \alpha_{i_2 j}|$. Thus, the heterogeneity of G_k consisting of N students with respect to S_j, is defined as

$$
HG_k(S_j) = \sum\nolimits_{i=2}^{N} D_j(P_i, P_{i-1}).
\tag{6}
$$

Here, the students in G_k have been sorted by values of skill proficiency in S_j, i.e., $\alpha_{1j} \leq \alpha_{2j} \leq \ldots \leq \alpha_{Nj}$. Then, the heterogeneity of G_k is computed as

$$
HG_k = \sum\nolimits_{j=1}^{V} HG_k(S_j).
\tag{7}
$$

One step further, the *heterogeneity of the solution* $\mathcal{G}$ is the average of the heterogeneity of all teams in the solution, i.e.,

$$
HG(\mathcal{G}) = \frac{\sum_{G_k \in \mathcal{G}} HG_k}{|\mathcal{G}|}.
\tag{8}
$$

Since not only heterogeneity but also the team balance, i.e., the difference of the heterogeneity among the teams, should be considered for the quality of the team formation solutions [18], we also define the *balance of the solution* $\mathcal{G}$ as

$$B(\mathcal{G}) = Variance(\{HG_k | \forall G_k \in \mathcal{G}\}). \tag{9}$$

That is, the teams in each solution should be as balanced as possible to ensure the fairness. Overall, the higher the solution heterogeneity is and the lower the solution balance is, the better the team formation result is.

Given this dissimilarity based objective, consisting of both solution heterogeneity $HG(\mathcal{G})$ and solution balance $B(\mathcal{G})$, we utilize the idea of clustering to solve the team formation problem. Intuitively, students could be first clustered using clustering algorithm, e.g., k-means [17], where features are their skill proficiency vectors. Then students of same cluster will be assigned to different teams. However,clusters under the classical k-means settings are often of different size, and this will have a negative effect on the team's heterogeneity. For instance, if there is a cluster with a very large size, which is bigger than the number of teams, according to the pigeon-hole principle , at least two students in the same cluster will be assigned to the same team. To address this problem, we think of an improved clustering method called *uniform k-means* to get uniform clusters with the same size. More specifically, in the process of k-means, the number of clusters is set to be $\lfloor \frac{U}{M} \rfloor$. Then, for every object(student), after calculating the distance between it and the center of every cluster, this student will be put into cluster which is not merely with shortest distance but also is not full, i.e., size of this cluster is no larger than M. After this, students are equally divided into $\lfloor \frac{U}{M} \rfloor$ clusters. Next, students in same cluster should be assigned into M different candidate teams. In this way, as students with similar skill proficiency will be put into different teams, the dissimilarity based objective can be easily achieved. The *U*niform *K*-means *B*ased algorithm (*UKB*), is summarized in Algorithm 1.

Algorithm 1. UKB: the uniform k-means based algorithm

Require:
    The set of U students $\mathcal{P} = \{P_1, P_2, \ldots, P_U\}$; The number of teams M;
Ensure:
    The set of teams $\mathcal{G} = \{G_1, G_2, \ldots, G_M\}$;
  1: Divide U students into $\lfloor \frac{U}{M} \rfloor$ clusters using uniform k-means;
  2: Determine the size of teams, each one with student number of $\lfloor \frac{U}{M} \rfloor$ or $\lfloor \frac{U}{M} \rfloor + 1$;
  3: Calculate the number of students that every team still needs;
  4: **for** each cluster c **do**
  5:     **while** c is not empty **do**
  6:         Put one student into every not-full team;
  7:         Calculate the number of students that every team still needs;
  8:     **end while**
  9: **end for**
 10: **return** $\mathcal{G}$;

Gain Based Objective. In addition to the dissimilarity based objective, another intuitive approach of measuring the quality of a team is to quantize the promotion of every student. Inspired by [1], we define a gain function to measure the promotion of students in terms of their skill proficiency.

In collaborative learning teams, the students can promote their skills through mutual exchanges and emulations. As a general rule, there are two factors which can influence the students' promotion. The first factor is the proficiency of each student, i.e., a student with higher proficiency is easier to promote than a student with lower proficiency. Another factor is the gap between her and the other students in the team, i.e., a student who collaborates with more capable team-mate can get more knowledge. According to these facts, we define a student's promotion in a skill as follows.

Suppose there is a non-empty team G_k with N students $G_k = \{P_1, P_2, \ldots, P_N\}$. G_k has a vector of maximum proficiency in every skill $a_k = \{a_{k1}, a_{k2}, \ldots, a_{kV}\}$, a_{kj} is the *maximum proficiency* in S_j among the students in G_k, i.e., $a_{kj} = MAX(\alpha_{1j}, \alpha_{2j}, \ldots, \alpha_{Nj})$. Then the *leader* of G_k in S_j, namely L_{kj} is the student with maximum proficiency in S_j,

$$L_{kj} = \{P_i | \alpha_{ij} = a_{kj}, i = 1, 2, \ldots, N\}. \tag{10}$$

Now, we put a new student P_i into G_k, if P_i is not the leader in S_j, i.e., $\alpha_{ij} < a_{kj}$, then P_i's *promotion* in S_j, $Q_j(P_i, G_k)$, is defined as:

$$Q_j(P_i, G_k) = (a_{kj} - \alpha_{ij}) \cdot \alpha_{ij}. \tag{11}$$

Here, $(a_{kj} - \alpha_{ij})$ is as the gap between P_i and L_{kj}, which is always positive. The definition is in conformity with the actual situation. For instance, there is a team with a leader who has a proficiency of 0.9 in a certain skill, suppose we put three new students A, B, C with proficiency of 0.2, 0.5, 0.8 respectively in the team. According to our definition, the promotions of them will be 0.14, 0.20, 0.08. A gets a small promotion of 0.14 because A has a low proficiency which brings bad influence for the promotion, C only gets 0.08 because there is only a little gap between C and the leader. Only B gets a big promotion of 0.20 due to B's higher proficiency and the bigger gap between B and the leader. For simplicity, we ignore the leader's promotion in her leading skill. Then, we define $Q(P_i, G_k)$, the *gain* of P_i as the overall promotion in every skill,

$$\begin{aligned} Q(P_i, G_k) &= \sum_{j=1}^{V} Q_j(P_i, G_k) \\ &= (a_k - \alpha_i) \cdot \alpha_i^T. \end{aligned} \tag{12}$$

Next, the average gain of G_k will be:

$$Q(G_k) = \frac{\sum_{P_i \in G_k} Q(P_i, G_k)}{|G_k|}. \tag{13}$$

Here, $|G_k|$ is the number of students in G_k. Finally, the *average gain of a solution* $\mathcal{G}$, $Q(\mathcal{G})$, can be defined as the average gain of all the teams,

$$Q(\mathcal{G}) = \frac{\sum_{G_k \in \mathcal{G}} Q(G_k)}{|\mathcal{G}|}. \tag{14}$$

We use the solution gain $Q(\mathcal{G})$ as the evaluative criterion for the quality of team formation solutions. We also propose a team formation algorithm which consists of two steps: First, the leader in each skill is chosen for each team; Then, all the non-leader students are put into teams according to their gains.

Specifically, $Q(\mathcal{G})$ is a monotone-increasing function of a_{kj}, so for maximizing $Q(\mathcal{G})$, the leader in each skill for each team should have proficiency as greater as possible. Given a set of U students and the number of teams M, the leader-choosing process is as follows. For each skill S_j, firstly we pick out M students $\mathcal{P} = \{P_1, P_2, \ldots, P_M\}$ with maximum proficiency. Secondly, for every student $P_i \in \mathcal{P}$, if P_i has been a leader in a team, then P_i will be the leader in S_j of this team; Otherwise, we choose a team G_k which still has no leader in S_j, and if G_k is not full, then P_i will be the leader in S_j of G_k, or else, a student in G_k with the maximum proficiency in S_j will be picked out as the leader.

After the leaders have been chosen, we show the way to put all non-leader students into teams. Assume all teams have the same size of λ, and the λ students in team G_k are presented by $P_{ki}, i = 1, 2, \ldots, \lambda$. Then $Q(\mathcal{G})$ will be

$$
\begin{aligned}
Q(\mathcal{G}) &= \frac{\sum_{G_k \in \mathcal{G}} Q(G_k)}{|\mathcal{G}|} \\
&= \frac{1}{|\mathcal{G}| \cdot \lambda} \cdot \sum_{G_k \in \mathcal{G}} \sum_{i=1}^{\lambda} Q(P_{ki}, G_k).
\end{aligned}
\tag{15}
$$

Obviously, $Q(\mathcal{G})$ will only be determined by $\sum_{G_k \in \mathcal{G}} \sum_{i=1}^{\lambda} Q(P_{ki}, G_k)$ since $\frac{1}{|\mathcal{G}| \cdot \lambda}$ is a constant. Also, $Q(\mathcal{G})$ increases with $Q(P_{ki}, G_k)$. As one student can only belong to one team, it is naturally to put P_i to G_k which can maximize $Q(P_i, G_k)$. In addition, since size of teams is limited, students with higher proficiency should be put into teams in priority. However, such team formation result violates the principle of heterogeneity because students with higher proficiency tend to be put into teams with higher maximum proficiency. To avoid this unbalanced result, we propose an algorithm which takes both the gain of students and the average level of teams into consideration, to get balanced teams.

To be specific, there are two factors determining the selection of groups for non-leader students. One is the gain of this student, another is the average level of the team. Here, we define LP_i, the level of a student P_i, as the average of P_i's proficiency in every skill, and LG_k, the *average level of a team* G_k, as the average level of all the students in team G_k,

$$
LP_i = \frac{\sum_{j=1}^{V} \alpha_{ij}}{V}, \quad LG_k = \frac{\sum_{P_i \in G_k} LP_i}{|G_k|}.
$$

Then we define a *balanced gain vector* $BG_i = \{BG_{i1}, BG_{i2}, \ldots, BG_{iM}\}$, where BG_{ik} is called the *balanced gain* of P_i if putting her to G_k,

$$
BG_{ik} = \frac{Q(P_i, G_k)}{LG_k}.
\tag{16}
$$

With the above definition, the entire process of *Balanced Gain Based algorithm* (*BGB*), is shown in Algorithm 2. In summary, we first choose the leaders

---

Algorithm 2. BGB: the balanced gain based algorithm

Require:
    The set of U students $\mathcal{P} = \{P_1, P_2, \ldots, P_U\}$; The number of teams M;
Ensure:
    The set of teams $\mathcal{G} = \{G_1, G_2, \ldots, G_M\}$;
 1: Determine the size of teams, each one with student number of $\lfloor \frac{U}{M} \rfloor$ or $\lfloor \frac{U}{M} \rfloor + 1$;
 2: Select the leaders in every skill for each team.
 3: Sort all the non-leader students by their levels LP in descending order;
 4: **for** each sorted students P_i **do**
 5:     Calculate the balanced gain vector BG_i;
 6:     Put student P_i in a not-full team G_k with maximum BG_{ik};
 7: **end for**
 8: **return** $\mathcal{G}$;

---

in each skill for each team, and then sort all the non-leader students by their levels LP in descending order. Next, for each sorted student P_i we put her into a not-full team G_k with maximum BG_{ik}. In this way, the higher level students will be put into relatively low level teams and vice versa, getting her a relatively high gain and keeping the heterogeneity of the teams, simultaneously.

4   Experiments

Firstly, we use the prediction of students' scores to evaluate the effectiveness of SDINA. Secondly, we make an expert evaluation to explore the effectiveness of features extracted by SDINA. At last, we evaluate the performance of our proposed team formation algorithms from various aspects.

4.1   Experimental Setup

Our experiments are conducted on three real-world datasets and two simulated ones. The real-world datasets contain two real private datasets and a public online dataset. The public dataset is Tatsuoka's fraction subtraction dataset [5], consisting of scores from middle school students on fraction subtraction problems. The two private datasets[2] are from two final math exams for high school students. Each of these three datasets is represented by a score matrix and also has multiple predefined skills. We denote these datasets as *FrcSub*, *Math1* and *Math2*. The brief summary is shown in Table 1. Figure 3 gives a brief preview of these datasets, where each column for each subfigure stands for a problem and each row for a student. The black element means the student is wrong in the problem, while white one means right. The two simulated datasets are made up of 500 and 1,000 students with 10 features[3]. The value of each feature is

---

[2] They will be publicly available after the paper acceptance.
[3] Unlike the real-world datasets, the simulated ones only consist of students with values between 0 and 1 on some features rather than students' test scores.

generated by random sampling from a uniform distribution of 0 to 1. We denote these two datasets as *SiData1* and *SiData2*.

All experiments are implemented by Matlab on a Core i5 3.10 Ghz machine with Window 7 and 4 GB memory.

Table 1. Datasets Summary.

Dataset	#Student	#Skill	#Problem
FrcSub	536	8	20
Math1	4,206	11	12
Math2	3,907	16	12

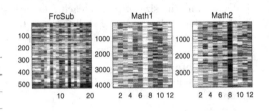

Fig. 3. The preview of the datasets.

4.2 Evaluation on Student Modeling

To demonstrate the effectiveness of *SDINA* on modeling students' skill proficiency, we conduct experiments of predicting students' scores (right as 1 and wrong as 0) on each problem. We perform 5-fold cross validation on the real-world datasets, i.e., 80 % of the students are randomly selected for training while the rest for testing. We consider two baseline approaches:

- *DINA* [4]: a cognitive diagnosis model which is detailed in Sect. 3.1.
- *PMF* [19]: a latent factor model, widely used in recommending system.

We record the best performance of each method by tuning their parameters, e.g., the latent dimension of PMF is set to be 10. As this task is actually a binary classification problem, *Accuracy* and *F1-measure* are used as evaluation metrics.

The experimental results are shown in Fig. 4. SDINA performs better than DINA and PMF in both Accuracy and F1 over all datasets. In Accuracy DINA performs better than PMF but in F1 PMF performs better than DINA. As a cognitive diagnosis model, SDINA is effective in modeling students' skill proficiency.

4.3 Evaluation on Feature Selection

After evaluating SDINA's effectiveness on student modeling, to demonstrate SDINA's effectiveness on team formation, first the performance of three different kinds of features on team formation are evaluated by educational experts, then teams based on these features are visually displayed for better illustration.

Specifically, we compare reasonablity of teams formed with the same algorithm based on three kinds of features, i.e., the continuous skill proficiency inferred by SDINA, the binary skill proficiency inferred by DINA and the raw examination scores[4]. Since conducting large-scale in-classroom experiments are

---

[4] The latent factor getting by PMF has not been used here since it's unexplainable.

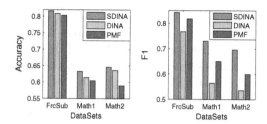

Table 2. Gold standard evaluation.

Feature	HR
Raw score	0.066
DINA	0.267
SDINA	**0.667**

Fig. 4. Performance on score prediction.

currently impractical, we choose to collect a *Gold Standard* [24] to evaluate effectiveness of team formation by various features. As educational experts, teachers are obviously able to give a relatively fair and convincing assessment with decades of educating experience. We will first specify the process of collecting a gold standard and then compare the performance of three kinds of teams.

To construct a gold standard, we simulate the team formation on real dataset and then ask corresponding teachers who are familiar with the chosen students, to evaluate the effectiveness of different methods. Specifically speaking, due to the labour cost of manual assessment, we first randomly draw five classes with 283 students in total, then we form teams with a fixed size (e.g. 5) by three different features based on UKB. And for each class, we randomly draw three formed teams for each method, that is, nine teams will be chosen and randomly ordered. Subsequently, we ask teachers of the five classes to pick three most reasonable teams for each class out of their understanding of students. So, nine of the forty-five formed teams are chosen and regarded as most effective.

Taking the gold standard as ground truth, we compute hit rate (HR) [6] for each method. Here this metric measures how closely the output of a method is to the gold standard and is defined as $HR_i = \frac{|T_i \cap GS|}{|GS|}$. Here, HR_i is the HR of the ith team formation method, T_i represents the teams formed by the ith method and GS means the teams picked by the gold standard. As is shown in Table 2, the effectiveness of team formation by SDINA greatly outperforms DINA and Raw score methods by a quantitative and more accurate cognitive diagnosis, and at the meanwhile DINA obtains more satisfying results than Raw score method.

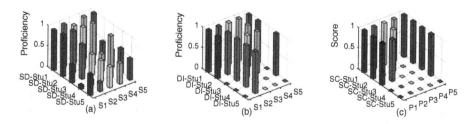

Fig. 5. An example of teams by using three kinds of features.

One step further, for better illustration about the difference between teams formed by three kinds of features, Fig. 5 shows a toy visualized example. We use UKB to form five-students teams on public dataset FrcSub. Figure 5(a) shows a team formed through continuous skill proficiency output by SDINA. Here, SD-Stu1 to SD-Stu5 represent five students and S1 to S5 represent five skills. Figure 5(b) shows a team formed through binary skill proficiency output by DINA. DI-Stu1 to DI-Stu5 represent students and S1 to S5 represent skills which are the same with Fig. 5(a). Figure 5(c) shows a team formed directly using students' score on every problem. P1 to P5 represent five problems respectively. Only five skills or problems are shown for better visualization. We can observe that team in Fig. 5(a) contains different levels of students, SD-Stu1 and SD-Stu2 are of highest ability. Others can promote skills following the lead of them. Since proficiency of students in each skill is also of great difference, students can promote their skills through learning from each other. Compared with other two kinds of features team formation using SDINA has better explanation. SDINA not only can support automatic team formation but also can provide guidance for manually forming teams.

4.4   Evaluation on Team Formation

In this subsection, we first fix features as continuous skill proficiency inferred by SDINA and evaluate performance of UKB and BGB on real-world datasets. Then, to demonstrate that UKB is not just effective with features of SDINA's output, we evaluate its performance on two simulated datasets. Quality of team formation solutions is evaluated with *Heterogeneity*, *Balance* and *Gain*, which are defined in Sect. 3.2. For Heterogeneity and Gain, higher value is better while for Balance, lower value is better. Our algorithms are compared with:

- *SPOS*: short for Semi-Pareto Optimal Set, which is proposed in [18].
- *RANDOM*: a standard random algorithm to form heterogeneity teams [2].

We should note that the algorithm in [1] is not chosen as a baseline, because it can only be applied to form teams for students with 1-dimensional ability.

Firstly, we perform team formation experiments on three real-world datasets and use the output of SDINA as input features for all the algorithms, to make sure grouping algorithms will be comparable. Figure 6 shows the experimental results. The subfigures in row 1 to row 3 represent three datasets and columns represent three measurements. The X-axis in each subfigure represents team size from 5 to 8 since the optimal team for collaborative learning should contain 5 to 8 students [10]. Please note that, if the student number is not evenly divisible by the team size, then actually team size here represents the basic student amount in every team, and there is at least one team which has one more student. The result shows the effectiveness of our proposed algorithms. In terms of the dissimilarity based objective, i.e., Heterogeneity and Balance, UKB outperforms the two baselines among all the team sizes in three datasets, and among the baselines, SPOS has a relatively good performance. Similarly, in terms of the gain based objective, BGB outperforms the two baselines in all cases.

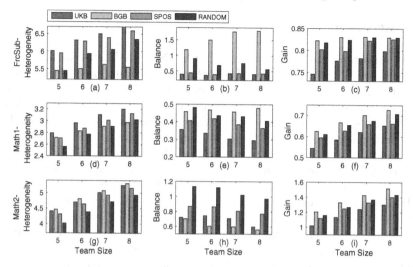

Fig. 6. Collaborative team formation performance on three real-world datasets.

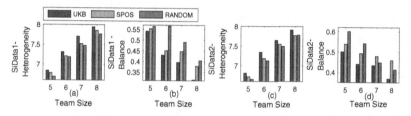

Fig. 7. Collaborative team formation performance on simulated datasets.

Table 3. Runtime(seconds).

Datasets	UKB	BGB	SPOS	RANDOM
FrcSub	1.936	0.174	1.811	**0.122**
Math1	595.2	**16.89**	750.3	22.7
Math2	664.2	**10.73**	620.4	19.5

Secondly, to demonstrate that UKB is not just effective with features of SDINA's output, we perform team formation experiments on datasets with simulated features. We can see from Fig. 7 that UKB has larger Heterogeneity and lower Balance among all team sizes on both datasets. We don't test BGB as it only focuses on features of students' skill proficiency.

Moreover, Table 3 shows runtime of each method to form five-students teams. BGB and RANDOM run much faster in general. For our methods UKB run faster than SPOS on Math1, BGB run faster than RANDOM on Math1 and Math2.

5 Discussion

The experimental results demonstrate that SDINA could better model students with continuous skill proficiency. Through more accurate analysis on students and following the existing research achievements, UKB and BGB could generate more effective collaborative learning teams for dissimilarity based objective and gain based objective. In the meanwhile, the team formation results are explainable, which makes the framework has more practical value.

Both stages of this general framework may be further improved. Firstly, we can employ SDINA on more data (e.g., the homework data) for feature extraction. Secondly, relationship between these two objectives could be studied and maybe the trade-off between them can be researched. Optimization methods should be tried to formulate an optimization problem for maximizing gain and heterogeneity, and minimizing balance. Thirdly, we plan to design more efficient solutions than UKB and we would like to consider more influence factors to get a more reasonable definition of students' promotion in BGB. Finally, we plan to apply this theoretical research in the real-world teaching and learning, e.g., we already served high schools where we collected the data. Indeed, given that modeling students' cognitive skills for collaborative learning has largely been neglected, there are many research directions remain to be explored.

6 Conclusion

In this paper, we designed a two-stage framework to exploit cognitive diagnosis for collaborative learning team formation. Firstly, we proposed a cognitive diagnosis model SDINA, which can automatically quantify students' skill proficiency in continuous values. Secondly, we proposed two objectives, the dissimilarity based objective and the gain based objective with heuristic algorithms to solve the team formation problem. At last, extensive experiments on several datasets demonstrated that our SDINA could model the students' skill proficiency more precisely and the proposed algorithms can help generate collaborative learning teams more effectively. We hope this work could lead to more future studies.

Acknowledgements. This research was partially supported by grants from the National Science Foundation for Distinguished Young Scholars of China (Grant No. 61325010), the Natural Science Foundation of China (Grant No. 61403358) and the Science and Technology Program for Public Wellbeing (Grant No. 2013GS340302). Qi Liu gratefully acknowledges the support of the Youth Innovation Promotion Association of CAS and acknowledges the support of the CCF-Intel Young Faculty Researcher Program (YFRP).

References

1. Agrawal, R., Golshan, B., Terzi, E.: Grouping students in educational settings. In: SIGKDD, pp. 1017–1026. ACM (2014)
2. Christodoulopoulos, C.E., Papanikolaou, K.: Investigation of group formation using low complexity algorithms. In: Proceeding of PING, Workshop, pp. 57–60 (2007)
3. De La Torre, J.: Dina model and parameter estimation: a didactic. J. Educ. Behav. Stat. **34**(1), 115–130 (2009)
4. De La Torre, J.: The generalized dina model framework. Psychometrika **76**(2), 179–199 (2011)
5. DeCarlo, L.T.: On the analysis of fraction subtraction data: the dina model, classification, latent class sizes, and the q-matrix. APM (2010)
6. Deshpande, M., Karypis, G.: Item-based top-N recommendation algorithms. ACM Trans. Inf. Syst. (TOIS) **22**(1), 143–177 (2004)
7. Desmarais, M.C.: Mapping question items to skills with non-negative matrix factorization. ACM SIGKDD Explor. Newsl. **13**(2), 30–36 (2012)
8. DiBello, L.V., Roussos, L.A., Stout, W.: 31a review of cognitively diagnostic assessment and a summary of psychometric models. Handb. Stat. **26**, 979–1030 (2006)
9. Embretson, S.E., Reise, S.P.: Item response theory for psychologists. Psychology Press, New York (2013)
10. Gall, M.D., Gall, J.P.: The discussion method. The psychology of teaching methods, (75 ppt 1), pp. 166–216 (1976)
11. Gibbs, G.: Learning in teams: a tutor guide. Oxford Centre for Staff and Learning Development (1995)
12. Gogoulou, A., Gouli, E., Boas, G., Liakou, E., Grigoriadou, M.: Forming homogeneous, heterogeneous and mixed groups of learners. In: Proceeding ICUM, pp. 33–40 (2007)
13. Graf, S., Bekele, R.: Forming heterogeneous groups for intelligent collaborative learning systems with ant colony optimization. In: Ikeda, M., Ashley, K.D., Chan, T.-W. (eds.) ITS 2006. LNCS, vol. 4053, pp. 217–226. Springer, Heidelberg (2006)
14. Hooper, S., Hannafin, M.J.: Cooperative cbi: The effects of heterogeneous versus homogeneous grouping on the learning of progressively complex concepts. J. Educ. Comput. Res. **4**(4), 413–424 (1988)
15. Hwang, G.-J., Yin, P.-Y., Hwang, C.-W., Tsai, C.-C., et al.: An enhanced genetic approach to composing cooperative learning groups for multiple grouping criteria. Educ. Technol. Soc. **11**(1), 148–167 (2008)
16. Lappas, T., Liu, K., Terzi, E.: Finding a team of experts in social networks. In: Proceedings of the 15th SIGKDD, pp. 467–476. ACM (2009)
17. Li, Q., Wang, P., Wang, W., Hu, H., Li, Z., Li, J.: An efficient K-means Clustering Algorithm on MapReduce. In: Bhowmick, S.S., Dyreson, C.E., Jensen, C.S., Lee, M.L., Muliantara, A., Thalheim, B. (eds.) DASFAA 2014, Part I. LNCS, vol. 8421, pp. 357–371. Springer, Heidelberg (2014)
18. Mahdi, B., Fattaneh, T.: A semi-pareto optimal set based algorithm for grouping of students. In: ICELET, pp. 10–13. IEEE (2013)
19. Mnih, A., Salakhutdinov, R.: Probabilistic matrix factorization. In: Advances in neural information processing systems, pp. 1257–1264 (2007)
20. Ounnas, A., Davis, H., Millard, D.: A framework for semantic group formation. In: ICALT, pp. 34–38. IEEE (2008)
21. Ozaki, K.: Dina models for multiple-choice items with few parameters considering incorrect answers. In: APM (2015)

22. Slavin, R.E.: Cooperative learning: theory, research, and practice, vol. 14. Allyn and Bacon, Boston (1990)
23. Smith, K.A., Sheppard, S.D., Johnson, D.W., Johnson, R.T.: Pedagogies of engagement: classroom-based practices. JEE **94**(1), 87–101 (2005)
24. Štajner, T., Thomee, B., Popescu, A.-M., Pennacchiotti, M., Jaimes, A.: Automatic selection of social media responses to news. In: 19th ACM SIGKDD, pp. 50–58. ACM (2013)
25. Toscher, A., Jahrer, M.: Collaborative filtering applied to educational data mining. In: KDD Cupp (2010)
26. Wu, R., Liu, Q., Liu, Y., Chen, E., Su, Y., Chen, Z., Hu, G.: Cognitive modelling for predicting examinee performance. In: Proceedings of the 24th International Conference on Artificial Intelligence, pp. 1017–1024. AAAI Press (2015)
27. Yannibelli, V., Amandi, A.: A deterministic crowding evolutionary algorithm to form learning teams in a collaborative learning context. Expert Syst. Appl. **39**(10), 8584–8592 (2012)

Advanced Applications(1)

Popular Route Planning with Travel Cost Estimation

Huiping Liu, Cheqing Jin$^{(\boxtimes)}$, and Aoying Zhou

School of Computer Science and Software Engineering, Institute for
Data Science and Engineering, East China Normal University, Shanghai, China
hpliu@stu.ecnu.edu.cn,{cqjin,ayzhou}@sei.ecnu.edu.cn

Abstract. With the increasing number of GPS-equipped vehicles, more
and more trajectories are generated continuously, based on which some
urban applications become feasible, such as route planning. In general,
route planning aims at finding a path from source to destination to meet
some specific requirements, i.e., the minimal travel time, fee or fuel consumption. Especially, some users may prefer popular route that has been
travelled frequently. However, the existing work to find the popular route
does not consider how to estimate the travelling cost. In this paper, we
address this issue by devising a novel structure, called popular traverse
graph, to summarize historical trajectories. Based on which an efficient
route planning algorithm is proposed to search the popular route with
minimal travel cost. The extensive experimental reports show that our
method is both effective and efficient.

1 Introduction

Route planning is important not only for our daily life, but also for business map
engines like Google and Bing Maps [1–4]. Although the shortest/fastest paths are
used commonly, they may be insufficient in some situations, while the popular
route that refers to a path being travelled frequently is sometimes important.
For example, drivers who travel in an unfamiliar city may prefer a popular path
which may be safer with better traffic condition and road quality, and a taxi
passenger may want to travel along a popular path in case of a roundabout
trip. Moreover, people care more about the travel cost, i.e., how long it takes
or how much it costs. An accurate travel cost estimation will improve people's
satisfaction. In practice, there are several popular paths with different travel
cost from source to destination, so a popular route with the minimal travel cost
which saves resource consumption (such as time, money, fuel) is a better choice.

Route planning or driving direction planning has been studied in recent years
and some influential works have been published. [3] proposes a framework to find
out the practically fastest route at a given departure time based on a landmark
graph learned from a large number of historical taxi trajectories. However, the
fastest route is not always popular, some shortcuts may reduce the travel time
but increase the risk and uncertainty of a trip. The work performs a two-stage
routing algorithm based on the graph to find the fastest route. The first stage

S.B. Navathe et al. (Eds.): DASFAA 2016, Part II, LNCS 9643, pp. 403–418, 2016.
DOI: 10.1007/978-3-319-32049-6_25

is to find the rough route represented by a sequence of landmarks whose travel time can be estimated by their model and the second stage is to find a practically detailed fastest route in the road network based on speed constraints. But the travel time of the detailed fastest route may be different from the estimated travel time at the first stage. Since there may exist several different paths between two landmarks and the estimated travel time is just the mean travel time of all possible paths. In most cases, the travel time of the detailed fastest route is less than the estimated travel time, which will cause an inaccurate travel time estimation of the route. Moreover, the model proposed in [5] to estimate the travel time of a given path by using the optimal route concatenation cannot be applied to route planning directly.

In this paper, we aim at finding the popular route with minimal travel cost from source to destination and estimating the travel cost for this route. We propose a framework to achieve this goal. Firstly, we construct a popular traverse graph based on the historical trajectories, where each node is a popular location, and each edge is a popular route between two locations. Subsequently, for each popular route in this graph, we use the minimum description length (MDL) principle [6] to model the travel cost of the routes. Finally, based on the graph, given a source-destination pair and a leaving time, we find the fastest popular route in consideration of the optimal route concatenation [5] for an accurate travel cost estimation. The contributions of this paper are summarized below.

- We propose a novel structure, called popular traverse graph, from trajectories without road network information, which contains the popular routes between locations.
- We present a parameter-free method using the minimum description length (MDL) principle to model the travel cost on each popular route in the graph.
- We devise an efficient routing algorithm which combines optimal route concatenation with route planning on the popular traverse graph.
- We have conducted extensive experiments upon a real dataset of millions of trajectories generated by more than 10000 taxis over a month in Beijing. The results show that our method is both effective and efficient.

The remainder of the paper is organized as follows. Section 2 reviews the related work. Section 3 describes some preliminary knowledge. Section 4 illustrates the popular traverse graph, and a way to model the travel cost for each popular route. Section 5 details the routing algorithm. Section 6 reports the evaluation and a brief conclusion is given in Sect. 7.

2   Related Work

Route planning has been widely investigated in recent years, including popular route planning and the fastest route planning.

Popular Route Planning: [7,8] discover the top-k possible popular routes that sequentially pass given locations from historical uncertain trajectories. [9] studies how to discover the most popular route between any two locations. The authors introduce a transfer network model by exploiting intersections and the popularity of transfer nodes on the transfer network. They infer the most popular routes according to the turning probability of each intersection on the network. [10] tries to find the time period-based most frequent path. It firstly constructs a footmark graph which is used to calculate the frequencies of candidate paths, then they retrieve the most frequent path in arbitrary time periods specified by the users on this graph. All the above studies try to find the popular routes without considering the travel cost, whereas our focus is to find a popular route with the minimal travel cost.

The Fastest Route Planning: [11] proposes a fundamental algorithm (Dijkstra's algorithm) to find the shortest path between two nodes in a graph and the A^* algorithm proposed by [12] boosts the searching performance with a heuristic estimation. In [1,2,13], the authors study how to find the fastest path on a time-dependent graph. [14] computes the fastest path on a road network by considering speed and driving patterns. Yuan et al. proposed a framework to find the fastest route from taxi trajectories [3,4]. In [3], they construct a landmark graph and based on which, a two-stage routing algorithm is performed to find the fastest route. In [4], traffic prediction is employed for optimization. However, the estimated travel time is not actually the travel time of the practical fastest route as [3,4] recommended. [15] studies stochastic skyline route queries in road networks with multiple travel costs. The authors provide the travel cost distribution given a source-destination pair with a leaving time. [5] proposes a model to estimate the travel time of a given path in a road network, but it's unsuitable for route planning from a source to a destination. Our work significantly differs from the above methods, because we aim at finding the popular route and estimating the travel cost.

3   Preliminary

We define some terms and the problem addressed in this paper.

Definition 1 (*Trajectory*). A *Trajectory Tr* is a time-ordered sequence of points generated by a moving object. Each point p consists of a geographic location $p.l$ and a timestamp $p.t$, i.e., $Tr : p_1 \rightarrow p_2 \rightarrow ... \rightarrow p_n$, where $p_{i+1}.t > p_i.t$ $(1 \leq i < n)$.

The trajectory [16,17] is a real reflection of the travelling behaviour of the moving object and provides us a possible path from the start to the end location if no road network information is provided.

Definition 2 (*Popular Route*). A *Popular Route* is a path with alternative condition holds: (1) A path that has been traversed at least τ times, where τ is a pre-defined threshold; (2) A path consists of multiple sub-paths and each sub-path satisfies condition (1).

A popular route is always a candidate for a trip, because most of people have chosen it. However, not all source-destination pair have a direct popular route (condition 1). For instance, a long route as a whole may not be frequently passed. In this case, it can still be treated as a popular route if all its sub-paths are popular routes (condition 2) [9].

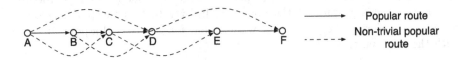

Fig. 1. An example of route concatenation

Definition 3 (*Popular Traverse Graph*). A *Popular Traverse Graph (PTG) G* = (*V, E*) is a directed graph where *V* is a set of popular locations and *E* is a set of popular routes between locations.

Since each edge in PTG is a popular route, the path between any two nodes is also popular by Definition 2.

Definition 4 (*Concatenation of Route*). Let $r : n_1 \rightarrow n_2 \rightarrow ... \rightarrow n_i$ denotes a route in the popular traverse graph G, where n_k ($1 \leq k \leq i$) is a node in G. Denote $|r|$ as the size of r which is the number of nodes it contains. Route $r^* : n_1 \rightarrow ... \rightarrow n_i$ is a concatenation of r if $|r^*| \leq |r|$.

The concatenation of a route means some consecutive road segments in the route are frequently traversed as a whole, and we can regard the consecutive road segments as a united road segment when we estimate the travel cost of this route. Since the united road segment covers the entire path, it reflects the traffic conditions of this whole path, including intersections, traffic lights and direction turns, which will improve the accuracy of the travel cost estimation.

Example 1. Figure 1 is a route on PTG. For a sub-route $r : A \rightarrow B \rightarrow C \rightarrow D$, if r is travelled frequently as a whole, then r is a popular route due to the condition (1) in Definition 2. We call r the *non-trivial popular route*. Then $r^* : A \rightarrow D$ is a concatenation of r (there are 4 in total for r). Obviously, only non-trivial popular route has different concatenations and each non-trivial popular route r has $|r| > 2$. Moreover, every sub-route r' of r with $|r'| > 2$ is also a non-trivial popular route, such as $A \rightarrow B \rightarrow C$ and $B \rightarrow C \rightarrow D$.

The next issue is how to find an optimal one from different concatenations for a route. [5] finds the optimal concatenation of a path by making an object function minimized. In this paper, we apply the object function: $\sum_{i=1}^{k} \frac{1}{n_{s_i}} Var(c_{s_i})$, where s_i is the ith segment of the path, $Var(c_{s_i})$ is the variance of the travel costs on s_i, and n_{s_i} is the number of trajectories that travelled on s_i.

Problem Definition: Given a popular traverse graph G and a route planning query with a source s, a destination d and a departure time t, we find the popular route with minimal travel cost in regard to the optimal route concatenation.

4   Popular Traverse Graph

This section first describes how to construct a popular traverse graph from trajectories without road network information, and then details the travel cost modelling of popular routes.

4.1   Constructing the Popular Traverse Graph

As the road network may be unavailable in some situations [8,9], it is infeasible to compute the travelling frequency of the roads to find the popular routes. Fortunately, it is still feasible to find the popular locations (been visited frequently) from the trajectories, so that the transitions between locations can be extracted. Finally, we discover the popular routes on each transition to construct the popular traverse graph. The major processes are listed below.

Popular locations mining. Since the trajectories consist of points, the popular locations come from the points of trajectories. To reduce the size of points, we only consider the end points of the trajectories which can reflect the locations where people usually start from or go to. Thus, the query source or destination has a high probability to locate at the locations. To find the popular locations, we first partition the points into different zones and then the DBSCAN clustering algorithm [18] is invoked to generate clusters for the points in each zone. Finally, k clusters with the maximal number of points are chosen as the popular locations.

Transitions extraction. We then search the transitions between locations by scanning the trajectory dataset. For each trajectory tr, we first map it to the popular locations, then the trajectory can be represented as $tr: l_1 \rightarrow l_2 \rightarrow ... \rightarrow l_n$, where l_i $(1 \leq i \leq n)$ is a popular location. For each $l_i \rightarrow l_{i+1}$ $(1 \leq i < n)$, we generate a transition from l_i to l_{i+1} if it does not exist, and we also keep the segment of tr belongs to this transition for popular routes discovering.

Popular routes discovering. There may exist several different paths for a transition that connects two popular locations. In order to get the popular paths on the transitions, we cluster the trajectory segments pertaining to the transition. As a result, the cluster with at least τ trajectories is treated as popular route. To tackle the ununiform rate trajectories, we apply Edit Distance with Projections (EDwP) proposed in [19] to measure the similarity between trajectories, and we use the density-based method [20] to cluster the trajectories.

Note that trajectory clustering can not only discover the popular routes but also detect the outliers in the trajectory set (i.e., the roundabout trips), which helps to improve the accuracy of the travel cost estimation.

4.2   Modeling Travel Cost Using the MDL Principle

It is challenging to estimate the travel cost for an arbitrary path [3,15,21], since the cost of a route at different time varied a lot. The major difficulty is how to partition the travel costs into different time slots. [15,21] divide the time slots statically. The VE-Clustering algorithm [3] attempts to partition a day into different time slots according to the travel costs by two clustering methods, namely V-clustering and E-clustering, and each with a threshold. However, it is hard to set the values of such global thresholds for all routes. We note that a good partition for the travel costs should meet the following two requirements: (1) *homogeneity*: stable travel costs in each time slot, and (2) *conciseness*: the significant difference (i.e., distribution) between two adjacent time slots, otherwise, there is no need to split them apart. Fortunately, although challenging, the minimum description length (MDL) principle [6] which is widely used in information theory is suitable to solve this problem. Hence, we propose a parameter-free algorithm by integrating the MDL principle.

Let's review MDL briefly [6]. The MDL cost consists of two components: $L(H)$ and $L(D|H)$, where $L(H)$ is the length of the description of the hypothesis, and $L(D|H)$ is the length of the description of data under the hypothesis, both in bits. To get the best hypothesis H to explain the data D, the value of $L(H) + L(D|H)$ must be minimized. In our work, H refers to the time partition and D is the set of travel costs along the popular route. Hence, $L(H)$ and $L(D|H)$ are defined formally below.

$$L(H) = \log_2(num) + \sum_{i=1}^{num} \lceil \log_2 \mathsf{span}(slot_i) \rceil \qquad (1)$$

$$L(D|H) = \sum_{i=1}^{num} \{ \lceil \log_2(N(slot_i) + 1) \rceil + \mathsf{Ent}(slot_i) \} \qquad (2)$$

where $\mathsf{Ent}(slot_i) = -\sum_{k=1}^{num_{cla}} \frac{N_k(slot_i)}{N(slot_i)} \log_2 \frac{N_k(slot_i)}{N(slot_i)}$. Equation (1) encodes the hypothesis of a partitioning, the first term describes the number of partitions (num) on time, the second term describes the span of each time slot ($slot_i$). Equation (2) describes data under the hypothesis. The first term encodes the number of the travel costs in $slot_i$[1]. The second term computes the information entropy to describe the stability of the slot, which needs to map the travel costs to different class first, each with a different cost level. For example, in travel time cost, the costs in seconds can map to minute level, that is costs from 0 to 120 s correspond to two classes, which are (0,60] and (60,120].

Obviously, $L(H)$ stands for *conciseness* and $L(D|H)$ for *homogeneity*. The more homogenic and concise, the better. However, *conciseness* and *homogeneity* are contradictory, hence we need to find the optimal trade-off by minimizing $L(H) + L(D|H)$. As it is expensive to compute the minimal value of $L(H) + L(D|H)$ due to too many potential partitionings, we devise an approximate

---

[1] To avoid the case that $N(slot_i) = 0$, we increase the value by 1.

Algorithm 1. approPartition(D)

---

1 $S \leftarrow \{[t_1, t_2]\}$, where t_1 and t_2 are the earliest and latest time in D;
2 $minCost \leftarrow MDL(S)$ // $MDL(S) = L(H) + L(D|H)$;
3 **while** *true* **do**
4 $\quad$ $s[t_1, t_2] \leftarrow$ a slot in S with the maximal value of $\mathsf{Ent}(\cdot)$;
5 $\quad$ $t \leftarrow$ arg $\min_{t \in [s.t_1, s.t_2]}(\mathsf{Ent}([s.t_1, t]) + \mathsf{Ent}([t, s.t_2]))$;
6 $\quad$ $S' \leftarrow (S - \{s\}) \cup \{[s.t_1, t], [t, s.t_2]\}$, $newCost \leftarrow MDL(S')$;
7 $\quad$ **if** $newCost < minCost$ **then**
8 $\quad\quad$ $minCost \leftarrow newCost$, $S \leftarrow S'$;
9 $\quad$ **else**
10 $\quad\quad$ **return** S;

---

solution instead, as shown in Algorithm 1. This algorithm accepts a travel costs set D as input. Initially, a set S only contains one time slot that covers the whole data set D (at line 1). Let $minCost$ denote the MDL cost of the current situation. We then probe the dataset S in greedy manner (at lines 3–10). At each time, we find a time slot s in S with the maximal entropy to split. Subsequently, the optimal splitting point t ($t \in [s.t_1, s.t_2]$) is found by minimizing the sum of entropies. Hence, the original time slot s in S is replaced by two new slots $[s.t_1, t]$ and $[t, s.t_2]$. The iteration will stop when $newCost \geq minCost$.

The time complexity of Algorithm 1 is $O(n \cdot k)$, where k is the number of time slots and n is the size of the dataset D, since it computes the MDL cost by scanning the whole dataset at each iteration. In this paper, we use the average travel cost in each time slot to represent the cost of it.

Example 2. Figure 2 illustrates an example of the popular traverse graph. The nodes A, B, C, D, E are popular locations and solid lines are the popular routes between them. The dash lines stand for the non-trivial popular routes mined from the trajectory dataset (see Sect. 5.2). For simplicity, we use a static number to represent the cost on each popular route and we assume that only one popular route exists on each transition.

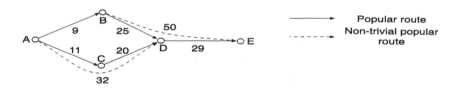

Fig. 2. An example of PTG

5   Route Planning on the Popular Traverse Graph

In this section, we introduce how to find the popular route with the minimal travel cost in consideration of the optimal route concatenation on the time-dependent popular traverse graph.

5.1   Routing Algorithm

Recall that we consider the optimal concatenation of route for accurate travel cost estimation, therefore the result route for a query should satisfy two conditions: (1) it's travel cost should be the cost of its optimal concatenation, (2) it spends minimal travel cost under condition (1). That is the route we expected has the optimal concatenation with the minimal travel cost.

Example 3. We show an example of route planning on PTG illustrated by Fig. 2. Suppose the non-trivial popular routes are also the optimal concatenations. Then the cost of route $A \rightarrow C \rightarrow D$ should be 32, since its optimal concatenation $A \rightarrow D$ costs 32, the same to $B \rightarrow D \rightarrow E$ which costs 50. So, the expected route from A to E should be $A \rightarrow B \rightarrow D \rightarrow E$ which has the minimal travel cost 59. An interesting observation is that the cost of sub-route on $A \rightarrow B \rightarrow D \rightarrow E$ may not be the minimal. For example, $A \rightarrow B \rightarrow D$ costs more than $A \rightarrow C \rightarrow D$ from A to D. That is the major difference from the shortest route planning.

A naive solution is to find all routes with optimal concatenation from source to destination, and choose the one with minimal travel cost. However, it's infeasible to enumerate all the possible routes with optimal concatenation, the cost of it is prohibitive. Instead, we propose a method to return the expected route more efficiently, as listed in Algorithm 2.

Given a PTG $G = (V, E)$ and the query $q = (s, d, t)$, Algorithm 2 returns the expected route. We use *cost*, *route* and *routeSet* for each node to keep the minimal cost, the optimal route and the routes have been considered from s respectively. In addition, we maintain a priority queue Q for nodes on PTG sorted by *cost* in ascending order. Initially we set $n.cost$ to maximum, $n.route$ to *null* and $n.routeSet$ to *empty* except for the source node s, and add s to Q (lines 1–3). Then we keep searching until Q is empty or the destination node d is settled (lines 4–14). At each iteration, we get the head node n of Q and find its outgoing popular routes set where each route is either an outgoing edge of n on PTG or a non-trivial popular route (line 6). For each outgoing route of n, we consider the entire route from s to v passed n. If the route has not been checked, we then find its optimal concatenation and the corresponding travel cost. We keep the route in $v.routeSet$ to avoid verifying twice (lines 7–11). If the new cost of v is less than its current cost, then we update v' cost and its route, and add v to Q if it's not in it (lines 12–14). In this way, we will get the route whose optimal concatenation has minimal travel cost by Lemma 1.

Lemma 1. *Algorithm 2 can return the expected route on popular traverse graph $G = (V, E)$ for a query $q = (s, d, t)$ if the route exists.*

Algorithm 2. routePlanning $(G = (V, E), s, d, t)$

---

1 **foreach** *For each n in V* **do**
2     $n.cost \leftarrow \infty$, $n.route \leftarrow null$, $n.routeSet \leftarrow \emptyset$;
3 $s.cost \leftarrow 0$, $s.route \leftarrow s$;
4 Create a new priority query Q, $Q.enqueue(s)$;
5 **while** $n \leftarrow Q.dequeue()$ and $n \neq NULL$ **do**
6   **return** $d.route$ *if* $n = d$;
7     $S \leftarrow \{r : n \rightarrow ... \rightarrow v | v$ is not settled and r is a non-trivial popular route or an outgoing edge of $n\}$;
8   **foreach** *route* $r : n \rightarrow ... \rightarrow v$ *in* S **do**
9        $r' \leftarrow n.route + r$;
10   **if** $r' \notin v.routeSet$ **then**
11          $r^* \leftarrow$ the optimal concatenation of r';
12          $v.routeSet \leftarrow v.routeSet \cup \{r'\}$;
13   **if** $r^*.cost < v.cost$ **then**
14             $v.cost \leftarrow r^*.cost$, $v.route \leftarrow r'$;
15             $Q.enqueue(v)$ if v is not in Q;

Proof. It is obvious that only the cost of optimal concatenations can be considered by Algorithm 2, and then we need to prove that the route returned by Algorithm 2 has minimal travel cost. We prove it by contradiction. Suppose there is a route whose optimal concatenation is $r' : s \rightarrow ... \rightarrow n_i \rightarrow d$ has less travel cost than the route r^* returned by Algorithm 2. Then the route $s \rightarrow ... \rightarrow n_i$ must cost less than r^*, too. That is n_i must have been settled before returning r^*, hence r' has been checked (by line 6). Since we choose the less cost optimal route concatenation, that is r^* has less cost than r', which is a contradiction. Thus the lemma is proved.

Example 4. We take Fig. 2 as the input PTG, and we find the expected route from A to E by Algorithm 2. We begin with A, and we check $A \rightarrow B$ and $A \rightarrow C$ whose optimal concatenation is themselves and $A \rightarrow C \rightarrow D$ with optimal concatenation $A \rightarrow D$. Then the cost of B, C and D will be 9, 10 and 32 respectively. We next settle B and check its outgoing routes, then the cost of E will be 59 with route $A \rightarrow B \rightarrow D \rightarrow E$. For C, since $A \rightarrow C \rightarrow D$ has been checked, we move to D whose cost is 32 with route $A \rightarrow C \rightarrow D$. Then we consider $A \rightarrow C \rightarrow D \rightarrow E$ whose cost is 61. Finally, we get the expected route $A \rightarrow B \rightarrow D \rightarrow E$ with cost 59.

Although Algorithm 2 can return the right route, the complexity is relatively high, especially in line 6 and line 10 when finding the non-trivial popular routes and computing the optimal concatenation of a route. We improve the procedures by introducing a suffix tree.

5.2   Indexing for Non-trivial Popular Routes

In this sub-section, we describe how to construct a suffix-tree-based index from trajectories dataset to quickly retrieve the non-trivial popular routes between nodes on PTG. For a route $r : n_1 \rightarrow n_2 \rightarrow ... \rightarrow n_k$, we define *prefix route* of r as the route $r' : n_1 \rightarrow ... \rightarrow n_i(1 < i < k)$, and *suffix route* of r as the route $r^* : n_i \rightarrow ... \rightarrow n_k(1 < i < k)$

For each trajectory, we first map it to a string of nodes on PTG, i.e., tr: $n_1 \rightarrow n_2 \rightarrow ... \rightarrow n_k$. For each $n_i \rightarrow n_{i+1}(1 \leq i < k)$, if it's not a popular route on PTG, then we know that $n_i \rightarrow n_{i+1}$ won't be a popular route or be a part of popular route, hence it will be deleted from tr. Now tr is split into several segments. For each segment $s : n_i \rightarrow ... \rightarrow n_j$, if $|s| > 2$, then it could be a non-trivial popular route. We create a path on the suffix-tree for each suffix route of s: $n_o \rightarrow ... \rightarrow n_j(i \leq o \leq j - 2)$ if it's not exists in the tree, otherwise, we increase its support. By this way, any path p with $|p| > 2$ in the suffix-tree whose support is no less than the threshold τ will be a non-trivial popular route. To reduce the size of the suffix-tree, we remove the paths with support less than τ. Given a starting node, we can find all the non-trivial popular routes in $O(1)$ with the reduced suffix-tree.

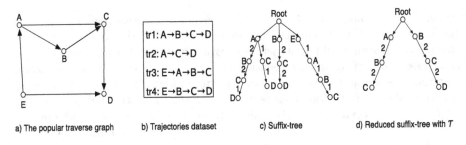

a) The popular traverse graph   b) Trajectories dataset   c) Suffix-tree   d) Reduced suffix-tree with T

Fig. 3. An example of constructing suffix-tree

Example 5. Figure 3 shows an running example of constructing a reduced suffix-tree based on the PTG and trajectories dataset with threshold $\tau = 2$. As we can see, there are two non-trivial popular routes in total which are $A \rightarrow B \rightarrow C$ and $B \rightarrow C \rightarrow D$.

5.3   Computing the Optimal Concatenation

As mentioned above, it is time-consuming when finding the optimal concatenation of a route with many different concatenations. In this paper, we pre-compute the optimal concatenations of the non-trivial popular routes on the reduced suffix-tree to improve the efficiency. For a non-trivial popular route $r : n_1 \rightarrow ... \rightarrow n_k$, there exist 2^{k-2} possible concatenations, we compute its optimal concatenation by using dynamic programming solution with complexity

$O(k^2)$. That is we calculate the minimal object function value (mentioned in Sect. 3) for each node by $opt_i = \arg\min_{1 \le j < i}(opt_j + (n_j \to n_i).obj)(1 < i \le k)$ with $opt_1 = 0$, where opt_i is the minimal object function value of each node, and obj is the object function value of the route.

With the optimal concatenations of non-trivial popular routes in the reduced suffix-tree, we compute the cost of the optimal concatenation of a route on the PTG, which is described in Algorithm 3. For each node in the route, we define $ocCost$ as the cost of the optimal concatenation from the source. We maintain a priority queue for the nodes by their order in the route. For each node n_i from the queue, we only consider the node n_j that n_i can reach as far as possible on the route through r', where r' is a non-trivial popular route whose optimal concatenation has been found or an outgoing edge of n_i whose optimal concatenation is itself, then we refine n_j by r' using Algorithm 4 and add it to the queue (lines 5–6). For the nodes (except for n_i and n_j) in r', we refine the nodes that can reach beyond r' by a non-trivial popular route and add them to the queue (lines 7–10). Finally, We return the result until we get to the ending node. By using the pre-computed results of non-trivial popular routes, we save lots of computing while refining the nodes. The complexity of Algorithm 3 will be $O(|r|)$ since the worst case is that every node on the route is in the queue.

Algorithm 3. ocOnPTG $(r : n_1 \to ... \to n_k)$

---

1 $n_1.ocCost \leftarrow 0$, $n_1.opt \leftarrow 0$, $n_i.opt \leftarrow \infty$ for $1 < i \le k$;
2 Initialize a priority queue Q, $Q.enqueue(n_1)$;
3 **while** $n_i \leftarrow Q.dequeue()$ and $n_i \ne NULL$ **do**
4   **return** $n_i.ocCost$ if $n_i = n_k$;
5      $r' \leftarrow \arg\max_{i < j \le k}|r : n_i \to ... \to n_j|$, where r is either a non-trivial popular route or an outgoing edge of n_i;
6   Refinement(r'), $Q.enqueue(n_j)$;
7   **foreach** $o = i + 1 ... j - 1$ **do**
8   **if** $r^p : n_o \to ... \to n_p(j < p \le k)$ is a non-trivial popular route **then**
9   refinement$(n_i \to ... \to n_o)$;
10   refinement(r^p), $Q.enqueue(n_p)$;

Algorithm 4. refinement $(r : n_i \to ... \to n_j)$

---

1 $obj \leftarrow n_i.opt + r.opt$;
2 **if** $obj < n_j.opt$ **then**
3      $n_j.opt \leftarrow obj$, $n_j.ocCost \leftarrow n_i.ocCost + r.ocCost$;

Example 6. Take Fig. 1 as an example. The nodes in the queue will be D, E, and F when compute the cost of the optimal concatenation from A to F.

6   Experiments

We report some experimental results in this section. Without loss of generality, in our experiments, we focus on travel time cost. All codes are written in Java, run at a computer with dual core 2.00 GHz CPU and 16 GB main memory.

We use a real-life dataset containing 7,122,320 paid trajectories generated by 13,007 taxicabs in Beijing from Oct. 1 to Oct. 31, 2013. The sampling rate is about 1 point per minute. The dataset is divided into two parts. The first part (the first 21 days) is used to construct a popular traverse graph ($\tau = 100$). The final PTG has 5,000 nodes and 33,357 popular routes, and the suffix-tree based on the PTG contains 210,338 different non-trivial popular routes. Such steps are implemented offline. The second part (the last 10 days) is used for evaluation. We randomly choose 30,000 trajectories from the dataset as the queries and regard the travel time as the ground truth. The length of the trajectories ranges from 3 km to 18 km, and the travel time varies from 3 min to 60 min.

As T-drive [3] is known to find the fastest route between two points at a departure time, we use it on PTG as the baseline method. Moreover, MDL+Dijkstra refers to our basic method that applies MDL to model the travel cost and Dijkstra algorithm on TPG to search route, and MDL+OC refers to our method that applies MDL and the optimal route concatenation.

6.1   Effectiveness

We use mean absolute error (MAE) and mean relative error (MRE) to measure the effectiveness of our method. Equation (3) describes the common definitions, where y_i is an estimate, $\widehat{y}_i$ is the ground truth and n is the number of samples.

$$MAE = \frac{\sum_{i=1}^{n} |y_i - \widehat{y}_i|}{n}, \quad MRE = \frac{\sum_{i=1}^{n} |y_i - \widehat{y}_i|}{\sum_{i=1}^{n} \widehat{y}_i} \tag{3}$$

Effectiveness of MDL. Table 1 compares MAE and MRE between MDL+Dijkstra and T-drive. Recall that T-drive needs to set two parameters in advance, one for V-Clustering and the other for E-Clustering. We consider 12 different parameter settings, and in the best case, MAE $= 74.9$ s and MRE $= 0.224$. Note that MDL+Dijkstra has MAE of 71.8 s and MRE of 0.214. Obviously, MDL+Dijkstra outperforms T-drive, since the MDL method we proposed can find an approximate optimal trade-off between homogeneity and conciseness on every popular route, which guarantees a good result. We use the best parameter setting which is 50 and 30 for T-drive in the following experiments.

Overall Effectiveness. We generate three groups of queries, each with top 10,000, 20,000 or 30,000 queries according to the length of trajectories in descending order, as shown in Table 2. Figure 4 shows the performance for these groups. Clearly, MDL+Dijkstra and MDL+OC behave better than T-drive, since MDL principle is suitable for travel cost modelling. Moreover, the performance gain

Table 1. Comparison of MDL+Dijkstra and T-drive

Methods	MDL+Dijkstra	T-Drive ($a=10, b=30, c=50, d=100$)											
		b,a	b,b	b,c	b,d	c,a	c,b	c,c	c,d	d,a	d,b	d,c	d,d
MAE (sec.)	**71.8**	75.3	75.0	76.1	77.9	75.5	**74.9**	76.1	77.9	75.6	75.0	76.2	78.2
MRE	**.214**	.225	.224	.227	.232	.225	**.224**	.227	.232	.226	.224	.227	.233

Table 2. Statistics of three query groups

Group	Size	Avg time (s)	Avg length (m)
A	10,000	435	6,226
B	20,000	369	5,266
C	30,000	335	4,751

of MDL+OC rises when the query distance increases. Since MDL+OC considers the optimal concatenation while routing and the longer the distance is, the more concatenations of the route we can find, which will benefit the travel time estimation. Although our method performs better on long distance query comparing to baselines, we cannot make sure that all queries follow the same route as we found, and the longer the distance is, the higher probability the real path will choose a different route. That's why the MRE increases with the increment of query distance.

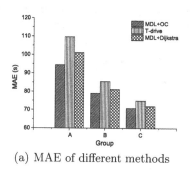

(a) MAE of different methods   (b) MRE of different methods

Fig. 4. Comparison of different methods on query groups

An interesting finding is that around 60–70 % queries follow the same route as we found in each group. Hence, we refine each group, and only reserve the ones along the same route as we returned, and generate three new groups (see Table 3). Figure 5 shows the performance on these groups. Compared with Fig. 4, the performance gain of our method rises. Clearly, MDL+OC outperforms all the baselines in terms of the two metrics, and it has significant advantage over the baselines in each query group. Figure 5(a) illustrates the MAE of different methods. As the length decreases, the MAE decreases, too. Conversely, in Fig. 5(b),

Table 3. Statistics of refined query groups

Group	Size	Avg time (s)	Avg length (m)	Percentage (%)
D	6,238	416	6,163	62.38
E	13,691	357	5,109	68.46
F	20,930	327	4,629	69.77

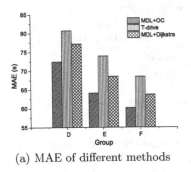 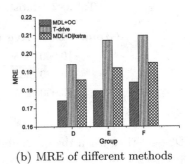

(a) MAE of different methods   (b) MRE of different methods

Fig. 5. Comparison of different methods on refined query groups

the MRE of all methods increases as the length of query decreases. This is because the shorter a path is, the more unstable its travel time could be, since the traffic conditions will have a big influence on it. On the other hand, a long distance means more concatenations, it helps to find a "better" optimal concatenation, therefore MDL+OC behaves better when the length increases.

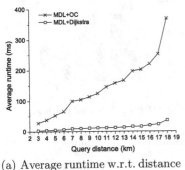 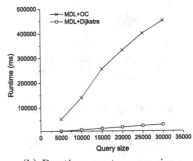

(a) Average runtime w.r.t. distance   (b) Runtime w.r.t. query size

Fig. 6. Runtime of different methods

6.2   Efficiency

Figure 6 shows the runtime on different query distance and query size. Since MDL+Dijkstra does not need to find the optimal concatenation, it runs much more faster than MDL+OC. Figure 6(a) shows that the runtime of MDL+OC increases as the query distance rises, since large distance will increase the number of concatenations and the non-trivial popular routes to be checked while routing. However, MDL+OC can return the result very quickly, far less than 1 s. Figure 6(b) illustrates the runtime of MDL+OC grows linearly as the query size increases.

7   Conclusion

In this paper, we propose a framework to find the popular route in consideration of travel cost estimation. We firstly construct a popular traverse graph by discovering the popular routes from historical trajectories, and then use MDL principle to model the travel cost on each popular route. Finally, we devise an efficient routing algorithm to find the popular route with minimal travel cost in terms of optimal concatenation. We evaluate our methods with extensive experiments, showing that our methods are both effective and efficient.

Acknowledgment. Our research is supported by the 973 program of China (No. 2012CB316203), NSFC (61370101, U1401256 and 61402180), Shanghai Knowledge Service Platform Project (No. ZF1213), Innovation Program of Shanghai Municipal Education Commission (14ZZ045), and Natural Science Foundation of Shanghai (No. 14ZR1412600).

References

1. Kanoulas, E., Du, Y., Xia, T., Zhang, D.: Finding fastest paths on A road network with speed patterns. In: Proceedings of ICDE (2006)
2. Ding, B., Yu, J.X., Qin, L.: Finding time-dependent shortest paths over large graphs. In: Proceedings of EDBT, pp. 205–216 (2008)
3. Yuan, J., Zheng, Y., Zhang, C., Xie, W., Xie, X., Sun, G., Huang, Y.: T-drive: driving directions based on taxi trajectories. In: Proceedings of SIGSPATIAL (2010)
4. Yuan, J., Zheng, Y., Xie, X., Sun, G.: Driving with knowledge from the physical world. In: Proceedings of KDD (2011)
5. Wang, Y., Zheng, Y., Xue, Y.: Travel time estimation of a path using sparse trajectories. In Proceedings of KDD (2014)
6. Grünwald, P.D., Myung, I.J., Pitt, M.A.: Advances in Minimum Description Length: Theory and Applications. MIT Press, Cambridge (2005)
7. Zheng, K., Zheng, Y., Xie, X., Zhou, X.: Reducing uncertainty of low-sampling-rate trajectories. In: Proceedings of ICDE (2012)
8. Wei, L.-Y., Zheng, Y., Peng, W.-C.: Constructing popular routes from uncertain trajectories. In: Proceedings of KDD (2012)
9. Chen, Z., Shen, H.T., Zhou, X.: Discovering popular routes from trajectories. In: Proceedings of ICDE (2011)
10. Luo, W., Tan, H., Chen, L., Ni, L.M.: Finding time period-based most frequent path in big trajectory data. In: Proceedings of SIGMOD (2013)

11. Dijkstra, E.W.: A note on two problems in connexion with graphs. Numerische mathematik **1**(1), 269–271 (1959)
12. Hart, P.E., Nilsson, N.J., Raphael, B.: A formal basis for the heuristic determination of minimum cost paths. IEEE Trans. Syst. Sci. Cybern. **4**(2), 100–107 (1968)
13. Kenneth, K.L., Halsey, E.: The shortest route through a network with time-dependent internodal transit times. J. Math. Anal. Appl. **14**(3), 493–498 (1966)
14. Gonzalez, H., Han, J., Li, X., Myslinska, M., Sondag, J.P.: Adaptive fastest path computation on a road network: a traffic mining approach. In: Proceedings of VLDB (2007)
15. Yang, B., Guo, C., Jensen, C.S., Kaul, M., Shang, S.: Stochastic skyline route planning under time-varying uncertainty. In: Proceedings of ICDE (2014)
16. Mao, J., Song, Q., Jin, C., Zhang, Z., Zhou, A.: Tscluwin: trajectory stream clustering over sliding window. In: Proceedings of DASFAA (2016)
17. Duan, X., Jin, C., Wang, X., Zhou, A., Yue, K.: Real-time personalized taxi-sharing. In: Proceedings of the DASFAA (2016)
18. Ester, M., Kriegel, H.-P., Sander, J., Xu, X.: A density-based algorithm for discovering clusters in large spatial databases with noise. In: Proceedings of KDD (1996)
19. Ranu, S., Deepak, P., Aditya, D. Telang, A., Deshpande, P., Raghavan, S.: Indexing and matching trajectories under inconsistent sampling rates. In: Proceedings of ICDE (2015)
20. Lee, J.-G., Han, J., Whang, K.-Y.: Trajectory clustering: a partition-and-group framework. In: Proceedings of SIGMOD (2007)
21. Balan, R.K., Nguyen, K.X., Jiang, L.: Real-time trip information service for a large taxi fleet. In: Proceedings of MobiSys (2011)

ETCPS: An Effective and Scalable Traffic Condition Prediction System

Dong Wang, Wei Cao, Mengwen Xu, and Jian Li[✉]

Institute for Interdisciplinary Information Sciences, Tsinghua University,
10084 Beijing, BJ, China
{wang-dong12,cao-w13,xmw12}@mails.tsinghua.edu.cn,
lijian83@mail.tsinghua.edu.cn

Abstract. Real-time prediction of the traffic condition is an important ingredient for a variety of applications. In this paper, we propose an *Ensemble based Traffic Condition Prediction System (ETCPS)* for predicting the traffic conditions of any roads in a city based on the current and historical GPS data collected from floating vehicles. We have observed two useful correlations in the traffic condition time series, which are the bases of our design. In order to exploit these two correlations for prediction, we propose two different models called *Predictive Regression Tree (PR-Tree)* and *Spatial Temporal Probabilistic Graphical Model (STPGM)*. Our best quality prediction is achieved by a careful ensemble of the two models. Our system provides high-quality prediction and can easily scale to very large datasets. We conduct extensive experimental evaluations with a large GPS data set collected from more than 12,000 taxis in Beijing during two months. The experimental results demonstrate the effectiveness, efficiency, and scalability of our system.

1 Introduction

Real-time prediction of the traffic condition becomes increasingly important. A well-performed traffic condition prediction system is the fundamental ingredient of various real applications. Examples include the traffic management [6], routing service [13], taxi ride sharing [8] etc. Such problem has been widely studied in recent years [1,10,11,15]. Generally, given the current and historical traffic conditions of the road network, our goal is to predict the traffic condition of each road after a few minutes or hours.

Most prior works on traffic condition prediction are based on the data generated by the road side loop sensors. However, such loop sensors are usually expensive and only embedded in highways and part of urban main roads. Alternatively, ubiquitous location based services enable us to collect a large volume of traffic data from GPS-embedded devices. Such GPS data provides valuable information for analyzing and predicting the traffic conditions. Despite there exist several researches and products for traffic prediction based on the GPS data, most of them only focused on the arterial roads and did not consider the urban roads.

© Springer International Publishing Switzerland 2016
S.B. Navathe et al. (Eds.): DASFAA 2016, Part II, LNCS 9643, pp. 419–436, 2016.
DOI: 10.1007/978-3-319-32049-6_26

In this paper, we study the efficient and scalable models for traffic condition prediction based on the GPS data collected from floating vehicles (taxis in our data). To make our exposition more concrete, we first illustrate several challenges in our problem.

- Large volume of GPS data has been generated routinely, especially for some metropolises such as New York or Beijing. Most prior works are based on probabilistic graphical models [3,5,9]. The state spaces explode in these algorithms under very large scale datasets. Thus, it takes a very long time to run the algorithms.
- The traffic conditions and their transition patterns (i.e., the patterns in which the traffic condition varies) for each road vary significantly under different time intervals. For example, if the traffic is in a jam during a peak hour, it usually lasts for a long time. However, if such congestion happens in a non-peak hour, the traffic usually become light soon. Such traffic pattern is changing over time. Prior works based on the Markov Chain and Hidden Markov Model (HMM) [5,9,11] can not capture such feature since the states of transition matrices are not related with time.
- The taxis sometimes slow down or even stop for picking or attracting the passengers. It is hard to distinguish whether such low travel speed is due to the congestion of the traffic. Such records may lead to erroneous estimations of the traffic condition.

To address the above challenges, we propose the *Ensemble based Traffic Condition Prediction System (ETCPS)*. Our system combines two different models called *Predictive Regression Tree (PR-Tree)* and *Spatial Temporal Probabilistic Graphical Model (STPGM)*. We summarize our technical contributions below:

- We present two useful observations in the traffic condition time series which are the bases of our design. We first present the correlations between the gaps of the traffic condition and its expected traffic condition. Then, we show the autocorrelations in the first order difference of the traffic condition series (See Sect. 2).
- We propose a regression tree based model called PR-Tree. PR-Tree can effectively capture the proposed correlations and thus predict the traffic conditions with a high accuracy. PR-Tree is very efficient on large scale datasets. Given a training set with 10^5 roads, it only takes 3.26 min to train a PR-Tree and the prediction of PR-Tree is real-time (See Sect. 5).
- We propose a probabilistic graphical model called STPGM. STPGM can capture the correlations between adjacent roads. It formulates the state transitions in different time intervals separately. Thus, the state space for STPGM is much smaller than the prior works [3,5,9]. On the other hand, STPGM captures different traffic patterns in different time intervals. We show that in the experiment STPGM is more efficient and accurate than the algorithms in prior works (See Sect. 6).
- We propose a prediction system called ETCPS which combines PR-Tree and STPGM. We evaluate our model with real dataset which consists of GPS

points generated by over 12,000 taxis collected in two months. It provides an experimental evidence that ETCPS is efficient, scalable in terms of supporting large size road networks, and achieves a high-quality prediction (See Sect. 7).

2 Preliminary

Road Network. We are given a data set consisting of GPS records of taxis. The GPS records of the j-th taxi is represented by $Tr_j = \{p_1, p_2, \ldots, p_{|Tr_j|}\}$. Each p_i represents a GPS record $(cid, time, location, speed)$ indicating the id of the j-th car, the time stamp when the record is generated, the latitude and longitude of the current location and the instantaneous speed respectively. We define a real urban road network as a directed graph $G = (V, E)$ where V is the set of nodes representing the terminal points of road segments and E is the set of road segments. A road segment r_i is a directed edge associated with a start point v_s, an end point v_e with length l_i. See Fig. 1 for an illustration. Utilizing the technique of map-matching [7], each GPS record p_i on the trajectory Tr_j can be located to a road segment r_i in which the car j is traveling on.

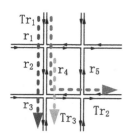

Fig. 1. Time cost

Table 1. Time cost (Million seconds)

Traj	Time (Intv)	Road segment	Speed (km/h)	Traj	Time (Intv)	Road segment	Speed (km/h)
Tr_1	34	r_1	56	Tr_2	34	r_1	60
Tr_1	35	r_2	60	Tr_2	35	r_2	58
Tr_1	35	r_3	61	Tr_2	35	r_4	58
Tr_3	35	r_2	15	Tr_2	36	r_5	60
Tr_3	35	r_3	60				

Traffic Condition. We define the traffic condition for a road segment r_i during a specific period as below. Given a GPS data set collected during D days, we split the period of D days into several intervals, and each time interval spans λ minutes. We assume that the traffic condition of a specific road segment remains unchanged in one interval. Such assumption is widely used in the transportation literature [11,15].

As each day has $M = \frac{60 \cdot 24}{\lambda}$ time intervals, for a GPS data set collected during D days, there are $T = M \cdot D$ time intervals. The t-th interval is $[t \cdot \lambda, (t+1) \cdot \lambda)$. For example, if we set $\lambda = 15, D = 31$, then we have $M = 96$, $T = 2976$, and the interval 34 is a time period from $8:30$ to $8:45$ in the first day.

By mapping each GPS record to a road segment, we consider the average speed of all the records observed in the t-th interval on a road segment. For example, in Table 1, the observed average speed for r_2 in the 35-th interval is $(60 + 58 + 15)/3$. However, some taxis may run at a very low speed or even stop for *boarding* or *balling* when the road is not congested. We regard such records as the noise which is eliminated in the pre-processing stage (see Sect. 8 for details). Then, the *traffic condition* of a road segment r_i in the t-th interval is defined as the average speed of all the GPS records observed in this road segment during the t-th interval, denoted as o_t^i. Note that for some road segments, there may not exist any GPS record in the t-th interval and thus we can not define the corresponding traffic condition. We explain how we deal with such case in Sect. 8. Currently, we simply assume o_t^i is well-defined for all i and t. Moreover, we use $\text{Org}^i = \{o_1^i, \dots, o_T^i\}$ to denote the traffic condition time series of road segment r_i.

Expected Traffic Condition. Note that the traffic conditions usually have the "daily pattern". For example, a road segment is usually in a jam during 6:00–9:00 each day whereas from 9:00 to 11:00 it is usually light. For the t-th interval, we define $t \bmod M$ as its *daily index*, i.e., it is the $t \bmod M$-th interval in its corresponding day. For example, if we set $M = 96$, then the 226-th interval represents the time period from $8:30$–$8:45$ in the third day and its daily index is $226 \bmod 96 = 34$. Let $A_t^i = \{o_{t'}^i | t' \equiv t \bmod M\}$ be the set of traffic conditions observed in road segment r_i during the $t \bmod M$-th interval for all days. For example, in Table 1, the 34-th interval is a time period from $8:30$ to $8:45$ on the first day. Then, A_{34}^i is the set of traffic conditions of the road segment r_i in all days from $8:30$ to $8:45$. We call the mean of A_t^i the *expected traffic condition* of r_i in time interval t, denoted as $a_t^i = \sum_{a \in A_t^i} a / |A_t^i|$. Essentially, the expected traffic condition a_t^i indicates the value that traffic conditions are usually around, in the $t \bmod M$-th interval of a day. We use $\text{Avg}^i = \{a_1^i, \dots, a_T^i\}$ to denote the expected traffic condition time series of the road segment r_i. Note that Avg^i is a periodic series and once we have the training data, a_t^i is always available for all $t \in Z$.

Problem Definition. Given the historical traffic conditions before time interval T, $\text{Org}^i = \{o_1^i, \dots, o_T^i\}$ for all i, our goal is to predict the traffic condition on the $T+1$-th interval o_{T+1}^i or even longer for each road segment r_i. For convenience, for any t, we use p_t to denote the predicted traffic condition in the time interval t.

3   Useful Observations

Most of prior works predict the future traffic conditions directly based on the traffic condition time series. However, it is difficult to extract the patterns in the traffic condition time series Org^i. We find that by transforming the Org^i into two different forms of time series, the new time series reveal very strong autocorrelations. We hope these observations can provide useful insight in further study of the travel condition prediction problem and related problems.

Expectation-Reality Gap. The traffic condition time series of the same road segment in each day usually exhibits strong periodic pattern which we refer to as the "daily pattern". We eliminate the daily pattern from the traffic condition series by subtracting the corresponding expected traffic condition from each of the traffic conditions. Specifically, we set $g_t^i = o_t^i - a_t^i$ and we thus obtain a new series $\text{Gap}^i = \{g_t^i | t = 1, \ldots, T\}$. Intuitively, if $g_t < 0$, it means that the traffic condition in the time interval t is more congested than usual. We find that there exists a strong correlation between g_{t+1} and g_t. Figures 2 and 3 show the scatter diagram of (o_t, o_{t+1}) and (g_t, g_{t+1}) of a specific road segment respectively. As we can see, by transforming the traffic condition series Org^i to the gap series Gap^i, we essentially extract the "pattern" of the traffic condition series.

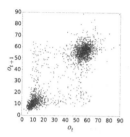 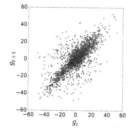 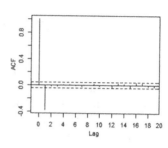

Fig. 2. o_t and o_{t+1} **Fig. 3.** g_t and g_{t+1} **Fig. 4.** ACF of Diff(Org)

First Order Difference of Traffic Condition Series. We use $\delta_t^i = o_t^i - o_{t-1}^i$ to represent the first order difference of traffic condition series, denoted as Diff(Org). We use ACF (Auto Correlation Function) to analyze the autocorrelation in the time series of δ_t^i. The autocorrelation of a random process describes the correlation between values of the process at different times with a time lag τ. Given a time series and time lag, ACF returns a value between $+1$ (total positive correlation) and -1 (total negative correlation) inclusive. If the absolute value of ACF is beyond ± 0.05, we usually think the time series is autocorrelated at time lag τ. In Fig. 4, we show the ACF value of the time series δ_t of a random road segment. The horizontal axis represents the time lag τ, and vertical axis represents the ACF value at lag τ. As the ACF value at lag $\tau = 1$ is far beyond the threshold -0.05, we conclude that there exists a correlation between δ_t and δ_{t+1}.

4   System Overview

The framework of our proposed traffic condition prediction system is illustrated in Fig. 5. We develop a system that utilizes the historical and real time taxi GPS records to estimate the current travel condition and predict the travel conditions in the next time intervals. It is composed of four major components: Pre-processing, Predictive Regression Tree Model (PR-Tree), Spatial Temporal Probabilistic Graphical Model (STPGM) and Ensemble.

In the pre-processing phase, first, we map match the GPS trajectories to road networks using the ST-Matching algorithm [11]. Then, we eliminate the

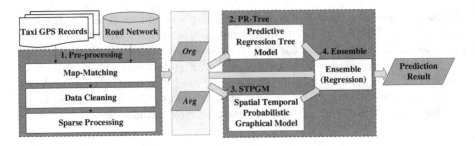

Fig. 5. Overview of system framework.

records which are under boarding or balling state. We then deal with the sparsity issure that no GPS record is observed for some roads during some time intervals. With the pre-processing, we thus obtain two time series Org and Avg as defined in Sect. 2. The details are presented in the experiment part (Sect. 8). Next, in Sect. 5, we use a regression tree based model called PR-Tree to predict the future traffic conditions based on our observed correlations. We further adopt a probabilistic graphical model called STPGM in Sect. 6 which captures both our observations and the correlations between the road segments. Finally, we combine two models in the ensemble stage as shown in Sect. 7. We show that combining two different models enhances the accuracy of the prediction in Sect. 8.

5   Predicting the Traffic Condition with PR-Tree

In this section, we define a regression tree based model called PR-Tree to predict the traffic condition of each road segment individually. We first describe the structure of PR-Tree in detail and how we predict the traffic condition on this tree in Sect. 5.1. Then in Sect. 5.2, we present the training algorithm of PR-Tree.

5.1   Description of PR-Tree

Recall that the time series Gap shows a strong autocorrelation as we claimed in Sect. 3. We can thus approximate g_{t+1} by an estimation $\hat{g}_{t+1}$ based on g_t and predict the traffic condition in the $t + 1$-th interval by $p_{t+1} = a_{t+1} + \hat{g}_{t+1}$ (the expected traffic condition a_{t+1} is always available as we claimed in Sect. 2). From Fig. 3, it is reasonable to set $\hat{g}_{t+1} = \theta \cdot g_t$ since the scatter diagram shows a nearly linear correlation. However, we find that the ratio g_{t+1}/g_t varies when g_t takes different values. For example, if g_t is closed to -10, g_{t+1} is usually around 1.2 times g_t whereas if g_t is closed to -8, g_{t+1} is usually around 1.4 times g_t. Motived by this, instead of estimating g_{t+1} by $\theta \cdot g_t$, we use a proper function $R(g_t)$ and estimate g_{t+1} by $g_t \cdot R(g_t)$.

Structure. To learn a proper function R, we propose a regression tree based model called PR-Tree. Specifically, PR-Tree splits the input space into several subspaces. Each subspace is associated with an output parameter θ. Given the input g_t, we find the subspace corresponding to g_t and return the corresponding θ as $R(g_t)$. Formally, each inner node of PR-tree has a splitting value and each leaf node has a output parameter θ. To find the corresponding subspace of g_t, we search on PR-Tree as follows. Initially, the current node is the root of PR-Tree. If g_t is less than or equal to the splitting value of the current node, we search the left child recursively. Otherwise, we search the right child. We perform such search until it reaches a leaf node and return the corresponding θ on the leaf node as $R(g_t)$. For simplicity, we use R to represent the corresponding PR-Tree.

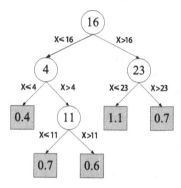

Fig. 6. An example of PR-Tree

We show an example of a PR-Tree in Fig. 6. The PR-Tree contains four inner nodes (the splitting value of these nodes are $\{4, 11, 16, 23\}$), and five leaf nodes (their values are $\{0.4, 0.7, 0.6, 1.1, 0.7\}$). We take $g_t = 5$ as the input. As the splitting value of the root node is 16 and $g_t \leq 16$, we search its left child recursively and finally reach a leaf node with output parameter $\theta = 0.7$.

Prediction. To predict the traffic condition in the time interval $t+1$, we simply set $\hat{g}_{t+1} = R(g_t) \cdot g_t$ and predict o_{t+1} by $p_{t+1} = a_{t+1} + R(g_t)$. Figure 6 shows an example. Given the current traffic condition $o_t = 45$, assuming the expected traffic condition on t and $t+1$ are $a_t = 40$, $a_{t+1} = 43$, we get $g_t = o_t - a_t = 5$. By taking g_t as the input of PR-Tree, we get $R(g_t) = 0.7$. Then, we estimate o_{t+1} by $a_{t+1} + R(g_t) \cdot g_t = 46.5$.

5.2 Training PR-Trees

First, we present the objective for training PR-Trees. Recall that we predict o_{t+1} as $p_{t+1} = a_{t+1} + R(g_t) \cdot g_t$. Given the training set $\text{Org}^i = \{o_1^i, \ldots, o_T^i\}$, our goal is to minimize the squared error $\sum_{t \in [1,T]} (p_{t+1} - o_{t+1})^2$. Equivalently, we need to find an optimal PR-Tree (function R^*) that

$$R^* = \underset{R}{\text{argmin}} \sum_{t \in [1,T]} (g_{t+1} - R(g_t) \cdot g_t)^2 \qquad (1)$$

Algorithm 1. PR-Tree Splitting (Split)

Require: Node *root*, Training sequence TR, cross validation sequence CV
Ensure: Update the PR-Tree.
1: $e_{TR} = f(TR, \text{out}(TR))$
2: $e_{min} = \infty$
3: **for** $i = 1, \ldots, |TR| - 1$ **do**
4:      $TR_l \leftarrow$ first i elements in TR
5:      $TR_r \leftarrow T \backslash TR_l$
6:      **if** $f(TR_l, \text{out}(TR_l)) + f(TR_r, \text{out}(TR_r)) < e_{min}$ **then**
7:          $e_{min} = f(TR_l, \text{out}(TR_l)) + f(TR_r, \text{out}(TR_r))$
8:          $TR_l^* = TR_l$, $TR_r^* = T \backslash TR_l^*$ ▷ update the best TR_l
9:      **end if**
10: **end for**
11: **If** $e_{min} > e_S - \gamma$ **return**
12: $root.lc \leftarrow$ a new node corresponds to TR_l^* ▷ split *root*
13: $root.rc \leftarrow$ a new node corresponds to TR_r^* ▷ split *root*
14: **if** $best_{CV} > Q(CV)$ **then**   ▷ qualify the splitted PR-Tree
15:      $best_{CV} = Q(CV)$ ▷ update the global best value
16:      Split($root.lc, TR_l^*, CV$), Split($root.rc, TR_r^*, CV$)
17:      set the splitting value of *root* as $max_{s \in TR_l^*}$ $s.u$ ▷ inner node
18: **else**
19:      $root.lc = $ None, $root.rc = $ None
20:      set the output value of *root* as $\text{out}(TR)$ ▷ leaf node
21:      **return**
22: **end if**

Our training algorithm is slightly different from the standard regression tree training algorithm. To train the PR-Tree, given the time series Gap $= \{g_1^i, \ldots, g_T^i\}$, we construct another sequence $S = \{(u, v) | u = g_t, v = g_{t+1}, \forall t = [1, T)\}$. Each element $s \in S$ indicates a pair of values (g_t, g_{t+1}). We use $s.u$ to denote the first value in pair s and $s.v$ to denote its second value. We sort S by increasing order of $s.u$. For any subsequence $S_x \subset S$ and any PR-Tree R, we define the cost of S_x as $Q(S_x) = \sum_{s \in S_x} (s.v - R(s.u) \cdot s.u)^2$, which represents the squared error if we use PR-Tree R to fit the set S_x.

Our training algorithm works as follows. During the training phase, each node corresponds to a subsequence of $S_x \subset S$. For a specific node, if it is an inner node, we use S_l, S_r to denote the corresponding subsequences of its left child and its right child respectively. Then, its splitting value is $max_{s \in S_l}$ $s.u$. Otherwise, it is a leaf node. We define $f(S_x, \alpha) = \sum_{s \in S_x} (s.v - \alpha \cdot s.u)^2$. The output θ of this leaf node is $argmin_\alpha$ $f(S_x, \alpha)$, denoted as $\text{out}(S_x)$.

Initially, we have a singleton tree. There is the only one node which corresponds to S. We split the PR-Tree recursively. For each node, there is a best splitter S_l^*, i.e.,

$$S_l^* = \underset{S_l}{\text{argmin}} \; \{f(S_l, \text{out}(S_l)) + f(S \backslash S_l, \text{out}(S \backslash S_l))\}.$$

We enumerate the first i elements of S_x as S_l ($S_r = S \backslash S_l$) to search the best splitter S_l^* (line 3 to line 10 in Algorithm 1). Note that since S is sorted and $f(S_l, \alpha)$ is the sum of quadratic terms which is still quadratic. To obtain the best splitter S_l^*, we can maintain the coefficients of $f(S_l, \alpha)$ and the minimum of the quadratic term can be calculated in $O(1)$ time. Each time when we enumerate a new subsequence, we only need to update the coefficients. Thus, we can obtain the best splitter in $O(|S|)$ time efficiently. We denote $S_r^* = S_x \backslash S_l^*$. If $f(S_l^*, \mathsf{out}(S_l^*)) + f(S_r^*, \mathsf{out}(S_r^*)) < f(S_x, \mathsf{out}(S_x)) - \gamma$, we split the current node into two child nodes with subsequences S_l^* and S_r^* respectively where γ is a threshold to be specified. Otherwise, we terminate the recursion.

The readers may notice that such splitting procedure may cause a serious overfitting problem, i.e., the PR-Tree keeps splitting until each node only contains a very short subsequence. To remedy this issue and reduce the generalization error, we split S into two parts, the training part TR and the cross validation part CV. We use TR to train PR-Tree, each time when a node is split, we qualify the current PR-Tree on the cross-validation set CV and check whether if $Q(CV)$ decreases. If the qualification on CV does not decrease, we undo the splitting operation (line 19 to line 21) and terminate the recursion. Otherwise, we split its children nodes recursively (line 14 to line 17). See Algorithm 1 for the pseudo code.

6   Predicting Traffic Condition with STPGM

Despite that the PR-Tree performs well in most of our data (which we show in Sect. 8), it does not consider the correlations between the road segments. Some roads are easily affected by its neighbors, the congestions of its neighbors usually lead to the congestion of its self in the next few time intervals. For such roads, PR-Tree does not perform well. Motivated by this, we propose a probabilistic graphical model called STPGM which is used in combination with the PR-Tree in our system.

We first construct a *spatial temporal probabilistic graph (STPG)* G_p which corresponds to a road network G. If a vehicle can travel from the road segment r_i to the road segment r_j (or from r_j to r_i) directly, we say that r_i and r_j are adjacent. We construct a vertex v_i in G_p which corresponds to a road segment r_i in G. We add an edge between v_i and v_j if and only if the road segments r_i and r_j are adjacent. For a specific v_i, we use $Neib(v_i)$ to denote all the adjacent vertices of v_i. Intuitively, the adjacent road segments affect each other much more significantly than the other road segments. Thus, each edge in G_p represents a "strong effectiveness" in the road network.

6.1   States of STPGM

We first discretize the traffic conditions into different states. Recall that as we claimed in Sect. 1, the traffic conditions and the transition patterns are very different not only at different road segments, but also at different time intervals.

However, for a specific road segment, we find that the traffic conditions and transition patterns are usually similar for the time intervals with the same daily index. For example, if the traffic is congested in $8:00$, it usually stays congested in next several time intervals. However, if the traffic is congested in $10:00$, the traffic becomes light in the next few minutes with a large probability. Motivated by this, we consider different time intervals separately and use the same state sets for the time intervals with the same daily index.

For a specific road segment r_i, instead of clustering all of its traffic conditions in series Org^i (which are widely used in the prior works [3,5,11,13]), we consider the traffic conditions under different daily index separately. Formally, we consider a specific daily index $l \in [M]$. Recall that $A_l^i = \{o_t^i | t \equiv l \bmod M\}$. We cluster the traffic condition set A_l^i into k clusters with K-Medoids where k is a parameter to be specified (see Sect. 8 for details). For example, if the daily index l corresponds to $8:30$–$8:45$ in a day, then we cluster the traffic conditions for all days during $8:30$–$8:45$. We use the center $c_{x,l}^i$ of each cluster to represent a state, and denote the set of the centers as $C_l^i = \{c_{1,l}^i, \ldots, c_{k,l}^i\}$. The state of the traffic condition in the time interval t is represented by its nearest center in $C_{[t \bmod M]}^i$, denoted as s_t^i. We show an example of a random selected road segment r_i where $C_{25}^i = \{44, 48, 52, 58\}$ and $C_{74}^i = \{15, 25, 32, 38\}$ (km/h). The time interval 25 corresponds to $6:00$–$6:15$ where the traffic is usually light and the time interval 74 corresponds to $18:30$–$18:45$ where the traffic is usually heavy.

6.2 Parameter Learning

We predict the traffic condition of a specific vertex (corresponds to a road segment) v_i based on the historical traffic conditions of itself and its neighbors. We assume that the traffic condition of v_i in the time interval $t+1$ is only related with the traffic conditions of v_i and $Neib(v_i)$ in the time interval t.

Formally, consider a vertex v_i. Let $\{v_i\} \cup Neib(v_i) = \{v_{i_1}, \ldots, v_{i_n}\}$ and the corresponding states in time interval t are $\{c_{x_i,t}^i, c_{x_{i_1},t}^{i_1}, c_{x_{i_2},t}^{i_2}, \ldots, c_{x_{i_n},t}^{i_n}\}$. Our goal is to learn the transition probability for all the possible states in $C_{(t+1) \bmod M}^i$, i.e.,

$$P(s_{t+1}^i = c_{x_i,t+1}^i | s_t^{i_1} = c_{x_{i_1},t}^{i_1}, s_t^{i_2} = c_{x_{i_2},t}^{i_2}, \ldots, s_t^{i_n} = c_{x_{i_n},t}^{i_n})$$

$$= \frac{P(s_{t+1}^i = c_{x_i,t+1}^i, s_t^{i_1} = c_{x_{i_1},t}^{i_1}, \ldots, s_t^{i_n} = c_{x_{i_n},t}^{i_n})}{P(s_t^{i_1} = c_{x_{i_1},t}^{i_1}, \ldots, s_t^{i_n} = c_{x_{i_n},t}^{i_n})} \quad (2)$$

For the prediction, it is unnecessary to compute the denominator, which we show in Sect. 6.3. As for the numerator, the state space in Eq. 2 explodes exponentially whereas the training data is relatively limited. It is not sufficient to estimate the numerator precisely. Thus, we approximate the numerator of Eq. 2 by

$$P(s_{t+1}^i = c_{i,t+1}^{x_i}) \prod_{j=1}^{n} P(s_{i_j}^t = c_{i_j,t}^{x_{i_j}} | s_{t+1}^i = c_{x_i,t+1}^i) \quad (3)$$

where $P(s_{i_j}^t = c_{i_j,t}^{x_{i_j}} | s_{t+1}^i = c_{x_i,t+1}^i)$ indicates that given the observed state in the time interval $t + 1$, the probability that the previous state of v_{i_j} is $c_{i_j,t}^{x_{i_j}}$.

We define the indicator function $I(s_t^i, c_{x,t}^i)$ which indicates that whether the state of the road segment r_i in the time interval t equals $c_{x,t}^i$. We use $N = \sum_{t' \equiv t \bmod M} I(s_{t'}^i, c_{x,t}^i)$ to represent the total days that the state of the road segment r_i in the $t \bmod M$-th interval of each day is $c_{x,t}^i$. Then, we calculate the probability $P(s_t^i = c_{x,t}^i)$ by the frequency $P(s_t^i = c_{x,t}^i) = N/D$. Similarly, for the term $P(s_{i_j}^t = c_{i_j,t}^{x_{i_j}} | s_{t+1}^i = c_{x_i,t+1}^i)$, we have

$$P(s_{i_j}^t = c_{i_j,t}^{x_{i_j}} | s_{t+1}^i = c_{x_i,t+1}^i) = \frac{\sum_{t' \equiv t \bmod M} \left(I(s_{t'+1}^i, c_{x_i,t+1}^i) \cdot I(s_{t'}^{i_j}, c_{x_{i_j},t}^{i_j}) \right)}{\sum_{t' \equiv t \bmod M} I(s_{t'+1}^{i_j}, c_{x_{i_j},t+1}^{i_j})}.$$

(4)

Thus, we get the approximation of the numerator of Eq. 2.

6.3 Prediction

Suppose the traffic conditions of the road network in time interval t are observed. We first construct the states for each road segment r_i. To predict the traffic condition of a road segment r_i, after obtaining the states of v_i and $Neib(v_i)$ in the time interval t, we use Eq. 2 to infer the probability of each state for v_i in the time interval $t + 1$. Then, we select the state with the largest probability as the predicted state and the corresponding cluster center as the predicted traffic condition. Note that as the denominator of Eq. 2 is a constant value when the states of v_i and $Neib(v_i)$ in the time interval t are given, it is actually unnecessary to compute this denominator.

7 Model Extensions

Ensemble. We find that in the experiment, the performances of PR-Tree and STPGM differ in different roads. Some roads are rarely affected by their neighbors, such as the arterial roads. For such roads, PR-Tree outperforms STGPM. However, as PR-Tree does not consider the correlations of the roads, STPGM performs better than PR-Tree for the roads which are highly affected by its neighbors, especially the roads that only few GPS records are observed. Our prediction for traffic condition in the $t + 1$-th interval is a linear combination of the previous traffic condition o_t^i, the prediction obtained by PR-Tree and STPGM. The weights of the linear combination is obtained by linear regression. We show that in the experiment, by combining the models, our system achieves a higher accuracy for the prediction.

Alternate of the Input Series. In fact, both the PR-Tree and STPGM are the models which capture the correlations in a time series. Recall that in the PR-Tree model, we use the time series Gap as the input. In STPGM, we use the traffic condition time series Org as the input. Essentially, we can use the any time series related with the traffic as the input of both models and predict the traffic condition in a proper way. For example, if we use the Org as the input of a PR-Tree, we actually try to approximate o_{t+1}^i by $o_t^i \cdot \theta(o_t^i)$ and we predict the traffic condition directly use $\theta(o_t^i)$. Similarly, we can use the Gap as the input of STPGM. Besides the proposed two series, we can also use the first order difference of Org (i.e., Diff(Org) as defined in Sect. 2) as our input or the traffic conditions filtered with Kalman filtering. The details are presented in Sect. 8.

8 Experimental Study

In this section, we evaluate the effectiveness and efficiency of the proposed models.

8.1 Experiment Setting

Data Set. In all experiments, we use the real dataset which consists of GPS records collected from 12,000 taxis from November 1st to December 31st in 2012[1]. The GPS data are map matched [7,14] to road network[2] of Beijing. We evaluate our algorithms on the data of November and December respectively. For each month, we divide the data set into the training set (1st - 24th), and the test set (25th - the last day). We distinguish two cases in our experiments: the standard case and the sparse case. For the standard case, we select 10812 road segments which contains more than 140 GPS records per day in average. In the sparse case, we select 101672 road segments in which the GPS records occurred in more than 10 time intervals per day in average. In all experiments, we focus on the time period from 6 : 00 to 24 : 00 in each day since there are only few GPS records observed during 00 : 00 to 6 : 00.

Measurement. We evaluate the performances of our models on the test data set by Mean Absolute Error (MAE), Mean Relative Error (MRE) and Mean Squared Error (MSE), i.e., $\text{MAE} = \frac{1}{|E|}\sum_{i=1}^{|E|}\sum_{t=1}^{T}|p_t^i - o_t^i|$, $\text{MRE} = \frac{1}{|E|}\sum_{i=1}^{|E|}\sum_{t=1}^{T}|p_t^i - o_t^i|/o_t^i$, $\text{MSE} = \frac{1}{|E|}\sum_{i=1}^{|E|}\sum_{t=1}^{T}(p_t^i - o_t^i)^2$. Recall that we evaluate our algorithms on the datasets of November and December respectively. For convenience, for each model, we use the mean of the errors on the two months as the final error. All the experiments are implemented parallelly with Python 2.7 and run on a service on Open Stack (Intel Xeon E312 CPU of 16 cores with 2.1 GHz for each core and 32 GB memory on Ubuntu 14.04 LTS operate system).

---

[1] This data can be downloaded in http://www.datatang.com/data/45888.
[2] This data can be downloaded in http://www.datatang.com/data/45422.

8.2 Pre-processing

Data Cleaning. In the data cleaning phase, we eliminate the GPS records for taxis which slow down or even stop for picking or attracting passengers. We distinguish two cases of such records. One is *boarding*, i.e., the passengers get on or get off the taxi. The other is *balling*, i.e., the taxis slow down or stop to attract guests who need taxis. For the boarding state, the speed of the taxi usually varies sharply in a short time. Therefore, once we detect such sharp variation of the speed, we eliminate such GPS records. To handle the balling state, for a specific road, we check the speeds of all taxis in this road in a specific time t. If the speeds of most taxis are relatively high, only few of the taxis are driving at a very low speed, we think such taxis are on the balling state and we eliminate the corresponding GPS records.

Deal with Sparsity. Recall that as we claimed in Sect. 2, some road segments may not contain any GPS record during the time interval t for some $t \in [T]$. Thus, the corresponding traffic condition o_t^i is not defined. To solve this issue, for the road segment r_i, if the GPS record set observed in the time interval t is not empty, we define $\bar{o}_t^i$ as the average speed of the GPS records in the t-th interval. Otherwise, we have $\bar{o}_t^i = -1$. Let $\overline{A_t^i} = \{\bar{o}_{t'}^i | t' \equiv t \bmod M \wedge \bar{o}_{t'}^i \neq -1\}$ indicate the traffic conditions during the $t \bmod M$-th interval in each day. We define $\bar{a}_t^i$ as the mean of $\overline{A_t^i}$ and the series Bias $= \{b_t = \bar{o}_t^i - \bar{a}_t^i | \forall \bar{o}_t^i \neq -1\}$. Then, for each pair of adjacent elements in Bias, we perform the linear interpolation to obtain the undefined b_i. For example, if $Bias = \{b_1 = 3, b_4 = 4.5, b_7 = 10.5\}$, we obtain a series $\{b_1 = 3, b_2 = 3.5, b_3 = 4, b_4 = 4.5, b_5 = 6.5, b_6 = 8.5, b_7 = 10.5\}$ after performing linear interpolation. Finally, we have that the traffic condition o_t^i is obtained by $o_t^i = \bar{a}_t^i + b_t$.

8.3 Performance Evaluation

Performances of Different Models. We present the evaluations of our models. We first compare our model with the baseline Avg, i.e., predict the traffic condition o_t^i by its expected value a_t^i. Furthermore, in the recent work, Yang et al. [11] proposed STHMM for traffic condition prediction which is based on a spatial temporal hidden markov model. We compare STHMM with our models as well.

The results are shown in Fig. 7a, b. As we can see, the baseline (Avg) performs worst in both cases. Despite that STHMM outperforms Avg in both cases, both of our models PR-Tree and STPGM perform better than STHMM in our data set. Moreover, in the standard case, PR-Tree performs better than STPGM as shown in Fig. 7a whereas in the sparse case STPGM performs better. By combining PR-Tree and STPGM, our system ETCPS achieves the best performs in both two cases.

Verifying the Observed Patterns. Recall that as we claimed in Sect. 7, any time series related with traffic can be taken as the input of both PR-Tree and STPGM, and predict the traffic condition in the proper way. To illustrate the

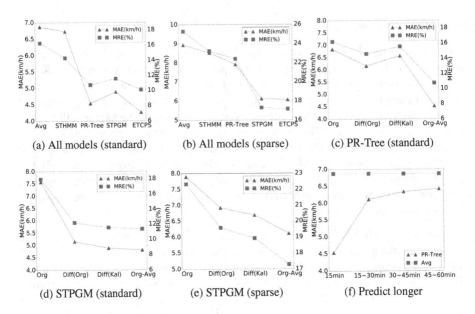

(a) All models (standard)   (b) All models (sparse)   (c) PR-Tree (standard)

(d) STPGM (standard)   (e) STPGM (sparse)   (f) Predict longer

Fig. 7. Performance analysis.

effects of the observations which we proposed in Sect. 3, we design four different experiments with different time series and evaluate each experiment on PR-Tree and STPGM respectively. The first two time series are Org and Gap = Org−Avg, as we used in Sects. 5 and 6. Then, we use the first order difference of Org as the input time series, denoted as Diff(Org). The t-th element in Diff(Org) is $o_{t+1} - o_t$. Furthermore, since the raw GPS records usually contain the noise such as the GPS drift, we use Kalman filtering to process the traffic condition series Org. We take the first order difference of the processed time series as the input as well, denoted as Diff(Kal).

We show the experimental results in Fig. 7c–e. Both PR-Tree and STPGM perform badly if we use Org as input directly. However, by using Diff(Org) and Gap instead, the performances improve significantly which verifies our observations.

Predict Longer Time Intervals. The PR-Tree model can be also used to predict the traffic conditions in the longer term. Given observations in interval t denoted as o_t, we first obtain the predicted traffic condition p_{t+1} and we take p_{t+1} as the "true traffic condition" in the time interval $t + 1$ and obtain p_{t+2}. Iteratively, we obtain the prediction after m time intervals p_{t+m}. In Fig. 7, we show the performance of PR-Tree in predicting the traffic condition in the next 0 to 60 min and comparing with the Avg method. As m increases, the performance becomes worse, but it is still better than Avg.

Effects of Time and Road Length. Figure 8 shows the effectiveness of our prediction across time. We plot the average mean squared error of travel speed

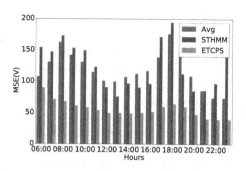

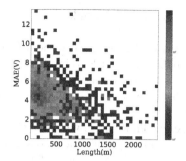

Fig. 8. MSE varies over day   **Fig. 9.** RMAE varies over road length

(MSE) for the baseline Avg, STHMM and ETCPS respectively during different hours for all days. The result shows that our system outperforms both the
baseline and STHMM.

To illustrate the effectiveness of the road length, in the Fig. 9, we show the
relation between MAE and the length of road segments. The result shows that
the road segments with longer length tend to have smaller MAE, i.e., our prediction performs better for the road segments with longer lengths.

Running Time. Since the predictions of both PR-Tree and STPGM are simple
which can be done in real time, we only present the running time for training
our models in Fig. 10 and Table 2. From Table 2, we can see that the training
time cost of PR-Tree is very small. It takes only 3.26 min to process 10^5 roads.
However, STPGM takes a much longer time to train as shown in Table 2. Especially for the state formulation phase, clustering the traffic conditions is time
costing. It takes 176.6 min to process the state formulation phase for 10^5 roads.
We stress that SHTMM applies a complicated state formulation algorithm and
the state space is much larger than STPGM. In our data set, the time consuming
of SHTMM is 1718 ms per road whereas even for STPGM, it only takes 13.3 ms
per road to train the model.

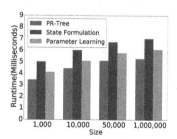

Fig. 10. Time cost

Table 2. Time cost (Minutes)

Size	PR-Tree	State Formulation	Parameter learning
10^3	0.04	1.72	0.22
10^4	0.46	17.46	2.09
$5*10^4$	2.14	88.1	10.09
10^5	3.26	176.6	19.61

9   Related Work

In this section, we review the related existing works. Most of prior works use the probabilistic models to predict the traffic conditions. Hunter et al. [4] formulated the traffic condition prediction in the arterial network to a maximum likelihood problem and estimated the travel time distributions based on the observed route travel times. Yeon et al. [12] estimated traffic conditions on a freeway using Discrete Time Markov Chains (DTMC). However these works assumed that the travel times on different road segments are independent without considering the correlation between the traffic conditions on different roads which may lead to incorrect prediction in the urban area [9].

To capture the correlations between road segments, Hofleitner et al. [3] formulated the transitions between states among adjacent road segments as a dynamic Bayesian network model and predicted the traffic conditions by an EM approach. However, it did not consider the efficiency on the large scale data. Yuan et al. [13] built a landmark graph based on the trajectories of taxis, where each node (entitled a landmark) indicates a road segment each edge indicates the aggregation of taxis commutes between two landmarks. They formulated the correlations and estimated the edge travel time distributions based on the landmark graph. However, as the landmarks are selected from the top-k frequently traversed road segments, many of road segments with sparse records can not be predicted.

The most related work with our model was proposed by Yang et al. [11]. They proposed an algorithm called STHMM which is a spatio temporal hidden markov model. They further presented an effective method to deal with the sparsity in the data. However, they did not consider the heterogeneity of transition patterns in different time intervals. In our experiment section (Sect. 8), we show that our model outperform STHMM in both the efficiency and accuracy. We stress that Chu et al. [2] considered the transition patterns in different time intervals and proposed a time-vary dynamic network. However their goal is to reveal the causal structure in a ring road system which differs from ours.

Furthermore, we stress two recent related works [1,10]. Wang et al. [10] presented an efficient algorithm to estimate the travel time of any path, based on sparse trajectories generated by taxi in recent time slots and in history, by using the tensor decomposition. Instead of predicting the traffic conditions, they studied the estimation of travel time for given travel paths in the current time slot. Asghari et al. [1] estimated the travel time distributions based on the historical sensor data. As their work studied the algorithm to find the most reliable route for the travel planning, it has a related but different scope.

10   Conclusion

We study the effective and scalable methods for traffic condition prediction. We propose an Ensemble based Traffic Condition Prediction System (ETCPS) which combines two novel models called Predictive Regression Tree (PR-Tree) and Spatial Temporal Probabilistic Graphical Model (STPGM). Our model is based on two useful observed correlations in the traffic condition data. Our system provides high-quality prediction and can easily scale to very large datasets. We conduct extensive experiments to evaluate our proposed models. The experimental results demonstrate that comparing with the existing methods, ETCPS is more efficient and accurate.

In the future, we plan to infer the traffic conditions by incorporating more features from heterogeneous data sources, such as the weather condition, POI information etc. Next, we will focus on the efficient way to deal with road segments which have extremely sparse trajectory records. Furthermore, we plan to try different ensemble methods to combine the different models in order to enhance the performance of the prediction.

Acknowledgment. This work was supported in part by the National Basic Research Program of China grants 2015CB358700, 2011CBA00300, 2011CBA00301, and the National NSFC grants 61033001, 61361136003.

References

1. Asghari, M., Emrich, T., Demiryurek, U., Shahabi, C.: Probabilistic estimation of link travel times in dynamic road networks. In: ACM SIGSPATIAL (2015)
2. Chu, V.W., Wong, R.K., Liu, W., Chen, F.: Causal structure discovery for spatio-temporal data. In: Bhowmick, S.S., Dyreson, C.E., Jensen, C.S., Lee, M.L., Muliantara, A., Thalheim, B. (eds.) DASFAA 2014, Part I. LNCS, vol. 8421, pp. 236–250. Springer, Heidelberg (2014)
3. Hofleitner, A., Herring, R., Abbeel, P., Bayen, A.: Learning the dynamics of arterial traffic from probe data using a dynamic Bayesian network. IEEE Trans. Intell. Transp. Syst. **13**(4), 1679–1693 (2012)
4. Hunter, T., Herring, R., Abbeel, P., Bayen, A.: Path and travel time inference from GPS probe vehicle data. NIPS Anal. Netw. Learn. Graphs **12**(1) (2009)
5. Kwon, J., Murphy, K.: Modeling freeway traffic with coupled HMMs. Technical report, University of California, Berkeley (2000)
6. Leontiadis, I., Marfia, G., Mack, D., Pau, G., Mascolo, C., Gerla, M.: On the effectiveness of an opportunistic traffic management system for vehicular networks. IEEE Trans. Intell. Transp. Syst. **12**(4), 1537–1548 (2011)
7. Lou, Y., Zhang, C., Zheng, Y., Xie, X., Wang, W., Huang, Y.: Map-matching for low-sampling-rate GPS trajectories. In: Proceedings of the 17th ACM SIGSPATIAL International Conference on Advances in Geographic Information Systems, pp. 352–361. ACM (2009)
8. Ma, S., Zheng, Y., Wolfson, O.: T-share: a large-scale dynamic taxi ridesharing service. In: 2013 IEEE 29th International Conference on Data Engineering (ICDE), pp. 410–421. IEEE (2013)
9. Ramezani, M., Geroliminis, N.: On the estimation of arterial route travel time distribution with Markov chains. Transp. Res. Part B: Methodol. **46**(10), 1576–1590 (2012)
10. Wang, Y., Zheng, Y., Xue, Y.: Travel time estimation of a path using sparse trajectories. In: Proceedings of the 20th ACM SIGKDD International Conference on Knowledge Discovery and Data Mining, pp. 25–34. ACM (2014)

11. Yang, B., Guo, C., Jensen, C.S.: Travel cost inference from sparse, spatio temporally correlated time series using Markov models. Proc. VLDB Endow. **6**(9), 769–780 (2013)
12. Yeon, J., Elefteriadou, L., Lawphongpanich, S.: Travel time estimation on a freeway using discrete time Markov chains. Transp. Res. Part B: Methodol. **42**(4), 325–338 (2008)
13. Yuan, J., Zheng, Y., Xie, X., Sun, G.: T-drive: enhancing driving directions with taxi drivers' intelligence. IEEE Trans. Knowl. Data Eng. **25**(1), 220–232 (2013)
14. Yuan, J., Zheng, Y., Zhang, C., Xie, X., Sun, G.Z.: An interactive-voting based map matching algorithm. In: Proceedings of the 2010 Eleventh International Conference on Mobile Data Management, pp. 43–52. IEEE Computer Society (2010)
15. Zheng, W., Lee, D.H., Shi, Q.: Short-term freeway traffic flow prediction: Bayesian combined neural network approach. J. Transp. Eng. **132**(2), 114–121 (2006)

Species Distribution Modeling via Spatial Bagging of Multiple Conditional Random Fields

Danhuai Guo, Yuanchun Zhou$^{(\boxtimes)}$, Yingqiu Zhu, and Jianhui Li

Computer Network Information Center, Chinese Academy of Sciences, Beijing, China
{guodanhuai,zyc,lijh}@cnic.cn, zhuyingqiu15@mails.ucas.ac.cn

Abstract. Satellite tracking technologies enable scientists to collect data of animal migrations and species habitats on a large scale. Modeling distributions of wild animals is of considerable use. It helps researchers to understand important ecological phenomena such as the spread of bird flu and climate changes. Species distribution modeling has been studied for a long time, however, most existing work provide solutions in a point-wise manner, ignoring the relevance between adjacent habitats, which may reflect an important dependency between nearby places. In this paper, we take the relevance into consideration, and then propose a novel method to model species habitats and predict possible distribution of wild animals by applying the Spatial Bagging of Multiple Conditional Random Fields(SBMCRFs) on remote-sensing data. To access the usability of our method, several experiments are implemented on a real world dataset of migratory birds from Qinghai Lake Reserve. The experiment results show that SBMCRFs outperforms the baselines significantly, and the relevance between nearby places is demonstrated to be an important factor in species distribution modeling.

Keywords: Species distribution modeling · Conditional random fields · Ensemble methods

1 Introduction

With the development of satellite tracking technology, geo-location and global positioning systems(GPS), many researchers apply them to find useful movement patterns. In biological research, tracking technology is used to observe animal migrations and collect a huge amount of detailed information of long-distance migratory species. Through analysis on collected datasets, many unanticipated patterns are found, which are proven to be of biological significance.

Due to the increasing size of trajectory data, it is difficult for ecologists to mine valuable patterns from the big size data. To help biologists, systems like MoveMine (Li et al. 2011) are developed to record and analysis the long-distance migration of wild animals. Moreover, several machine learning algorithms, including logistic regression and max entropy model are adopted to model the species distribution using the ecological niche features and position features. Researchers extract environmental features of each place and predict

© Springer International Publishing Switzerland 2016
S.B. Navathe et al. (Eds.): DASFAA 2016, Part II, LNCS 9643, pp. 437–450, 2016.
DOI: 10.1007/978-3-319-32049-6_27

the probability that animals may stay there. However, they model habitats in a point-wise manner and fail to take the dependencies between nearby places into considerations. As flocks moving between different habitats quite frequently, the suitable living places should be spatially correlated. In order to solve this problem, Conditional Random Fields(CRFs) is an optional method to model the spatial distribution of migratory animals.

CRFs is an undirected discriminative probabilistic graphical model and performs well in structured prediction. It combines the idea of graphical models and discriminative models and exploits the long dependent interactions. We first transform the geospatial space into a collection of grids separated by longitudinal and latitudinal parallel lines with equal cell size. The computation of prediction is implemented according to the grids. By using CRFs, both the ecological information of an individual grid and the relevance between grids are exploited. As the grids are not directed, it will be confusing to manually choose a direction to transform the grids in to a sequence. In this work, each direction has a corresponding separate model and the final result is an aggregation of all models. Finally, a regression model is introduced to better balance the weight of each CRFs.

The main contributions of our paper lies in two aspects:

Species Distribution Modeling: We propose a novel method, using SBMCRFs to solve the problem of species distribution modeling, which integrate the information from both a specific habitat and the relevance between nearby habitats to support the prediction.

Real evaluation: We evaluate our methods using a large-scale real GPS dataset. The experimental results demonstrate that CRFs improves the prediction accuracy and outperforms several baselines. Moreover, experiments on behavior prediction prove that both the richness and the utilities of environmental resources are useful for training the classifiers.

The remaining of the paper is organized as follows: In Sect. 2, we present some related work such as modeling movement patterns of wild species and applications of CRFs. The preliminary process of our method is introduced in Sect. 3. And the proposed distribution prediction method is described in Sect. 4. Detail descriptions about CRFs and our species distribution model are given in this section. Then relevant features the model used are listed in Sect. 5. In Sect. 6, the experiments are illustrated to show the effectiveness of the proposed methods. Finally, Sect. 7 concludes the paper and introduces possible future work.

2   Related Work

Pearson (2007) and Elith et al. (2009) surveyed a series of distribution modeling algorithms. The most cited works related to species distribution modeling are the methods proposed by Phillips et al. (2006), which adopted the Maximum Entropy (Maxent) model to predict the probability of a place to be the habitat. Caruana et al. (2006) proposed to select the most important environmental features to understand the species distribution. Various models including

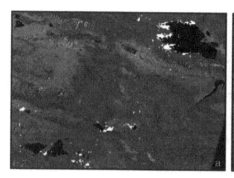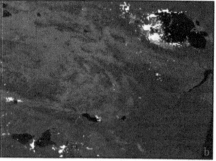

Fig. 1. a. Remote Sensing Image and Species Distribution in Qinghai-Tibet Plateau Area, where the white points indicate the places which have been visited by animals. The data are obtained from a GPS dataset recording the migration of Bar-headed goose from Mar 2007 to Oct 2007; and b. The Ruddy Shelduck from Mar 2008 to Oct 2008

Zheng et al. (2010), Li et al. (2011), Tang et al. (2009) and Tang et al. (2010) are proposed to analyze the movement patterns based on the trajectory data mining. They focus on mining the moving patterns from observation data and fail to make a prediction on the moving routes if the outside condition changes. Environmental features, which have a significant impact on moving objects, are not taken into their consideration. In this paper, we incorporate environmental features extracted from remote sensing data into the trajectory dataset to find potential habitats for wild species.

Conditional Random Fields(CRFs) is a discriminative graphical model which has been proven to be useful in a wide range of applications including natural language processing (Lafferty et al. 2001), computer vision (Phillips et al. 2006), text mining and bio-informatics. CRFs can perform better than traditional Hidden Markov Model(HMM) or Max-Entropy Model(MEMM), because it takes neighbor relationships and long distant dependent information into consideration. The discriminative model can obtain decision boundaries without calculating prior probabilities, which are hard to confirm.We apply CRFs to the species distribution prediction to make full use of information from neighbouring habitats. In addition, we use the graph structure in the discriminative condition model (Kumar and Hebert, 2006) for our 2-D graphical model.

3 Preliminary

Figure 1 illustrates the distribution of Bar-headed goose and the Ruddy Shelduck while they were breeding in Qinghai Lake. It can be observed that in some places, like left lakeshore of the Qinghai Lake, bird appears denser than the rest of the places. If these areas can be predicted in advance, scientists would gain more access to wild animals and make full preparation for wild animal protection. A challenge is that it is hard to predict these areas point by point, since the

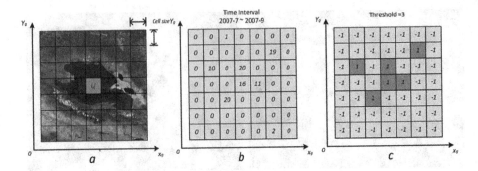

Fig. 2. Example of 2-D geographic model

scale of the distribution is too large while the dense parts take only a small portion of the whole district. Considering the sparseness, we divide the map into a set of rectangular grids as displayed in Fig. 2.

The remote sensing data on the specific area can be divided according to geographic grids. Certain longitudes and latitudes are set to define grids. The grid size is not fixed and various settings are tested in the experiments.

In Fig. 2, the first figure illustrates the grids divided by the longitudinal and latitudinal parallel lines. Then we calculate the number of bird occurrences in each grid. The second figure shows the statistical results. The calculation is repeated with a defined time interval and the time interval is from July 2007 to September 2007 in this case. The third figure shows labels that the prediction requires after preprocessing. First we set a threshold to classify whether a specific area is dense or sparse in terms of migration activities. Then grids with counts larger than the threshold are labeled as positive (+1) while the rest labeled as negative (-1).

4   Proposed Approach

Different from most existing work which only focus on the features of the point to be predicted, in this work, we leverage the dependencies of labels between the neighbouring grids. Meanwhile, the distribution modeling is transformed from a point-wise prediction to a sequence prediction. However, there is not an original sequence order between different grids, so we need find a direction to build the sequence. Obviously, the choosing of direction counts much in the prediction.

Figure 3 illustrates a toy example. Since the grids are divided by the longitudinal and latitudinal parallel lines, a grid can be viewed as a successive node of a sequence that have eight possible directions. As shown in the example, when computing the probability of point p, sequential dependencies from $a \to p$, $b \to p$, $c \to p$, $d \to p$, $e \to p$, $f \to p$, $g \to p$ and $h \to p$ are considered. Here we turn to statistical learning techniques to solve the problem. A separate model is trained for each direction, then we linearly integrate individual models. Furthermore, the eight models are combined by voting.

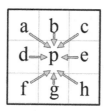

Fig. 3. A toy example. A grid can be viewed as a successive node of a sequence that have eight possible directions.

4.1 Individual Conditional Random Fields

In this section, we will describe the individual Conditional Random Fields in our model. Here we define X as the random variable over the observations and Y as the labels to be predicted. In this work, Y can take only two values, positive and negative, indicating a place is frequently visited by birds or not.

The CRFs model is an undirected graphical model, firstly proposed for text sequence segmentation and labeling (Lafferty et al. 2001). Let $G = (V, E)$ be an undirected graph where the node set E can be divided into two parts, X contains the observations and Y contains the corresponding labels. V is the vertex set of the graph. With the predefined graph structure G, the CRFs can model the conditional possibility $p(y|x)$ by

$$P(y|x) = \frac{1}{Z} \prod_{c \in C} \varphi_c(x_c, y_c) \tag{1}$$

where C is the set of cliques in the graph G, x_c and y_c is the set of x and y in the clique. φ_c denotes the potential function with positive value and Z is the normalization factor where

$$Z = \sum_y \prod_{c \in C} \varphi_c(x_c, y_c) \tag{2}$$

Traditionally, the potential feature function can be constructed in any form $f_k(x, y)$ for different applications. Each feature function $f_k(x, y)$ denotes a numeral feature on cliques c. So the $P(y|x)$ can be expressed as

$$P(y|x) = \frac{\exp(\sum_k \lambda_k f_k(x, y))}{\sum_y \exp(\sum_k \lambda_k f_k(x, y))} \tag{3}$$

where the parameter λ is the weight for the corresponding feature function. The feature functions are association potential functions. Each feature function reflects an specific influence on label y_i from feature x, or influences from labels of neighbor grids N_i. Thus, we have

$$P(y|x) = \frac{1}{Z(x)} \exp\{\sum_i \beta g(x, y_i) + \sum_{i \in S} \sum_{j \in N_i} \gamma f_{ij}(y_i, y_j, x)\} \tag{4}$$

where the parameters β, γ balance the relative contribution of the two feature functions. $g_i(x, y)$ denotes the effect from x while f_{ij} denotes that from neighbor grids. The potential feature functions can be in different forms. In this work, logistic function is adopted since the ecological variables are in high dimensions and correlated. We defined the feature function as

$$g(x, y_i) = \log(\sigma(y_i w^T \phi_i(x))) \tag{5}$$

$$f_{ij}(y_i, y_j, x) = \log \sigma((y_i y_j) u^T \phi_i(\mu_{ij})) \tag{6}$$

where σ is the logistic sigmoid function. The $\phi_i(x)$ is the basis function, and gaussian form is chosen in this paper. The $\mu_{ij}(\phi_i(x), \phi_j(x))$ is a new feature vector, which is concatenated by $[\phi_i(x), \phi_j(x))]$. w and u are the parameter for the correspond logistic feature function. The parameter of our CRF model are $\Theta = \{w, u, \gamma, \beta\}$

4.2 Model Inference with Constraint

With a given set of label data $(x^l, y^l)_1^n$, the model parameter Θ can be learned by maximizing the regularized log-loss of the training data using iterative searching algorithms, such as gradient descent. But the key problem in both learning and inference is the computation of the normalization factor Z. A simple measure is using the pseudolikelihood to reduce the computation cost (Kumar et al. 2006). However, the max-pseudolikelihood methods often cause over-fitting. Thus we add the L2 regularization term for parameter w, u. Finally, the model learning problem is to find the parameters maximizing the pseudolikelihood, i.e.,

$$\Theta = argmaxL(y|x) = \sum_l \sum_{i \in S} \ln P(y_i^l | y_{N_i}^l, x^l) - \frac{\lambda_1}{2} w^T w - \frac{\lambda_2}{2} u^T u \tag{7}$$

where the

$$P(y_i^l | y_{N_i}^l, x^l) = \frac{1}{z_i} \exp\{\beta g_i(x, y) + \sum_{j \in N_i} \gamma f_{ij}(y_i, y_j, x)\} \tag{8}$$

and new normalization factor is

$$z_i = \sum_{y_i \in \{-1, +1\}} \exp\{\beta g_i(x, y) + \sum_{j \in N_i} \gamma f_{ij}(y_i, y_j, x)\} \tag{9}$$

Another problem is that pseudolikelihood function is not convex. Moreover, initial values are needed in searching global maximum for w, u. Here we first assume that the feature function g, f are independent. Then the initial value for the parameters are computed though traditional maximum log-likelihood method. For example, the log-likelihood for association potential $g_i(x, y)$ can be expressed as

$$L_1(y|w, x) = \sum_l \sum_{i \in S} \log(\sigma(y_i^l w^T \phi_i(x^l))) \tag{10}$$

we then can get its second deviant hessian as

$$\nabla^2 L_1 = \sum_l^n \sum_{i \in S} y_i^{l^2} \phi_i(x^l) \phi_i(x^l) \sigma(y_i^l w^T \phi_i(x^l))(1 - \sigma(y_i^l w^T \phi_i(x^l))) \qquad (11)$$

Now w reaches global maximum though Newton's methods. The value of u can be obtained in a similar way. After getting initial values, equations listed below can be used to compute the Θ by gradient descent method.

$$\frac{\partial L}{\partial w} = \sum_l^n \sum_{i \in S} \{\beta m_1 - \frac{\sum_y m_1 \exp(E_i(x^l, y_i^l))}{Z_i}\} - \lambda_1 w \qquad (12)$$

$$\frac{\partial L}{\partial u} = \sum_l^n \sum_{i \in S} \{\gamma m_2 - \frac{\sum_y m_2 \exp(E_i(x^l, y^l))}{Z_i}\} - \lambda_2 u \qquad (13)$$

$$\frac{\partial L}{\partial \beta} = \sum_l^n \sum_{i \in S} \{g_i(x^l, y^l) - \frac{\sum_y \beta \alpha_i(x^l, y^l) \exp(E_i(x^l, y^l))}{Z_i}\} \qquad (14)$$

$$\frac{\partial L}{\partial \gamma} = \sum_l^n \{m_3 - \frac{\sum_y m_3 \exp(E_i(x^l, y^l))}{Z_i}\} \qquad (15)$$

where

$$E_i(x^l, y^l) = g_i(x^l, y^l) + \sum_{j \in N_i} f_{ij}(y_i^l, y_j^l, x) \qquad (16)$$

$$m_1 = y_i^l \phi_i(x^l)(1 - \sigma(y_i^l w^T \phi_i(x^l))) \qquad (17)$$

$$m_2 = \sum_{j \in N_i} y_i^l \phi_i(\mu_{ij}^l)(1 - \sigma(y_i^l y_j^l u^T \phi_i(\mu_{ij}^l))) \qquad (18)$$

$$m_3 = \sum_{i \in S} \sum_{j \in N_i} f_{ij}(y_i^l, y_j^l, x^l) \qquad (19)$$

Thus, the task is to find the most possible label for the given observation data with the conditional possibility defined above, i.e.,

$$y^{opt} = argmax \sum_{i \in S} \exp\{\beta g_i(x, y) + \sum_{j \in N_i} \gamma f_{ij}(y_i, y_j, x)\} \qquad (20)$$

It is obvious that we need to compute the expectation of features over the models and search over all possible assignments of the label to reach the maximum. A simple exhaustive search would be prohibitively expensive. There are two major approaches to compute the marginal probability or conditional probability i.e., exact inference method and approximate inference method. In this work, we use an iterative technique named iterated conditional modes (ICM, Besag 1986) to solve the problem, which is an application of coordinate-wise gradient ascent. Finally, the label serves as a reference for distribution prediction. A grid with a positive label means that animals have a relatively high possibility to gather in the corresponding area. Furthermore, the combination of labels of nearby grids reveals the most probable moving direction of animal species, which also contributes to biological researches.

5   Features

In order to model the species distribution of wild animals, ecological niche features are extracted to characterize habitats. In this paper, a set of ecological features are compiled, as described in Table 1.

Table 1. List of ecological features from the satellite remote sensing data

Feature	Source	Unit	Resolution
Temperature	MODIS(MOD11A2)	°C	1 km
NDVI	MODIS(MOD13A2)		1 km
Elevation	SRTM-DEM	meter	0.09 km
Distance(Water)	MODIS(MOD12Q1)	meter	1 km
Distance(Grassland)	MODIS(MOD12Q1)	meter	1 km
Distance(Forest)	MODIS(MOD12Q1)	meter	1 km
Distance(Wetland)	MODIS(MOD12Q1)	meter	1 km
Distance(Farmland)	MODIS(MOD12Q1)	meter	1 km
Distance(Bareland)	MODIS(MOD12Q1)	meter	1 km
Distance(Ice)	MODIS(MOD12Q1)	meter	1 km
Distance(Bush)	MODIS(MOD12Q1)	meter	1 km

In addition, several open data sources are employed in this work. We first refine the remote-sensing images with the Geospatial Data Abstraction Library (GDAL) and the ArcGIS 9.2 Dataset proposed by Environmental Systems Research Institute. The Moderate Resolution Imaging Spectroradiometer (MODIS) dataset and Shuttle Radar Topography Mission (SRTM) dataset are adopted to obtain ecological features, including temperature, Normalized Difference Vegetation Index (NDVI), elevation and the richness of water, grassland, forests, wetland, farmland, bareland, ice and bushes. A detailed description is given below.

5.1   Temperature

Temperature is derived from the MODIS Land Surface Temperature (LST) product (MOD11A2).

5.2 NDVI

Normalized Difference Vegetation Index (NDVI) is derived from the MODIS product which is named as MOD13A2. The value denotes whether a specific region is covered with living vegetation. Generally, NDVI represents the vegetation condition for regions. The value ranges from -1.0 to +1.0, where a negative value indicates a lack of plants while a positive one indicates the prosperity of vegetables.

5.3 Elevation

Elevation is from the dataset of SRTM, which is the digital topographic database of Earth with resolution of 90 m.

5.4 Distance

In this work, we calculate the distance between the center of a specific grid and the nearest water, wetland, farmland, grassland, bare land, forests, ice and bushes. Here distance measures the richness of the natural resources for a specific grid area. The information is obtained from MOD12Q1 product of MODIS.

6 Experiment

In this section, we will describe our dataset and settings of our experiments. Experiment results are illustrated and discussed later.

6.1 Dataset

The dataset is collected from an on-site study conducted at the Qinghai Lake National Nature Reserve, Qinghai Province, China, from March 2007 to December 2009. By equipping transmitters on birds, the satellite tracks the positions of migratory birds. Until 2010, over one million migration records were collected. The data from Bar-Headed goose, Ruddy-shelduck and Great Black-headed Gull are used in the experiments.

6.2 Experiment Settings

Two experiments are performed in this work. The first one is a comparison between our model and several existing baselines including BIOCLIM (Busby, 1991), DOMAIN (Carpenter et al. 1993), MAHAL (Etherington et al. 2009), Generalized Linear Model Logistic Regression (GLM logistic), Generalized Linear Model Poisson Regression (GLM Poisson), Generalized Additive Model (GAM), Maximum Entropy (Maxent), Support Vector Machine (SVM), Random Forest and Boost Regression Tree (BRT, Elith et al. 2008).

Area Under Curve (AUC, Pencina et al. 2008) is adopted to measure the effectiveness of models above.

The baseline models are given with the same data and features as our model. For each baseline model, the original data is inputted after several preprocessings that are aimed at satisfying the requests and constrains of the model. Several common prediction methods, such as Logistic Regression, which is widely used in both research fields and industry, are included in the baseline models. Moreover, classifiers like binary SVM are exploited to provide comparable performances. For an individual area on the map, the prediction of species distribution can be considered as a classification problem, namely to classify whether there would appear a large mount of animals at the specific time or not.

After the first experiment, performances of all models on the same geographical and ecological data are obtained. Each model outputs a set of labels corresponding to areas on the map, which describes the possible distribution. To access the effects of the models, we conduct an evaluation on their performances mainly using AUC as the measurement.

As the grid size is manually set and may influence the performance of the models, the second experiment is implemented to investigate its impact. We choose five algorithms and vary the grid size. Changes of results are recorded.

6.3   Results

Figure 4 shows the results of the models on different kinds of birds. Our proposed model outperforms the baselines on the Bar-headed Goose data and Ruddy Shelduck data. The Maxent algorithm provide the best performance on average. Our method can be viewed as a maximum entropy prediction on sequence data, and the result proves that the spatial bagging is useful on identification of the habitats and distribution prediction.

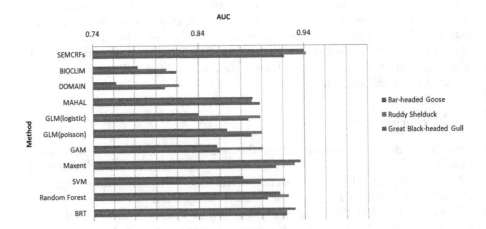

Fig. 4. The experimental results in terms of AUC of different models

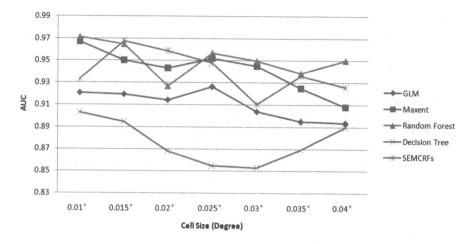

Fig. 5. The AUC of different models with a varying cell size

Figure 5 illustrates the performance of models with different grid sizes. For the simplicity of computation, the grid is set as squares. And decision tree is added in the experiment. The random forest is significantly better than decision tree, which proves that the ensemble of multiple models is more effective on this task. But the AUC of the proposed model is not stable as is shown in the figure, a smooth method may be needed to improve the effectiveness.

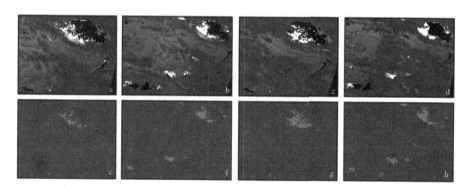

Fig. 6. Results of Species Distribution Prediction using SBMCRFs. From left to right, first two columns are the predicated spatial distribution of Bar-headed goose in 2008 and 2009. And last two columns are the predicated spatial distribution of Ruddy Shelducks

As is shown in Fig. 5, the size of grid actually has an impact on the performance of the model. Consider the sparsity of species distribution provided by the given data, the manually set grid size is acceptable and provides preferable effects.

448   D. Guo et al.

Figure 6 shows an example of the prediction results. In the first row there are four maps displaying the predicated species distribution. The maps illustrate the spatial distribution of bar-headed geese and Ruddy Shelducks. On each map, the white spots denote that wild animals would gather in the corresponding areas with considerable possibilities. With labels provided by our model, areas where species are possible to appear are highlighted. As is shown in the map, animals are likely to gather near the lake, which is consistent with the real situation. The prediction results may provide some interesting findings for biologists. The results can imply several unexpected areas where animals gather, which are significant for biological research. The second row consists of maps that display the density distribution of the prediction. The density distribution represents a more detailed description.

7 Conclusion and Future Works

In this work, we propose a novel method to model habitats and predict species distribution using Spatial Bagging of Multiple Conditional Random Fields. Conditional Random Fields is introduced to extract auxiliary information from neighbouring habitats, which positively supports the prediction. In practice of biological researches, we divided maps into grids and label them according to a serious of computations as illustrated above. Based on those labels, we identify the possible habitats and potential moving direction of animal species. The experimental results show that the proposed method provide better performance than baselines. Moreover, the introduction of relevance between nearby areas is demonstrated to be of significance.

Although our method performs well on the given data, there remains some processings to be improved. In order to expand our method to a general method, the most important phase is the selection of the grid size. During the processing on the original geographical and ecological data, the size of a grid cell has a significant impact on the distribution of calculated values. The attributes of grid cells are influenced and limited by the grid size. Either too large or too small of a grid would lead to an apparent loss of accuracy of the modeling. To fit different dataset, a hierarchical size set can provide improvements on both effect and efficiency. Instead of using an uniform size, we can set large sizes for the sparse parts of the data while set samll sizes for the dense parts. For further optimization, techniques such as krigging and spatio-temporal variograms can be used. It is predictable that they can provide access to higher precision of measurements on geographical and ecological data. In the future, we will continuously test our method with different datasets and improve the effectiveness though constant optimizations.

Acknowledgments. This work is partly supported by the Natural Science Foundation of China (NSFC) under Grant No. 41371386 and 91224006.

References

Busby, J.: BIOCLIM a bioclimate analysis and prediction system. Plant Prot. Q., 6 (1991)

Carpenter, G., Gillison, A.N., Winter, J.: DOMAIN: a flexible modelling procedure for mapping potential distributions of plants and animals. Biodivers. Conserv. **2**(6), 667–680 (1993)

Etherington, T.R., Ward, A.I., Smith, G.C., Pietravalle, S., Wilson, G.J.: Using the Mahalanobis distance statistic with unplanned presence only survey data for bio-geographical models of species distribution and abundance: a case study of badger setts. J. Biogeogr. **36**(5), 845–853 (2009)

Elith, J., Leathwick, J.R., Hastie, T.: A working guide to boosted regression trees. J. Anim. Ecol. **77**(4), 802–813 (2008)

Pencina, M.J., D'Agostino, R.B., Vasan, R.S.: Evaluating the added predictive ability of a new marker: from area unde the ROC curve to reclassification and beyond. Stat. Med. **27**(2), 157–172 (2008)

Zheng, V.W., Zheng, Y., Xie, X., Yang, Q. Collaborative location, activity recommendations with GPS history data. In: Proceedings of the 19th International Conference on World Wide Web, pp. 1029–1038. ACM (2010)

Li, Z., Han, J., Ji, M., Tang, L.A., Yu, Y., Ding, B., Kays, R.: Movemine: mining moving object data for discovery of animal movement patterns. ACM Trans. Intell. Syst. Technol. (TIST) **2**(4), 37 (2011)

Tang, M., Zhou, Y., Li, J., Wang, W., Cui, P., Hou, Y., Yan, B.: Exploring the wild birds migration data for the disease spread study of H5N1: a clustering and association approach. Knowl. Inf. Syst. **27**(2), 227–251 (2011)

Tang, M.J., Zhou, Y.C., Cui, P., Wang, W., Li, J., Zhang, H., Hou, Y.S., Yan, B.P.: Discovery of migration habitats and routes of wild bird species by clustering and association analysis. In: Huang, R., Yang, Q., Pei, J., Gama, J., Meng, X., Li, X. (eds.) ADMA 2009. LNCS, vol. 5678, pp. 288–301. Springer, Heidelberg (2009)

Tang, M.J., Wang, W., Jiang, Y., Zhou, Y., Li, J., Cui, P., Liu, Y., Yan, B.: Birds bring flues? mining frequent and high weighted cliques from birds migration networks. In: Kitagawa, H., Ishikawa, Y., Li, Q., Watanabe, C. (eds.) DASFAA 2010. LNCS, vol. 5982, pp. 359–369. Springer, Heidelberg (2010)

Pearson, R.G.: Species distribution modeling for conservation educators and practitioners. Lessons in Conservation (LinC) Developing the capacity to sustain the earth's diversity, 54 (2007)

Elith, J., Leathwick, J.R.: Species distribution models: ecological explanation and prediction across space and time. Annu. Rev. Ecol. Evol. Syst. **40**, 677–697 (2009)

Caruana, R., Elhawary, M., Munson, A., Riedewald, M., Sorokina, D., Fink, D., Hochachka, W.M., Kelling, S.: Mining citizen science data to predict orevalence of wild bird species. In: Proceedings of the 12th ACM SIGKDD International Conference on Knowledge Discovery and Data Mining, pp. 909–915. ACM, NewYork (2006)

Kumar, S., Hebert, M.: Discriminative random fields. Int. J. Comput. Vision **68**(2), 179–201 (2006)

Lafferty, J., McCallum, A., Pereira, F.C.: Conditional random fields,: Probabilistic models for segmenting and labeling sequence data. In: Proceedings of the Eighteenth International Conference on Machine Learning, ICML 01, pp. 282–289. CA, USA, Morgan Kaufmann Publishers Inc, San Francisco (2001)

Phillips, S.J., Anderson, R.P., Schapire, R.E.: Maximum entropy modeling of species geographic distributions. Ecol. Model. **190**(3), 231–259 (2006)

Besag, J.: On the statistical analysis of dirty pictures. J. R. Stat. Soc. Ser. B (Methodol.) **48**(3), 259–302 (1986)

Real-Time Personalized Taxi-Sharing

Xiaoyi Duan[1], Cheqing Jin[1]($\boxtimes$), Xiaoling Wang[1], Aoying Zhou[1], and Kun Yue[2]

[1] School of Computer Science and Software Engineering,
Institute for Data Science and Engineering,
East China Normal University, Shanghai, China
dollyxiaoyi@gmail.com, {cqjin,xlwang,ayzhou}@sei.ecnu.edu.cn
[2] School of Information Science and Engineering,
Yunnan University, Kunming, China
kyue@ynu.edu.cn

Abstract. Taxi-sharing is an efficient way to improve the utility of taxis by allowing multiple passengers to share a taxi. It also helps to relieve the traffic jams and air pollution. It is common that different users may have different attitudes towards the taxi-sharing scheduling plan, such as the fee to be paid and the additional time to the destination. However, this property has not been paid enough attention to in the traditional taxi-sharing systems – the traditional focus is how to decrease the travel distance. We study the problem of personalized taxi-sharing in this paper, with the consideration of each passenger's preference in payment, travel time and waiting time. We first define the satisfaction degree of each party involved in the scheduling plan, based on which two goals are defined to evaluate the overall plan, including MaxMin and MaxSum. Subsequently, we devise a two-phase framework to deal with this problem. The statistical information gathered during the offline phase will be used to hasten query processing during the online phase. Experimental reports upon the real dataset illustrate the effectiveness and efficiency of the proposed method.

1 Introduction

The rapid increment of vehicles in large cities brings with some serious social issues, such as traffic jams and air pollution [1–3]. Taxi is an important means of transport because it is convenient for people to go to the destination quickly. However, the benefits of taxi have not been fully exploited since one taxi is allowed to take only one passenger or one group of passengers, leaving some seats unoccupied. Consequently, it is meaningful to devise a taxi-sharing system that allows more than one group of passengers to share one taxi, so as to relieve the effects of traffic jams and air pollution.

Actually, some attempts have been made to fulfill the idea of vehicle sharing. For example, some Web sites (e.g., AApinche) allow uses to publish their travel plan or to find a matched partner from the existing posts. However, it usually takes a long time before users can find their matches [4,5]. Recently, some works

S.B. Navathe et al. (Eds.): DASFAA 2016, Part II, LNCS 9643, pp. 451–465, 2016.
DOI: 10.1007/978-3-319-32049-6_28

focus on devising a ride-sharing strategy, which allows multiple groups of passengers sharing one vehicle [6,7]. The T-share system aims at finding a plan to minimize the total travel distance for each request [8]. Unfortunately, none of the above works consider the customized preference of each user. Our pruning rules fully take account of users' preferences.

In general, a successful schedule plan should satisfy each party in this plan. Otherwise, this plan may not go into reality. There are three parties in a taxi-sharing plan: driver, taxi rider, and waiting passenger. Taxi rider is the passenger who has already ridden on the taxi, and waiting passenger is the one who is hailing a taxi [9]. Each user has personalized requirement about the satisfaction. Someone is sensitive to the fee, but insensitive to the driving time, while someone else may be insensitive to the fee but wants to go to the destination as soon as possible. Therefore, a customized taxi-sharing plan needs to fully consider users' preferences and then we designed the plan.

However, it is challenging to deal with personalized taxi-sharing due to the following reasons. (i) *Real-time service requirement.* It is critical to send the response to users as soon as possible. However, it is non-trivial to achieve the goal due to the huge amount of taxis and the dynamic properties of taxis. (ii) *Personalized preference.* Similar route is not the only factor to be concerned in taxi sharing, because customers' service preferences are too diverse to compose a sharing plan, such as sharing purpose and requirements to partners. (iii) *Different route matching.* As most users in our system tend to have different origins and destinations, it is not likely to find a taxi rider with the same origin and destination. In order to enhance the sharing chance, we need to relax the matching condition. To handle the aforementioned challenges, we propose a novel framework in this paper, which contains two phases: *offline* and *online*. The goal of the *offline* phase is to compute some statistical information to support online computation. For example, we can estimate the upper and lower bounds of the driving time between two points. During the *online* phase, some pruning rules are devised to hasten query processing.

The contributions are summarized below.

- We formalize the personalized taxi-sharing issue which fully considers users' preferences. We define *satisfaction* degree of a person (driver, waiting passenger, or taxi rider) and two global targets, MaxSum and MaxMin, to measure the quality of the overall plan.
- We deal with the real-time taxi-sharing issue efficiently to support the scenario with huge number of taxis and passengers. To hasten query processing of online phase, we precompute several operators during offline phase and use pruning rules to save significant part of computations.
- We conduct extensive experiments on the real trajectory dataset to validate the effectiveness and efficiency of the proposed method.

The remainder of this paper is organized as follows. Section 2 introduces the preparatory work of data model and problem statement. Section 3 introduces the framework of our new method. Section 4 describes the offline phase that precompute some statistics. Section 5 describes the online phase to process queries efficiently by using pruning rules. Section 6 reports experimental results upon real data sets. Section 7 reviews related work. A brief conclusion is given in the last section.

2 Preliminaries

Data Model: A typical taxi-sharing system involves three parties, including taxi driver (shorten as TD), taxi rider (shorten as TR), and waiting passenger (shorten as WP)[1]. Since TR and WP may have different origins and destinations, taxi sharing plans can be diverse. Assuming a WP plans to go from O to D, and a TR is traveling from A to B, there exist two possible scheduling plans to share this taxi, namely WPF (Waiting Passenger First) and TRF (Taxi Rider First). As illustrated in Fig. 1, the route in WPF plan is $AODB$, as the taxi goes to D at first after picking up WP at O, while the route in TRF plan is $AOBD$, as the taxi goes to B at first instead. In this paper, we focus on the scenario that a taxi is shared by at most two groups of passengers, and leave the scenario that multiple groups sharing one taxi as a piece of future work.

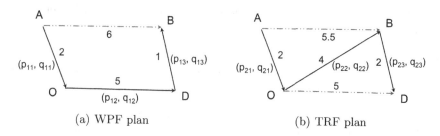

(a) WPF plan   (b) TRF plan

Fig. 1. Taxi-sharing Plans (WPF and TRF)

Pricing Strategy: The pricing strategy varies a lot in different cities, e.g., the unit price may change by different time and places. Here, we consider a model for the illustration purpose where the per-kilometer price is fixed, denoted as k. The detail of pricing strategy is described below.

Let $\overrightarrow{D}_{WPF} = (\mathsf{rdis}(A,O), \mathsf{rdis}(O,D), \mathsf{rdis}(D,B))$ denote a distance vector in the WPF plan, where $\mathsf{rdis}(l_1, l_2)$ denotes the road network distance between two locations l_1 and l_2. Let $\overrightarrow{p_1} = k \cdot (p_{11}, p_{12}, p_{13})$ and $\overrightarrow{q_1} = k \cdot (q_{11}, q_{12}, q_{13})$ ($\forall i$, $p_{1i}, q_{1i} \in [0,1]$) denote the payment share for WP and TR in the WPF plan respectively (see Fig. 1(a)). Then, the payment for WP (denoted as M_{WP}), TR (denoted as M_{TR}), and the income for TD (denoted as M_{TD}) are computed as: $M_{WP} = \overrightarrow{p_1} \cdot \overrightarrow{D}_{WPF}$, $M_{TR} = \overrightarrow{q_1} \cdot \overrightarrow{D}_{WPF}$, and $M_{TD} = M_{WP} + M_{TR}$. To ensure TD always gets more income, the parameters satisfy: $\forall i$, $p_{1i} + q_{1i} \geq 1$.

---

[1] Note that TR or WP can also be a group of people rather than one person.

Symmetrically, let $\overrightarrow{D}_{TRF} = (\text{rdis}(A,O), \text{rdis}(O,B), \text{rdis}(B,D))$ denote the distance vector in the TRF plan. Let $\overrightarrow{p_2} = k \cdot (p_{21}, p_{22}, p_{23})$ and $\overrightarrow{q_2} = k \cdot (q_{21}, q_{22}, q_{23})$ ($\forall i,\ 0 \le p_{2i},\ q_{2i} \le 1$) denote the payment share for WP and TR respectively (see Fig. 1(b)). Accordingly, $M_{WP} = \overrightarrow{p_2} \cdot \overrightarrow{D}_{TRF}$, $M_{TR} = \overrightarrow{q_2} \cdot \overrightarrow{D}_{TRF}$, and $M_{TD} = M_{WP} + M_{TR}$.

Example 1. Table 1 illustrates a pricing strategy for Fig. 1. For example, in the WPF plan, the payments of WP and TR are computed as: $M_{WP} = 2 \times (0.5 \times 2 + 0.6 \times 5 + 0.5 \times 1) = 9$, $M_{TR} = 2 \times (0.6 \times 2 + 0.6 \times 5 + 0.6 \times 1) = 9.6$.

Table 1. An example of pricing strategy based on Fig. 1

Plan	$\overrightarrow{p}$	$\overrightarrow{q}$	M_{WP}	M_{TR}	M_{TD}
WPF	$2 \times (0.5, 0.6, 0.5)$	$2 \times (0.6, 0.6, 0.6)$	9	9.6	18.6
TRF	$2 \times (0.5, 0.6, 1)$	$2 \times (0.6, 0.6, 0)$	10.8	7.2	18

Note: k is 2 RMB/km.

Preference and Satisfaction: The personalized preference of each user (either WP or TR) is defined below.

Definition 1 (Preference). *The preference of a user (either WP or TR) is defined as (α, β), where α controls user preference to the time cost, and β controls user preference to the payment. $\alpha \ge 0$, $\beta \ge 0$, and $\alpha + \beta = 1$.*

Furthermore, we use $(\alpha_{WP}, \beta_{WP})$ and $(\alpha_{TR}, \beta_{TR})$ to denote the preference of WP and TR respectively. If $\alpha_{WP} > \beta_{WP}$, WP prefers to save more time. If $\alpha_{TR} < \beta_{TR}$, TR prefers to save money, even though the travel time may be extended.

Generally, the "satisfaction" degree of each party (either TR or WP) must reflect the change with/without taxi-sharing.

Definition 2 (Satisfaction). *For passengers (WP or TR), satisfaction is defined as $S = \alpha\Delta_T + \beta\Delta_P$, where Δ_T denotes the time difference between sharing taxi ans non-sharing taxi, and Δ_P denotes the payment difference between sharing and non-sharing. For drivers (TD), satisfation is defined as income difference between sharing taxi and non-sharing taxi.*

Specifically, the satisfaction degrees of WP (S_{WP}), TR (S_{TR}) and TD (S_{TD}) are computed as:

- **WP's Satisfaction:** $S_{WP} = \alpha_w \Delta_T + \beta_w \Delta_P$, where (α_w, β_w) denotes WP's preference in time and payment.
- **TR's Satisfaction:** $S_{TR} = \alpha_r \Delta_T + \beta_r \Delta_P$, where (α_r, β_r) denotes TR's preference in time and payment.

– **TD's Satisfaction:** $S_{TD} = (I - I')$, where I is the real gain in taxi-sharing, and I' is the gain when only carrying with one passenger along the same route.

A user is satisfied with this plan only if *satisfaction* is greater than zero. Note that there exists a bit difference when computing Δ_T and Δ_P between TR and WP. Originally, TR can go to the destination without carrying WP. But in a taxi-sharing plan, he must pick up WP at first, so that the time is extended. Under such situation, $\Delta_T < 0$. Different from TR, WP has two choices, either sharing a taxi with WP, or just waiting for a new taxi. But it is unclear whether a new taxi will come there quickly or not. We use the average waiting time to estimate the time of the second choice by using historical statistics. Under such a condition, the value of Δ_T is definitely not negative.

The *satisfaction* degrees defined above illustrate how a party treats the scheduling plan. A scheduling plan is valid only when all of three parties feel satisfied with that plan, i.e., $S_{TR} \geq 0 \wedge S_{WP} \geq 0 \wedge S_{TD} \geq 0$. This condition can also be shorten as $S_{TR} \geq 0 \wedge S_{WP} \geq 0$ since $S_{TD} \geq 0$ always holds.

Query Definition: The final task is to select a best choice out of all valid scheduling plans. Let's consider two valid plans with satisfactions as (S_{TR}, S_{WP}, S_{TD}) and $(S'_{TR}, S'_{WP}, S'_{TD})$. If $(S_{TR} \geq S'_{TR} \wedge S_{WP} \geq S'_{WP} \wedge S_{TD} \geq S'_{TD})$, the former is better. Otherwise, there exists no clear winner. Hence, we define two queries for this issue, including MaxSum and MaxMin.

Definition 3 (MaxSum Query). *Given a set of scheduling plans, return a valid plan with maximal value of the sum of satisfactions.*

Definition 4 (MaxMin Query). *Given a set of scheduling plans, return a valid plan with maximal value of the minimal satisfaction among three.*

Example 2. Consider the satisfaction degrees in Table 2. The MaxSum query will return the WPF plan, while the MaxMin query will return the TRF plan.

Table 2. An example of MaxSum and MaxMin

Plan	S_{TR}	S_{WP}	S_{TD}	MaxSum	MaxMin
WPF	1.68	12.62	2.8	17.1	1.68
TRF	2.48	11.24	2	15.72	2

3 The Framework

A typical taxi-sharing system manages a large number of taxis. The position of each taxi will be reported to the system continuously. When a WP sends a request to the system, the system will select a most suitable taxi for WP immediately.

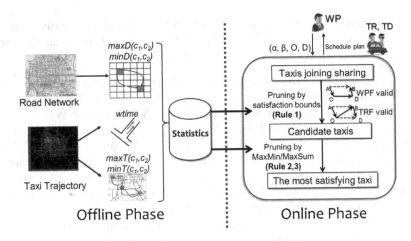

Fig. 2. Framework of real-time personalized taxi sharing

Figure 2 illustrates the framework, which contains two phases, *offline* phase for statistics and *online* phase for query processing. With the help of statistical information stored in the *offline* phase, the query processing can be significantly improved, avoiding comparing all taxis running in the system.

Offline Phase. This phase aims at providing statistical information to support pruning rules. Recall that the satisfaction is actually the tradeoff of the traveling distance and the travel time, but to compute the road network distance and travel time precisely is infeasible. Hence, we attempt to estimate them in the offline manner. The road network helps to compute the distance between two points. Their travel time and waiting time of a road are estimated based on taxi trajectory records. More details will be presented in Sect. 4.

Online Phase. The goal of this phase is to select a taxi most suitable for a request launched by a WP. Given all taxis that are willing to join sharing, we devise several pruning rules so that the candidate taxis to be further evaluated are significantly smaller than the original set of taxis. By using the statistics generated from the offline phase, each pruning rule can be evaluated in $O(1)$ time. In comparison, the cost to evaluate a taxi needs to find the shortest path between two points, which is expensive when the distance between these two points is far. See Sect. 5 for the details about this phase.

4   Offline Phase

During the offline phase, we generate statistical information to support the pruning rules (to be introduced in the next section). The road network is divided into cells with equal sizes. We need to compute the following five operators based on historical trajectories. The first four operators use cells of source point and the destination point as input, while the last one is for a pick-up or drop-off point in the space. Each operator returns a value that is pre-computed and stored in the system.

- minD(c, c'). It returns a value no greater than the minimal distance of the shortest path between two points in two cells, c and c', respectively.
- maxD(c, c'). It returns a value no smaller than the maximal distance of the shortest path between two points in two cells, c and c', respectively.
- minT(c, c'). It estimates the minimal travel time between two arbitrary points in two cells, c and c', respectively.
- maxT(c, c'). It estimates the maximal travel time between two arbitrary points in two cells, c and c', respectively.
- wtime(p). It returns the average waiting time for the road segment where the pick-up or drop-off point p locates, which is calculated by the average time intervals between two unoccupied taxis passing by.

For any two cells c and c', the shortest path between them must start and end at the border of each cell. Let $P(c)$ and $P(c')$ denote the sets of the points that intersect with roads on the border of cells c and c' respectively. Then, minD is computed as $\min_{p \in P(c), p' \in P(c')}$ rdis(p, p'), where rdis(p, p') denotes the road network distance of the shortest path between two points p and p'. The common ways to implement rdis(p, p') includes Dijkstra's algorithm [10] and the A* algorithm [11].

To compute maxD precisely is more complex than minD, since the starting and ending points for the corresponding path may be located inside of two cells, no longer at the borders. Let $d(c, p)$ denote the maximal distance of the shortest path from point p to any point in c. In general, computing $d(c, p)$ is cheap since cell c is set small. In this way, we return $\min_{p \in P(c), p' \in P(c')}($rdis$(p, p') + d(c, p) + d(c', p'))$.

It is infeasible to predict minT or maxT precisely only based on road network. It is affected by some other factors, such as the number of traffic lights in the path. Moreover, it is inaccurate to use the maximal allowed speed to estimate minT, since in most cases the driving speed is lower than the maximal allowed speed. Hence, we attempt to use the historical trajectory database. Let ttime1(c, c', tr) denote the travel time during trajectory tr leaving c and entering c'. We return $\min_{tr}$ ttime1(c, c', tr) for the minT operator. For the case where there exists no historical trajectory passing through c and c', we can (i) search for another cell c'', and return $\min_{c'', tr, tr'}($ttime1$(c, c'', tr) + $ttime1$(c'', c', tr'))$, or (ii) return the time by using the maximal allowed speed.

We return $\max_{tr}$ ttime2(c, c', tr) to estimate maxT based on historical trajectories. It records the time when tr first comes in c and finally leaves c'. Compared with $\max_{tr}$ ttime1(c, c', tr), this new subroutine considers the travel time in each cell. For the case when there exists no historical trajectory passing through c and c', we return: $\min_{c''}(\max_{tr}$ ttime2$(c, c'', tr) + \max_{tr'}$ ttime2$(c'', c', tr'))$.

The goal of the last operator, wtime, is to estimate the average time to wait for a vacant taxi. We match the pick-up point p to the road, and then compute the average number of unoccupied taxis passing during a certain period.

For example, if 10 unoccupied taxis have gone through the road in one hour, it costs approximately $3 = (\frac{60}{10 \times 2})$ min to wait for a vacant taxi in average.

5   Online Phase

The goal of the online phase is to select one taxi to match a WP's request. To enhance the performance, some pruning rules are devised to avoid expensive computation. The following contents are organized as below. We first introduce the bound analysis for two plans in Sects. 5.1 and 5.2. Subsequently, three pruning rules, along with the overall algorithm, are described in Sect. 5.3.

5.1   Bound Analysis of the WPF Plan

We first study the WPF plan (Fig. 1(a)). Assume a WP wants to go from O to D, and a TD happens to be at A, taking a TR to B at that time. The route in the WPF plan will be AODB. We analyze the lower and upper bounds of satisfactions for three parties below.

Waiting Passenger. WP's satisfaction is defined as: $S_{WP} = \alpha_w \Delta_T + \beta_w \Delta_P$ (Definition 2), where Δ_T represents the time difference between waiting for an unoccupied taxi at O and waiting for the taxi traveling from A to O. The first item is estimated by wtime(O), while the second item is within $(\min T(A, O), \max T(A, O))$. Hence, Δ_T is in a range of (wtime(O) $-$ maxT(A, O), wtime $-$ minT(A, O)).

Δ_P is computed as: $\Delta_P = k \cdot \text{rdis}(O, D) - M_{WP}$, where $k \cdot \text{rdis}(O, D)$ denotes the WP's payment from O to D without sharing, and M_{WP} is the payment with taxi-sharing, i.e., $M_{WP} = p_{11}\text{rdis}(A, O) + p_{12}\text{rdis}(O, D) + p_{13}\text{rdis}(D, B)$. Note that WP may also need to pay for the segment DB though he will leave at D according to our pricing strategy. For any two points P and P', $\min D(P, P') \leq \text{rdis}(P, P') \leq \max D(P, P')$. Hence, the upper bound of the payment is $U_{M_{WP}} = p_{11}\max D(A, O) + p_{13}\max D(D, B)$, and the lower bound is $L_{M_{WP}} = p_{11}\min D(A, O) + p_{13}\min D(D, B)$. Accordingly, the upper bound (U^w_{WPF}) and the lower bound (L^w_{WPF}) of a WP's satisfactions are defined below.

$$U^w_{WPF} = \alpha_w \text{wtime}(O) + \beta_w(k - p_{12})\text{rdis}(O, D) - \alpha_w \min T(A, O) - \beta_w L_{M_{WP}}$$
$$L^w_{WPF} = \alpha_w \text{wtime}(O) + \beta_w(k - p_{12})\text{rdis}(O, D) - \alpha_w \max T(A, O) - \beta_w U_{M_{WP}}$$

Taxi Rider. TR's satisfaction is defined as: $S_{TR} = \alpha_r \Delta_T + \beta_r \Delta_P$ (Definition 2). Δ_T is the difference of travel time between the original route AB and the new route $AODB$. Hence, the lower bound of Δ_T is minT(A, B) $-$ maxT(A, O) $-$ maxT(O, D) $-$ maxT(D, B), and the upper bound of Δ_T is maxT(A, B)$-$minT(A, O)$-$minT(O, D)$-$minT(D, B). $\Delta_P = k \cdot \text{rdis}(A, B) - M_{TR}$, where $k \cdot \text{rdis}(A, B)$ denotes the payment for the segment AB without sharing, and the payment for sharing is $M_{TR} = q_{11}\text{rdis}(A, O) + q_{12}\text{rdis}(O, D) + q_{13}\text{rdis}(D, B)$. Similar to those of WP, the upper and lower payment bounds of

TR are computed as $U_{MTR} = q_{11}\mathsf{maxD}(A,O) + q_{12}\mathsf{rdis}(O,D) + q_{13}\mathsf{maxD}(D,B)$, $L_{MTR} = q_{11}\mathsf{minD}(A,O) + q_{12}\mathsf{rdis}(O,D) + q_{13}\mathsf{minD}(D,B)$. Accordingly, the upper bound (U_{WPF}^r) and the lower bound (L_{WPF}^r) of a TR's satisfactions are defined below.

$$U_{WPF}^r = \alpha_r\mathsf{maxT}(A,B) + k\beta_r\mathsf{maxD}(A,B)$$
$$- \alpha_r[\mathsf{minT}(A,O) + \mathsf{minT}(O,D) + \mathsf{minT}(D,B)] - \beta_r L_{MTR} \qquad (1)$$
$$L_{WPF}^r = \alpha_r\mathsf{minT}(A,B) + k\beta_r\mathsf{minD}(A,B)$$
$$- \alpha_r[\mathsf{maxT}(A,O) + \mathsf{maxT}(O,D) + \mathsf{maxT}(D,B)] - \beta_r U_{MTR} \qquad (2)$$

Taxi Driver. TD's satisfaction is defined as $I - I'$ (Definition 2). If launching taxi-sharing, the driver's income is the sum of TR's payment and WP's payment, i.e., $I = M_{TR} + M_{WP}$, and $I' = k \cdot (\mathsf{rdis}(A,O) + \mathsf{rdis}(O,D) + \mathsf{rdis}(D,B))$ when he travels along the same route. Thus, the upper bound (U_{WPF}^d) and the lower bound (L_{WPF}^d) of a driver's satisfactions are defined below.

$$U_{WPF}^d = \lambda_1\mathsf{maxD}(A,O) + \lambda_2\mathsf{rdis}(O,D) + \lambda_3\mathsf{maxD}(D,B) \qquad (3)$$
$$L_{WPF}^d = \lambda_1\mathsf{minD}(A,O) + \lambda_2\mathsf{rdis}(O,D) + \lambda_3\mathsf{minD}(D,B) \qquad (4)$$

where $\lambda_1 = p_{11} + q_{11} - k$, $\lambda_2 = p_{12} + q_{12} - k$, and $\lambda_3 = p_{13} + q_{13} - k$.

5.2 Bound Analysis of the TRF Plan

Assuming a WP wants to go from O to D, and a TD is at A, taking a TR to B at that time. The route in the TRF plan will be AOBD. We analyze the lower and upper bounds of satisfactions for the three parties.

Waiting Passenger. First, Δ_T for WP is the time difference between her original travel time of OD (plus the waiting time $\mathsf{wtime}(O)$) and the new travel time of $AOBD$. Second, Δ_P is the payment difference between $k \cdot \mathsf{rdis}(O,D)$ and $M_{WP} = p_{21}\mathsf{rdis}(A,O) + p_{22}\mathsf{rdis}(O,B) + p_{23}\mathsf{rdis}(D,B)$. The upper and lower bounds of payment are $U_{M_{WP}}$ and $L_{M_{WP}}$. Thus, the upper bound (U_{TRF}^w) and the lower bound (L_{TRF}^w) of a WP's satisfactions are defined below.

$$U_{TRF}^w = \alpha_w[\mathsf{wtime}(O) + \mathsf{maxT}(O,D)] + k\beta_w\mathsf{rdis}(O,D)$$
$$- \alpha_w[\mathsf{minT}(A,O) + \mathsf{minT}(O,B) + \mathsf{minT}(B,D)] - \beta_w L_{M_{WP}} \qquad (5)$$
$$L_{TRF}^w = \alpha_w[\mathsf{wtime}(O) + \mathsf{minT}(O,D)] + k\beta_w\mathsf{rdis}(O,D)$$
$$- \alpha_w[\mathsf{maxT}(A,O) + \mathsf{maxT}(O,B) + \mathsf{maxT}(B,D)] - \beta_w U_{M_{WP}} \qquad (6)$$

Taxi Rider. The route for TR changes from AB to AOB, so that Δ_T rises. In addition, the payment change Δ_P is computed as: $M_{TR} - k \cdot \mathsf{rdis}(A,B)$, where $M_{TR} = \mathsf{rdis}(A,O)q_{21} - \mathsf{rdis}(O,B)q_{22}$. Let U_{MTR} and L_{MTR} denote the upper and lower bounds of payment. Thus, the upper bound (U_{TRF}^r) and the lower bound (L_{TRF}^r) of a TR's satisfactions are defined below.

$$U_{TRF}^r = \alpha_r\mathsf{maxT}(A,B) + k\beta_r\mathsf{maxD}(A,B)$$
$$- \alpha_r(\mathsf{minT}(A,O) + \mathsf{minT}(O,B)) - \beta_r L_{MTR} \qquad (7)$$
$$L_{TRF}^r = \alpha_r\mathsf{minT}(A,B) + k\beta_r\mathsf{minD}(A,B)$$
$$- \alpha_r(\mathsf{maxT}(A,O) + \mathsf{maxT}(O,B)) - \beta_r U_{MTR} \qquad (8)$$

Algorithm 1. Schedule(w)

---

1 $C \leftarrow \emptyset, \tau \leftarrow 0, C' \leftarrow \emptyset$;
2 **foreach** *registered taxi r* **do**
3      $WPFvalid \leftarrow$ **false**; $TRFvalid \leftarrow$ **false**;
4   **if** $\min(U_{WPF}^w, U_{WPF}^r) > 0$ **then**
5          $\lfloor$ $WPFvalid \leftarrow$ **true**;
6   **if** $\min(U_{TRF}^w, U_{TRF}^r) > 0$ **then**
7          $\lfloor$ $TRFvalid \leftarrow$ **true**;
8   **if** $WPFvalid$ **or** $TRFvalid$ **then**
9          $C \leftarrow C \cup \{r\}, \alpha_1 \leftarrow 0, \alpha_2 \leftarrow 0, \beta_1 \leftarrow 0, \beta_2 \leftarrow 0$;
10   **if** $WPFvalid =$ **true then**
11             $\lfloor$ $\alpha_1 \leftarrow \min(L_{WPF}^w, L_{WPF}^r, L_{WPF}^d); \beta_1 \leftarrow L_{WPF}^w + L_{WPF}^r + L_{WPF}^d$;
12   **if** $TRFvalid =$ **true then**
13             $\lfloor$ $\alpha_2 \leftarrow \min(L_{TRF}^w, L_{TRF}^r, L_{TRF}^d); \beta_2 \leftarrow L_{TRF}^w + L_{TRF}^r + L_{TRF}^d$;
14   **if** *the query is MaxMin* **and** $\max(\alpha_1, \alpha_2) > \tau$ **then**
15             $\lfloor$ $\tau \leftarrow \max(\alpha_1, \alpha_2)$;
16   **else if** *the query is MaxSum* **and** $\max(\beta_1, \beta_2) > \tau$ **then**
17             $\lfloor$ $\tau \leftarrow \max(\beta_1, \beta_2)$;

18 **foreach** *registered taxi* $r \in C$ **do**
19   **if** *the query is MaxMin* **and**
    $(\min(U_{WPF}^w, U_{WPF}^r, U_{WPF}^d) > \tau$ $\parallel$ $\min(U_{TRF}^w, U_{TRF}^r, U_{TRF}^d) > \tau)$ **then**
20         $\lfloor$ $C' \leftarrow C' \cup \{r\}$;
21   **if** *the query is MaxSum* **and**
    $(U_{WPF}^w + U_{WPF}^r + U_{WPF}^d > \tau$ $\parallel$ $U_{TRF}^w + U_{TRF}^r + U_{TRF}^d > \tau)$ **then**
22         $\lfloor$ $C' \leftarrow C' \cup \{r\}$;

23 **foreach** *candidate taxi* $r \in C'$ **do**
24     $\lfloor$ Compute the satisfactions of three parties;
25 **return** a taxi with the biggest minimal (total) satisfaction to MaxMin (MaxSum);

Taxi Driver. The route for TD is $AOBD$. Thus, the upper bound (U_{TRF}^d) and the lower bound (L_{TRF}^d) of a TD's satisfactions are defined below.

$$U_{TRF}^d = \lambda_1 \mathsf{maxD}(A, O) + \lambda_2 \mathsf{maxD}(O, B) + \lambda_3 \mathsf{maxD}(B, D) \tag{9}$$

$$L_{TRF}^d = \lambda_1 \mathsf{minD}(A, O) + \lambda_2 \mathsf{minD}(O, B) + \lambda_3 \mathsf{minD}(B, D) \tag{10}$$

where $\lambda_1 = p_{21} + q_{21} - k$, $\lambda_2 = p_{22} + q_{22} - k$, and $\lambda_3 = p_{23} + q_{23} - k$.

5.3 Pruning Rules and Query Processing

We then design some pruning rules. In any feasible plan, the satisfaction degrees for all three parties must be positive. Note that TD's satisfaction is always positive under our pricing strategy settings. Hence, the rule is as below.

Pruning Rule 1. *One taxi with* $\min(U^w_{WPF}, U^r_{WPF}) < 0$ *or* $\min(U^w_{TRF}, U^r_{TRF})$ < 0 *is improper.*

All improper taxis can be filtered out by using Pruning Rule 1. However, the computation cost is still high if there exist a large number of suitable taxis. Therefore, we propose other two pruning rules to filter a significant part of suitable taxis, which respectively correspond to MaxMin and MaxSum queries. According to the bound analysis, the upper and lower bounds of two target satisfaction values are $\max(U^w_{WPF}, U^r_{WPF}, U^d_{WPF})$ and $\min(L^w_{WPF}, L^r_{WPF}, L^d_{WPF})$ respectively. Hence, let τ denote the maximal lower bound for all taxis. Then, any taxi with the upper bound below τ can be filtered safely. The same filtering process is used for TRF strategy. The pruning rule is defined formally below.

Pruning Rule 2. *The MaxMin query returns a taxi with the greatest minimal satisfaction among all of three parties. A taxi with* $\min(U^w_{WPF}, U^r_{WPF}, U^d_{WPF}) <$ τ *or* $\min(U^w_{TRF}, U^r_{TRF}, U^d_{TRF}) < \tau$ *can be pruned safely.*

Pruning Rule 3. *The MaxSum query returns a taxi with the greatest sum of satisfactions. A taxi with* $\min(U^w_{WPF} + U^r_{WPF} + U^d_{WPF}) < \tau$ *or* $\min(U^w_{TRF} +$ $U^r_{TRF} + U^d_{TRF}) < \tau$ *can be pruned safely.*

Algorithm 1 describes details on selecting the most suitable taxi by using the above pruning rules. At first, the algorithm scans all registered taxis to construct a candidate set C by using Pruning Rule 1. In addition, the threshold τ is computed for further pruning (at lines 2–17). Subsequently, it scans all candidate taxis in C to construct C' by using Pruning Rules 2 and 3 (at lines 18–22). At last, it scans each taxi in C' to obtain the final result (at lines 23–24). The most proper plan for each kind of query is returned to WP (at lines 25).

6   Experiments

We report some experimental results in this section. The offline phase is conducted on a Hadoop cluster. All codes for online processing are written in Java, and run on a computer with 16 GB RAM and Intel Xeon 2.00 GHz CPU.

Road Network: We utilize the road network of Beijing, which is in the range of 30,000 m in width and length. It is partitioned into 30 * 30 cells, each with 1,000 m in width and length.

Taxi Trajectory: We use a real taxi trajectory set containing 13,000 Beijing taxis over one month (October 2013). The number of occupied/unoccupied taxis varies a lot in time and place. We sample every hour within a day to count all taxis and occupied taxis. About 4,000 taxis in average are occupied during the rush hours, and 90 % of occupied taxis move around urban districts, while around 1,000 taxis are occupied at mid-night. We assume that all occupied taxis join taxi-sharing. Therefore, the number of occupied taxis (TR) varies from 1,000 to 4,000 in the experiments. Since the requests are processed in order, the number of passengers does not affect the query time. The parameter setting of the pricing strategy are listed in Table 1.

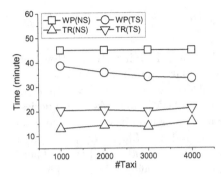 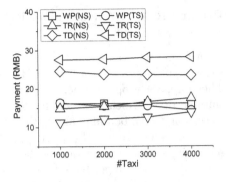

Fig. 3. Time comparison   **Fig. 4.** Payment/income comparison

6.1 Effectiveness

We evaluate the effectiveness of the proposed taxi-sharing method. First, we compare the travel time and payment between taxi-sharing (TS) and non-sharing (NS) methods after randomly selecting requests from the trips in the real dataset. The number of TRs ranges from 1,000 to 4,000, and the number of WPs is set to 1,000. The preferences of TRs and WPs are generated uniformly in (0,1). Figure 3 illustrates the average time for WP/TR with and without sharing taxi. Figure 4 compares payment of WP/TR and income of TD in TS and NS methods. Note that the travel time for WP in two methods contains the waiting time. When using TS strategy, the travel time for WP is reduced even if WP plans to wait for TR, because WP's waiting time decreases. WP may pay more money in taxi-sharing scenario, the average payment for WPs is less than NS as the amount of TRs rises. Although TR spends more time on traveling, the average payment is significantly reduced. In addition, more occupied taxis make WP and TR both save more time or money. TDs' income rises by more than 15 % compared to NS in all situations, due to the payment sharing strategy (Table 1). In fact, TD can earn more if any of the parameters in Table 1 rises.

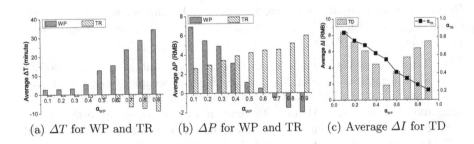

(a) ΔT for WP and TR   (b) ΔP for WP and TR   (c) Average ΔI for TD

Fig. 5. Parameter settings comparison

We then evaluate the influence of preference parameters, assuming there are 2,000 TRs. α_{WP} and α_{TR} are generated uniformly in (0,1). We compare the average travel time difference (ΔT), average payment difference (ΔP) from NS to TS for WP/TR, and TD's average income difference (ΔI) from TS to NS, as reported in Fig. 5. In Fig. 5(a), higher α_{WP} results in more ΔT for WP, which means WP saves more time in TS. In Fig. 5(b), lower α_{WP} leads to more ΔP, which indicates WP saves more money. Figure 5(c) illustrates two metrics: (1) TD's average ΔI. For all preference settings, TD can earn more in TS compared to NS. Moreover, biased preferences of passengers lead to more income. (2) Average α_{TR}. If WP and TR have the opposite preferences in time (money), they are more likely to share one taxi. For example, the time-preferred WP is matched with the payment-preferred TR.

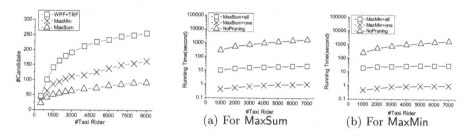

Fig. 6. Pruning effects   (a) For MaxSum   (b) For MaxMin

Fig. 7. Efficiency comparison

6.2 Efficiency

We then evaluate the performance of pruning rules. Figure 6 illustrates the performance of three pruning rules, including WPF+TRF (Rule 1), MaxMin (Rule 2), and MaxSum (Rule 3). Recall that WPF+TRF is the basis of MaxMin and MaxSum. By WPF+TRF, approximately 97 % of taxis are safely filtered, e.g., the number of candidates is around 250 when there are nine thousand taxi riders. After employing MaxMin or MaxSum, 36 % or 60 % more candidates are removed safely, e.g., only 160 or 95 candidates are left when $|TR| = 9,000$. Moreover, the pruning power will be strengthened when the number of taxi riders increases. For example, when $|TR| = 3,000$, around 2 % of taxis (60) are left for processing. But when $|TR| = 9,000$, around 1 % of taxis (90) are left.

The use of pruning rules deeply influences the running time of the proposed methods. Since computing the shortest path for each candidate costs much time, we use two schemes to fulfill this goal. The first is to precisely compute the satisfactions for all candidates (+all). The second is to randomly choose a candidate every time, and try to evaluate whether this one satisfies the requirement or not (+one). If not, we repeat to choose another one. Figure 7(a) and (b) show that MaxMin and MaxSum pruning methods both scale significantly better than NoPruning method, because NoPruning method needs to precisely compute each taxi. Note that MaxMin+one and MaxSum+one methods can answer the query in a second.

7   Related Work

The problem of ride-sharing has been studied intensively in recent years. There are several popular methods for ride-sharing, mainly including mobile applications (e.g., Uber), and the agencies providing ride-sharing service (e.g., Lyft). [12] proposes the dynamic ride-sharing problem. [13] discusses the dynamic slugging problem with the vehicle-capacity and delay bounded constraints. However, they require drivers' original routes not to be changed and trips to be arranged in advance. Later works propose new ride-sharing problems. The work [14] summarizes different optimization goals for ride-sharing: to minimize system-wide vehicle-miles, to minimize system-wide travel time and to maximize the number of participants. [15] creates a dynamic ride-sharing community service architecture, which combines ITS, ride-sharing and social network. We also develop a passenger matching system [16] to find passengers with similar locations and estimate additional time and payment. [17] proposes a scheme that allows detours and four sharing patterns. Existing ride-sharing methods [12,13,15] apply to some of them, but our work contains all of them.

Taxi sharing is a more popular way of ride-sharing in realtime service. Research efforts on analysis of taxi-sharing have been made to address the objective issue by considering different constraints. For example, [18] considers the vehicle capacity constraint, [19] considers the waiting time constraint, and [8] considers time and money constraints. Our work considers all these constraints and the pricing strategy of [8] can be considered as a special case of our pricing method. Other works focus on how to schedule passengers. [20] builds a dynamic and distributed taxi-sharing system which processes every passenger's request on each end of taxi. [19] allows more than two groups of passengers sharing one taxi, but it simply filters out taxis outside the waiting time constraint. [8] aims to obtain global minimal additional incurred travel distance which is an NP-complete problem. So it utilizes a greedy algorithm to find taxi-sharing candidates by predicting taxis' future positions, which is a very uncertain task. To the best of our knowledge, our work first takes into account of different people's customized preferences for taxi-sharing, in purpose of saving time and payment.

8   Conclusion

We have studied the taxi-sharing problem on a parameterized real-time taxi-sharing system for user satisfaction. We develop a method to maximize people's satisfaction with their personalized needs. Experimental results show the efficiency and effectiveness of our method. Future work includes: (1) allowing more than two groups of passengers to share one taxi, (2) creating indexes for passengers' moving positions to speedup online processing, and (3) making estimation of travel time more accurate by using probability distribution function instead of scalar.

Acknowledgement. Our research is supported by the 973 program of China (No. 2012CB316203), NSFC (U1401256, 61370101, U1501252, 61402180 and 61472345), Shanghai Knowledge Service Platform Project (No. ZF1213), Innovation Program of Shanghai Municipal Education Commission(14ZZ045), and Natural Science Foundation of Shanghai (No. 14ZR1412600).

References

1. Brunekreef, B., Holgate, S.T.: Air pollution and health. Lancet **360**, 1233–1242 (2002)
2. Morency, C.: The ambivalence of ridesharing. Transportation **34**, 239–253 (2007)
3. Chan, N.D., Shaheen, S.A.: Ridesharing in North America: past, present, and future. Transp. Rev. **32**, 93–112 (2012)
4. Baldacci, R., Maniezzo, V., Mingozzi, A.: An exact method for the car pooling problem based on Lagrangean column generation. Oper. Res. **52**, 422–439 (2004)
5. Attanasio, A., Cordeau, J., Ghiani, G., Laporte, G.: Parallel tabu search heuristics for the dynamic multi-vehicle dial-a-ride problem. Parallel Comput. **30**, 377–387 (2004)
6. Gidofalvi, G., Pedersen, T.B., Risch, T., Zeitler, E.: Highly scalable trip grouping for large-scale collective transportation systems. In: EDBT, pp. 678–689 (2008)
7. Ferguson, E.: The rise and fall of the American carpool: 1970–1990. Transportation **24**, 349–376 (1997)
8. Ma, S., Zheng, Y., Wolfson, O.: T-share: a large-scale dynamic taxi ridesharing service. In: ICDE, pp. 410–421 (2013)
9. Song, L., et al.: TaxiHailer: a situation-specific taxi pick-up points recommendation system. In: Bhowmick, S.S., Dyreson, C.E., Jensen, C.S., Lee, M.L., Muliantara, A., Thalheim, B. (eds.) DASFAA 2014, Part II. LNCS, vol. 8422, pp. 523–526. Springer, Heidelberg (2014)
10. Dijkstra, E.W.: A note on two problems in connexion with graphs. Numerische Mathematik **1**, 269–271 (1959)
11. Zeng, W., Church, R.L.: Finding shortest paths on real road networks: the case for a*. IJGIS **23**, 531–543 (2009)
12. Agatz, N., Erera, A., Savelsbergh, M., Wang, X.: Sustainable passenger transportation: dynamic ride-sharing. Technical report, ERIM Report Series Research in Management (2010)
13. Ma, S., Wolfson, O.: Analysis and evaluation of the slugging form of ridesharing. In: SIGSPATIAL/GIS, pp. 64–73 (2013)
14. Agatz, N., Erera, A., Savelsbergh, M., Wang, X.: Optimization for dynamic ridesharing: a review. EJOR **223**, 295–303 (2012)
15. Fu, Y., Fang, Y., Jiang, C., Cheng, J.: Dynamic ride sharing community service on traffic information grid. In: ICICTA, vol. 2, pp. 348–352. IEEE (2008)
16. Duan, X., Jin, C., Wang, X.: POP: a passenger-oriented partners matching system. In: 31st IEEE ICDE Workshops 2015, pp. 117–118 (2015)
17. Furuhata, M., Dessouky, M., et al.: Ridesharing: the state-of-the-art and future directions. Transp. Res. Part B Methodol. **57**, 28–46 (2013)
18. Tao, C.C.: Dynamic taxi-sharing service using intelligent transportation system technologies. In: 2007 WiCOM, pp. 3209–3212 (2007)
19. Huang, Y., Jin, R., Bastani, F., Wang, X.S.: Large scale real-time ridesharing with service guarantee on road networks. CoRR abs/1302.6666 (2013)
20. d'Orey, P., Fernandes, R., Ferreira, M.: Empirical evaluation of a dynamic and distributed taxi-sharing system. In: ITSC, pp. 140–146 (2012)

Author Index

Printed in the United States
By Bookmasters